CLASSICAL
ELECTROMAGNETISM

REVISED SECOND EDITION

CLASSICAL
ELECTROMAGNETISM
REVISED SECOND EDITION

Jerrold Franklin
TEMPLE UNIVERSITY

Dover Publications
Garden City, New York

Bibliographical Note

This Dover edition, first published in 2017 and updated in 2023, is a revised
edition of the work first published by Pearson / Addison-Wesley in 2005.

Library of Congress Cataloging-in-Publication Data

Names: Franklin, Jerrold, author.
Title: Classical electromagnetism / Jerrold Franklin (Temple University).
Description: Revised Second edition. | Garden City, New York : Dover Publications,
 [2017] | Includes bibliographical references.
Identifiers: LCCN 2017031035| ISBN 9780486813714 | ISBN 0486813711
Subjects: LCSH: Electromagnetism—Textbooks.
Classification: LCC QC760 .F74 2017 | DDC 537—dc23
LC record available at https://lccn.loc.gov/2017031035

Manufactured in the United States of America
81371104   2023
www.doverpublications.com

Contents

Preface to the Second Edition

This second edition includes a number of improvements to the first edition in clarifying and extending a number of the derivations, but it serves the same purpose of providing a good learning experience for students. My intention has been to write a comprehensive textbook for the first-year electromagnetism course. It could also serve as supplementary reading for graduate students to help them achieve a better understanding of the subject matter, or for advanced undergraduate students who want to go into more detail than is generally covered in the undergraduate course. I hope it will prove to be helpful for students, and will also be a good textbook from which professors can teach.

Preface to the First Edition

This text is designed for first year Graduate Students who generally will have had at least one, and probably two, earlier undergraduate courses in Electromagnetism. Because the text starts each topic at a fundamental level and works up, it could be also used by a good student with little prior knowledge of Electromagnetism, but enough background in mathematics to feel comfortable.

The book tells the story of Electromagnetism (EM), from its 19th century beginnings to its present place, somewhere in the 21st century. It starts with simple expositions of Coulomb's law and the magnetic law of Biot-Savart. At each stage in the development it demonstrates how the subject could be defined on a more fundamental basis. This continues through the laws of Gauss and Ampere, to the unifying partial differential equations of Maxwell, and the ensuing electromagnetic radiation. Then it is shown how the principles of Special Relativity *require* the unification of Electricity and Magnetism that had been achieved earlier by Maxwell, and extends this to the unification of Space and Time. In the last Chapter, the presentation of Electromagnetism as a Quantum Gauge Theory demonstrates that EM is a manifestation of the general principle of Local Gauge Invariance in Quantum Mechanics. A bit more Quantum Mechanics is introduced to show that Classical Electromagnetism is a limiting form of Quantum Electrodynamics (QED).

As an afterword, it is shown how the generalization of Gauge Invariance to

coupled fields leads to the unification of the Electromagnetic and Weak Interactions, and the conjectured unification of these interactions with the Strong
Interaction (Grand Unified Theory), ultimately completing the unification of
physics started by Maxwell in the 19th Century.

The mathematics is completely unified with the physics at each stage in the
text. There are no separate Mathematical Appendices or Flyleaf Formulas to
memorize. The appropriate mathematics is learned more easily in the context
of a real physics application. The mathematical concepts are developed as they
are needed for the physics. This makes the learning process for both math and
physics more natural and, I hope, more interesting. As with the physics, the
mathematics is introduced first on a basic level, but ends up at the high level
needed for a good development of the physics of Electromagnetism.

My own Electromagnetism was learned first from Sears 1st Edition, and
then from Panofsky & Phillips. (Information for each book is given in the
Bibliography.) Over the years I have used P & P, Jackson, and Good &
Nelson as texts for the course I taught. The remarkably interesting books
by Mel Schwartz and Landau & Lifschitz were also used for reference by me
and my students. Consciously or subconsciously, much of the material in
this text draws on those texts (and also on Maxwell's own remarkable 1873
Treatise). But it is not an amalgam. You will find many new treatments and
new insights. I have tried to write a new text that is enjoyable to learn from,
and to teach from, while maintaining a high level of rigor and reaching a deep
level of understanding.

The chapters should not all be covered at the same rate. The earlier chapters, especially chapters 1-4, are somewhat of a review and could be covered
more quickly. But they are important, with some new ways of looking at old
EM. A reasonable break in a one year course would be to complete at least
chapters 1-8 in the first semester. It would also be reasonable to go as far as
section 9.5 (Magnetic Energy).

A word must be added about units. The two texts I learned my early
Electromagnetism from, Sears 1st Edition (a pioneer in SI, then called Georgi
units) and Panofsky and Phillips, each used Systeme Internationale (SI) units.
What I learnt was the roadblock that SI units place in the unification of
Electricity with Magnetism. (A fuller discussion is in the last section of this
book.) I know this book goes against the trend for the use of SI units in all new
textbooks, and thank my publisher for permitting it. I hope that my book will
go some way against this stream, at least for advanced texts. I use Gaussian
units for most of the book, then dispense with the conversion constant c after
Relativity, finally getting to natural units in the last chapter. This evolution of
units follows the course of unification of Electricity with Magnetism, then with
the Weak Interaction, and maybe more. My recommendation for numerical
calculations is to use Gaussian (or natural) units throughout any calculation
(putting SI quantities into Gaussian using Appendix A). Then, at the end,

Gaussian quantities can be put into SI units if desired.

The use of SI units beclouds the obvious connections between the **E** and **D** fields, and between the **B** and **H** fields, as well as precluding a simple relativistic unification of the **E** and **B** fields. Beyond this, SI units make the extension of Classical Electromagnetism to QED, and the unification with Weak Interactions particularly unwieldy. In fact, SI units fly in the face of all the advances in the unification of physics of the past 150 years. I should also mention the introduction (in SI units) of two misnamed constants (ϵ_0 and μ_0) that have no physical meaning and serve only to complicate EM for beginning students as well as working physicists. Enough about units. I hope you enjoy and learn from my book.

Jerrold Franklin

Chapter 1

Foundations of Electrostatics

1.1 Coulomb's Law

Historically, the quantitative study of electrostatics began with **Coulomb's law**, which is illustrated in Fig. 1.1.

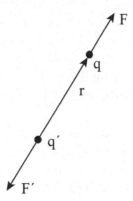

Figure 1.1: Coulomb's law for the force between two charges.

This law states that the electric force between two point charges is inversely proportional to the square of the distance between them and directly proportional to the product of the charges, with the direction of the force being along the straight line connecting the two charges. In these respects, the Coulomb force between two point charges is similar to the gravitational force between two point masses. An important difference between the two force laws is that the electric charge comes in two signs, with the force between like charges being repulsive, and that between two opposite charges being attractive.

Coulomb's law can be written

$$\mathbf{F} = \frac{kqq'\hat{\mathbf{r}}}{r^2},\tag{1.1}$$

giving the force on charge q due to charge q' in terms of the unit vector $\hat{\mathbf{r}}$ which specifies the direction from q' to q. The unit vector $\hat{\mathbf{r}}$ is a dimensionless

1

vector defined as a vector divided by its magnitude, $\hat{\mathbf{r}} = \mathbf{r}/|r|$, so Coulomb's law can also be written as

$$\mathbf{F} = \frac{kqq'\mathbf{r}}{r^3}. \tag{1.2}$$

This form of Coulomb's law is more useful in vector operations.

Because Coulomb's law is a proportionality, a constant k is included in Eqs. (1.1) and (1.2). This constant can be chosen to define the unit of electric charge. Unfortunately, several different definitions have been used for the unit of charge, and some care is required in treating the units consistently. We take time now to discuss some of the different systems.

The simplest choice, in terms of Coulomb's law, is to set k=1, and use Coulomb's law to define the unit of electric charge. This leads to the electrostatic unit (**esu**) of charge called the **statcoulomb**, which can be defined in words by **The electrostatic force between two charges, each of one statcoulomb, a distance one centimeter apart, is one dyne.** More simply stated, if all distances and forces are in cgs (centimeter-gram-second) units, then the charge in Coulomb's law (with k=1) is in statcoulombs.

Another choice for the definition of the unit of electric charge uses MKS (meter-kilogram-second) units in Coulomb's law with the constant k being given by $k \simeq 9 \times 10^9$. This defines the unit of charge called the **coulomb**, which could be defined in words by **The electrostatic force between two charges, each of one coulomb, a distance one meter apart, is 9×10^9 Newtons.**

The coulomb is somewhat more familiar than the statcoulomb because the common unit of current, the ampere, is defined as one coulomb per second. Possibly for this reason, the coulomb, and the form of Coulomb's law with $k \simeq 9 \times 10^9$ was adopted as part of the Systeme International (SI) system of units, and has gained almost universal usage in elementary physics textbooks. However the study and understanding of electrostatics is considerably simpler using esu units with k=1. The SI system is also particularly awkward to use for relativistic or quantum formulations of electromagnetism. For this reason, we will consistently use esu units as part of what is called the **Gaussian system** of units relating the esu and emu systems of units, which will be introduced in Chapter 7. Conversions between the Gaussian system and the SI system are given in Appendix A.

In the SI system, Coulomb's law is further complicated by introducing a new quantity ϵ_0 defined by

$$k = \frac{1}{4\pi\epsilon_0}. \tag{1.3}$$

The use of the $1/4\pi$ is said to "rationalize" the units because it makes some later equations (such as Gauss's law) simpler.

The constant ϵ_0 is sometimes called "the permittivity of free space". This terminology is unfortunate because, in the theory of Quantum Electrodynamics (QED), which is the foundation theory for Classical Electromagnetism, a frequency dependent permittivity does arise (vacuum polarization) that has nothing to do with ϵ_0. Actually, if $\frac{1}{4\pi\epsilon_0}$ is converted from SI to Gaussian units, it is equal to c^2, the square of the speed of light.

The constant c was originally introduced as the ratio between the unit of charge (esu) used in electric phenomena and the unit (emu) used in magnetic phenomena. Some years after its introduction, Maxwell showed that c was, in fact, the speed of light in vacuum. (See Chapter 10.) Herman Minkowski then showed in Einstein's theory of Special Relativity (See Chapter 14) that space and time just referred to different directions in a completely symmetric space-time manifold.

So the constant c is just the conversion constant between the space axes and the time axis (just like the conversion between miles and feet in an American topographical map). This means that c is no longer a constant to be measured, but a specified number used to define the meter in terms of the second. This is the modern definition of the meter, which is defined so that light travels exactly 299,792,458 meters in one second. In cgs units, then

$$c = 2.99792458 \times 10^{10} \text{ cm/sec}, \tag{1.4}$$

which is the number we will use in this text.

The defined value for c is very close to 3×10^{10} in magnitude (equal to three significant figures), and we will generally use the value 3 for conversions, with the understanding that the more accurate value could be used if greater accuracy were desired. (Whenever the numbers 3 or 9 appear in conversion equations, the more accurate value could be substituted.) The conversion numbers are also changed by various powers of 10 related to the difference between cgs and MKS units, as well as a mismatch in relating the ampere to the emu unit of current, the abampere. (Ten amperes equal one abampere.) The esu unit of charge and the emu unit of charge are related by

$$1 \text{ abcoulomb(emu)} = c \text{ statcoulombs(esu)}, \tag{1.5}$$

which leads to the connection

$$1 \text{ coulomb} = 3 \times 10^9 \text{ statcoulombs}. \tag{1.6}$$

between the units of charge in the SI and esu (or Gaussian) systems.

As an example of the connection between the units, the magnitude of the charge on an electron is given in esu units by

$$e = 4.80 \times 10^{-10} \text{ statcoulomb}, \tag{1.7}$$

or

$$e = 1.60 \times 10^{-19} \text{ coulomb} \tag{1.8}$$

in SI units. It can be seen from the large negative power of 10 required in either case, that neither system of units is really appropriate for elementary

4

CLASSICAL ELECTROMAGNETISM

particle, nuclei, atomic, or molecular physics (microscopic physics) where the electron charge is the relevant unit of charge. In Chapter 16, we will discuss other systems of units that are more appropriate for those cases.

The above discussion may seem confusing, but it is a confusion we have to live with.

The Coulomb's law force on each of the two charges is proportional to the product of the two charges, and each force is along their common axis. Thus Coulomb's law satisfies Newton's third law of equal and opposite forces. We could try to use the third law as a theoretical basis for the symmetrical appearance of the two charges, and the common action line of the forces, but Newton's third law is not a sturdy basis on which to build. Although it is satisfied in electrostatics, we will see in Chapter 7 that it is violated by the magnetic force between moving charges.

A better principle is the conservation of linear and angular momentum. Conservation of linear momentum requires the two Coulomb forces to be equal and opposite. Conservation of angular momentum requires the two forces to be along the same action line. So we see that it is from these two conservation laws that the two forces in Coulomb's law are colinear, and the charges appear symmetrically.

The fact that the force is proportional to the first power of either charge is called **linearity**. A related, but logically somewhat more extended assumption (verified by experiment), is that the Coulomb force due to two charges, located at different points, on a third is the vector sum of the two individual forces. This is called **the superposition principle** for the electric force. Extended to the force due to several charges, the superposition principle leads to the form

$$\mathbf{F} = \sum_n \frac{(\mathbf{r} - \mathbf{r}_n)qq_n}{|\mathbf{r} - \mathbf{r}_n|^3},$$ (1.9)

for the force on a point charge q at $\mathbf{r}$ due to other point charges q_n located at points $\mathbf{r}_n$. We can see from Eq. (1.9) that the $\mathbf{r}/r^3$ form of Coulomb's law is more convenient than using $\hat{\mathbf{r}}/r^2$, because the unit vector for $(\mathbf{r} - \mathbf{r}_n)$ would be awkward to use.

1.2 The Electric Field

At this point, it becomes useful to find the force in two stages by introducing the concept of the **electric field E**, defined by

$$\mathbf{F} = q\mathbf{E}$$ (1.10)

for the force on a point charge q due to any collection of other charges. With this definition, Coulomb's law for the electric field due to a point charge q is

$$\mathbf{E} = \frac{q\hat{\mathbf{r}}}{r^2}.$$ (1.11)

The electric field at a point $\mathbf{r}$ due to a number of point charges q_n, located at positions $\mathbf{r}_n$, is given by

$$\mathbf{E}(\mathbf{r}) = \sum_n \frac{(\mathbf{r} - \mathbf{r}_n)q_n}{|\mathbf{r} - \mathbf{r}_n|^3}. \tag{1.12}$$

The effect of Eq. (1.9) can now be accomplished in two steps by first using Eq. (1.12) to find $\mathbf{E}$, and then Eq. (1.10) to give the force on the point charge q located at $\mathbf{r}$. Although introduced in this way as a mathematical convenience, we will see (as often happens in physics) that the electric field has important physical significance on its own, and is not merely a mathematical construct.

Equation (1.10) defines the electric field $\mathbf{E}$ at the point $\mathbf{r}$ in the presence of the charge q. However care must be exercised if Eq. (1.10) is to be used to measure the electric field that existed at $\mathbf{r}$ before the charge q was introduced. That is because the introduction of the charge q can polarize any nearby matter, changing the field at point $\mathbf{r}$. This polarization can even produce an $\mathbf{E}$ field where none existed before the introduction of charge q. (This is a common phenomenom in static electricity, causing lightning as well as other effects.) For this reason, the use of a test charge to measure a pre-existing electric field is accomplished by

$$\mathbf{E}_0 = \lim_{q \to 0} \frac{\mathbf{F}}{q}, \tag{1.13}$$

where $\mathbf{E}_0$ is the electric field that was present before the test charge was introduced.

We have thus far limited our considerations to point charges. In principle, this is all that is needed because it is believed that all charges appear as point charges of value $\pm e$ for leptons and $\pm\frac{2}{3}e$ or $\pm\frac{1}{3}e$ for quarks (the constituents of strongly interacting matter). However, the sum in Eq. (1.12) would have of the order of 10^{23} terms for macroscopic objects, and be impossible to use. For this reason the concept of a continuous charge distribution as an abstraction of a huge number of point particles is introduced. That is, in Eq. (1.12), the sum on n is first taken over a large number of the point charges q_n that is still small compared with the total number of charges in a macroscopic sample. This leads to clusters of charge Δq_i each having n_i charges, so

$$\Delta q_i = \sum_{n=n_{i-1}}^{n_i} q_n. \tag{1.14}$$

The number of charges in each cluster can be large, and yet all of the charges in a cluster still be at about the same point, since the total number of charges is huge for a macroscopic sample. (For instance, one million atoms are contained in a cube 10^{-6} cm on a side.) This means that a very large number of point charges looks like, and can be well aproximated by, a still large collection of effective point charges Δq_i. Then the electric field will be given by

$$\mathbf{E}(\mathbf{r}) = \sum_i \frac{(\mathbf{r} - \mathbf{r}_i)\Delta q_i}{|\mathbf{r} - \mathbf{r}_i|^3}. \tag{1.15}$$

In the limit that the number of charge clusters becomes infinite, (In this case, "infinity" is of the order of 10^{20}.) and the net charge in each cluster approaches zero compared to the total charge, the sum approaches an integral over charge differentials dq', and Eq. (1.15) is replaced by

$$\mathbf{E}(\mathbf{r}) = \int \frac{(\mathbf{r} - \mathbf{r}')dq'}{|\mathbf{r} - \mathbf{r}'|^3}. \tag{1.16}$$

In Eq. (1.16), there are two different position vectors, which we refer to as the source vector $\mathbf{r}'$ and the field vector $\mathbf{r}$.

The form of the differential charge element dq' depends on the type of charge distribution. The integral operator $\int dq'$ becomes

$$\int dq' = \int \lambda(\mathbf{r}')dl' \text{ for a linear charge density } \lambda, \tag{1.17}$$

$$\int dq' = \int \sigma(\mathbf{r}')dA' \text{ for a surface charge density } \sigma, \tag{1.18}$$

$$\int dq' = \int \rho(\mathbf{r}')d\tau' \text{ for a volume charge density } \rho, \tag{1.19}$$

where dl', dA', and $d\tau'$ are differentials of length, area, and volume, respectively. Equation (1.16) can also be extended to point charges with the understanding that

$$\int dq' f(\mathbf{r}') = \sum_n q_n f(\mathbf{r}_n) \text{ for point charges } q_n \text{ at positions } \mathbf{r}_n. \tag{1.20}$$

The charge distribution on which $\mathbf{E}$ acts can also be considered continuous, and then the force on the continuous charge distribution would be

$$F = \int dq \, \mathbf{E}(\mathbf{r}), \tag{1.21}$$

with $\int dq$ given as in Eqs. (1.17-1.20).

Equation (1.16) gives the static electric field for any charge distribution. The term 'static', as used here, means that all time derivatives of the charge distribution are zero, or are neglected if the charges are moving. But using Eq. (1.16) is a 'brute force' method that often requires complicated integration, and may not be a practical way to find $\mathbf{E}$. (With the use of modern computers, it has become more practical to sometimes just do these integrals on the computer.)

There are several simple geometries for which the use of symmetry simplifies the integrals, and some examples of these are given in the problems at the end of this chapter. Aside from these simple cases, better methods are usually needed to find the electric field.

1.3   Electric Potential

The work done by the electric field in moving a charge q from a point A to a point B along a path C is given by

$$W_C = q \int_{AB_C} \mathbf{E} \cdot \mathbf{dr}, \tag{1.22}$$

where the notation $\int_{AB_C}$ means that the displacement $\mathbf{dr}$ is along the path defined by the curve C from A to B.

The definition of a conservative force field is one for which the net work done around any closed path is zero. We now show that this is true for the $\mathbf{E}$ field of a point charge. This follows because the integrand of Eq. (1.22) can be written as the perfect differential $d(-1/r)$ when $\mathbf{E}$ is given by Coulomb's law:

$$d\left(\frac{-1}{r}\right) = \frac{d(r)}{r^2} = \frac{d[(\mathbf{r} \cdot \mathbf{r})]^{\frac{1}{2}}}{r^2} = \frac{\mathbf{r} \cdot \mathbf{dr}}{r^3}, \tag{1.23}$$

with the final form in Eq. (1.23) being the integrand for the net work on a unit charge using Coulomb's law for $\mathbf{E}$. This result can be extended to the $\mathbf{E}$ given by Eq. (1.15) or Eq. (1.16), because each of these are just linear sums of Coulomb's law for a single point charge. Thus

$$\oint \mathbf{E} \cdot \mathbf{dr} = 0 \tag{1.24}$$

for any static electric field integrated around any closed path, and $\mathbf{E}$ is said to be a **conservative field**. (The notation $\oint$ indicates that the line integral is taken around a closed path.)

It follows that the work done on a unit charge by a conservative field in moving from point A to point B is independent of the path taken. This can be seen in Fig. 1.2 by picking any two points A and B on a closed path, and breaking the closed path integral into an integral from A to B along path C_1, followed by an integral from B to A along path C_2. Then

$$0 = \oint \mathbf{E} \cdot \mathbf{dr} = \int_{AB_1} \mathbf{E} \cdot \mathbf{dr} + \int_{BA_2} \mathbf{E} \cdot \mathbf{dr} = \int_{AB_1} \mathbf{E} \cdot \mathbf{dr} - \int_{AB_2} \mathbf{E} \cdot \mathbf{dr}, \tag{1.25}$$

so

$$\int_{AB_1} \mathbf{E} \cdot \mathbf{dr} = \int_{AB_2} \mathbf{E} \cdot \mathbf{dr}. \tag{1.26}$$

Since the $\int_{AB} \mathbf{E} \cdot \mathbf{dr}$ is independent of the path, it can be written as a scalar function ψ of only the endpoint positions.

$$\int_{AB} \mathbf{E} \cdot \mathbf{dr} = \psi(\mathbf{r}_A, \mathbf{r}_B). \tag{1.27}$$

Then, by the property of an integral that

$$\int_{AB_C} = \int_{AP_C} + \int_{PB_C} \tag{1.28}$$

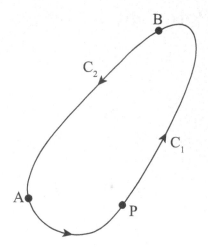

Figure 1.2: Closed integration path for $\oint \mathbf{E} \cdot \mathbf{dr}$.

for any point P on the path C, it follows that ψ must be the difference

$$\psi(\mathbf{r}_A, \mathbf{r}_B) = \phi(\mathbf{r}_A) - \phi(\mathbf{r}_B) \tag{1.29}$$

of a scalar function of position $\phi(\mathbf{r})$ evaluated at the two positions . The scalar function $\phi(\mathbf{r})$ is defined as the **electric potential**, related to the electric field by

$$\phi(\mathbf{r}_B) - \phi(\mathbf{r}_A) = -\int_{\mathbf{r}_A}^{\mathbf{r}_B} \mathbf{E} \cdot \mathbf{dr}. \tag{1.30}$$

The integral defining the difference in potentials is independent of the path chosen from $\mathbf{r}_A$ to $\mathbf{r}_B$. The potential ϕ defined in this way, with the minus sign before the integral in Eq. (1.30), has the physical significance of being the potential energy per unit charge due to the electric field.

Where practicable it is convenient to take point A to be at infinity. Then

$$\phi(\mathbf{r}) = \int_{\mathbf{r}}^{\infty} \mathbf{E} \cdot \mathbf{dr}. \tag{1.31}$$

This integral represents the work done on a unit charge by the electric field in moving the charge from the point $\mathbf{r}$ to infinity, which is equal to the work we would have do to move the charge from infinity to $\mathbf{r}$.

For a point charge, the integration in Eq. (1.31) is easily done using Eq. (1.23). This results in Coulomb's law for the potential of a point charge:

$$\phi = \frac{q}{r}. \tag{1.32}$$

In the same way,

$$\phi(\mathbf{r}) = \sum_n \frac{q_n}{|\mathbf{r} - \mathbf{r}_n|} \tag{1.33}$$

for a collection of point charges, and

$$\phi(\mathbf{r}) = \int \frac{dq'}{|\mathbf{r} - \mathbf{r}'|} \tag{1.34}$$

for continuous charge distributions.

Since the electric potential is the potential energy per unit charge, its units in the esu system are ergs per esu of charge. A more convenient energy unit for microphysics (atomic, molecular, nuclear, or elementary particle) is the **electron volt** (eV) which is the potential energy that a particle with the electron charge ($e = 4.8 \times 10^{-10}$ esu) has in a potential of one volt (SI units). The esu **statvolt(unit)** equals 300 volts, so the esu result for the potential energy of a particle with charge e should be multiplied by 300 to give its energy in eV. For instance the potential energy of an electron, a distance of 0.529 Å from the proton in a hydrogen atom is

$$U = -\frac{e^2}{r} = -300 \times \frac{4.8 \times 10^{-10}}{0.529 \times 10^{-8}} = -27.2\,\text{eV}. \tag{1.35}$$

Note that there is only one factor of e in the numerical calculation. That is because the second e is included in the definition of the eV unit. The unit Å (pronounced Angstrom) equals 10^{-8} cm, and is a convenient unit for atomic and molecular physics because the typical atomic size is about 1 Å.

When using the electron volt energy unit, it is convenient to also give particle masses in energy units (anticipating relativity), even for low velocities. For instance, the velocity of an electron of kinetic energy T=13.6 eV can be given as

$$\frac{v}{c} = \left[\frac{2T}{mc^2}\right]^{\frac{1}{2}} = \left[\frac{27.2}{.511 \times 10^6}\right]^{\frac{1}{2}} = 7.3 \times 10^{-3}, \tag{1.36}$$

where we have used $mc^2 = 0.511$ MeV for the electron.

1.3.1 Potential gradient

It is usually easier to find the potential than it is to calculate the electric field by integrating over the charge distribution. What is needed then is a way to determine the electric field from the potential (i.e., the inverse process of Eq. (1.31). We begin by first considering the relation between the potential ϕ and the field **E**.

A potential field is conveniently pictured by means of equipotentials, that is, surfaces along which ϕ is constant. A common example (in two dimensions) shown in Fig. 1.3 is a topographic map where the lines of equal altitude are equipotentials of the gravitational field. No work is done in moving along an equipotential, so the direction of **E** ('line of force') is everywhere perpendicular to the equipotential. Mathematically, that perpendicular direction is defined as the direction of the **gradient** of the potential. In our topographic example, this direction is the steepest direction up the hill. Experimentally, this would be opposite the direction a ball would roll if placed at rest on the hillside.

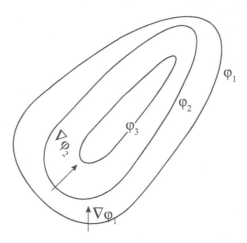

Figure 1.3: Equipotentials ϕ_i and gradients $\boldsymbol{\nabla}\phi_i$.

The magnitude of the gradient is defined to be the rate of change of the potential with respect to distance in the direction of maximum increase. For infinitesimal dispacements, an equipotential surface in three dimensions can be approximated by its tangent plane so the change in a scalar field in an infinitesimal displacement $\mathbf{dr}$ will vary as the cosine of the angle between the direction of maximum gradient and $\mathbf{dr}$. Then a vector gradient (written as $\mathbf{grad}\phi$) can be defined by

$$d\phi(\mathbf{r}) = \mathbf{dr}\cdot\mathbf{grad}\phi. \tag{1.37}$$

From Eq. (1.37), it can be seen that $\mathbf{grad}\phi$ determines how much the scalar field ϕ will change when you move a short distance $\mathbf{dr}$. The definition of $\mathbf{grad}\phi$ by Eq. (1.37) may seem a bit indirect, but Eq. (1.37) can be used to give the direct definition that

$$\mathbf{grad}\phi = \hat{\mathbf{n}}\frac{d\phi}{|\mathbf{dr}|}, \tag{1.38}$$

where the unit vector $\hat{\mathbf{n}}$ is in the direction of maximum increase of ϕ, and $\mathbf{dr}$ is taken in that direction of maximum increase. The direction of the gradient will always be in the direction of maximum increase and perpendicular to the equipotentials of the scalar function.

The rate of change of the scalar field in a general direction $\hat{\mathbf{a}}$, that is not necessarily the direction of maximum change, can be defined by choosing $\mathbf{dr}$ in that direction and dividing both sides of Eq. (1.37) by the magnitude $|\mathbf{dr}|$. This is called the **directional derivative** of ϕ, defined by $\hat{\mathbf{a}}\cdot\mathbf{grad}\phi$ for the rate of change of ϕ in the direction $\hat{\mathbf{a}}$.

To get a slightly different feel for the gradient of a scalar field, we consider its application in a specific coordinate system. (Up to now, we have used no specific coordinate system. For the most part, we will continue that practice since it permits more generality and introduces less algebraic complexity.)

In Cartesian (x, y, z) coordinates, an infinitesimal displacement is given by

$$\mathbf{dr} = \hat{\mathbf{i}}dx + \hat{\mathbf{j}}dy + \hat{\mathbf{k}}dz. \tag{1.39}$$

Then, Eq. (1.37) defining the gradient can be written

$$
\begin{aligned}
d\phi(x, y, z) &= (\hat{\mathbf{i}}dx + \hat{\mathbf{j}}dy + \hat{\mathbf{k}}dz) \cdot \mathbf{grad}\phi \\
&= (\mathbf{grad}\phi)_x dx + (\mathbf{grad}\phi)_y dy + (\mathbf{grad}\phi)_z dz
\end{aligned} \tag{1.40}
$$

At the same time, the differential of the function $\phi(x, y, z)$ of three variables is given by

$$d\phi(x, y, z) = \partial_x\phi dx + \partial_y\phi dy + \partial_z\phi dz. \tag{1.41}$$

Note that we will use the notation ∂_x rather than the more cumbersome $\frac{\partial}{\partial x}$ to represent the partial derivative in the x direction.

Comparing Eqs. (1.40) and (1.41) for the same differential, and using the fact that the displacements dx, dy, dz are independent and arbitrary, we see that

$$\mathbf{grad}\phi = \hat{\mathbf{i}}\partial_x\phi + \hat{\mathbf{j}}\partial_y\phi + \hat{\mathbf{k}}\partial_z\phi. \tag{1.42}$$

Equation (1.42) holds only in Cartesian coordinates. The form of $\mathbf{grad}\phi$ is somewhat more complicated in other coordinate systems.

Equation (1.42) can be written in terms of a vector differential operator $\boldsymbol{\nabla}$, given in Cartesian coordinates by

$$\boldsymbol{\nabla} = \hat{\mathbf{i}}\partial_x + \hat{\mathbf{j}}\partial_y + \hat{\mathbf{k}}\partial_z. \tag{1.43}$$

Then, the gradient of a scalar field can be written as

$$\mathbf{grad}\phi = \boldsymbol{\nabla}\phi. \tag{1.44}$$

The representation of $\mathbf{grad}$ by the vector differential operator $\boldsymbol{\nabla}$ (usually called $\mathbf{del}$) is not limited to Cartesian coordinates, although its form is not given by Eq. (1.43) for other coordinate systems. We will derive a coordinate independent definition of $\boldsymbol{\nabla}$ later.

From Eq. (1.30), relating ϕ to $\mathbf{E}$, we see that the integrand in Eq. (1.30) is the differential

$$d\phi = -\mathbf{E} \cdot \mathbf{dr}. \tag{1.45}$$

This is just the definition of the gradient of ϕ, so comparing Eq. (1.45) with Eq. (1.37), we see that

$$\mathbf{E} = -\boldsymbol{\nabla}\phi. \tag{1.46}$$

This permits us to find $\mathbf{E}$ once ϕ has been determined. In words, Eq. (1.46) states that the electric field is the negative gradient of the potential, or, more simply, "E equals minus Del phi."

The actual calculation of $\boldsymbol{\nabla}\phi$ can be made using a coordinate system, but it is usually better to use the definition of the gradient in Eq. (1.38) to find it for various functions of the position vector $\mathbf{r}$ directly. We start with r, the magnitude of $\mathbf{r}$, treated as a scalar field. Its maximum rate of change is in the $\hat{\mathbf{r}}$ direction, and its derivative in that direction is dr/dr=1. So

$$\boldsymbol{\nabla} r = \hat{\mathbf{r}}. \tag{1.47}$$

Next, we consider any scalar function, f(r), of the magnitude of $\mathbf{r}$. The direction of maximum rate of change of f(r) will also be $\hat{\mathbf{r}}$, and its derivative in that direction is df/dr. So

$$\boldsymbol{\nabla} f(r) = \hat{\mathbf{r}}\frac{df}{dr}. \tag{1.48}$$

Applying this result to Coulomb's law for the potential due to a point charge gives

$$\mathbf{E} = -\boldsymbol{\nabla}\left(\frac{q}{r}\right) = -q\hat{\mathbf{r}}\frac{d}{dr}\left(\frac{1}{r}\right) = \frac{q\hat{\mathbf{r}}}{r^2}, \tag{1.49}$$

so we have derived Coulomb's law for $\mathbf{E}$ from Coulomb's law for ϕ.

The same derivation works for the electric field of a collection of point charges or a continuous charge distribution. For a continuous distribution, this becomes

$$\begin{aligned} \mathbf{E}(\mathbf{r}) &= -\boldsymbol{\nabla}\int\frac{dq'}{|\mathbf{r}-\mathbf{r}'|} \\ &= -\int dq'\boldsymbol{\nabla}\frac{1}{|\mathbf{r}-\mathbf{r}'|} \\ &= \int\frac{(\mathbf{r}-\mathbf{r}')dq'}{|\mathbf{r}-\mathbf{r}'|^3}, \end{aligned} \tag{1.50}$$

where the $\boldsymbol{\nabla}$ can be taken under the integral because it is a partial derivative that does not act on $\mathbf{r}'$. We have also used the fact that $(\mathbf{r}-\mathbf{r}')$ is just $\mathbf{r}$ with the origin displaced by the constant vector $\mathbf{r}'$, so

$$\boldsymbol{\nabla}\left(\frac{1}{|\mathbf{r}-\mathbf{r}'|}\right) = -\frac{(\mathbf{r}-\mathbf{r}')}{|\mathbf{r}-\mathbf{r}'|^3}. \tag{1.51}$$

1.4   Gauss's Law

A method for using symmetry to find $\mathbf{E}$ without integrating over the charge distribution is given by **Gauss's law**. We first derive Gauss's law from Coulomb's law for a single point charge. We start with the surface integral,

$$\oint d\mathbf{A}\cdot\mathbf{E} = q \oint d\mathbf{A}\cdot\frac{\mathbf{r}}{r^3}, \tag{1.52}$$

of the normal component of $\mathbf{E}$ over a closed surface surrounding the point charge, as shown in Fig. 1.4.

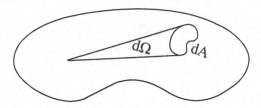

Figure 1.4: The solid angle $d\Omega$ subtended by the surface differential dA for Gauss's law.

The vector differential of area, $d\mathbf{A}$, is an infinitesimal surface element of magnitude dA. Since it is infinitesimal, it approaches a plane surface, tangent to the closed surface. By convention, its vector direction is along the outward normal to the closed surface. The integrand of the surface integral in Eq. (1.52) can be recognized as the definition of the solid angle subtended by the differential surface element $d\mathbf{A}$, as can be seen on Fig. 1.4.

$$d\Omega = \frac{\hat{\mathbf{r}}\cdot d\mathbf{A}}{r^2}. \tag{1.53}$$

Then the surface integral can be written as

$$\oint d\mathbf{A}\cdot\mathbf{E} = q \oint d\Omega = 4\pi\,q, \tag{1.54}$$

with the factor 4π arising as the magnitude of the total solid angle of any closed surface.

If the point charge q were located outside the closed surface, then the surface integral would be zero. This can be seen in Fig. 1.5.

If a plane surface is made to cut the closed integration surface into two parts, then the integrated solid angle over each part of the surface will equal in magnitude the solid angle subtended by the part of the plane surface inside the closed surface. But the two solid angles will be of opposite sign and just cancel. Thus the integral over the closed surface will be $4\pi q$ for a point charge inside the surface, and zero for a point charge outside the surface.

For a collection of point charges, only those inside the surface will contribute to the integral, and we have Gauss's law

$$\oint d\mathbf{A}\cdot\mathbf{E} = 4\pi\,Q_{\text{enclosed}}, \tag{1.55}$$

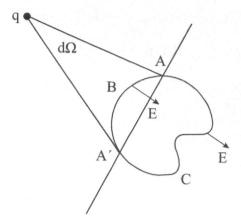

Figure 1.5: Gauss's law for charge outside a closed suface. The solid angle $d\Omega$ subtended by each of the curved surfaces ABA' and ACA' are equal to the solid angle subtended by the plane AA'.

where Q_{enclosed} is the net charge within the surface. Since a continuous charge distribution is really a very large number of point charges, Gauss's law for a continuous distribution is also given by Eq. (1.55), with

$$Q_{\text{enclosed}} = \int \rho(\mathbf{r})d\tau, \qquad (1.56)$$

where the integral is over the volume enclosed by the closed surface.

For SI units, Gauss' law is written

$$\oint d\mathbf{A}{\cdot}\mathbf{E} = \frac{1}{\epsilon_0}Q_{\text{enclosed}}, \text{ SI} \qquad (1.57)$$

without the 4π, but introducing ϵ_0. Actually the 4π in Eq. (1.55), the Gaussian form of Gauss's law, is a good reminder that it comes from the integral over all solid angle.

Gauss's law provides a powerful and simple method to find $\mathbf{E}$ whenever there is enough symmetry to enable the surface integral to be done without integration. But, if use of Gauss's law requires a complicated surface integration, then another method should be used to find $\mathbf{E}$.

1.4.1   Examples of Gauss's law

Point charge

As a simple example of Gauss's law, we use it to derive Coulomb's law for a point charge, demonstrating the steps used in application of Gauss's law. The first step is to recognize the symmetry of the charge configuration, which, in the case of an isolated point charge is spherical symmetry about the point charge. The type of symmetry dictates the **Gaussian surface** to be used for the surface integral. For the point charge, this is a sphere, of any radius r, centered at the point charge so as to make use of the symmetry, as shown in Fig. 1.6. Note that the Gaussian surface is just a mathematical surface that need not be (and usually isn't) any physical surface of the problem.

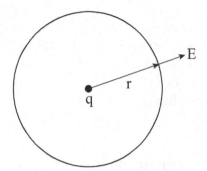

Figure 1.6: Gaussian sphere for a single point charge.

Next the symmetry is used to make simplifying observations about the **E** field at the Gaussian surface. For the point charge, we first observe that **E** must be in the radial direction with respect to the charge. This follows, most simply from the principle of insufficient reason. That is, looking at Fig. 1.6, there is no more reason for **E** to be directed to the right of radial than to the left, since there is no reference point other than the charge to define left or right. Another, more mathematical, derivation of the radial direction of **E** is the fact that only one vector, the position vector **r**, can be defined for this geometry. Thus the vector function **E** must be given only in terms of this vector and can only be $\mathbf{E}(\mathbf{r}) = E(\mathbf{r})\hat{\mathbf{r}}$. (This reasoning will be used later in the book to simplify other vector integrations.)

The next observation is that the radial **E** must have the same magnitude for all points on the sphere, because all such points are equivalent, given the spherical symmetry. These two observations (that **E** is radial and constant in magnitude around the spherical surface) lead to the simplification of Gauss's law for this case that

$$\oint \mathbf{E}\cdot d\mathbf{A} = E \oint dA = 4\pi r^2 E = 4\pi q. \qquad (1.58)$$

The $\oint dA$ in Eq. (1.58) was done by just observing that it is the surface area of the sphere. Then dividing by $4\pi r^2$, gives

$$\mathbf{E}(\mathbf{r}) = \frac{q\hat{\mathbf{r}}}{r^2}, \qquad (1.59)$$

which is just Coulomb's law for the electric field.

Because Gauss's law is derived from Coulomb's law, and Coulomb's law for the electric field can be derived from Gauss's law, the two laws are mathematically equivalent. Either one could be chosen as the starting point for electrostatics. Historically, Coulomb's law was discovered first, but Gauss's law is more general in the sense that Coulomb's law is just one of the simple applications of Gauss's law. Also, as we will see later in discussing Faraday's 'ice bucket' experiment, Gauss's law can be verified to greater accuracy than Coulomb's law.

However, Gauss's law by itself does not lead to the Coulomb law of force, Eq.(1.1). Starting from Gauss's law, the additional assumption that the charge appearing in $\mathbf{F} = q\mathbf{E}$ is the same (i.e., interchangeable) with the charge appearing in Gauss's law. This was implicitly assumed in Coulomb's force law, leading to Newton's third law for the Coulomb force. The equivalence of these two charges, the **source charge** in Gauss's law, and the **force charge** in $\mathbf{F} = q\mathbf{E}$, must be an additional assumption in classical electromagnetism.

1.4.2   Spherically symmetric charge (and mass) distributions

You may have noticed in the derivation of Coulomb's law from Gauss's law, that the fact that the charge was a point charge never entered. The only property that was needed for the charge distribution was that it was spherically symmetric and entirely contained within the Gaussian sphere. Thus the derivation would work just as well for any such charge distribution, and we have the result that the electric field outside any spherically symmetric charge distribution is the same as that for a point charge of the same net charge, located at the center of spherical symmetry. This means, for instance, that a uniformly charged sphere, or a uniformly charged shell (or collection of shells) would have the same electric field (and, therefore, the same electric potential) beyond the charge distribution as a point charge, and be indistinguishable from a point charge. Of course, inside the charge distribution, the electric field would be modified, and not fall off like $1/r^2$.

This result would have been very important in the development of gravitational theory (which is mathematically equivalent to electrostatics), had Newton known Gauss's law. Newton formulated his law of gravitation for point masses, just as the later Coulomb's law was formulated for point charges. However, Newton knew that the Earth and the Moon were not point masses. How could the point mass formula work so well for extended masses?

Newton predated Gauss so could be excused for not knowing Gauss's law. The only formulation he had for the force on an extended object B exerted by an extended object A was (We use Coulomb's law, but the arguments would be the same for Newton's law of gravity.)

$$\begin{aligned}
\mathbf{F_B} &= \int \rho_B(\mathbf{r_B})\mathbf{E}_A(\mathbf{r_B})d\tau_B \\
&= \int d\tau_B \int d\tau_A \rho_B(\mathbf{r_B})\rho_A(\mathbf{r_A})\frac{(\mathbf{r_B} - \mathbf{r_A})}{|\mathbf{r_B} - \mathbf{r_A}|^3}.
\end{aligned} \qquad (1.60)$$

The double integral in Eq. (1.60) satisfies Newton's third law nicely, but is difficult to integrate directly, even for two uniformly charged spheres. The double integral does not look anything like the force between two point charges (point masses for Newton). Poor Newton worked for many years. He actually used a complicated geometrical argument (given in his *Principia*) to show that the force between two spheres, each with spherically symmetric density, was exactly the same as the force between two point masses. So all his relatively simple equations for point masses also worked for large physical objects, as long as they were spherically symmetric.

We now show how Newton could have saved a lot of work had he known Gauss's law. Consider the force between the two extended, but spherically symmetric and non-overlapping, charge distributions, A and B, shown in Fig. 1.7a. The electric field due to A would appear at sphere B like that of a point charge, by the Gauss's law argument given above. So, the $\mathbf{E_A}(\mathbf{r_B})$ in Eq.(1.60) would be that of a point charge, but this would still leave an integral over the extended object B. However we now know that the force on B is the same as if the extended object A were a point charge, as shown in Fig. 1.7b.

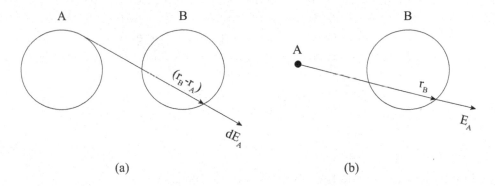

(a)   (b)

Figure 1.7: Gauss's law for two uniformly charged spheres. By Gauss's law, the sphere A in Fig. (a) can be replaced by a point charge in Fig. (b).

We use Newton's third law to observe that the force on the (equivalent) point charge A due to the extended charge B is the same as that on B due to A. In calculating the force on the point charge A, we use Gauss's law again to effectively replace the extended object B by a point charge, so the force on A is the same as that due to a point charge B. We thus see that the force between A and B is the same as that between two point charges, without performing any integral.

Line charge

Gauss's law can be applied to other simple symmetries. For an infinitely long (In practical terms, "infinitely long" means length $\gg r_\perp$, and "$\gg$" will usually mean about a factor of 10, although $\gg$ is often sneaked in for lower ratios.) straight line charge with linear charge density λ, the Gaussian surface is a cylinder of arbitrary length L and radius $r_\perp$, coaxial with the line charge, as shown in Fig. 1.8.

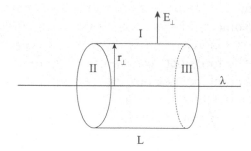

Figure 1.8: Gaussian surface for a line charge.

From the axial symmetry, **E** is everywhere directed straight out from the line charge, and can depend only on the perpendicular distance from the line charge, $\mathbf{r}_\perp$. The integral for Gauss's law is over three surfaces, the curved surface of the cylinder (I), and the two end caps (II and III). The integrals over the endcaps vanish because **E** is always parallel to the surface of the endcaps. $E_\perp$ is constant on the curved surface, and can be taken out of the integral, which becomes just the curved surface area. Then, Gauss's law becomes

$$\oint \mathbf{E} \cdot d\mathbf{A} = E_\perp \int_I dA = 2\pi r_\perp L E_\perp = 4\pi\lambda L, \tag{1.61}$$

so

$$E_\perp = \frac{2\lambda}{r_\perp}. \tag{1.62}$$

The potential for the line charge is the radial integral of $E_\perp$

$$\phi(\mathbf{r}) = -2\lambda \int_r^{r_0} \frac{dr}{r} = 2\lambda \ln(r/r_0). \tag{1.63}$$

$E_\perp$ falls off too slowly for the integral for ϕ to converge at infinity, and an infinite amount of work would have to be done to bring a charge from infinity to finite distances for the infinite line charge. Therefore, an arbitrary finite point, r_0, has been chosen for us to set $\phi(r_0) = 0$. As with spherical symmetry, the Gauss's law derivation for the wire would also hold outside any axially symmetric charge distribution that is uniform along its axis. So, **E** outside any such charge distribution is the same as that for a uniform line charge.

Infinite plane

We next look at an infinite plane sheet with a constant surface charge density σ. The symmetry of the infinite plane sheet requires that **E** be perpendicular to the plane of the sheet, and not depend on position parallel to the sheet. The appropriate Gaussian surface, shown in Fig. 1.9, is a **Gaussian pillbox**, a flat box with identical parallel ends (I and II) of arbitrary area A and any shape. (Gauss meant a flat box used to carry healing pills, and not the deadly military fort of the same general shape.) The pillbox is oriented parallel to the sheet of charge and is bisected by the sheet. **E** is parallel to the side (III) of the pillbox, so the only contribution to the Gauss's law integral is from the two flat surfaces.

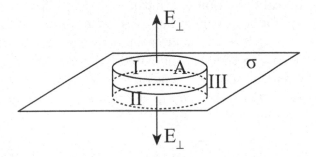

Figure 1.9: Gaussian pillbox for an infinite plane.

From the symmetry of the sheet, **E** has the same magnitude, but opposite direction, on either side of the sheet. Gauss' law then leads to

$$\oint \mathbf{E}{\cdot}\mathbf{dA} = 2E_\perp \int_I dA = 2E_\perp A = 4\pi\sigma A, \tag{1.64}$$

so

$$E_\perp = 2\pi\sigma, \tag{1.65}$$

for an infinite sheet of charge with constant surface charge density σ. Note that **E** is independent of the distance from the infinite plane sheet, as long as the sheet still looks infinite.

The above derivation can be modified slightly to give the discontinuity in **E**, when a charged surface is crossed. For the discontinuity, we consider infinitesimal distances on either side of the surface so that any continuous surface looks like a plane. Then, using the Gaussian pillbox in Fig. 1.9, the discontinuity in **E** is given by

$$\mathbf{\hat{n}}{\cdot}\mathbf{E_1} - \mathbf{\hat{n}}{\cdot}\mathbf{E_2} = 4\pi\sigma, \tag{1.66}$$

where $\mathbf{\hat{n}}$ is the unit vector normal to the plane, pointing from region 1 to region 2. Equation (1.65) is a special case of this result when the only source of **E** is the surface charge.

1.5   The Variation of E

We have seen that the variation of a scalar field ϕ is determined by its gradient, so $d\phi = -\mathbf{dr}{\cdot}\boldsymbol{\nabla}\phi$. A vector field can have two different types of variation. It can vary along its direction, for example like the velocity field, **v**, of a stream as the slope gets steeper. The vector field can also vary across its direction, as when the velocity is faster in the middle of the stream than near the edges. How can these two variations be measured?

1.5.1   Divergence

We now give a physical definition of what is called the divergence of a vector field. The increase of a vector field along its direction is shown in Fig. 1.10. A measure of the strength of the field is the density of **lines of force** in the figure, with the increase in the field indicated by increasing lines of force. We construct a mathematical volume V enclosed by a surface S, as shown in the figure. The increase in **E** can be seen in the figure as more lines of **E** leaving the volume than entering it.

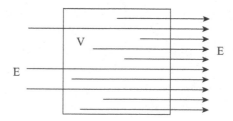

Figure 1.10: Divergence of lines of **E**. More lines leave the volume than enter it.

A quantitative measure of the excess of lines leaving the volume is given by the integral $\oint_S \mathbf{E} \cdot \mathbf{dA}$. This integral can be used to define an average **divergence** (written as "div")of the lines of the vector field. That is

$$\langle \text{div } \mathbf{E} \rangle_V = \frac{1}{V} \oint_S \mathbf{E} \cdot \mathbf{dA}, \tag{1.67}$$

where the notation $\langle \text{div } \mathbf{E} \rangle_V$ denotes the average of div **E** over the volume V. The value of div **E** at a point can be defined by shrinking the integral about the point, so that

$$\text{div } \mathbf{E} = \lim_{V \to 0} \frac{1}{V} \oint_S \mathbf{E} \cdot \mathbf{dA} \tag{1.68}$$

gives the divergence of the vector field at a point (if the limit exists), and is a measure of its rate of increase along the direction of the vector field.

As we did with the gradient, we now show what the divergence would look like in Cartesian coordinates. Figure 1.11 shows an infinitesimal volume (a parallelepipid in Cartesian coordinates) of dimensions $\Delta x \times \Delta y \times \Delta z$, that will shrink to zero at the point x, y, z. The surface integral in the definition of div**E** is over the six faces of the parallelepipid, I-VI, so the integral can be written as

$$\text{div } \mathbf{E} = I + II + III + IV + V + VI, \tag{1.69}$$

where I indicates the integral over face I, and similarly for the other faces.

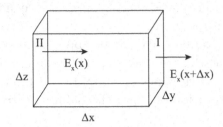

Figure 1.11: Volume element in Cartesian coordinates.

We concentrate first on faces I and II, each parallel to the y-z plane. In the limit as both Δy and Δz approach zero, the integral over face I approaches $E_x(x + \Delta x, y, z)\Delta y \Delta z$, and that over face II approaches $-E_x(x, y, z)\Delta y \Delta z$, provided that E_x is continuous at the point x,y,z. So, for these two faces

$$
\begin{aligned}
I + II &= \lim_{\Delta x, \Delta y, \Delta z \to 0} \frac{[E_x(x + \Delta x, y, z)\Delta y \Delta z - E_x(x, y, z)\Delta y \Delta z]}{\Delta x \Delta y \Delta z} \\
&= \lim_{\Delta x \to 0} \frac{[E_x(x + \Delta x, y, z) - E_x(x, y, z)]}{\Delta x} \\
&= \partial_x E_x,
\end{aligned}
\tag{1.70}
$$

where the last step follows from the definition of the partial derivative.

The integrals over the other four faces are done in the same way, leading to similar results, with the substitutions $x \to y$ and then $x \to z$, so

$$
\text{div } \mathbf{E} = \partial_x E_x + \partial_y E_y + \partial_z E_z \tag{1.71}
$$

in Cartesian coordinates.

The form of Eq. (1.71) suggests that div $\mathbf{E}$ could be written as a dot product

$$
\text{div } \mathbf{E} = \boldsymbol{\nabla} \cdot \mathbf{E}, \tag{1.72}
$$

of the vector differential operator $\boldsymbol{\nabla}$ with $\mathbf{E}$. Equation (1.72) can be stated in words as "divergence $\mathbf{E}$ equals **Del dot E**".

The representation of div by the vector differential operator $\boldsymbol{\nabla}\cdot$ is not limited to Cartesian coordinates, although its explicit form is not given by Eq. (1.71) for other coordinate systems. We will derive a coordinate independent definition of $\boldsymbol{\nabla}$ shortly.

Equation (1.67) defines the average of the divergence over a finite volume. Using the definition of a volume average that

$$
\langle \boldsymbol{\nabla} \cdot \mathbf{E} \rangle_V = \frac{1}{V} \int_V \boldsymbol{\nabla} \cdot \mathbf{E} d\tau, \tag{1.73}
$$

Eq. (1.67) can be rewritten as

$$
\int_V \boldsymbol{\nabla} \cdot \mathbf{E} d\tau = \oint_S d\mathbf{A} \cdot \mathbf{E}. \tag{1.74}
$$

In this form, it is called the **divergence theorem**. Our derivation of the divergence theorem has been so simple because we have effectively defined the average divergence for a finite volume by the divergence theorem, and then shown that this average divergence approaches other definitions of the divergence at a point as the volume shrinks to the point.

The definition of the divergence given by Eq. (1.68) can be used to evaluate the divergence of the position vector. The definition gives

$$\mathbf{\nabla \cdot r} = \lim_{V \to 0} \frac{1}{V} \oint \mathbf{r \cdot dA}$$

$$= \lim_{V \to 0} \frac{1}{V} \oint R^3 d\Omega = 3, \qquad (1.75)$$

where we have used the fact that

$$V = \oint d\Omega \int_0^R r^2 dr = \frac{1}{3} \oint R^3 d\Omega. \qquad (1.76)$$

(Note that the R in the integral $\oint R^3 d\Omega$ refers to the distance from the origin to the bounding surface, and can be a function of angle.)

We next calculate the divergence of the electric field in Coulomb's law to get

$$\mathbf{\nabla \cdot} \left[\frac{q\mathbf{r}}{r^3} \right] = q \frac{\mathbf{\nabla \cdot r}}{r^3} + q\mathbf{r \cdot \nabla} \frac{1}{r^3}$$

$$= \frac{3q}{r^3} - \frac{3q\mathbf{r \cdot \hat{r}}}{r^4} = 0, \quad r \neq 0. \qquad (1.77)$$

The restriction $r \neq 0$ is necessary because both terms in Eq. (1.77) are singular at $r = 0$.

In Eq. (1.77) we have demonstrated the procedure for applying the vector differential operator $\mathbf{\nabla}$ to a combination of functions. We have made use of the fact that the operator $\mathbf{\nabla}$ has two distinct properties:

1. $\mathbf{\nabla}$ is a differential operator.

2. $\mathbf{\nabla}$ is a vector.

Because $\mathbf{\nabla}$ is a differential operator, it acts on functions one at a time, just as in d(uv)=udv+vdu. We also follow the convention that the differential operator acts only on functions to its right, so the order in which $\mathbf{\nabla}$ appears must be to the left of the functions it acts on and to the right of the other fucntions. As a vector, $\mathbf{\nabla}$ must behave, in any expansion, like any other vector.

The utilization of both of these properties can be seen in Eq. (1.77). There are two terms, because $\mathbf{\nabla}$ acts separately on the $\mathbf{r}$ and on the $1/r^3$, and, in each term, the $\mathbf{\nabla}$ remains dotted with the $\mathbf{r}$, and to the left of the function it acts on. In every case, strict adherence to these two properties of $\mathbf{\nabla}$ will lead to the correct evaluation of vector derivatives.

1.5.2 Dirac delta function

We now investigate the behavior of $\boldsymbol{\nabla}\cdot[\mathbf{r}/r^3]$ at $r = 0$. In fact, something quite dramatic happens at the origin to the divergence of $\mathbf{r}/r^3$, as can be seen by applying the divergence theorem to $\mathbf{r}/r^3$:

$$\int d\tau \boldsymbol{\nabla}\cdot\left[\frac{\mathbf{r}}{r^3}\right] = \oint \frac{d\mathbf{A}\cdot\mathbf{r}}{r^3} = \oint d\Omega = 4\pi. \tag{1.78}$$

So, even though $\boldsymbol{\nabla}\cdot[\mathbf{r}/r^3]$ vanishes at all but one point, its volume integral is not zero. This property is consistent with one definition of the **Dirac delta function** in three dimensions:

$$\begin{aligned} \int_V d\tau \delta(\mathbf{r}) &= 1, &&\text{if r = 0 inside V,} \\ \int_V d\tau \delta(\mathbf{r}) &= 0, &&\text{if r} \neq 0 \text{ inside V.} \end{aligned} \tag{1.79}$$

Then

$$\boldsymbol{\nabla}\cdot\left[\frac{\mathbf{r}}{r^3}\right] = 4\pi\delta(\mathbf{r}). \tag{1.80}$$

We can now take the divergence of the Coulomb electric field, including the origin, to get

$$\boldsymbol{\nabla}\cdot\left[\frac{q\mathbf{r}}{r^3}\right] = 4\pi q\delta(\mathbf{r}). \tag{1.81}$$

To apply the divergence theorem to integrals over continuous charge distributions, we extend Eq.(1.80) to

$$\boldsymbol{\nabla}\cdot\left[\frac{(\mathbf{r}-\mathbf{r}')}{|\mathbf{r}-\mathbf{r}'|^3}\right] = 4\pi\delta(\mathbf{r}-\mathbf{r}'), \tag{1.82}$$

where the constant vector $\mathbf{r}'$ just shifts the origin of coordinates from $\mathbf{0}$ to $\mathbf{r}'$.

In any integral over a volume V containing the point $\mathbf{r}' = \mathbf{r}$, the region of integration can be shrunk to an infinitesimal volume surrounding $\mathbf{r}$. Then the integral including $\delta(\mathbf{r}-\mathbf{r}')$ has the property

$$\begin{aligned} \int_V \delta(\mathbf{r}-\mathbf{r}')f(\mathbf{r}')d\tau' &= f(\mathbf{r}), &&\text{if } \mathbf{r}' = \mathbf{r} \text{ inside V,} \\ \int_V \delta(\mathbf{r}-\mathbf{r}')f(\mathbf{r}')d\tau' &= 0, &&\text{if } \mathbf{r}' \neq \mathbf{r} \text{ inside V,} \end{aligned} \tag{1.83}$$

provided that $\lim_{\mathbf{r}'\to\mathbf{r}} f(\mathbf{r}')$ exists. So, an integral with a delta function is the simplest integral to do. The integration just involves evaluating the rest of the integrand at the point where the argument of the delta function vanishes.

Equation (1.83) is a somewhat better definition of the Dirac delta function than Eq.(1.79) because it permits a more general definition of the delta function with respect to a class of functions $f(\mathbf{r})$. Then, Eq.(1.79) follows from this definition if $f(\mathbf{r}')$ is chosen to be 1.

We must empasize that the Dirac delta function is not a mathematical function in the strict sense. In fact, as a function, it does not make sense. It would vanish everywhere, except where it was not defined (loosely speaking, "infinite"). That is why we have been careful, in either definition, to define the delta function only in terms of its property in integrals.

When we write it in equations where we do not integrate, such as Eq.(1.82), it is always with the understanding that the delta function will only be given physical meaning in a subsequent integration. That is, in equations like Eq.(1.82), the delta function is just an indication of how to perform a pending integration.

Applying the divergence to the Coulomb integral for the electric field of a volume distribution of charge leads to

$$
\begin{aligned}
\boldsymbol{\nabla}{\cdot}\mathbf{E}(\mathbf{r}) &= \boldsymbol{\nabla}{\cdot}\int \frac{(\mathbf{r}-\mathbf{r}')\rho(\mathbf{r}')d\tau'}{|\mathbf{r}-\mathbf{r}'|^3} \\
&= \int \rho(\mathbf{r}')d\tau'\boldsymbol{\nabla}{\cdot}\left[\frac{(\mathbf{r}-\mathbf{r}')}{|\mathbf{r}-\mathbf{r}'|^3|}\right] \\
&= 4\pi \int \rho(\mathbf{r}')d\tau'\delta(\mathbf{r}-\mathbf{r}').
\end{aligned}
\tag{1.84}
$$

Doing the delta function integral gives

$$
\boldsymbol{\nabla}{\cdot}\mathbf{E} = 4\pi\rho
\tag{1.85}
$$

for any continuous charge distribution.

Equation (1.85) has been derived starting from Coulomb's law, which was the historical order of development. However, the theory of electrostatics could also start with the partial differential equation given by Eq.(1.85). Then, Gauss's law can be derived by applying the divergence theorem

$$
\oint_S \mathbf{E}{\cdot}\mathbf{dA} = \int_V \boldsymbol{\nabla}{\cdot}\mathbf{E}d\tau = 4\pi \int_V \rho(\mathbf{r})d\tau = 4\pi Q_{\text{enclosed}}.
\tag{1.86}
$$

And, as we have shown, Gauss's law can be used to derive Coulomb's law for a point charge. So we see there are three, mathematically equivalent "starting points" for electrostatics (Coulomb's law, Gauss's law, $\boldsymbol{\nabla}{\cdot}\mathbf{E} = 4\pi\rho$).

Since Gauss's law can be derived from $\boldsymbol{\nabla}{\cdot}\mathbf{E} = 4\pi\rho$, and $\boldsymbol{\nabla}{\cdot}\mathbf{E} = 4\pi\rho$ can be derived from Gauss's law (by following the steps of Eq.(1.85) backward), they are mathematically equivalent. For this reason, the equation $\boldsymbol{\nabla}{\cdot}\mathbf{E} = 4\pi\rho$ is sometimes called "the differential form of Gauss's law". However, the two laws represent quite different physical manifestations.

Equation (1.85) can be put in terms of the electric potential ϕ, leading to **Poisson's equation**

$$
\boldsymbol{\nabla}{\cdot}(\boldsymbol{\nabla}\phi) = \nabla^2\phi = -4\pi\rho.
\tag{1.87}
$$

Equation (1.87) introduces the **Laplacian** differential operator ∇^2 defined by the application of the divergence to the gradient of a scalar. In Cartesian coordinates, the Laplacian operator is given by

$$\nabla^2 = \partial_x^2 + \partial_y^2 + \partial_z^2, \qquad (1.88)$$

but it is more complicated in other coordinate systems. The homogeneous form of Poisson's equation, with the source function $\rho = 0$,

$$\nabla^2\phi = 0 \qquad (1.89)$$

is called **Laplace's equation.**

1.5.3   Curl

Next, we look at how **E** can vary across its direction, and we give a physical definition of the **curl** of a vector field.

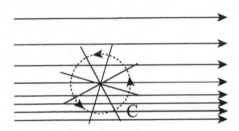

Figure 1.12: Velocity field with curl. The current increases going down on the figure causing the paddle wheel to rotate counterclockwise.

Figure 1.12 shows a vector field having such a variation, with the density of lines being proportional to the strength of the field. If this were a velocity field, such as the current of water in a stream, this variation could be measured experimentally by placing a paddle wheel in the stream as shown in the figure. Then the rotation of the paddle wheel would be a measure of the variation of the vector field. This can be done without getting wet by calculating a line integral around a typical closed curve C, as shown on the figure.

This integral can be used to define an average value of the variation (called **curl**) over a surface S bounded by the curve C. The average **curl** is defined by

$$\langle \hat{\mathbf{n}}\cdot\mathbf{curl}\,\mathbf{E}\rangle_S = \frac{1}{S}\oint_C d\mathbf{r}\cdot\mathbf{E}, \qquad (1.90)$$

where $\hat{\mathbf{n}}$ is the unit vector normal to the surface S at any point. Note that, by this definition, the combination $S\langle\hat{\mathbf{n}}\cdot\mathbf{curl}\,\mathbf{E}\rangle_S$ does not depend on the shape of the surface S, but only on the bounding path C. Since the variation will be different in different directions, it is the average value of the normal component of **curl** that is defined by Eq.(1.90).

The positive sign for the direction of n̂ is taken by convention to be the **boreal** direction. That is, if integral around the contour C is taken in the direction of the rotation of the earth, then the north pole is in the positive direction as shown on Fig. 1.13a. This is also stated as the **right hand rule**: If the integral around the contour C is taken in the direction that the four fingers of the right hand curl as they tend to close, then the right thumb points in the positive direction for n̂, as shown in Fig. 1.13b. This will be our general sign convention relating the direction of integration around a closed curve and the positive direction of the normal vector to any surface bounded by the curve.

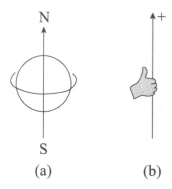

(a)   (b)

Figure 1.13: (a) Boreal direction on the globe. (b) Right-hand rule for positive direction.

The value of **curl E** at a point can be defined by starting with a smooth surface through the point and taking the limit as the curve bounding the surface shrinks about the point, and the enclosed surface shrinks to zero area. This gives the definition of **curl** at a point:

$$(\mathbf{curl\,E})_n = \lim_{S \to 0} \frac{1}{S} \oint_C \mathbf{dr}\cdot\mathbf{E}. \tag{1.91}$$

As the curve C shrinks to a point, the smooth surface approaches its tangent plane at the point, and $(\mathbf{curl\,E})_n$ in Eq.(1.90) represents the component of curl in the direction of the normal vector n̂ to the tangent plane.

With **curl E** defined by Eq.(1.91), it is possible to find its specific form in Cartesian coordinates. To find **curl E** at the point x,y,z, we consider an infinestimal rectangle of dimension Δx by Δy parallel to the x-y plane, as shown in Fig. 1.14. The line integral around the rectangle consists of four parts, so

$$\lim_{S \to 0} \frac{1}{S} \oint_C \mathbf{dr}\cdot\mathbf{E} = I + II + II + IV. \tag{1.92}$$

As Δy approaches zero, we can make the replacement

$$\int_y^{y+\Delta y} f(y')dy' \to f(y)\Delta y, \tag{1.93}$$

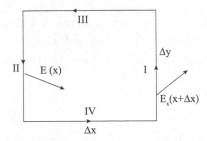

Figure 1.14: Differential surface for curl in Cartesian coordinates.

provided that f(y) is continuous at y. Then the contribution to Eq.(1.91) from sides I and II is

$$
\begin{aligned}
I + II &= \lim_{\Delta x, \Delta y \to 0} \frac{[E_y(x + \Delta x, y, z)\Delta y - E_y(x, y, z)\Delta y]}{\Delta x \Delta y} \\
&= \lim_{\Delta x \to 0} \frac{[E_y(x + \Delta x, y, z) - E_y(x, y, z)]}{\Delta x} \\
&= \partial_x E_y.
\end{aligned}
\tag{1.94}
$$

The procedure for sides III and IV is the same, with the interchange of x and y, and a change of overall sign, leading to

$$
(\mathbf{curl\,E})_z = \partial_x E_y - \partial_y E_x,
\tag{1.95}
$$

for the z component of **curl E**.

The other components of **curl** follow in the same way with the cyclic substitution $x \to y$, $y \to z$, $z \to x$, as illustrated in Fig. 1.15. With these substi-

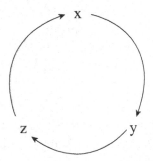

Figure 1.15: Cyclic order of $x \to y \to z$.

tutions, Eq. (1.95) can be extended to

$$
(\mathbf{curl\,E})_z = \partial_x E_y - \partial_y E_x, \quad \text{and cyclic.}
\tag{1.96}
$$

The "and cyclic" in Eq.(1.96) means that it stands for three equations, the one written and two others that follow by cyclic substitution. As a mnemonic, the positive term in any component of **curl** is written in cyclic order, and then the second term is in the opposite order with a minus sign. The form of Eq.(1.96) suggests that **curl E** can be written as a cross product of the derivative operator ∇ and **E**:

$$\mathbf{curl\,E} = \nabla \times \mathbf{E}, \qquad (1.97)$$

although the simple form of Eq. (1.96) only holds in cartesian coordinates.

We can use the definition of **curl** in Eq.(1.91) to show that the **curl** of any gradient is zero:

$$\nabla \times \nabla \phi = 0. \qquad (1.98)$$

This is because any gradient is a conservative vector field and the closed contour integral on the right hand side of Eq.(1.91) vanishes. As a mnemonic, the fact that $\nabla \times \nabla \phi = 0$ can be thought of as following from the vanishing of the cross product of a vector with itself. But that is not strictly the case for vector differential operators, and an important counter-example is given as a problem.

Since the position vector **r** is a gradient ($\nabla r^2 = 2\mathbf{r}$), it follows that

$$\nabla \times \mathbf{r} = 0. \qquad (1.99)$$

Also,

$$\nabla \times \mathbf{E} = 0 \qquad (1.100)$$

follows, because **E** is a gradient.

As an example of vector derivative operations, we now explicitly take $\nabla \times \mathbf{E}$ to show that it vanishes. First, for the Coulomb field of a unit point charge:

$$\begin{aligned} \nabla \times \mathbf{E} &= \nabla \times \left[\frac{\mathbf{r}}{r^3} \right] \\ &= \frac{1}{r^3} \nabla \times \mathbf{r} - \mathbf{r} \times \nabla \frac{1}{r^3} \\ &= 0 + 3\mathbf{r} \times \frac{\hat{\mathbf{r}}}{r^4} = 0. \end{aligned} \qquad (1.101)$$

We do not have to specify $r \neq 0$ in Eq.(1.101) because each term is explicitly zero for any **r**, and, more safely, we already know that the curl of any gradient is zero everywhere. For a continuous charge distribution, the same result ($\nabla \times \mathbf{E} = 0$) follows by bringing the $\nabla \times$ operation inside the volume integral, as was done in Eq. (1.85) for the divergence.

Equation (1.90) defines the average of **curl** over a finite surface. That equation can be rewritten as

$$\int_S d\mathbf{A} \cdot \nabla \times \mathbf{F} = \oint_C d\mathbf{r} \cdot \mathbf{F} \qquad (1.102)$$

for any vector **F**. In this form, it is called **Stokes' theorem** relating the integral of the curl of a vector over a surface S to the line integral of the vector around the closed curve, C, bounding the surface. From Stokes' theorem it follows that if $\boldsymbol{\nabla}\times\mathbf{F} = 0$ everywhere in a region, then

$$\oint_C \mathbf{dr}\cdot\mathbf{F} = 0 \qquad (1.103)$$

around any closed path in the region.

We now have seen the general properties of the electric field vector $\mathbf{E}$. It is **conservative, derivable from a potential**, and **irrotational**. These properties have different physical manifestations, but we have shown them to be mathematically equivalent. The connection between the properties of $\mathbf{E}$ are illustrated below.

The arrows indicate that each property is directly derivable from the adjacent property. The result that $\mathbf{E} = -\boldsymbol{\nabla}\phi$ does not follow directly from $\boldsymbol{\nabla}\times\mathbf{E} = 0$, but follows in two steps as shown.

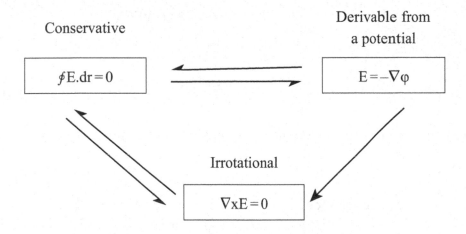

1.6   Summary of Vector Calculus

This section is a summary of the application the vector differential operator $\boldsymbol{\nabla}$. Many of the results listed have been derived earlier in the text, and others follow from the basic definitions of gradient, divergence, and curl.

1.6.1   Operation by $\boldsymbol{\nabla}$

In most applications, it is useful to know the operation of $\boldsymbol{\nabla}$ on simple functions of the position vector $\mathbf{r}$. These results are summarized below:

1. $\boldsymbol{\nabla} f(r) = \hat{\mathbf{r}}\frac{df}{dr}$

2. $(\mathbf{p}\cdot\boldsymbol{\nabla})\mathbf{r} = \mathbf{p}$

3. $\nabla(\mathbf{p}\cdot\mathbf{r})|_{\mathbf{p}\,\text{constant}} = \mathbf{p}$

4. $\nabla\cdot\mathbf{r} = 3$

5. $\nabla\times\mathbf{r} = 0$

6. $\nabla\cdot\left(\frac{\mathbf{r}}{r^3}\right) = 4\pi\delta(\mathbf{r}).$

Operations on combinations of functions of position can be simplified by using the two distinct properties of ∇:

1. ∇ is a differential operator.

2. ∇ is a vector.

We give several examples (that will be used later in the text) of the use of these two properties of ∇:

$$
\begin{aligned}
\nabla\left[\frac{\mathbf{p}\cdot\mathbf{r}}{r^3}\right]_{\mathbf{p}\,\text{constant}} &= \frac{1}{r^3}\nabla(\mathbf{p}\cdot\mathbf{r}) + (\mathbf{p}\cdot\mathbf{r})\nabla\left(\frac{1}{r^3}\right) \\
&= \frac{\mathbf{p}}{r^3} - \frac{3\hat{\mathbf{r}}}{r^4}(\mathbf{p}\cdot\mathbf{r}) \\
&= \frac{\mathbf{p} - 3(\mathbf{p}\cdot\hat{\mathbf{r}})\hat{\mathbf{r}}}{r^3}, \quad r \neq 0,
\end{aligned}
\tag{1.104}
$$

where we have taken the combination $(\mathbf{p}\cdot\mathbf{r})/r^3$ as the product of two scalar functions.

Also,

$$
\begin{aligned}
\nabla\times(\mathbf{A}\times\mathbf{r})_{\mathbf{A}\,\text{constant}} &= \mathbf{A}(\nabla\cdot\mathbf{r}) - (\mathbf{A}\cdot\nabla)\mathbf{r} \\
&= 3\mathbf{A} - \mathbf{A} \\
&= 2\mathbf{A}.
\end{aligned}
\tag{1.105}
$$

In this example of the curl of a vector product, we have used the algebraic vector identity

$$\mathbf{a}\times(\mathbf{b}\times\mathbf{c}) = \mathbf{b}(\mathbf{a}\cdot\mathbf{c}) - \mathbf{c}(\mathbf{a}\cdot\mathbf{b}). \tag{1.106}$$

This identity should be kept firmly in memory as the **bac minus cab** rule, using the mnemonic 'a cross b cross c equals bac minus cab'. Also note that, since $\mathbf{A}$ was a constant vector, each term in the expansion of $\nabla\times(\mathbf{A}\times\mathbf{r})_{\mathbf{A}\,\text{constant}}$ was written in the order $\mathbf{A}\,\nabla\,\mathbf{r}$ in Eq,. (1.105).

The bac minus cab rule is used in deriving the following two vector identities:

$$\nabla\times(\mathbf{A}\times\mathbf{B}) = \mathbf{A}(\nabla\cdot\mathbf{B}) - (\mathbf{A}\cdot\nabla)\mathbf{B} - \mathbf{B}(\nabla\cdot\mathbf{A}) + (\mathbf{B}\cdot\nabla)\mathbf{A} \tag{1.107}$$

$$\nabla(\mathbf{A}\cdot\mathbf{B}) = \mathbf{A}\times(\nabla\times\mathbf{B}) + (\mathbf{A}\cdot\nabla)\mathbf{B} + \mathbf{B}\times(\nabla\times\mathbf{A}) + (\mathbf{B}\cdot\nabla)\mathbf{A}. \tag{1.108}$$

We go through the derivation of these identities step by step to illustrate the method.

For $\nabla \times (\mathbf{A} \times \mathbf{B})$, the derivative operation of ∇ initiallly leads to two terms, one with $\mathbf{A}$ constant and one with $\mathbf{B}$ constant:

$$\nabla \times (\mathbf{A} \times \mathbf{B}) = \nabla \times (\mathbf{A} \times \mathbf{B})|_{\mathbf{A}\,\text{constant}} + \nabla \times (\mathbf{A} \times \mathbf{B})|_{\mathbf{B}\,\text{constant}}. \tag{1.109}$$

These two terms are treated separately. For $\mathbf{A}$ constant, the first step is to write the three vectors in the order $\mathbf{A}\,\nabla\,\mathbf{B}$, since ∇ only acts on $\mathbf{B}$. This is written down twice, anticipating the use of the bac minus cab rule. The second step is to write the bac minus cab equation above the line, so this intermediate step looks like:

$$
\begin{aligned}
\mathbf{a} \times (\,\mathbf{b} \times \mathbf{c}) \quad &= \quad \mathbf{b}(\mathbf{a}\cdot\mathbf{c}) - \mathbf{c}(\mathbf{a}\cdot\mathbf{b}) \\
\nabla \times (\mathbf{A} \times \mathbf{B})|_{\mathbf{A}\,\text{constant}} &= \quad \mathbf{A}(\nabla\cdot\mathbf{B}) - (\mathbf{A}\cdot\nabla)\mathbf{B}
\end{aligned} \tag{1.110}
$$

Then, we relate the a, b, and c of the bac minus cab rule to the vectors ∇, $\mathbf{A}$, and $\mathbf{B}$, so $\mathbf{a} \sim \nabla$, $\mathbf{b} \sim \mathbf{A}$, $\mathbf{c} \sim \mathbf{B}$. This association tells us where to put the dots and crosses, in order to preserve the vector algebra. This third step leaves

$$
\begin{aligned}
\mathbf{a} \times (\mathbf{b} \times \mathbf{c}) \quad &= \quad \mathbf{b}\,(\mathbf{a}\cdot\mathbf{c}) - \mathbf{c}(\mathbf{a}\cdot\mathbf{b}) \\
\nabla \times (\mathbf{A} \times \mathbf{B})|_{\mathbf{A}\,\text{constant}} &= \quad \mathbf{A}(\nabla\cdot\mathbf{B}) - (\mathbf{A}\cdot\nabla)\mathbf{B},
\end{aligned} \tag{1.111}
$$

with the signs determined by bac minus cab.

There will be two more terms for $\mathbf{B}$ constant. These could be found by repeating the same process as when $\mathbf{A}$ was held constant, but there is a simpler way. We note that the original expression $\nabla \times (\mathbf{A} \times \mathbf{B})$ changes sign under the interchange $\mathbf{A} \rightleftharpoons \mathbf{B}$. So, the two terms with $\mathbf{B}$ constant just follow by interchanging $\mathbf{A}$ and $\mathbf{B}$ with a sign change in Eq. (1.111), which gives the four terms in the full identity.

For $\nabla(\mathbf{A}\cdot\mathbf{B})$, we consider it to be the bac of bac minus cab, so the intermediate step for $\mathbf{A}$ constant is

$$
\begin{aligned}
\mathbf{b}(\mathbf{a}\cdot\mathbf{c}) \quad &= \quad \mathbf{a} \times (\mathbf{b} \times \mathbf{c}) + \mathbf{c}(\mathbf{a}\cdot\mathbf{b}) \\
\nabla(\mathbf{A}\cdot\mathbf{B})|_{\mathbf{A}\,\text{constant}} &= \quad \mathbf{A} \quad \nabla \quad \mathbf{B} \qquad \mathbf{A}\nabla\mathbf{B}
\end{aligned} \tag{1.112}
$$

The left hand side of Eq. (1.112) shows that the appropriate identification is $\mathbf{a} \sim \mathbf{A}$, $\mathbf{b} \sim \nabla$, $\mathbf{c} \sim \mathbf{B}$, which tells us how to put in the dots and crosses on the right hand side, so

$$\nabla(\mathbf{A}\cdot\mathbf{B})|_{\mathbf{A}\,\text{constant}} = \mathbf{A} \times (\nabla \times \mathbf{B}) + (\mathbf{A}\cdot\nabla)\mathbf{B}. \tag{1.113}$$

The other two terms with $\mathbf{B}$ constant follow by just interchanging $\mathbf{A}$ and $\mathbf{B}$ since the original expression $\nabla(\mathbf{A}\cdot\mathbf{B})$ is symmetric in $\mathbf{A}$ and $\mathbf{B}$.

The expression $(\mathbf{A}\cdot\boldsymbol{\nabla})$ in some of the above equations represents the magnitude of $\mathbf{A}$ times the **directional derivative** which was defined in Section 1.3. In this case, the directional derivative is acting on the vector $\mathbf{B}$ rather than a scalar. Although the directional derivative is a scalar operator, the direction of $(\mathbf{A}\cdot\boldsymbol{\nabla})\mathbf{B}$ is not usually that of $\mathbf{B}$. It is useful to know that the action of the operator $(\mathbf{A}\cdot\boldsymbol{\nabla})$ acting on the postion vector $\mathbf{r}$ just gives the vector $\mathbf{A}$. That is

$$(\mathbf{A}\cdot\boldsymbol{\nabla})\mathbf{r} = \mathbf{A}. \tag{1.114}$$

This can be shown from Eq. (1.113), using the relations $\boldsymbol{\nabla}\times\mathbf{r} = \mathbf{0}$ and $\boldsymbol{\nabla}(\mathbf{A}\cdot\mathbf{r})_{\mathbf{A}\,\text{constant}} = \mathbf{A}$. Note that although $\mathbf{A}$ need not be a constant vector for Eq. (1.114) to hold, we can treat $\mathbf{A}$ as constant in using Eq. (1.113).

The reader is urged not to memorize identities like Eqs. (1.107) and (1.108), but to develop facility in deriving them on the spot as needed. With practice, you can eventually leave out penciling in the bac minus cab rule above the line, but dispensing with this crutch too early in the game will lead to error. Mind your dots and crosses!

1.6.2   Integral theorems

We have derived the divergence theorem, Eq. (1.74), and Stokes' theorem, Eq. (1.102), relating integrals over vectors and their derivatives. In this section, we derive other useful integral theorems from these two theorems.

We first use the divergence theorem to derive two theorems, first derived by George Green in 1828. Green's theorems are of great use in solving the vector differential equations of electrostatics. Applying the divergence theorem to the combination $\phi\boldsymbol{\nabla}\psi$ of two scalar fields gives

$$\oint_S d\mathbf{A}\cdot[\phi\boldsymbol{\nabla}\psi] = \int_V \boldsymbol{\nabla}\cdot[\phi\boldsymbol{\nabla}\psi]d\tau. \tag{1.115}$$

Then, expanding $\boldsymbol{\nabla}\cdot[\phi\boldsymbol{\nabla}\psi]$ using the derivative property of $\boldsymbol{\nabla}$ gives **Green's first theorem**:

$$\oint_S d\mathbf{A}\cdot[\phi\boldsymbol{\nabla}\psi] = \int_V [\phi\nabla^2\psi + (\boldsymbol{\nabla}\phi)\cdot(\boldsymbol{\nabla}\psi)]d\tau. \tag{1.116}$$

Interchanging ϕ and ψ in Eq. (1.116), and subtracting leads to **Green's second theorem**:

$$\oint_S d\mathbf{A}\cdot[\phi\boldsymbol{\nabla}\psi - \psi\boldsymbol{\nabla}\phi] = \int_V [\phi\nabla^2\psi - \psi\nabla^2\phi]d\tau. \tag{1.117}$$

Green's second theorem could also be derived by applying the divergence theorem to the combination $[\phi\boldsymbol{\nabla}\psi - \psi\boldsymbol{\nabla}\phi]$.

Also starting with the divergence theorem, we can derive a **gradient theorem** relating the integral over a volume V of the gradient of a scalar to the integral over the bounding surface S of the scalar function:

$$\int_V d\tau\boldsymbol{\nabla}\phi = \oint_S d\mathbf{A}\phi. \tag{1.118}$$

The proof of this theorem follows by first dotting the left hand side of Eq. (1.118) by an arbitrary constant vector $\mathbf{k}$

$$
\begin{aligned}
\mathbf{k} \cdot \int_V d\tau \boldsymbol{\nabla}\phi &= \int_V d\tau \boldsymbol{\nabla} \cdot [\mathbf{k}\phi], \quad \text{(since } \mathbf{k} \text{ is constant)} \\
&= \oint_S d\mathbf{A} \cdot \mathbf{k}\phi, \quad \text{(by the divergence theorem)} \\
&= \mathbf{k} \cdot \oint_S d\mathbf{A}\phi \quad \text{(since } \mathbf{k} \text{ is constant)}.
\end{aligned}
\tag{1.119}
$$

Now, since $\mathbf{k}$ is an arbitrary vector, Eq.(1.119) must hold for any component of the vectors (in any coordinate system), and the vector equation (1.118) follows.

A **curl theorem**:

$$
\int_V d\tau \boldsymbol{\nabla} \times \mathbf{E} = \oint_S d\mathbf{A} \times \mathbf{E}
\tag{1.120}
$$

can be proved in the same way as was the gradient theorem by dotting with a constant vector $\mathbf{k}$:

$$
\begin{aligned}
\mathbf{k} \cdot \int_V d\tau \boldsymbol{\nabla} \times \mathbf{E} &= \int_V d\tau \mathbf{k} \cdot (\boldsymbol{\nabla} \times \mathbf{E}) \\
&= \int_V d\tau \boldsymbol{\nabla} \cdot (\mathbf{E} \times \mathbf{k}) \\
&= \oint_S d\mathbf{A} \cdot (\mathbf{E} \times \mathbf{k}) \\
&= \mathbf{k} \cdot \oint_S d\mathbf{A} \times \mathbf{E}.
\end{aligned}
\tag{1.121}
$$

Since $\mathbf{k}$ is an arbitrary vector, it follows, as before, that the vector equation (1.120) must follow from Eq.(1.121).

In the steps in Eq. (1.121), we have moved $\mathbf{k}$ in and out of integrals and derivatives, because it is a constant vector. Notice that, at each step, the vector character of the equation is preserved. In two of the steps, we have used the cyclic property of the **triple scalar product** $\mathbf{A} \cdot (\mathbf{B} \times \mathbf{C})$ that

$$
\mathbf{A} \cdot (\mathbf{B} \times \mathbf{C}) = \mathbf{B} \cdot (\mathbf{C} \times \mathbf{A}).
\tag{1.122}
$$

This identity holds for the vector derivative $\boldsymbol{\nabla}$ as long as it still acts as a derivative on the appropriate variables. In the above example, the rearrangement of the vector order in the triple scalar product could only be done because $\mathbf{k}$ is a constant vector. (Another useful mnemonic for rearranging the triple scalar product is the fact that it is unchanged by simple interchange of the dot and the cross.)

Taking the limit of the divergence, gradient, or curl theorem as the volume shrinks to a point, we can provide (as promised) a definition of $\boldsymbol{\nabla}$ that is independent of any coordinate system:

$$
\boldsymbol{\nabla} = \lim_{V \to 0} \frac{1}{V} \oint_S d\mathbf{A}.
\tag{1.123}
$$

In Eq.(1.123), the vector derivative operator $\boldsymbol{\nabla}$ is defined in terms of the vector integral operator on the right hand side of the equation. Then, any operation of $\boldsymbol{\nabla}$ can be calculated by the corresponding vector operation by the integral operator.

From Stokes' theorem, we can derive a **Stokes' theorem for the gradient**:

$$\int_S d\mathbf{A} \times (\boldsymbol{\nabla}\phi) = \oint_C d\mathbf{r}\phi. \tag{1.124}$$

The proof of this theorem is similar to that for the gradient and curl theorems. We dot the left hand side of Eq.(1.124) by an arbitrary constant vector $\mathbf{k}$:

$$
\begin{aligned}
\mathbf{k} \cdot \int_S d\mathbf{A} \times (\boldsymbol{\nabla}\phi) &= \int_S \mathbf{k} \cdot [d\mathbf{A} \times (\boldsymbol{\nabla}\phi)] \\
&= \int_S d\mathbf{A} \cdot [(\boldsymbol{\nabla}\phi) \times \mathbf{k}] \\
&= \int_S d\mathbf{A} \cdot [\boldsymbol{\nabla} \times (\mathbf{k}\phi)] \\
&= \oint_C d\mathbf{r} \cdot (\mathbf{k}\phi) \\
&= \mathbf{k} \cdot \oint_C d\mathbf{r}\phi. \tag{1.125}
\end{aligned}
$$

And again, the vector equation follows because $\mathbf{k}$ is an arbitrary vector.

1.7 Problems

1. Four point charges, each of charge q and mass m are located at the four corners of a square of side L.

   (a) Find the magnitude of the force on one of the charges.

   (b) Use the force in part (a) to find the velocity of one of the charges a long time after the four charges are released from rest in the original configuration.

2. Four point charges, each of charge q are fixed at the four corners of a square of side L. Find the electric field a distance z above the plane of the square on the perpendicular axis of the square.

3. Show that the electric field a distance r from a long straight wire with a uniform linear charge density λ is given by

$$\mathbf{E} = \frac{2\lambda \hat{\mathbf{r}}}{r}. \tag{1.126}$$

4. Two long parallel wires, each with uniform linear charge density λ, are a distance $\mathbf{a}$ apart. The origin of coordinates is the midpoint between the two wires.

   (a) Find the electric field in terms of the vectors $\mathbf{r}$ and $\mathbf{a}$.

   (b) Write down the x and y components of $\mathbf{E}$, taking $\mathbf{a}$ in the x direction.

5. (a) Find the electric field a distance z along the axis of a uniformly charged ring of charge Q and radius R.

   (b) Integrate the field for a ring to find the electric field a distance z on the axis of a uniformly charged disk of charge Q and radius R.

   (c) Find the electric field of the charged disk for
   (1) $z = 0$,
   (2) $z >> R$. (Expand the square root.)

6. A straight wire of length L has a uniformly distributed charge Q. Find the electric field on the axis of the wire a distance z from the center of the wire for the cases

   (a) $z > L/2$.

   (b) $-L/2 < z < L/2$. (Hint: Use symmetry for part b.)

7. A straight wire of length L has a uniformly distributed charge Q. Find E_x and E_y a distance d from the wire in the configuration shown in the figure below. Express your answers in terms of the angles θ_1 and θ_2.

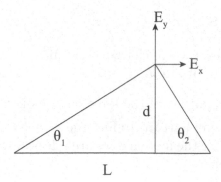

8. (a) Find the potential a distance d from the wire in terms of the angles θ_1 and θ_2 shown in the figure above.

   (b) Use the potential to find E_x and E_y at the same point.

9. Four point charges, each of charge q, are fixed at the four corners of a square of side L.

   (a) Find the potential a distance z above the plane of the square on the perpendicular axis of the square.

   (b) Use the potential to find the electric field at the same point.

10. A straight wire of length L has a uniformly distributed charge Q.

   (a) Find the potential on the axis of the wire a distance $z>L/2$ from the center of the wire.

   (b) Use the potential to find the electric field at the same point.

11. (a) Find the potential a distance z along the axis of a uniformly charged ring of charge Q and radius R.

   (b) Integrate the potential for a ring to find the potential a distance z along the axis of a uniformly charged disk of charge Q and radius R.

   (c) Use the potentials in parts a and b to find the corresponding electric fields.

12. (a) Integrate the potential for a ring to find the potential a distance $d>R$ from the center of a uniformly charged spherical shell of charge Q and radius R.

   (b) Integrate the potential for a spherical shell to find the potential a distance $d>R$ from the center of a uniformly charged sphere of charge Q and radius R.

   (c) Use the potentials in parts (a) and (b) to find the corresponding electric fields.

13. (a) Use Gauss's law to find the electric field inside and outside a uniformly charged solid sphere of charge Q and radius R.

    (b) Integrate **E** to find the potential inside and outside the solid sphere.

14. (a) Use Gauss's law to find the electric field inside and outside a uniformly charged hollow sphere of charge Q and radius R.

    (b) Integrate **E** to find the potential inside and outside the hollow sphere.

15. (a) Use Gauss's law to find the electric field inside and outside a long straight wire of radius R with uniform charge density ρ.

    (b) Integrate **E** to find the potential inside and outside the wire with the boundary condition $\phi(R) = 0$.

16. Consider a large sheet of aluminum foil of area A with a total charge Q, assumed to be uniformly distributed on the surfaces of the aluminum sheet.

    (a) Calculate the electric field close to the foil and far from any edge.

    (b) Show that using the formula for the electric field for a sheet of charge gives the same answer as using the formula for the electric field just outside a conductor, when each formula is applied to this problem.

17. For a screened potential $\phi(r) = qe^{-\mu r}/r$,

    (a) find the electric field.

    (b) find the charge distribution that produces this potential.

    (c) show that Gauss's law is satisfied by your answers. (Use a spherical Gaussian surface and be careful about the origin.)

18. Find the Curl and Divergence (for $r \neq 0$) of each of the vector fields:
    (a) $\mathbf{F} = (\mathbf{r}\times\mathbf{p})(\mathbf{r\cdot p})$, (b) $\mathbf{G} = (\mathbf{r\cdot p})^2\mathbf{r}$.
    The vector **p** is a constant vector.

19. Show that the operator $\mathbf{L} = -i\mathbf{r}\times\nabla$ satisfies $\mathbf{L}\times\mathbf{L}\phi = i\mathbf{L}\phi$. [Hint: Let $\mathbf{E} = -\nabla\phi$. Expand $(\mathbf{r}\times\nabla)\times(\mathbf{r}\times\mathbf{E})$ in bac-cab.]

Chapter 2

Further Development of Electrostatics

2.1 Conductors

A conductor is defined as a continuous material in which some charge is free to move. This **free charge** can be either conduction electrons (usually valence electrons that can leave their original atom and move through the conductor) or excess electrons placed on the conductor. The free charge can also be effective positive charges due to an absence of electrons. Most of the charge in any material is **bound charge** that is bound in the atoms and molecules of the matter, and cannot move through the material. When we talk about the "charge on a conductor" we refer to the excess negative or positive free charge, and not to either the bound charge or the free conduction electrons. If an electric field **E** is applied to the conductor, the free charges will be pushed around by the force of the field until the field inside the conductor is everywhere zero, at which time the charges stop moving, and a static situation is reached.

The conductor does not have to be a perfect conductor. That is, even if there is resistance to the movement of the charges, they will keep moving until the field is zero. For a good conductor, the time by which the movement ceases, called the **relaxation time**, is of the order of nanoseconds. For a poorer conductor, the relaxation time will be somewhat longer, but is usually very short. For electrostatics, the assumption is made that this relaxation has taken place so that **E**=0 everywhere inside a conductor. This is the basic property of a conductor from which all of its other properties follow.

We can use the fact that **E**=0 inside a conductor to derive a number of properties of a conductor in electrostatics. We start by considering a solid conductor with no cavities. As we have seen above, the defining property of a conductor in electrostatics is

1. **E=0 everywhere inside a conductor.**

From the equation $\nabla\cdot\mathbf{E} = 4\pi\rho$ with $\mathbf{E}=0$, it follows that

2. There is no charge inside a conductor. Any excess charge on a conductor must be on the surface.

Since $\mathbf{E}=0$ inside the conductor, no work is done in moving a test charge from one part of the conductor to another, and therefore

3. The entire conductor is an equipotential.

Since the conductor's surface is also an equipotential, it follows that, just outside a conductor, the field parallel to the conductor (E_{parallel}) vanishes, so

4. E just outside a conductor is normal to the surface of the conductor.

If we apply Gauss's law to the surface of a conductor, using a Gaussian pillbox of cross-sectional area A, as shown in Fig. 2.1, we see that the electric field passes through only one face of the pillbox since $\mathbf{E}=0$ on the inside the conductor.

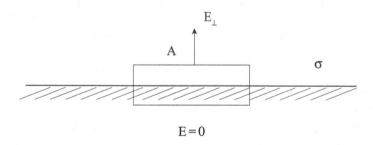

Figure 2.1: Gaussian pillbox for the surface of a conductor. E exits only through the top face of the pillbox.

Therefore, the surface integral of $\mathbf{E}$ is

$$\oint \mathbf{dA}\cdot\mathbf{E} = E_\perp A, \tag{2.1}$$

The pillbox can be made small enough so that charge inside the pillbox can be approximated by

$$\int \sigma dA = \sigma A. \tag{2.2}$$

Thus, from Gauss's law,

5. Just outside a conductor, $\mathbf{E}_\perp = 4\pi\sigma$.

Note that this value of **E** is twice as large as that for a flat sheet of charge, because, for a conductor, the **E** field can go out of the pillbox through only one face.

Now we consider conductors which have one or more cavities inside. If a cavity inside a conductor is empty of charge, then the net charge on the surface of the cavity must be zero from Gauss's law applied to a surface just outside the cavity, and inside the conductor. It turns out that a stronger condition holds. The charge density on the surface of the cavity is identically zero. We first show that the **E** field inside the cavity is zero. We show this by using the fact that $\nabla \cdot \mathbf{E} = 0$ inside the cavity, and apply the divergence theorem to the product $\phi \mathbf{E}$ as follows:

$$\int \nabla \cdot [\phi \mathbf{E}] d\tau = \oint d\mathbf{A} \cdot [\phi \mathbf{E}]$$

$$\int [\mathbf{E} \cdot \nabla \phi + \phi \nabla \cdot \mathbf{E}] d\tau = 4\pi \oint \phi \sigma dA$$

$$-\int \mathbf{E} \cdot \mathbf{E} d\tau + 0 = 4\pi \phi \oint \sigma dA$$

$$\int \mathbf{E} \cdot \mathbf{E} d\tau = 0. \qquad (2.3)$$

In the above derivation, we used the fact that ϕ was constant to take it out of the surface integral. The remaining surface integral of σ is just the net charge on the surface which is zero. Since the remaining volume integral of **E·E** vanishes, and the integrand is never negative, the integrand must be identically zero everywhere in the region of integration. Thus we have

6. E=0 everywhere inside a vacant cavity inside a conductor.

Also, since $E_\perp = 4\pi\sigma$,

7. The surface charge density is identically zero on the surface of a vacant cavity inside a conductor.

The fact that **E=0** inside a cavity in a conductor, independently of what is happening outside the conductor, makes such a cavity a shield for screening out unwanted fields. This is the basis for the **Faraday cage**, constructed of metal screens to provide a field free region in a laboratory. It also makes a closed metal automobile (but not a convertible) a safe haven during a lightning storm.

If there is charge inside the cavity, even if the net charge inside is zero, there will be an **E** field inside the cavity, and, by Gauss's law,

8. The net charge on the surface of a cavity inside a conductor is the negative of the net charge within the cavity.

Then, if a charge Q is placed within a cavity in a neutral conductor, a net charge $-Q$ will result on the surface of the cavity, requiring (by conservation of charge), a net charge $+Q$ to be added to the outer surface of the conductor. Then, if the charge Q inside the cavity touches the cavity wall, it will neutralize the charge $-Q$ on the inner surface, leaving a completely neutral cavity surface, and a net charge Q still added to the outer surface of the conductor. The entire conductor will still be an equipotential, but will be at a higher potential than before the charge Q was introduced into the cavity.

The above procedure is the basis for several interesting physical phenomena. In the **Faraday ice bucket** experiment the charging procedure described above is repeated many times for a metal bucket with a metal cover enclosing it. Of course a small hole is necessary to permit the charge to be introduced into the cavity. No matter how much charge is inserted into the cavity, none remains on the inner surface. Measuring this lack of charge gives an accurate test of Gauss's law, which was the basis for the conclusion that no charge would be left on the inner surface. More sophisticated experiments of this type have given the most accurate determination that the exponent in Coulomb's law is, in fact, 2, since that fact was the original basis for Gauss's law.

The **Van de Graaff generator** also uses the fact that all charge deposited on the inner surface of a conductor must end up on the outer surface, leaving the inner surface uncharged. For the generator, a moving belt carries the charge through a small opening in a spherical conducting shell, with a brush transferring the charge from the moving belt to the inside surface of the shell. Since the inner surface remains uncharged, this process can continue indefinitely. The potential, V, of the sphere of radius R increases as more charge winds up on the outer surface, with $V = Q/R$. The process is only limited by the eventual breakdown of the insulating property of the air, as the electric field at the surface, given by $E = Q/R^2 = V/R$, reaches the limit at which a spark discharge through the air will occur. This method is used in Vandegraff accelerators to achieve high voltages and strong electric fields to accelerate charged particles. The procedure is also used at Science Museums to produce 'artificial lightning' between two such generators, of opposite signs.

Since the electric field is identically zero inside the part of a conductor surrounding a cavity, regardless of any movement of the charge within the cavity, any movement of this charge can have no effect beyond the conductor. Because of this,

9. The electric field outside a conductor is not affected by the movement of charge within a cavity within the conductor.

This means that a Faraday cage can also be used to keep unwanted electrical effects from the region outside the cage.

2.2 Electrostatic Energy

Since the static electric field conserves energy, the potential energy of a charge distribution is equal to the work required to assemble the charge distribution. The potential energy of a single charge q at position $\mathbf{r}$ in an electric potential $\phi(\mathbf{r})$ is given by

$$U = q\phi(\mathbf{r}). \tag{2.4}$$

In Eq. (2.4), $\phi(\mathbf{r})$ is the potential due to charge distributions other than the point charge q itself. Thus, electrostatic 'self energy' (which would be infinite) is excluded.

The potential energy of two point charges would be

$$U_2 = q_2\phi_1(\mathbf{r_2}) = \frac{q_2 q_1}{r_{21}}, \tag{2.5}$$

where $\phi_1(\mathbf{r_2})$ is the potential at the point $\mathbf{r_2}$ due to the charge q_1, and $r_{21} = |\mathbf{r_2} - \mathbf{r_1}|$ is the magnitude of the vector from q_1 to q_2. As additional charges are added, the work required to bring the ith charge from infinity to its final location is determined by the potential due to the i-1 previous charges:

$$\delta U_i = q_i\phi_{i-1}(\mathbf{r_i}) = q_i\sum_{j=1}^{i-1}\frac{q_j}{r_{ij}}. \tag{2.6}$$

Then, the potential energy for a collection of N point charges is

$$U_N = \sum_{i=2}^{N}\delta U_i = \sum_{i=2}^{N}\sum_{j=1}^{i-1}\frac{q_i q_j}{r_{ij}}. \tag{2.7}$$

This potential energy can be written in somewhat simpler notation as

$$U_N = \sum_{j<i}^{N}\frac{q_i q_j}{r_{ij}}. \tag{2.8}$$

The form of the function being summed in Eq. (2.8) is symmetric with respect to the indices i and j. This means we can interchange i and j in the sum, and then

$$U_N = \sum_{j<i}^{N}\frac{q_i q_j}{r_{ij}} = \sum_{i<j}^{N}\frac{q_j q_i}{r_{ji}}. \tag{2.9}$$

Adding the two sums in Eq. (2.9) gives a sum over all i and j with j$\neq$i that is just twice U_N. Then,

$$U_N = \frac{1}{2}\sum_{j\neq i}^{N}\frac{q_i q_j}{r_{ij}}, \tag{2.10}$$

with the factor $\frac{1}{2}$ entering because of the double counting in the sum.

This analysis can be extended to continuous charge distributions by introducing the limit $\Delta q_i \to dq'$ as was done in section 1.1, leading to

$$U = \frac{1}{2} \int dq \int dq' \frac{1}{|\mathbf{r} - \mathbf{r}'|} \tag{2.11}$$

or

$$U = \frac{1}{2} \int \phi(\mathbf{r}) dq. \tag{2.12}$$

The integral in Eq. (2.12) is over all space, but need only be carried out where the charge distribution is non-zero.

Equation (2.12) can also be derived completely in the framework of a continuous charge distribution. Consider a final charge distribution given by $\rho(\mathbf{r})$. We can think of arriving at $\rho(\mathbf{r})$ by starting with zero charge everywhere and building up the the final $\rho(\mathbf{r})$ by infinitesimal increments $\delta\rho'_n(\mathbf{r})$, where $\delta\rho'_n(\mathbf{r})$ has the same spatial dependence as $\rho(\mathbf{r})$. At each stage, the increase in the potential energy is

$$\delta U_n = \int d\tau \phi'_n \delta\rho'_n, \tag{2.13}$$

where ϕ'_n is the potential produced by the charge distribution ρ' already present at that stage. Both ρ' and ϕ' are functions of $\mathbf{r}$. Then the final potential energy is given by

$$U = \int d\tau \sum_n \phi'_n \delta\rho'_n \to \int d\tau \int_0^{\rho(\mathbf{r})} d\rho' \phi'. \tag{2.14}$$

Now we make the important observation that the equations determining ϕ' from ρ' are linear. That is any fractional increase in ϕ' is equal to the fractional increase in ρ'. This means that

$$\frac{\delta\phi'}{\phi'} = \frac{\delta\rho'}{\rho'}, \tag{2.15}$$

so

$$\rho'\delta\phi' = \phi'\delta\rho'. \tag{2.16}$$

Then it follows that

$$\delta(\rho'\phi') = \rho'\delta\phi' + \phi'\delta\rho' = 2\phi'\delta\rho', \tag{2.17}$$

and Eq. (2.14) can be written as

$$U = \frac{1}{2} \int d\tau \int_0^{\rho\phi} d(\rho'\phi') = \frac{1}{2} \int \rho(\mathbf{r})\phi(\mathbf{r}) d\tau. \tag{2.18}$$

Surface or line charge densities could be treated in the same way, so the potential energy can be given as

$$U = \frac{1}{2} \int \phi(\mathbf{r}) dq, \tag{2.19}$$

in agreement with our previous result. In this derivation, the factor of $\frac{1}{2}$ is seen to come from the linearity property of electrostatics.

Direct application of Eq. (2.19) would be difficult in the presence of conductors because the charge distribution cannot be specified for a conductor. The charge adjusts to make the conductor an equipotential. However, since each conductor is an equipotential, the potential ϕ can be taken outside the integral, and the contribution of each conductor will be given by $\frac{1}{2}QV$. Then, equation (2.19) can be extended to

$$U = \frac{1}{2} \int \phi(\mathbf{r})dq + \frac{1}{2} \sum_i Q_i V_i, \tag{2.20}$$

where the integral is taken over all space excluding conductors, and the sum is over all conductors.

The potential energy can also be given in terms of the electric field $\mathbf{E}$, by using $\boldsymbol{\nabla}\cdot\mathbf{E} = 4\pi\rho$ to eliminate ρ in Eq. (2.20) (Assuming a volume distribution of charge.) This gives

$$U = \frac{1}{2} \int \phi \left[\frac{\boldsymbol{\nabla}\cdot\mathbf{E}}{4\pi}\right] d\tau + \frac{1}{2} \sum_i Q_i V_i. \tag{2.21}$$

Then, we introduce a total divergence to the volume integration in Eq. (2.21):

$$U = \frac{1}{8\pi} \int \{\boldsymbol{\nabla}\cdot[\phi\mathbf{E}] - \mathbf{E}\cdot\boldsymbol{\nabla}\phi\} d\tau + \frac{1}{2} \sum_i Q_i V_i. \tag{2.22}$$

Next, we apply the divergence theorem to the first integral in Eq. (2.22), and use $\mathbf{E} = -\boldsymbol{\nabla}\phi$ in the second integral to get

$$U = \frac{1}{8\pi} \sum_j \oint_j \mathbf{dA}\cdot[\phi\mathbf{E}] + \frac{1}{8\pi} \oint_R \mathbf{dS}\cdot(\phi\mathbf{E}) + \frac{1}{8\pi} \int_V \mathbf{E}\cdot\mathbf{E}d\tau + \frac{1}{2} \sum_i Q_i V_i. \tag{2.23}$$

The first term in Eq. (2.23)) is a sum over all conductors. It arises because the surface of each conductor is a boundary surface for the volume integration. The surface integral, $\oint_R$, is over a large surface, S_R, that can be taken as a sphere of radius R, in the limit as $R \to \infty$.

We assume that the average over solid angle of the product $\phi\mathbf{E}$ goes to zero faster than $1/R^2$, so the integral over S_R vanishes as $R \to \infty$. (This is not true for radiation fields, in which case the entire derivation must be done differently.) The volume integral of $\mathbf{E}\cdot\mathbf{E}$ is over all space, not including conductors.

For the surface integrals summed over conductors in Eq. (2.23), the potential ϕ is constant on each conductor, and can be taken out of each integral. This gives

$$\begin{aligned}
U &= \frac{1}{8\pi} \int E^2 d\tau + \frac{1}{8\pi} \sum_i V_i \oint_i \mathbf{dA}\cdot\mathbf{E} + \frac{1}{2} \sum_i Q_i V_i \\
&= \frac{1}{8\pi} \int E^2 d\tau - \frac{1}{2} \sum_i V_i \oint_i \sigma dA + \frac{1}{2} \sum_i Q_i V_i.
\end{aligned} \tag{2.24}$$

The minus sign in the surface integral of σ arises because the direction of the vector surface differential $\mathbf{dA}$ is into the metal. The last two terms in Eq. (2.24) cancel, leaving

$$U = \frac{1}{8\pi} \int E^2 d\tau, \qquad (2.25)$$

where the volume integral is over all space except conductors, where the integrand would vanish anyway.

When some of the charge distribution consists of point charges, the integrals in either Eq. (2.19) or Eq. (2.25) must be treated with special care. The energy given by either integral represents the total energy required to assemble the charge distribution. For a point charge, this would be infinite. In doing these integrals, this infinite 'self energy' should be excluded, just as its in the sum over charges in Eq. (2.10) at the start of our derivation.

In Eq. (2.19), this is done by taking the potential ϕ in the integral to be from all charges except the point charge at that point. For instance, for point charges q_1 and q_2 located at positions $\mathbf{r_1}$ and $\mathbf{r_2}$, Eq. (2.19) should be given as

$$U = \frac{1}{2} \int \phi(\mathbf{r}) dq = \frac{1}{2}[q_1\phi_2(\mathbf{r_1}) + q_2\phi_1(\mathbf{r_2})] = \frac{q_1 q_2}{r_{12}}. \qquad (2.26)$$

In using Eq. (2.25) to find the potential energy, the square of the electric field must be taken so as to exclude the square of the 'self-field' of any point charge. For the case of the two point charges, the electric field is

$$\mathbf{E}(\mathbf{r}) = \frac{q_1(\mathbf{r} - \mathbf{r_1})}{|\mathbf{r} - \mathbf{r_1}|^3} + \frac{q_2(\mathbf{r} - \mathbf{r_2})}{|\mathbf{r} - \mathbf{r_2}|^3}. \qquad (2.27)$$

In the integral over E^2 to find the potential energy, only the cross term in the square of Eq. (2.27) is used, so

$$U = \frac{q_1 q_2}{4\pi} \int \frac{(\mathbf{r} - \mathbf{r_1}) \cdot (\mathbf{r} - \mathbf{r_2})}{|\mathbf{r} - \mathbf{r_1}|^3 |\mathbf{r} - \mathbf{r_2}|^3} d\tau. \qquad (2.28)$$

We will evaluate this integral to show that it gives the same potential energy as Eq. (2.26). The second factor under the integral can be written as a gradient

$$U = -\frac{q_1 q_2}{4\pi} \int \frac{(\mathbf{r} - \mathbf{r_1})}{|\mathbf{r} - \mathbf{r_1}|^3} \cdot \nabla \left[\frac{1}{|\mathbf{r} - \mathbf{r_2}|} \right] d\tau. \qquad (2.29)$$

Then we form a divergence under the integral:

$$\int \left\{ \nabla \cdot \left[\frac{(\mathbf{r} - \mathbf{r_1})}{|\mathbf{r} - \mathbf{r_1}|^3 |\mathbf{r} - \mathbf{r_2}|} \right] - \frac{1}{|\mathbf{r} - \mathbf{r_2}|} \nabla \cdot \left[\frac{(\mathbf{r} - \mathbf{r_1})}{|\mathbf{r} - \mathbf{r_1}|^3} \right] \right\} d\tau. \qquad (2.30)$$

Now we apply the divergence theorem to the first integral. This gives an integral over the surface of a sphere of radius R enclosing the volume of integration

$$\int \nabla \cdot \left[\frac{(\mathbf{r} - \mathbf{r_1})}{|\mathbf{r} - \mathbf{r_1}|^3 |\mathbf{r} - \mathbf{r_2}|} \right] d\tau = \oint R^2 d\Omega \left[\frac{\hat{\mathbf{r}} \cdot (\mathbf{r} - \mathbf{r_1})}{|\mathbf{r} - \mathbf{r_1}|^3 |\mathbf{r} - \mathbf{r_2}|} \right]. \qquad (2.31)$$

The area of the sphere is $4\pi R^2$, while the integrand $\sim 1/R^3$. Therefore, this surface integral goes to zero like $1/R$ in the limit as $R \to \infty$.

Our procedure has had the effect of converting the integral in Eq. (2.28) to

$$U = \frac{q_1 q_2}{4\pi} \int \frac{1}{|\mathbf{r} - \mathbf{r_2}|} \boldsymbol{\nabla} \cdot \left[\frac{(\mathbf{r} - \mathbf{r_1})}{|\mathbf{r} - \mathbf{r_1}|^3} \right] d\tau. \tag{2.32}$$

The divergence term in this integral is $4\pi\delta(\mathbf{r} - \mathbf{r_1})$, so

$$\begin{aligned}
U &= q_1 q_2 \int \frac{\delta(\mathbf{r} - \mathbf{r_1})}{|\mathbf{r} - \mathbf{r_2}|} d\tau \\
&= \frac{q_1 q_2}{|\mathbf{r_1} - \mathbf{r_2}|}.
\end{aligned} \tag{2.33}$$

We would like to emphasize the usefulness of the procedure we used in converting the integral in Eq. (2.28) to that in Eq. (2.32). For a general scalar function $\phi(\mathbf{r})$ and vector function $\mathbf{F}(\mathbf{r})$, the steps are

$$\begin{aligned}
\int \mathbf{F} \cdot \boldsymbol{\nabla}\phi \, d\tau &= \int [\boldsymbol{\nabla} \cdot (\mathbf{F}\phi) - \phi \boldsymbol{\nabla} \cdot \mathbf{F}] d\tau \\
&= \oint \mathbf{dA} \cdot \mathbf{F}\phi - \int \phi \boldsymbol{\nabla} \cdot \mathbf{F} d\tau.
\end{aligned} \tag{2.34}$$

If the original volume integral is over all space, then the surface integral can be taken over a large sphere of radius $R \to \infty$. If the limit $\lim_{r\to\infty}(r^2 F\phi) = 0$, the surface integral will vanish, leaving

$$\int \mathbf{F} \cdot \boldsymbol{\nabla}\phi \, d\tau = -\int \phi \boldsymbol{\nabla} \cdot \mathbf{F} d\tau. \tag{2.35}$$

We will call this procedure **the divergence trick**, and will use it throughout the text. Students of calculus will recognize it as just integration by parts in three dimensions.

2.3 Electric dipoles

2.3.1 Fields due to dipoles

The Coulomb's law integral of Eq. (1.34) gives the electric potential of a given charge distribution, but it is only practical to perform this integral for simple charge distributions. Also, for a general charge distribution, a different volume integral must be evaluated for each point where the potential is desired. For these reasons, it is useful to expand the denominator of the integrand, $|\mathbf{r} - \mathbf{r'}|^{-1}$, in a factorized form. This is feasible for finite charge distributions that vanish outside some region. Then r' will always be smaller than some maximum distance d.

To generate what is called the electric dipole approximation, we use the first two terms of the vector form of a three dimensional Taylor expansion:

$$f(\mathbf{r}) = f(\mathbf{0}) + \mathbf{r}\cdot[\boldsymbol{\nabla} f(\mathbf{r})]_{\mathbf{r}=0} + \dots \tag{2.36}$$

Expanding the denominator of the Coulomb integral as a function of $\mathbf{r}'$ in this way gives

$$\frac{1}{|\mathbf{r}-\mathbf{r}'|} = \frac{1}{r} + \frac{\mathbf{r}\cdot\mathbf{r}'}{r^3} + O\left(\frac{r'^2}{r^3}\right). \tag{2.37}$$

This is an expansion in powers of (r'/r). When the field point r is far from the charge distribution so that the ratio (r'/r) is always small, the higher order terms may be neglected, leading to the dipole approximation.

Putting this approximate form into the Coulomb integral leads to

$$\phi(\mathbf{r}) = \frac{1}{r}\int dq' + \frac{\mathbf{r}}{r^3}\cdot\int \mathbf{r}'dq'. \tag{2.38}$$

The **electric dipole moment** of a charge distribution is defined by

$$\mathbf{p} = \int \mathbf{r}dq, \tag{2.39}$$

and then Eq. (2.38) can be written as

$$\phi(\mathbf{r}) = \frac{q}{r} + \frac{\mathbf{r}\cdot\mathbf{p}}{r^3}. \tag{2.40}$$

The first term in Eq. (2.40) is the same as the potential of a point charge q equal to the total charge of the charge distribution. The second term is called the **dipole potential** due to the **electric dipole moment p**. Consistent with this terminology, the first term is sometimes called the **monopole potential** due to the **electric monopole moment** q. Of course, Eq. (2.40) only makes sense when the field point r is outside a sphere of radius d, beyond which there is no charge, and the dipole approximation improves as the ratio d/r decreases.

If the total charge vanishes, then the dipole moment is independent of the origin of the integration coordinate in Eq. (2.39). As a practical matter, the origin is usually chosen inside the charge distribution so as to keep the ratio (d/r) as small as possible.

If the total charge does not vanish, then **p** will depend on where the origin of the integration coordinate is chosen, and the origin can always be chosen so as to make **p** = **0**. This is usually what is done for a non-vanishing total charge, unless other considerations (such as position in a crystal, or a desire to pick the origin at the center of mass if this differs from the 'center of charge') prevent this.

An important class of charge distributions are those called 'pure dipoles' where the total charge vanishes and higher terms in the expansion either vanish or are negligible. The abstraction of a point dipole of negligible size is often useful. Then r can be vanishingly small, and Eq. (2.40) will still apply. The dipole potential is singular at the origin, but for now we will treat pure dipoles leaving out the singular behavior at the origin.

An elementary electric dipole is often defined (especially in elementary texts) as consisting of two point charges +q and -q, a vector distance **L** apart, as in Fig. 2.2, in the limit as $L \to 0$, with the product qL remaining finite. Then Eq. (2.39) gives

$$\mathbf{p} = \lim_{L \to 0} q\mathbf{L}. \tag{2.41}$$

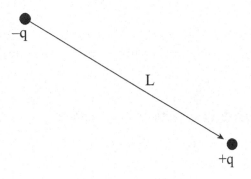

Figure 2.2: Model of an elementary dipole.

It is sometimes useful to visualize a pure electric dipole in this form to get a general idea of the direction and effect of its dipole fields.

The electric dipole field is given by the negative gradient of the dipole potential

$$\mathbf{E_p} = -\boldsymbol{\nabla} \left[\frac{\mathbf{p} \cdot \mathbf{r}}{r^3} \right]. \tag{2.42}$$

This is the gradient of the product of two scalars, so the dipole field is

$$\mathbf{E_p} = \frac{3(\mathbf{p} \cdot \hat{\mathbf{r}})\hat{\mathbf{r}} - \mathbf{p}}{r^3}. \tag{2.43}$$

The dipole electric field is conservative, since it is derivable from a potential, but it is not central since $\mathbf{r} \times \mathbf{E} \neq \mathbf{0}$.

The energy of a dipole in an external electric potential can be found starting with the general relation

$$U = \int \phi dq \tag{2.44}$$

for the potential energy of a charge distribution in an external potential $\phi(\mathbf{r})$. Note that Eq. (2.44) does not have the $\frac{1}{2}$ of Eq. (2.12), because the potential in Eq. (2.44) is due only to charge other than the charge distribution being considered.

We expand ϕ in a Taylor series:

$$\phi(\mathbf{r}) = \phi(0) + \mathbf{r} \cdot \nabla \phi(\mathbf{r})|_{\mathbf{r}=\mathbf{0}} + ..., \qquad (2.45)$$

and keep only the first two terms, in keeping with the dipole approximation. Putting this expansion into the integral for the potential energy gives

$$U = \phi(0) \int dq + [\nabla \phi(\mathbf{r})|_{\mathbf{r}=\mathbf{0}}] \cdot \int \mathbf{r} dq. \qquad (2.46)$$

The first term is the energy ($q\phi$) due to the electric charge, and the second term is the energy of the dipole

$$U_p = -\mathbf{p} \cdot \mathbf{E}. \qquad (2.47)$$

The negative sign for the dipole energy corresponds to the fact that a dipole will tend to minimize its energy by lining up in the direction of an applied electric field.

2.3.2   Forces and torques on dipoles

The force on an electric dipole in an external electric field can be found by taking the negative gradient of the dipole part of the potential energy:

$$\mathbf{F_p} = \nabla(\mathbf{p} \cdot \mathbf{E}). \qquad (2.48)$$

This shows that there will be no force on a dipole in a constant electric field.

The force on a dipole can be given in a different form by applying bac minus cab to $\nabla(\mathbf{p} \cdot \mathbf{E})$:

$$\begin{aligned} \mathbf{F_p} &= \nabla(\mathbf{p} \cdot \mathbf{E}) \\ &= \mathbf{p} \times (\nabla \times \mathbf{E}) + (\mathbf{p} \cdot \nabla)\mathbf{E} \\ &= (\mathbf{p} \cdot \nabla)\mathbf{E}. \end{aligned} \qquad (2.49)$$

This shows that the force on a dipole will be in the direction of the directional derivative of the electric field. Either Eq. (2.48) or Eq. (2.49) can be used to find the force on a dipole, but Eq. (2.48) is usually easier to use.

The force on a dipole can also be found by expanding the external electric field in a Taylor's expansion

$$\mathbf{E}(\mathbf{r}) = \mathbf{E}(0) + (\mathbf{r} \cdot \nabla)\mathbf{E}(\mathbf{r})|_{\mathbf{r}=\mathbf{0}} + ... \qquad (2.50)$$

Keeping the first two terms in the expansion gives for the force on a charge distribution

$$\begin{aligned} \mathbf{F} &= \int \mathbf{E}(\mathbf{r})dq \\ &= \int [\mathbf{E}(0) + (\mathbf{r} \cdot \nabla)\mathbf{E}(\mathbf{r})|_{\mathbf{r}=\mathbf{0}}]dq \\ &= q\mathbf{E} + (\mathbf{p} \cdot \nabla)\mathbf{E}, \end{aligned} \qquad (2.51)$$

in agreement with Eq. (2.49).

As an example, we find the force on a dipole a distant $\mathbf{r}$ from a point charge q.

$$\mathbf{F_{pq}} = q\mathbf{\nabla}\left[\frac{\mathbf{p}\cdot\mathbf{r}}{r^3}\right] = q\left[\frac{\mathbf{p} - 3(\mathbf{p}\cdot\hat{\mathbf{r}})\hat{\mathbf{r}}}{r^3}\right]. \tag{2.52}$$

This is just equal and opposite to the force on the charge q that would be given by the electric field of the dipole in Eq. (2.43), in agreement with Newton's third law.

The torque tending to rotate a dipole in an electric field can be found from the energy expression by considering a rotation of the dipole through an infinitesimal angle $\mathbf{d\theta}$. ($\mathbf{d\theta}$ is a vector along the axis of rotation with the magnitude of the infinitesimal angle $d\theta$.) Then

$$dU = -(\mathbf{E}\cdot\mathbf{dp}) = -\boldsymbol{\tau}\cdot\mathbf{d\theta}, \tag{2.53}$$

where $\boldsymbol{\tau}$ represents the vector torque on the dipole.

The infinitesimal change in the dipole moment $\mathbf{p}$ due to the rotation $\mathbf{d\theta}$, as for any rotation of a vector, is given by

$$\mathbf{dp} = \mathbf{d\theta}\times\mathbf{p}, \tag{2.54}$$

Then we can write

$$dU = -\boldsymbol{\tau}\cdot\mathbf{d\theta} = -\mathbf{E}\cdot(\mathbf{d\theta}\times\mathbf{p}) = -\mathbf{d\theta}\cdot(\mathbf{p}\times\mathbf{E}), \tag{2.55}$$

where we have used the cyclic property of the triple scalar product. Since the rotation $\mathbf{d\theta}$ is arbitrary, it can be removed from the second and fourth parts of Eq. (2.55), leading to

$$\boldsymbol{\tau} = \mathbf{p}\times\mathbf{E}. \tag{2.56}$$

Application of the right hand rule shows that this torque tends to align the dipole along the electric field.

The torque on a dipole can also be found by integrating the differential torque over a charge distribution

$$\boldsymbol{\tau} = \int \mathbf{r}\times\mathbf{E}(\mathbf{r})dq, \tag{2.57}$$

and keeping only the first (constant) term in the Taylor expansion, so that $\mathbf{E}(0)$ can be taken out of the integral. Then

$$\boldsymbol{\tau} = \left[\int \mathbf{r}dq\right]\times\mathbf{E}(0) = \mathbf{p}\times\mathbf{E}, \tag{2.58}$$

in agreement with Eq. (2.56).

The 'spin torque' tending to rotate a dipole $\mathbf{p}$ a distance $\mathbf{r}$ from a point charge q is given by

$$\boldsymbol{\tau}_{\text{spin}} = \frac{q\mathbf{p}\times\mathbf{r}}{r^3}. \tag{2.59}$$

There will also be an 'orbital torque' on the dipole for motion about the point charge given by

$$\tau_{\text{orbital}} = \mathbf{r} \times \mathbf{F}_{pq} = \frac{q\mathbf{r} \times \mathbf{p}}{r^3}, \tag{2.60}$$

where we have used Eq. (2.52) for the force on the dipole. The two torques cancel so there is no net torque on the charge-dipole system, and total angular momentum (spin+orbital) is conserved.

In calculating the orbital torque, we used the vector $\mathbf{r}$ from the point charge to the dipole so that there would be no orbital torque on the point charge. But the orbital torque on the combined charge-dipole system does not depend on what point the torques are calculated from because we have shown that the two forces (on the charge and on the dipole) are equal and opposite, forming a force couple. The motion of the charge-dipole system starting from rest would be a complicated combination of orbital motion about each other with an oscillating spin motion of the dipole.

The potential energy of two dipoles, $\mathbf{p}$ and $\mathbf{p}'$, a distance $\mathbf{r}$ apart, oriented as shown in Fig. 2.3, is given by

$$U_{\mathbf{pp}'} = -\mathbf{p} \cdot \mathbf{E}_{\mathbf{p}'} = \frac{\mathbf{p} \cdot \mathbf{p}' - 3(\mathbf{p} \cdot \hat{\mathbf{r}})(\mathbf{p}' \cdot \hat{\mathbf{r}})}{r^3}. \tag{2.61}$$

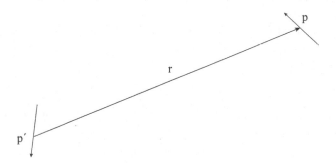

Figure 2.3: System of two dipoles.

The force between the two dipoles is the negative gradient of this potential energy. There are four different scalar functions of $\mathbf{r}$ in Eq. (2.61) (Recall that $\hat{\mathbf{r}} = \mathbf{r}/r$.). We take the gradient of each in turn, so the force on a dipole $\mathbf{p}'$ a distance $\mathbf{r}$ from a dipole $\mathbf{p}$ is

$$\mathbf{F}_{\mathbf{p}'\mathbf{p}} = \frac{3(\mathbf{p} \cdot \hat{\mathbf{r}})\mathbf{p}' + 3(\mathbf{p}' \cdot \hat{\mathbf{r}})\mathbf{p} + 3(\mathbf{p} \cdot \mathbf{p}')\hat{\mathbf{r}} - 15(\mathbf{p} \cdot \hat{\mathbf{r}})(\mathbf{p}' \cdot \hat{\mathbf{r}})\hat{\mathbf{r}}}{r^4}. \tag{2.62}$$

The force on $\mathbf{p}$ due to $\mathbf{p}'$ follows by interchanging $\mathbf{p}$ and $\mathbf{p}'$, and letting $\hat{\mathbf{r}} \rightarrow -\hat{\mathbf{r}}$. We see from Eq. (2.62) that the dipole-dipole force satisfies Newton's third law.

2.3.3 Dipole singularity at r=0

So far we have avoided the point $\mathbf{r}=\mathbf{0}$ in discussing dipoles. Here we derive the electric field of a point electric dipole taking careful account of this singularity. To facilitate the calculation, we make use of the result from Chapter 1 that

$$\nabla \cdot \left(\frac{\mathbf{r}}{r^3}\right) = 4\pi\delta(\mathbf{r}). \tag{2.63}$$

The electric field of a point electric dipole, taking into account this representation of the delta function is given by

$$
\begin{aligned}
\mathbf{E}(\mathbf{r}) &= -\nabla\phi(\mathbf{r}) = -\nabla\left(\frac{\mathbf{p}\cdot\mathbf{r}}{r^3}\right) \\
&= -\frac{\mathbf{p}}{r^3} - (\mathbf{p}\cdot\mathbf{r})\nabla\left(\frac{1}{r^3}\right) \\
&= -\frac{\mathbf{p}}{r^3} - (\mathbf{p}\cdot\hat{\mathbf{r}})\hat{\mathbf{r}}\left[\mathbf{r}\cdot\nabla\left(\frac{1}{r^3}\right)\right] \\
&= -\frac{\mathbf{p}}{r^3} - (\mathbf{p}\cdot\hat{\mathbf{r}})\hat{\mathbf{r}}\left[\nabla\cdot\left(\frac{\mathbf{r}}{r^3}\right) - \frac{\nabla\cdot\mathbf{r}}{r^3}\right] \\
&= \frac{3(\mathbf{p}\cdot\hat{\mathbf{r}})\hat{\mathbf{r}} - \mathbf{p}}{r^3} - 4\pi\hat{\mathbf{r}}(\hat{\mathbf{r}}\cdot\mathbf{p})\delta(\mathbf{r}).
\end{aligned}
\tag{2.64}
$$

Equation (2.64) shows the dipole electric field with its singular behavior at the origin. This singular behavior at the origin adds a 'contact term' to the potential energy of two dipoles as given by Eq. (2.61), so

$$U_{\mathbf{pp}'} = \frac{\mathbf{p}\cdot\mathbf{p}' - 3(\mathbf{p}\cdot\hat{\mathbf{r}})(\mathbf{p}'\cdot\hat{\mathbf{r}})}{r^3} + 4\pi(\mathbf{p}\cdot\hat{\mathbf{r}})(\hat{\mathbf{r}}\cdot\mathbf{p}')\delta(\mathbf{r}). \tag{2.65}$$

The contact term in the potential energy is not too important in classical physics where the likelihood of two dipoles touching is remote, but it must be taken into account in quantum mechanics whenever the wave function of the two dipoles does not vanish for $r=0$. This only happens for s-wave (orbital angular momentum zero) states, and then the wave function is spherically symmetric. The potential energy is usually integrated with the wave function squared, so the delta function enters averaged over all solid angle.

The average over solid angle is defined as

$$\langle(\hat{\mathbf{p}}\cdot\hat{\mathbf{r}})(\hat{\mathbf{r}}\cdot\hat{\mathbf{p}}')\rangle = \frac{1}{4\pi}\int d\Omega(\hat{\mathbf{p}}\cdot\hat{\mathbf{r}})(\hat{\mathbf{r}}\cdot\hat{\mathbf{p}}'). \tag{2.66}$$

This average over angles will come up a lot, so we evaluate it in some detail here. We first take the constant vector $\hat{\mathbf{p}}'$ out of the integral, and write

$$\langle(\hat{\mathbf{p}}\cdot\hat{\mathbf{r}})(\hat{\mathbf{r}}\cdot\hat{\mathbf{p}}')\rangle = \frac{1}{4\pi}\mathbf{I}\cdot\hat{\mathbf{p}}', \tag{2.67}$$

where $\mathbf{I}$ is a vector given by

$$\mathbf{I} = \int d\Omega (\hat{\mathbf{p}} \cdot \hat{\mathbf{r}}) \hat{\mathbf{r}}. \tag{2.68}$$

Now we note a very useful property of a vector integral like $\mathbf{I}$ that depends on only one constant vector. The integral can only be in the $\hat{\mathbf{p}}$ direction, since no other vector direction can be defined. Thus we can write $\mathbf{I}$ as

$$\mathbf{I} = I\hat{\mathbf{p}}. \tag{2.69}$$

Then,

$$\langle (\hat{\mathbf{p}} \cdot \hat{\mathbf{r}})(\hat{\mathbf{r}} \cdot \hat{\mathbf{p}}') \rangle = (\hat{\mathbf{p}} \cdot \hat{\mathbf{p}}')I, \tag{2.70}$$

where the scalar integral I is given by

$$I = \int d\Omega (\hat{\mathbf{p}} \cdot \hat{\mathbf{r}})(\hat{\mathbf{r}} \cdot \hat{\mathbf{p}}). \tag{2.71}$$

Choosing the $\hat{\mathbf{p}}$ direction as the z-axis, the angular integral becomes

$$I = \int d\Omega \cos^2 \theta = \frac{4\pi}{3}, \tag{2.72}$$

so the average over angles is

$$\langle (\hat{\mathbf{p}} \cdot \hat{\mathbf{r}})(\hat{\mathbf{r}} \cdot \hat{\mathbf{p}}') \rangle = \frac{1}{3}(\hat{\mathbf{p}} \cdot \hat{\mathbf{p}}'). \tag{2.73}$$

The potential energy, averaged over angles, is then given by

$$\langle U_{\mathbf{pp}'} \rangle = \frac{4\pi}{3}(\mathbf{p} \cdot \mathbf{p}')\delta(\mathbf{r} - \mathbf{r}'). \tag{2.74}$$

Note that the non-singular part of U vanishes in the average over solid angle, so Eq. (2.74) gives the full potential energy for a spherically symmetric quantum state of two electric dipoles.

2.4 Electric Quadrupole Moment

If we extend the three dimensional Taylor's expansion of $|\mathbf{r} - \mathbf{r}'|^{-1}$ beyond the first two terms, we get

$$\frac{1}{|\mathbf{r} - \mathbf{r}'|} = \frac{1}{r} + \frac{\mathbf{r} \cdot \mathbf{r}'}{r^3} + \frac{1}{2}\mathbf{r}'\mathbf{r}' : \left[\boldsymbol{\nabla}' \boldsymbol{\nabla}' \frac{1}{|\mathbf{r} - \mathbf{r}'|} \right]_{\mathbf{r}'=0} + O\left(\frac{r'^3}{r^4} \right). \tag{2.75}$$

We can use this expansion to generate an expansion of the potential in powers of $1/r$ given by

$$\phi(\mathbf{r}) = \phi_0(\mathbf{r}) + \phi_1(\mathbf{r}) + \phi_2(\mathbf{r}) + O\left(\frac{1}{r^4} \right) \tag{2.76}$$

with each order of potential in Eq. (2.76) coming from the corresponding term in Eq. (2.75). The terms in Eq. (2.76) are identified as the **monopole**, **dipole**, and **quadrupole** potentials, respectively.

In Eq. (2.75) we have introduced the new notation $(\mathbf{rr} : \nabla\nabla)$, which is the double dot product of two dyadics. We digress for a bit now to present the basics of dyadic algebra, which will prove to be useful throughout the text.

2.4.1   Dyadics

We first give the definition of a **dyad**, which is the direct product of two vectors written without a dot or a cross as $\mathbf{AB}$. The meaning of $\mathbf{AB}$ is given by the following operations with a third vector $\mathbf{C}$:

$$\mathbf{AB\cdot C} = \mathbf{A(B\cdot C)}, \qquad \mathbf{C\cdot AB} = \mathbf{(C\cdot A)B} \qquad (2.77)$$

$$\mathbf{AB\times C} = \mathbf{A(B\times C)}, \qquad \mathbf{C\times AB} = \mathbf{(C\times A)B}, \qquad (2.78)$$

with $\mathbf{A(B\times C)}$ and $\mathbf{(C\times A)B}$ being new dyads. The double dot operation in Eq. (2.75) has the following meaning:

$$\mathbf{AB : CD} = \mathbf{A\cdot(B\cdot C)D} = \mathbf{(A\cdot D)(B\cdot C)}. \qquad (2.79)$$

That is, we first dot $\mathbf{B}$ with $\mathbf{CD}$ resulting in the vector $\mathbf{(B\cdot C)D}$, and then $\mathbf{A}$ is dotted into $\mathbf{(B\cdot C)D}$.

A dyad can be represented by a matrix. This is usually done in Cartesian coordinates, with the matrix elements defined as

$$[\mathbf{AB}]_{ij} = A_i B_j, \quad (i, j = 1, 2, 3), \qquad (2.80)$$

where we have used the notation that $i=1,2,3$ corresponds to x, y, z. Equation (2.80) shows that, although the 3×3 matrix representation of a dyad has nine matrix elements, they depend on only six parameters.

A **dyadic** is a more general object, which we represent as $[\mathbf{D}]$. The dyadic has nine independent matrix elements defined by

$$D_{ij} = \hat{\mathbf{n}}_i\cdot[\mathbf{D}]\cdot\hat{\mathbf{n}}_j, \qquad (2.81)$$

where $\hat{\mathbf{n}}_1$, $\hat{\mathbf{n}}_2$, $\hat{\mathbf{n}}_3$ are the three basic unit vectors $(\hat{\mathbf{i}}, \hat{\mathbf{j}}, \hat{\mathbf{k}})$ in Cartesian coordinates). A dyadic can be expanded as sums of dyads. The standard expansion is in terms of the basic unit vectors:

$$[\mathbf{D}] = \Sigma_{i,j=1}^3 D_{ij}\hat{\mathbf{n}}_i\hat{\mathbf{n}}_j. \qquad (2.82)$$

If the coordinate axes were not orthogonal, the unit vectors in Eq. (2.81) and Eq. (2.82) would not be the same, and the two sets of D_{ij} could also be different. However, we shall use only orthogonal coordinate systems in this text, for which the $\hat{\mathbf{n}}_i$ and D_{ij} are the same in each equation.

An important dyadic is the **unit dyadic** $\hat{\hat{n}}$, defined by its dot product with a general vector $\mathbf{A}$:

$$\hat{\hat{n}} \cdot \mathbf{A} = \mathbf{A} \cdot \hat{\hat{n}} = \mathbf{A}. \tag{2.83}$$

Expanded in the basic unit vectors, the unit dyadic is given by

$$\hat{\hat{n}} = \Sigma_{i=1}^{3} \hat{n}_i \hat{n}_i. \tag{2.84}$$

Its matrix representation is the unit matrix with elements given by

$$[\hat{\hat{n}}]_{ij} = \delta_{ij}. \tag{2.85}$$

We are now ready to continue to the third term of the Taylor expansion. Applications of dyadics will become clearer as we use them in the text.

2.4.2   Quadrupole dyadic

We first evaluate the dyadic $\mathbf{\nabla}' \mathbf{\nabla}' \frac{1}{|\mathbf{r}-\mathbf{r}'|}$ in Eq. (2.75)

$$
\begin{aligned}
\mathbf{\nabla}' \mathbf{\nabla}' \frac{1}{|\mathbf{r} - \mathbf{r}'|} &= \mathbf{\nabla}' \left[\frac{(\mathbf{r} - \mathbf{r}')}{|\mathbf{r} - \mathbf{r}'|^3} \right] \\
&= \frac{3(\mathbf{r} - \mathbf{r}')(\mathbf{r} - \mathbf{r}')}{|\mathbf{r} - \mathbf{r}'|^5} - \frac{\hat{\hat{n}}}{|\mathbf{r} - \mathbf{r}'|^3}.
\end{aligned}
\tag{2.86}
$$

Note that the operator $\mathbf{\nabla}'$ acts only on the vector $\mathbf{r}'$ with $\mathbf{r}$ being considered a constant vector here. In the second term above we have used the identity

$$\mathbf{\nabla} \mathbf{r} = \hat{\hat{n}}, \tag{2.87}$$

which follows from the property of the directional derivative that

$$(\mathbf{p} \cdot \mathbf{\nabla}) \mathbf{r} = \mathbf{p}. \tag{2.88}$$

Setting $\mathbf{r}' = \mathbf{0}$ in Eq. (2.86), we find

$$\left[\mathbf{\nabla}' \mathbf{\nabla}' \frac{1}{|\mathbf{r} - \mathbf{r}'|} \right]_{\mathbf{r}'=0} = \frac{3\hat{\mathbf{r}}\hat{\mathbf{r}} - \hat{\hat{n}}}{r^3}. \tag{2.89}$$

Using this result in Eq. (2.75) the quadrupole potential is given by

$$\phi_2(\mathbf{r}) = \left(\frac{3\hat{\mathbf{r}}\hat{\mathbf{r}} - \hat{\hat{n}}}{2r^3} \right) : \int \mathbf{r}' \mathbf{r}' dq' \tag{2.90}$$

The dyadic $(3\hat{\mathbf{r}}\hat{\mathbf{r}} - \hat{\hat{n}})$ has the property that

$$(3\hat{\mathbf{r}}\hat{\mathbf{r}} - \hat{\hat{n}}) : \hat{\hat{n}} = 0. \tag{2.91}$$

This follows from the identity

$$\hat{\mathbf{n}} : \hat{\mathbf{n}} = 3, \tag{2.92}$$

that is the trace of the unit dyadic is 3. Using this result, we can modify the integrand in Eq. (2.90) by subtracting the dyadic $\frac{1}{3}\hat{\mathbf{n}}$, so the quadrupole potential can be written

$$\phi_2(\mathbf{r}) = \left(\frac{3\hat{\mathbf{r}}\hat{\mathbf{r}} - \hat{\mathbf{n}}}{2r^3}\right) : \int (\mathbf{r}'\mathbf{r}' - \frac{1}{3}r'^2\hat{\mathbf{n}})dq' \tag{2.93}$$

To put this is into a more compact form, we define a **quadrupole dyadic** as

$$[\mathbf{Q}] = \frac{1}{2}\int (3\mathbf{r}\mathbf{r} - r^2\hat{\mathbf{n}})dq. \tag{2.94}$$

Then, the quadrupole potential can be written as

$$\phi_2(\mathbf{r}) = \frac{(3\hat{\mathbf{r}}\hat{\mathbf{r}} - \hat{\mathbf{n}}) : [\mathbf{Q}]}{3r^3} \tag{2.95}$$

From the property (2.91), it follows that

$$\hat{\mathbf{n}} : [\mathbf{Q}] = 0, \tag{2.96}$$

so Eq. (2.95) can also be written

$$\phi_2(\mathbf{r}) = \frac{\hat{\mathbf{r}}\hat{\mathbf{r}} : [\mathbf{Q}]}{r^3}. \tag{2.97}$$

Either Eq. (2.95) or Eq. (2.97) for the quadrupole potential can be used, depending on which is more convenient in any particular application.

The quadrupole dyadic can be represented by a matrix using Eq. (2.81). The quadrupole dyadic transforms under rotations like a second rank tensor (see Chapter 14). Since it is real and symmetric, it can be diagonalized in a rotation by the same procedure as finding the principle axes of the tensor of inertia for a rigid body. If the orthogonal principal axes are chosen as the coordinate axes, then the quadrupole dyadic can be represented by the diagonal matrix

$$[\mathbf{Q}] = \begin{bmatrix} Q_x & 0 & 0 \\ 0 & Q_y & 0 \\ 0 & 0 & Q_z \end{bmatrix} \tag{2.98}$$

with the trace constraint

$$Q_x + Q_y + Q_z = 0. \tag{2.99}$$

Very often the symmetry of the charge distribution results in two of the diagonal quadrupole elements being equal. These are usually chosen to be Q_x and Q_y, and the diagonal matrix representation of the quadrupole is written

$$[\mathbf{Q}] = Q_0 \begin{bmatrix} -\frac{1}{2} & 0 & 0 \\ 0 & -\frac{1}{2} & 0 \\ 0 & 0 & 1 \end{bmatrix}. \tag{2.100}$$

The quantity Q_0 is referred to as **the quadrupole moment** of the charge distribution. It is given in Cartesian coordinates by

$$Q_0 = \int \frac{(2z^2 - x^2 - y^2)}{2} dq, \tag{2.101}$$

and in spherical coordinates (See Section 4.2.) by

$$Q_0 = \int \frac{(3\cos^2\theta - 1)}{2} r^2 dq. \tag{2.102}$$

We can see from Eq. (2.101) that Q_0 will generally be positive for an elongated charge distribution (football) and negative for a squashed distribution (pancake).

The symmetric quadrupole can also be written in dyadic notation as

$$[\mathbf{Q}] = \frac{1}{2} Q_0 [3\hat{\mathbf{k}}\hat{\mathbf{k}} - \hat{\mathbf{n}}], \tag{2.103}$$

where $\hat{\mathbf{k}}$ is the symmetry axis of the quadrupole. This dyadic form is useful for vector calculations.

Substituting the dyadic form of $[\mathbf{Q}]$ into Eq. (2.97), the potential for a symmetric quadrupole can be written as

$$\phi = \frac{Q_0}{2r^3} [3(\hat{\mathbf{k}} \cdot \hat{\mathbf{r}})^2 - 1]. \tag{2.104}$$

The electric field is the negative gradient of this potential, given by

$$\mathbf{E} = \frac{3Q_0}{2r^4} [5(\hat{\mathbf{k}} \cdot \hat{\mathbf{r}})^2 \hat{\mathbf{r}} - 2(\hat{\mathbf{k}} \cdot \hat{\mathbf{r}})\hat{\mathbf{k}} - \hat{\mathbf{r}}]. \tag{2.105}$$

As with the dipole, it is often useful to consider the abstraction of a 'point quadrupole' in the limit as the size of the charge distribution shrinks to a point while retaining a finite quadrupole moment. One elementary example of a point quadrupole is two point charges +q a distance L apart with a third collinear point charge -2q at their midpoint. Another simple example is a point charge +q at the center of a circular uniform line charge -q at radius L. In each case, taking the limit $L \to 0$ while keeping the product qL^2 constant defines a point quadrupole. These quadrupole models are shown in Fig. 2.4.

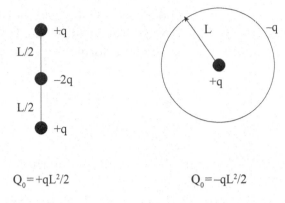

$Q_0 = +qL^2/2$ $Q_0 = -qL^2/2$

Figure 2.4: Two models of a quadrupole.

Although the abstraction of a point quadrupole is a useful concept, quadrupole moments generally occur only for extended objects (even if small), in contrast to dipole moments which can occur in point objects like elementary particles. Consequently, we will not investigate the complicated question of the singularity at the origin for point quadrupoles, and all of our quadrupole formulas are for the region $r>0$.

The energy of a quadrupole in an external electric potential can be found from the relation

$$U = \int \phi dq. \tag{2.106}$$

The charge distribution producing the external potential should be be well outside the charge distribution of the quadrupole, so we can expand the external potential, ϕ, to second order in $\mathbf{r}$:

$$\phi(\mathbf{r}) = \phi(0) - \mathbf{r}\cdot\mathbf{E}(0) - \frac{1}{2}\mathbf{r}\mathbf{r} : [\boldsymbol{\nabla}\mathbf{E}(\mathbf{r})]_{\mathbf{r}=0} + ... \tag{2.107}$$

Putting this expansion into the integral for the potential energy gives for the quadrupole energy (omitting the charge and dipole energy terms)

$$U_Q = -\frac{1}{2} \int \mathbf{r}\mathbf{r} : [\boldsymbol{\nabla}\mathbf{E}(\mathbf{r})]_{\mathbf{r}=0} dq. \tag{2.108}$$

We can subract the term $\frac{1}{3}\hat{\mathbf{n}} : \boldsymbol{\nabla}\mathbf{E}(\mathbf{r})$ from the integrand in Eq. (2.108) because

$$\hat{\hat{\mathbf{n}}} : \boldsymbol{\nabla}\mathbf{E}(\mathbf{r}) = \boldsymbol{\nabla}\cdot\mathbf{E}(\mathbf{r}) = 0 \tag{2.109}$$

for the external electric field $\mathbf{E}$. Also, The dyadic $[\boldsymbol{\nabla}\mathbf{E}(\mathbf{r})]_{\mathbf{r}=0}$ is a constant and can be taken out of the integral. The remaining integral is then proportional to the quadrupole moment, so the result for the quadrupole energy is

$$U_Q = -\frac{1}{3}[\mathbf{Q}] : [\boldsymbol{\nabla}\mathbf{E}]. \tag{2.110}$$

We see that the energy of a quadrupole depends on the gradient of the electric field, and that the energy would be zero in a constant field.

For a symmetric quadrupole in diagonal form, the energy can be written as

$$U_Q = -\frac{1}{6}Q_0[3\hat{\mathbf{k}}\hat{\mathbf{k}} - \hat{\mathbf{n}}] : [\boldsymbol{\nabla}\mathbf{E}]$$

$$= -\frac{1}{2}Q_0(\hat{\mathbf{k}}\cdot\boldsymbol{\nabla})(\hat{\mathbf{k}}\cdot\mathbf{E}). \tag{2.111}$$

We have again used the fact that $\boldsymbol{\nabla}\cdot\mathbf{E}$ vanishes for the external field $\mathbf{E}$ to eliminate the $\hat{\mathbf{n}} : [\boldsymbol{\nabla}\mathbf{E}]$ term in Eq. (2.111).

The force on a quadrupole in an external field is given by the negative gradient of the energy:

$$\mathbf{F_Q} = -\boldsymbol{\nabla}U_Q = \frac{1}{3}\boldsymbol{\nabla}([\mathbf{Q}] : [\boldsymbol{\nabla}\mathbf{E}]). \tag{2.112}$$

The force on a quadrupole can also be found from the expression

$$\mathbf{F} = \int \mathbf{E}dq \tag{2.113}$$

by expanding $\mathbf{E}$ to second order in $\mathbf{r}$. This leads to

$$\mathbf{F} = q\mathbf{E} + (\mathbf{p}\cdot\boldsymbol{\nabla})\mathbf{E} + \frac{1}{2}\int\{\mathbf{r}\mathbf{r} : [\boldsymbol{\nabla}\boldsymbol{\nabla}\mathbf{E}(\mathbf{r})]_{\mathbf{r=0}}\}dq \tag{2.114}$$

(The quantity $\boldsymbol{\nabla}\boldsymbol{\nabla}\mathbf{E}$ is a **three dyadic**, which is a straightforward generalization of the two dyadics we have considered up to now.) We can subtract $\frac{1}{3}\hat{\mathbf{n}}$ from the $\mathbf{r}\mathbf{r}$ term in the integrand in Eq. (2.114) because

$$\hat{\mathbf{n}} : [\boldsymbol{\nabla}\boldsymbol{\nabla}\mathbf{E}] = \nabla^2\mathbf{E} = -\boldsymbol{\nabla}(\nabla^2\phi) = 0 \tag{2.115}$$

for the external electric field. Then, the quadrupole part of the force in Eq. (2.114) becomes

$$\mathbf{F_Q} = \frac{1}{3}[\mathbf{Q}] : [\boldsymbol{\nabla}\boldsymbol{\nabla}\mathbf{E}] \tag{2.116}$$

The two different forms of the force on a quadrupole given by Eqs. (2.112) and (2.116) can be shown to lead to the same force. We show this by writing the force in Eq. (2.112) as

$$\overset{b\,(\quad a\quad\cdot\ c)}{\mathbf{F_Q} \;=\; \frac{1}{3}[\boldsymbol{\nabla}([\mathbf{Q}]\cdot\boldsymbol{\nabla})\cdot\mathbf{E}].} \tag{2.117}$$

The quantity $([\mathbf{Q}]\cdot\boldsymbol{\nabla})$ can be considered as the vector a in bac -cab, so we can write

$$\overset{a\quad\times\,(b\times\,c)\;+\quad(a\quad\cdot b)\ c}{\mathbf{F_Q} \;=\; \frac{1}{3}\{([\mathbf{Q}]\cdot\boldsymbol{\nabla})\times(\boldsymbol{\nabla}\times\mathbf{E}) + [([\mathbf{Q}]\cdot\boldsymbol{\nabla})\cdot\boldsymbol{\nabla}]\,\mathbf{E}\}}$$

$$= \frac{1}{3}[\mathbf{Q}] : [\boldsymbol{\nabla}\boldsymbol{\nabla}\mathbf{E}], \tag{2.118}$$

which is the same as Eq. (2.116)

For a symmetric quadrupole given by Eq. (2.103), the two different forms, Eq. (2.112) and Eq. (2.116), for the quadrupole force reduce to

$$\mathbf{F_Q} = \frac{1}{2}Q_0\boldsymbol{\nabla}(\hat{\mathbf{k}}\hat{\mathbf{k}}:\boldsymbol{\nabla}\mathbf{E}) \qquad (2.119)$$

and

$$\mathbf{F_Q} = \frac{1}{2}Q_0(\hat{\mathbf{k}}\!\cdot\!\boldsymbol{\nabla})(\hat{\mathbf{k}}\!\cdot\!\boldsymbol{\nabla})\mathbf{E}, \qquad (2.120)$$

respectively. Either of these equivalent forms for the force can be used.

2.4.3   Multipole expansion

The electric monopole, dipole, and quadrupole moments are the first three terms in what is called the **multipole expansion** of a charge distribution, leading to a multipole expansion of the potential. The dipole term and the quadrupole term in the expansion are particularly important if the lower terms in the expansion vanish because of some symmetry. If a lower term is present, then it will usually be much larger and swamp the next term in the expansion, although sometimes the different angular distribution and symmetry of the potential of the higher multipole still make it important.

Each term in the expansion should be considerably smaller than the previous terms or the expansion will not converge rapidly enough to make it useful. The ratio d/r where d is a characteristic size of the charge distribution and r is the distance to where the potential or field is measured is a measure of how well the multipole expansion will converge. For atoms and nuclei the ratio d/r is usually so small that only the first non-vanishing multipole moment will be important.

The next term in the multipole expansion of the potential would be the octupole term of order d^3/r^4. However this term, and all higher terms would be very complicated using dyadics. In chapter 4, the multipole expansion will be developed using Spherical Harmonics which are readily applicable to any order of multipole.

2.5 Problems

1. A conducting shell of arbitrary (but smooth) shape has an electric field E just outside a point on the sphere. Show that, if a small hole is drilled in the shell at that point, the electric field just inside the hole is $E/2$.

2. Four point charges, each of charge q and mass m are located at the four corners of a square of side L.

    (a) Find the potential energy of this configuration

    (b) Use conservation of energy to find the velocity of one of the charges a long time after the four charges are released from rest in the original configuration.

3. Find the potential energy of a uniformly charged sphere of charge Q and radius R by integrating

    (a) $\rho\phi$.

    (b) E^2.

4. Find the potential energy of a uniformly charged shell of charge Q and radius R by integrating

    (a) $\rho\phi$.

    (b) E^2.

5. A conducting shell with charge q and radius A is located inside, and at the center, of a conducting shell with charge $-q$ and radius B. Find the potential energy of this system by integrating

    (a) $\rho\phi$.

    (b) E^2.

6. Find the dipole moment of

    (a) a straight wire of length L with a linear charge density $\lambda(z) = \lambda_0 z/L, |z| < L/2$.

    (b) a hollow sphere of radius R with a surface charge distribution $\sigma(\theta) = (q/R^2)\cos(\theta)$.

7. (a) Find the spin torque on a dipole $\mathbf{p}_2$ a distance $\mathbf{r}$ from a dipole $\mathbf{p}_1$.

    (b) Find the spin torque on the dipole $\mathbf{p}_1$ due to $\mathbf{p}_2$.

    (c) Show that the net torque on the two dipoles is zero when the force couple on the dipoles is taken into account.

8. Two dipoles $\mathbf{p}_1$ and $\mathbf{p}_2$ are a distance $\mathbf{r}$ apart. $\mathbf{p}_1$ is parallel to $\mathbf{r}$, and $\mathbf{p}_2$ is perpendicular to $\mathbf{r}$, as shown in the figure.

   (a) Find the force on each dipole. Use the unit vectors $\hat{\mathbf{r}}$, $\hat{\mathbf{p}}_1$ and $\hat{\mathbf{p}}_2$ to specify direction in your answer.

   (b) Find the torque on each dipole.

   (c) Show that the net force and torque on the system are zero.

   (d) Find the energy of this system.

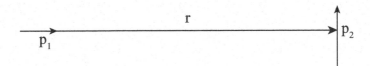

9. Calculate the quadrupole moment for

   (a) a uniformly charged needle of length $\mathbf{L}$ and charge q.

   (b) a uniformly charged disk of radius R and charge q.

   (c) a spherical shell of radius R with surface charge distribution $(q/R^2)\cos^2\theta$.

10. Calculate the quadrupole moment for

   (a) two point charges $+q$ a distance L apart with a third point collinear charge $-2q$ at their midpoint.

   (b) a point charge $+q$ at the center of a circular uniform line charge $-q$ at radius L.

11. Use Eq. (2.120) to find the quadrupole force on a nucleus with quadrupole moment Q_0 a distance $\mathbf{r}$ from an electron of charge $-e$. Show that this force is equal and opposite to the force on the electron in the electric quadrupole field of the nucleus.

12. (a) Find the potential energy of an electric dipole $\mathbf{p}$ a distance $\mathbf{r}$ from a symmetric electric quadupole Q_0 with symmetry axis $\hat{\mathbf{k}}$.

   (b) Evaluate the potential energy for the case when $\mathbf{p}$ and $\hat{\mathbf{k}}$ are parallel (or antiparallel), and each perpendicular to $\mathbf{r}$.

   (c) Evaluate the potential energy for the case when $\mathbf{p}$ and $\hat{\mathbf{k}}$ are parallel (or antiparallel), and each parallel to $\mathbf{r}$.

13. (a) Find the potential energy of two symmetric quadupoles, each of magnitude Q_0, a distance $\mathbf{r}$ apart. The quadupoles have symmetry axes $\hat{\mathbf{k}}$ and $\hat{\mathbf{q}}$, respectively.

(b) Evaluate the potential energy for the case when $\hat{\mathbf{k}}$ and $\hat{\mathbf{q}}$ are parallel (or antiparallel), and each perpendicular to $\mathbf{r}$.

(c) Evaluate the potential energy for the case when $\hat{\mathbf{k}}$ and $\hat{\mathbf{q}}$ are parallel (or antiparallel), and each parallel to $\mathbf{r}$.

14. Write Eqs. (2.111), (2.112), (2.116) in Cartesian coordinates. Take the symmetry axis in the z direction.

Chapter 3

Methods of Solution in Electrostatics

3.1   Differential Form of Electrostatics

In chapter 1, we started the study of electrostatics with Coulomb's law. But, in the presence of conductors, the electric field cannot be found directly from the Coulomb law integral of Eq. (1.16) because the surface charge distribution on conductors cannot be specified until the electric field is known. In this chapter, we resolve this difficulty by using differential equations that follow from Coulomb's law to formulate electrostatic theory. These basic equations are Eq. (1.80) for the divergence, and Eq. (1.92) for the curl of the electric field:

$$\boldsymbol{\nabla}\cdot\mathbf{E} = 4\pi\rho, \tag{3.1}$$
$$\boldsymbol{\nabla}\times\mathbf{E} = 0. \tag{3.2}$$

That is, the charge density ρ is the source of the electric field, and the electric field is irrotational.

In fact, these two equations form a more logical starting point for the study of electrostatics than does the historical order of starting from Coulomb's law for point charges. We present a scenario for this order of development here. The first experimental element of electrostatics is that objects have a force on them beyond that of gravity. This new **electric force** is proportional to a quantity called the **electric charge** q of the object. The force per unit charge on the object then defines a vector field called the **electric field**, so

$$\mathbf{F} = q\mathbf{E} \tag{3.3}$$

is the force on a point charge. Then experiments could be performed demonstrating that the electric force on a charge is a conservative force, so the vector field $\mathbf{E}$ is a conservative field, and is therefor irrotational, leading to Eq. (3.2).

The remaining question is how is the electric field $\mathbf{E}$ produced? We have shown that a vector field can vary through either its curl or its divergence. Thus, the remaining possible variation of the electric field must be given by equating its divergence to a scalar source field ρ'. The esu system of units puts the factor 4π into the divergence equation here, leading to Eq. (3.1) above. This choice of units makes most of the electrostatics in Chapter 1 simpler. Gauss's law for $\mathbf{E}$ could be used instead of the divergence equation, since the two methods are mathematically equivalent by the divergence theorem

A key assumption in electrostatics is that the source field ρ' that produces the electric field is identical to the charge field ρ that the electric field acts on. This is similar to the key assumption in Newtonian gravity that the source field of gravity is the same as the matter field that gravity acts on. For electrostatics this assumption can be justified by invoking Newton's third law (or more generally, conservation of linear momentum).

3.1.1   Uniqueness theorem

To develop all of electrostatics from equations (3.1) and (3.2) for the divergence and curl of $\mathbf{E}$, we must first show that these two equations, with appropriate boundary conditions, lead to unique solutions for the electric field, even in the presence of conductors. The sets of boundary conditions that lead to a unique solution will be developed during the the proof of the **uniqueness theorem**.

We prove the uniqueness theorem by assuming the existence of two distinct solutions of Eqs. (3.1) and (3.2), and then showing that their difference must vanish everywhere. We call these two solutions $\mathbf{E}_1$ and $\mathbf{E}_2$. Each solution satisfies Eqs. (3.1) and (3.2)

$$\boldsymbol{\nabla}\cdot\mathbf{E}_1 \;=\; 4\pi\rho \tag{3.4}$$
$$\boldsymbol{\nabla}\times\mathbf{E}_1 \;=\; 0 \tag{3.5}$$
$$\boldsymbol{\nabla}\cdot\mathbf{E}_2 \;=\; 4\pi\rho \tag{3.6}$$
$$\boldsymbol{\nabla}\times\mathbf{E}_2 \;=\; 0. \tag{3.7}$$

Since each electric field has the same source function, their difference defined by

$$\mathcal{E} = \mathbf{E}_1 - \mathbf{E}_2 \tag{3.8}$$

satifies the homogeneous equations

$$\boldsymbol{\nabla}\cdot\mathcal{E} \;=\; 0 \tag{3.9}$$
$$\boldsymbol{\nabla}\times\mathcal{E} \;=\; 0. \tag{3.10}$$

From Eq. (3.10), we see that $\mathcal{E}$ can be given as the gradient of a scalar field

$$\mathcal{E} = -\boldsymbol{\nabla}\psi \tag{3.11}$$

which, from Eq. (3.9), ψ satisfies Laplace's equation

$$\nabla^2\psi = 0. \tag{3.12}$$

Since each electric field $\mathbf{E}_1$ and $\mathbf{E}_2$ is separately irrotational , they each are the gradient of their own scalar fields ϕ_1 and ϕ_2, with

$$\psi = \phi_1 - \phi_2. \tag{3.13}$$

We next apply the divergence theorem (or Green's first theorem) to the combination $\psi\nabla\psi$, leading to

$$\int_V [(\nabla\psi)\cdot(\nabla\psi) + \psi\nabla^2\psi]d\tau = \oint_S (\psi\nabla\psi)\cdot\mathbf{dA}. \tag{3.14}$$

Since $\nabla^2\psi = 0$ and $\mathcal{E} = -\nabla\psi$, Eq. (3.14) becomes

$$\int_V |\mathcal{E}|^2 d\tau = -\oint_S \psi\mathcal{E}\cdot\mathbf{dA}. \tag{3.15}$$

Now any boundary condition on the surface S that makes the surface integral in Eq. (3.15) vanish will make $\int_V |\mathcal{E}|^2 d\tau$ vanish. Since the integrand $|\mathcal{E}|^2$ is never negative, the vanishing of the integral means that $\mathcal{E}$ must vanish throughout V, proving the uniqueness theorem inside the volume V.

If there are no conductors, then the surface S can be taken as a large sphere approaching infinite radius, and, in this limit, the volume integral will be over all space. In this case, the required boundary condition is that the possible electric fields all have the same asymptotic behavior (at least to order $1/R^2$). We will see that this asymptotic condition is generally satisfied.

In the presence of conductors, we must take the surface S to be composed of the discontinuous combination of the surface of each conductor plus one outer bounding surface. The outer bounding surface will either be a mathematical surface at infinity (**exterior problem**) or a finite surface enclosing the entire volume (**interior problem**).

The volume V is the volume of space excluding the conductors and bounded by the outer surface. Equation (3.15) then becomes

$$\int_V |\mathcal{E}|^2 d\tau = -\sum_i \oint_{S_i} \psi\mathcal{E}\cdot\mathbf{dA}, \tag{3.16}$$

where the sum is on all the surfaces described above. Even without conductors, we may choose to specify boundary conditions on arbitrary surfaces, in which case, Eq. (3.16) would also include those surfaces in the S_i sum. A typical volume of integration with several discontinuous boundary surfaces is shown in Fig. 3.1. Such a volume, looking like Swiss cheese, is said to be 'multiply connected'.

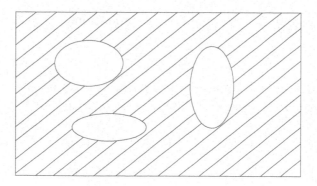

Figure 3.1: Multiply connected "Swiss cheese" volume with several bounding surfaces.

The simplest boundary condition to make the surface integrals in Eq. (3.16) vanish is that ψ vanish on all surfaces S_i. This means that $\phi_1 = \phi_2$ on each surface, which is the

Dirichlet Boundary Condition: The potential ϕ is specified on the surface.

The surface integrals will also vanish if the normal component of $\mathcal{E}$, defined by $\mathcal{E}_\perp = \hat{\mathbf{n}} \cdot \mathcal{E}$, vanishes on the surface. This means that $\mathbf{E}_{1\perp} = \mathbf{E}_{2\perp}$ on each surface, which is the

Neumann Boundary Condition: The normal component of the electric field, $\mathbf{E}_\perp$, is specified on the surface.

We have thus proved the

Uniqueness Theorem of Electrostatics: The electric field in a volume V is uniquely specified by the solution to $\nabla \cdot \mathbf{E} = 4\pi\rho$ and $\nabla \times \mathbf{E} = 0$, if either the potential ϕ or the normal component of the electric field $E_\perp$ is specified on each boundary surface of the volume.

The potential as well as the electric field will be unique except for the case of completely Neumann boundary conditions. In that case, an arbitrary constant could be added to the potential. This ambiguity is usually resolved by specifying the potential at one point in space.

There are other boundary conditions besides these two that lead to the vanishing of the surface integrals, and thus to uniqueness of the solutions, but the Dirichlet and Neumann conditions are the two that apply for most problems in electrostatics. The uniqeness theorem only proves the uniqueness of a possible solution. We have not yet proved the necessary existence of a solution. This will be done later by explicit construction of a solution to Eqs. (3.1) and (3.2) satisfying either the Dirichlet or Neumann boundary conditions.

The boundary conditions can be applied separately on each surface. That is Dirichlet on some surfaces and Neumann on others, and even can be Dirichlet on part of one surface and Neumann on the rest. But one or the other condition must be specified on the entire surface of the volume.

Care must be taken not to overspecify the boundary conditions. For instance, specifying both ϕ and $E_\perp$ on the same surface will lead to no solution, unless they are fortuitously consistent. That is, specifying the potential on the surface is sufficient to lead to a unique solution for ϕ from which $\mathbf{E}$ could be calculated.

Another case of overspecification could arise if the Neumann boundary condition is specified on the entire surface, including the bounding surface. Since Gauss's law follows by applying the divergence theorem to Eq. (3.1), the integral of $E_\perp$ over the entire surface is constrained by

$$\oint_S \mathbf{E} \cdot d\mathbf{A} = 4\pi \int_V \rho \, d\tau. \tag{3.17}$$

For the exterior problem, this constraint will be satisfied on the bounding surface at infinity by any solution to the problem. For the interior problem, $E_\perp$ on the bounding surface (or at least one other surface) must be chosen to satisfy Eq. (3.17), or there will be no static solution. (There will be a solution to the time dependent equation.) An example of this was the requirement we noted in Chapter 2 that the surface charge on the wall of a cavity in a conductor must be just the negative of any charge enclosed within the cavity. If any part of any surface has the Dirichlet boundary condition, then the constraint of Eq. (3.17) will be satisfied by any solution.

So far, the surfaces of the volume V could be arbitrary surfaces on which the boundary conditions are specified. If any surface is the surface of a conductor, then that surface must be an equipotential, with a constant value over the surface of the conductor. Also, the value of $E_\perp$ on the surface of the conductor is given by

$$E_\perp = 4\pi\sigma. \tag{3.18}$$

Then the surface integrals in Eq. (3.16) can be easily done, since the constant potential function ψ can be taken out of the integral. The remaining surface integral just equals the net charge on the conductor. Thus each surface integral will vanish if either the potential or net charge is specified on each conductor. This is the

Uniqueness Theorem of Electrostatics in the Presence of Conductors: The electric field in a volume V is uniquely specified by the solution to $\boldsymbol{\nabla} \cdot \mathbf{E} = 4\pi\rho$ and $\boldsymbol{\nabla} \times \mathbf{E} = 0$, if either the potential or charge is specified on each conductor.

Note that only the net charge is needed for each conductor. This is important, because the actual surface charge distribution cannot be known for a conductor until the problem is solved. Because of this, the Neumann boundary condition will not be a useful boundary condition for conductors, since $E_\perp$ cannot be specified for a conductor. Only the Dirichlet boundary condition with ϕ being constant on the surface of the conductor will be useful in solving problems in electrostatics with conducting surfaces. We will see later how this affects the solution of problems where the charge, and not the potential, is specified on a conductor.

The uniqueness theorem is often stated in terms of the potential as a solution of Poisson's equation.

Uniqueness Theorem for the Electrostatic Potential: The potential in a volume V is uniquely specified (up to an additive constant for a pure Neumann boundary condition) by the solution to $\nabla^2\phi = -4\pi\rho$, if either the potential or its normal derivative is specified on each surface of the volume, or if the potential or charge is specified on each conductor.

Having the uniqueness theorem means that any method of solution is valid as long as the resulting potential satisfies Poisson's equation and the boundary conditions. Even guessing a solution is alright, if it works. As an example of this, we look again at a vacant cavity in a conductor, which we treated in chapter 2. Since the surface of the cavity is an equipotential, we take as our solution the simple choice of that constant potential throughout the cavity. Since this satifies Poisson's equation for zero charge and the boundary condition, it is the unique solution. Then it follows that $\mathbf{E} = 0$ within the cavity and the surface charge density vanishes on the surface of the cavity. You may recall that it was not easy to prove the vanishing of the surface charge distribution in Chapter 2 before we had the uniqeness theorem.

In the remainder of this chapter, we consider various methods of solution of Poisson's and Laplace's equations with confidence that, if we find any solution satisfying the boundary conditions, it will be the unique, correct solution.

3.2   Images

In this section we illustrate the use of image charges to satisfy the boundary conditions for problems in which the conductors have useful symmetry properties.

3.2.1   Infinite grounded plane

As a first example, we solve the problem of a point charge q a distance d from an infinite grounded conducting plane. (The term **grounded** means that the potential is zero on the plane.) As shown in Fig. 3.2, the volume of interest is the space above the plane in the figure. Taking the vector position of the point charge as $\mathbf{d} = d\hat{\mathbf{k}}$, the physical region is the half space $z \geq 0$.

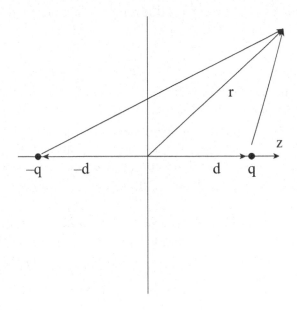

Figure 3.2: Point charge a distance d from a grounded plane.

The solution of this problem using Coulomb's law would be the field of the point charge plus the field due to the induced surface charge distribution on the grounded plane, but this is not known beforehand. Instead, we use an image charge behind the plane to reproduce the boundary condition that $\phi = 0$ on the surface z=0. As an educated guess, using the symmetry of the plane, we choose a charge -q a distance d behind the plane as shown in Fig. 3.2. It is easy to see that the potential due to the original charge q and its image -q is zero on the plane z=0.

Since the image charge is outside of the physical region, it does not enter Poisson's equation for $z \geq 0$. Thus, with the addition of this image charge to the original charge, Poisson's equation is satisfied for $z \geq 0$ and the boundary condition that $\phi(z = 0) = 0$ is satisfied. Therefor, the potential and electric field of the two charges are the unique solutions to the problem for $z > 0$. The region $z < 0$ is outside of the physical region for this problem, so whatever happens there is irrelevant.

The potential for $z > 0$ is that of the two point charges, and can be written

$$\phi(\mathbf{r}) = \frac{q}{|\mathbf{r} - \mathbf{d}|} - \frac{q}{|\mathbf{r} + \mathbf{d}|}, \quad z \geq 0. \tag{3.19}$$

The electric field is the corresponding field of the two point charges. The actual distribution of charges is a surface charge density on the grounded plane, but the potential and electric field in the region above the plane due to this surface charge is correctly reproduced by using the single image charge.

The image charge solution can be used to answer other questions about this configuration. For instance, the force on the point charge q a distance d from the grounded plane is actually caused by the distributed surface charge, but it can be given by the force on q due to the image charge -q a distance 2d away as

$$\mathbf{F} = -\frac{q^2\hat{\mathbf{d}}}{4d^2} \qquad (3.20)$$

This result is an example of a force acting on a charge even though there was no electric field before the charge was introduced. The electric field that causes the force on the charge is due to the surface charge induced on the plane by the point charge q. That this force is due to an induced charge distribution can be recognized by the dependence of the force on q^2. In the limit $q \to 0$ the ratio F/q will go to zero indicating that there was no pre-existing electric field.

Once the problem has been solved using the image charge, the actual surface charge distribution on the grounded plane can be found from $E_\perp$ at the plane. This is just the electric field of the two charges, so the surface charge density on the plane is given by

$$\sigma(\rho) = \frac{1}{4\pi}E_\perp = -\frac{qd}{2\pi(\rho^2 + d^2)^{\frac{3}{2}}}, \qquad (3.21)$$

where ρ (not to be confused with the volume charge density) is used here as the distance in the plane from the z-axis.

Although it gives the correct potential and electric field for $z \geq 0$, some care must be used to not take the image charge too seriously. If we use the image charge solution to calculate the work done in moving the charge q from infinity to the distance d, this gives

$$W = \int_\infty^d \mathbf{F} \cdot \mathbf{dr} = q^2 \int_\infty^d \frac{dr}{4r^2} = -\frac{q^2}{4d}. \qquad (3.22)$$

The negative of this work should be (and is) the potential energy U of the configuration. However, using the charge q and its image in Eq. (2.5) for the energy of two point charges gives

$$U = -\frac{q^2}{2d} \quad \text{(image charge result)}. \qquad (3.23)$$

Our use of the image charge seems to have given two different answers for the potential energy of the point charge and the conducting plane!

The resolution of this seeming paradox is that Eq. (3.23) is wrong because it is not appropriate to use the image charge in Eq. (2.5) for the potential energy of two point charges. There really is no point charge -q a distance 2d from the original charge q, although placing it there does correctly reproduce the potential and electric field for $z \geq 0$.

For integrating over the charge distribution to get the potential energy, the actual charge distribution must be used. Doing so in Eq. (2.12) leads to the correct answer

$$U = \frac{1}{2} \int \phi dq = \frac{1}{2} q \phi_{q'}(\mathbf{d}) = -\frac{q^2}{4d}, \tag{3.24}$$

with only the original point charge q contributing to the integral. The integral over the actual surface charge distribution vanishes because the potential on the grounded plane is zero.

3.2.2   Conducting sphere

Figure 3.3 shows a point charge q a distance d from the center of a grounded conducting sphere of radius a. The sphere could be either solid or hollow. In fact, the solution outside does not depend in any way on the internal construction of the sphere, if its outer surface is a conductor. We know that the potential inside the conducting sphere must be constant, but are interested in solving for the potential outside the sphere.

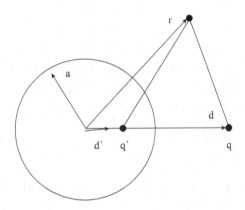

Figure 3.3: Point charge a distance d from a grounded sphere of radius a.

We place an image charge inside the sphere so that this image charge does not affect Poisson's equation in the physical region of interest $r > a$. The magnitude q' and location $\mathbf{d'}$ of the image charge are chosen to make the spherical surface at r=a (in the absence of the conductor) an equipotential of potential zero. By symmetry, we have placed the image charge along the axis defined by the location of the point charge q, as shown in Fig. 3.3.

The potential outside the sphere due to the two charges q and q' is given by

$$\phi(\mathbf{r}) = \frac{q}{|\mathbf{r} - \mathbf{d}|} + \frac{q'}{|\mathbf{r} - \mathbf{d}'|}, \quad r \geq a. \tag{3.25}$$

We want to find q' and $\mathbf{d}'$ to make this potential zero when r=a. This gives the condition

$$\phi(\mathbf{r})|_{r=a} = \frac{q}{|a\hat{\mathbf{r}} - d\hat{\mathbf{d}}|} + \frac{q'}{|a\hat{\mathbf{r}} - d'\hat{\mathbf{d}}|} = 0. \tag{3.26}$$

If we factor the quantity a/d out of the second denominator in Eq. (3.26), we can write

$$\phi(\mathbf{r})|_{r=a} = \frac{q}{|a\hat{\mathbf{r}} - d\hat{\mathbf{d}}|} + \frac{q'(d/a)}{|d\hat{\mathbf{r}} - (dd'/a)\hat{\mathbf{d}}|} = 0. \tag{3.27}$$

Now if we take

$$d' = a^2/d, \tag{3.28}$$

the second denominator in Eq. (3.27) will be $|d\hat{\mathbf{r}} - a\hat{\mathbf{d}}|$, and will be equal to the first denominator by the law of cosines. The numerators of Eq. (3.27) will cancel if we choose

$$q' = -qa/d. \tag{3.29}$$

Thus picking d' and q' as given by Eqs. (3.28) and (3.29) will make the spherical surface $r = a$, an equipotential with potential zero and satisfy the boundary condition of this problem. This gives the potential outside the sphere as

$$\phi(\mathbf{r}) = \frac{q}{|\mathbf{r} - \mathbf{d}|} - \frac{q}{|d\hat{\mathbf{r}} - r\hat{\mathbf{d}}|} = 0, \quad d \geq a. \tag{3.30}$$

The potential outside a grounded conducting sphere in the presence of a point charge q is that of the two point charges q and q' at the positions $\mathbf{d}$ and $\mathbf{d}'$. The image charge can be used to find the force on the original charge q to be

$$\mathbf{F} = \frac{qq'\hat{\mathbf{d}}}{(d - d')^2} = -\frac{q^2 a\hat{\mathbf{d}}/d}{(d - a^2/d)^2} = -\frac{q^2 a\mathbf{d}}{(d^2 - a^2)^2}. \tag{3.31}$$

The solution for a point charge outside a grounded sphere can be used as the starting point for the solution of several related problems. The solution for the case of a point charge inside a grounded hollow sphere is exactly the same as that for the charge outside. The steps above for the outside case never required d to be greater than a, so Eqs. (3.25) - (3.31) also apply for d less than a. Then the image charge would be outside the sphere and the solution would apply inside the hollow sphere.

Another case is that of a point charge outside an isolated neutral conducting sphere. For this, we place an additional point charge

$$q'' = -q' \tag{3.32}$$

at the center of the sphere. Because it is at the center, the surface of the sphere will still be an equipotential, now of potential

$$V = \frac{-q'}{a} = \frac{q}{d},$$ (3.33)

while the net image charge inside the sphere will be zero. Application of Gauss's law just outside the surface of the sphere would give the same result as if the net surface charge was zero. The potential outside is now that of the original charge and the two image charges. The force on the original charge is given by

$$\mathbf{F} = -\frac{q^2 a \mathbf{d}}{(d^2 - a^2)^2} + \frac{q^2 a \mathbf{d}}{(d^4)},$$ (3.34)

which is less than that for the grounded sphere. The case of a point charge outside an isolated conducting sphere of charge q_0 can be solved by taking the image charge at the origin to be

$$q'' = q_0 - q'.$$ (3.35)

The situation is a little different for a point charge inside an isolated hollow conducting sphere that originally had a charge q_0. There could be no additional image charge for this case. An image charge could not be placed inside the sphere without changing Poisson's equation, and an additional image charge could not be placed outside and keep the sphere as an equipotential. The solution would be exactly the same as for the charge inside a grounded sphere, except for an additive constant

$$\phi_0 = \frac{(q_0 + q)}{a}$$ (3.36)

due to the charges inside and on the sphere. This would be like the case discussed in connection with the Newton ice bucket experiment. The electric field inside a conducting sphere would not depend on the excess charge on the sphere, since that charge would all be on the outside surface

3.3   Separation of Variables for Laplace's Equation

3.3.1   Cartesian coordinates

The method of **separation of variables** is useful for solving Laplace's equation when the boundary surfaces (or conductors) have shapes that correspond to constant coordinate surfaces in some coordinate system. The simplest case is Cartesian coordinates (x,y,z) for which Laplace's equation is

$$[\partial_x^2 + \partial_y^2 + \partial_z^2]\phi(x, y, z) = 0.$$ (3.37)

As an example, we apply the method of separation of variables to the Dirichlet problem of finding the potential inside a hollow rectangular box of dimensions $A \times B \times C$ with its six bounding surfaces kept at specified potentials. We first consider a simplified problem with five sides grounded and the sixth side kept at a specified potential, as shown in Fig. 3.4.

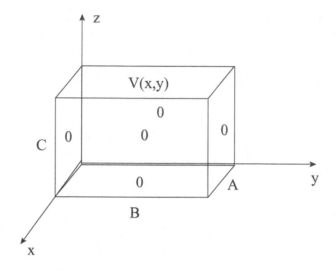

Figure 3.4: A rectangular box with 5 sides grounded, and the top kept at potential $V(x, y)$.

The boundary conditions for this problem are

$$\phi(x, y, C) = V(x, y), \tag{3.38}$$

and

$$\phi(0, y, z) = \phi(A, y, z) = \phi(x, 0, z) = \phi(x, B, z) = \phi(x, y, 0) = 0. \tag{3.39}$$

The method of separation of variables is to assume that a solution can be written as a product of three factors, each of which is a function of only one variable. That is, we take as an **ansatz** (trial solution) the form

$$\phi(x, y, z) = X(x)Y(y)Z(z). \tag{3.40}$$

If all six of the sides (the entire surface) were grounded, then the unique solution would be the trivial solution $\phi = 0$. If the potential on any part of the bounding surface is not zero, this trivial zero solution is excluded and we can divide Laplace's equation by ϕ. Then we substitute our ansatz into this equation

$$\frac{[\partial_x^2 + \partial_y^2 + \partial_z^2] X(x)Y(y)Z(z)}{X(x)Y(y)Z(z)} = \frac{X''}{X} + \frac{Y''}{Y} + \frac{Z''}{Z} = 0. \tag{3.41}$$

In deriving Eq. (3.41), we have used the property of a partial derivative that it acts only on its particular variable with the other variables being considered constant. Once the derivative acts on a function [such as $X(x)$] of only one variable , we have replaced the partial derivative $\partial_x X(x)$ by the total derivative $\frac{dX}{dx}$ for which we have used the standard notation X'.

Now, we move one of the terms in Eq. (3.41) to the other side so that it can be rewritten as

$$\frac{Y''}{Y} + \frac{Z''}{Z} = -\frac{X''}{X}. \tag{3.42}$$

Looking at Eq. (3.42), we see that the left hand side does not depend on x, so the right hand side, which can only depend on x, must be a constant. A similar argument holds for the left hand side of the equation as a function of y and z. This means each side of the equality must be the same constant. Then we can write an ordinary differential equation for $X(x)$

$$X'' = \lambda_x X, \tag{3.43}$$

where λ_x is a constant which, at this stage is arbitrary.

Equation (3.43) has the two possible solutions

$$X(x) = e^{\pm\sqrt{\lambda_x}x}. \tag{3.44}$$

The boundary conditions on $X(x)$ are that

$$X(0) = X(A) = 0, \tag{3.45}$$

so $X(x)$ must vanish in two places. This is only possible if the solution for X is oscillatory, which requires $\lambda_x < 0$. Then the general oscillatory solution for X can be written as

$$X(x) = \alpha \sin(\sqrt{-\lambda_x}x) + \beta \cos(\sqrt{-\lambda_x}x). \tag{3.46}$$

The boundary condition that $X(0) = 0$ requires the constant β to be zero. The boundary condition $X(A) = 0$ requires

$$\sin(\sqrt{-\lambda_x}A) = 0, \tag{3.47}$$

so

$$\sqrt{-\lambda_x}A = m\pi, \quad m = 1, 2, ... \tag{3.48}$$

The result is that the constant λ_x must satisfy the condition

$$\lambda_x = -\frac{m^2\pi^2}{A^2}, \tag{3.49}$$

where m is any positive integer.

The procedure we have just gone through is called solving an **eigenvalue problem**. The eigenvalue problem consists of a homogeneous differential equation, given here by Eq. (3.43) , and boundary conditions on the solution, given here by Eq. (3.45). Every term of a homogeneous equation has the same power of the unknown function, and a homogeneous equation always has zero as a possible solution.

In an eigenvalue problem with properly constituted homogeneous boundary conditions, a non-zero solution is only possible for certain characteristic values of the constant λ multiplying the solution on the right hand side of Eq. (3.43). These proper values of λ are called the **eigenvalues** of the problem. (In German "eigen" means proper or characteristic.) Here the eigenvalues are given by Eq. (3.49).

Without the boundary conditions, the constant λ multiplying the solution is arbitrary and could have any value. The differential equation is an **eigenvalue equation** if the boundary conditions are sufficient to restrict λ to its eigenvalues. The differential equation together with sufficient boundary conditions to restrict λ is called an eigenvalue problem.

A similar eigenvalue procedure can be carried out for the function $Y(y)$, which also must vanish in two places. Then Eq. (3.41) becomes an ordinary differential equation for the function $Z(z)$:

$$Z'' = \lambda_z Z, \tag{3.50}$$

with the condition

$$\lambda_z = \frac{m^2\pi^2}{A^2} + \frac{n^2\pi^2}{B^2}. \tag{3.51}$$

Equation (3.50) is not an eigenvalue equation because $Z(z)$ is not specified at both ends of its range, and the constant λ_z cannot be adjusted, but is determined by prior results. Note that, since the three constants λ_x, λ_y, and λ_z add up to zero, they can't all be negative. This means that there cannot be any non-zero solution with all six surfaces having zero potential, as we have already seen using the uniqueness theorem.

The single boundary condition on $Z(z)$ is

$$Z(0) = 0, \tag{3.52}$$

while $Z(C)$ is not yet determined. Therefor the only possible solution for $Z(z)$ is

$$Z(z) = \alpha_{mn} \sinh(\gamma_{mn}z), \tag{3.53}$$

where α_{mn} is arbitrary, and we have introduced the constant

$$\gamma_{mn} = \sqrt{\lambda_z} = \pi\sqrt{\frac{m^2}{A^2} + \frac{n^2}{B^2}}. \tag{3.54}$$

We now have all three functions X, Y, and Z, and the solution of Laplace's equation satisfying the boundary condition that five of the surfaces be grounded can be written as

$$\phi_{mn}(x, y, z) = \alpha_{mn} \sin\left(\frac{m\pi x}{A}\right) \sin\left(\frac{n\pi y}{B}\right) \sinh(\gamma_{mn} z). \quad (3.55)$$

The subscript mn in Eq. (3.55) refers to the two positive integers m and n characterizing the solution.

We see that there is a doubly infinite set of solutions that satisfy Laplace's equation and (so far) five of the boundary conditions of this problem. Since Laplace's equation is a homogeneous differential equation, any linear combination of solutions is a solution. Then, the most general solution that satisfies the five homogeneous ($\phi = 0$) boundary conditions is

$$\phi(x, y, z) = \sum_{mn} \alpha_{mn} \sin\left(\frac{m\pi x}{A}\right) \sin\left(\frac{n\pi y}{B}\right) \sinh(\gamma_{mn} z). \quad (3.56)$$

Now we are in a position to satisfy the last boundary condition that $\phi(x, y, C) = V(x, y)$. Applying this boundary condition to Eq. (3.56) gives

$$V(x, y) = \sum_{mn} \alpha_{mn} \sin\left(\frac{m\pi x}{A}\right) \sin\left(\frac{n\pi y}{B}\right) \sinh(\gamma_{mn} C). \quad (3.57)$$

This equation can be recognized as a double **Fourier sine expansion** of the function V(x,y).

3.3.2 Fourier series

We give a brief review of the expansion of a function f(x) in a Fourier sine series. The set of functions $\sin(n\pi x/L)$, with n being a positive integer, form a set of orthogonal functions satisfying

$$\int_0^L \sin(n\pi x/L) \sin(n'\pi x/L) dx = \frac{L}{2} \delta_{nn'}. \quad (3.58)$$

The **Kronecker delta** in Eq. (3.58) satisfies the conditions

$$\delta_{nn'} = 1, \quad n = n'; \qquad \delta_{nn'} = 0, \quad n \neq n'. \quad (3.59)$$

The expansion of a function $f(x)$ defined on an interval $(0, L)$ in a Fourier sine series is given by

$$f(x) = \sum_n \alpha_n \sin\left(\frac{n\pi x}{L}\right). \quad (3.60)$$

The expansion coefficients α_n can be determined by multiplying both sides of this expansion by one of the expansion functions $\sin\left(\frac{n'\pi x}{L}\right)$ and integrating from 0 to L. This gives

$$
\begin{aligned}
\int_0^L \sin\left(\frac{n'\pi x}{L}\right) f(x)dx &= \sum_n \int_0^L \alpha_n \sin\left(\frac{n'\pi x}{L}\right)\sin\left(\frac{n\pi x}{L}\right)dx \\
&= \sum_n \alpha_n (L/2)\delta_{nn'} \\
&= \frac{L}{2}\alpha_{n'}.
\end{aligned} \tag{3.61}
$$

Thus the expansion coefficient α_n is given by

$$
\alpha_n = \frac{2}{L}\int_0^L \sin\left(\frac{n\pi x}{L}\right)f(x)dx. \tag{3.62}
$$

We must call your attention to one step that was slipped into the first line of Eq. (3.61). (Did you notice it?) This was the interchange of order

$$
\int \sum \rightarrow \sum \int. \tag{3.63}
$$

That is the integral of the sum was performed by integrating term by term before completing the infinite sum. Term by term integration generally improves the convergence of an infinite sum, so its use here is justified and causes no problem. Later, going the other way will turn out to have interesting consequences.

The expansion coefficients in the double Fourier sine expansion in Eq. (3.57) are given by a double application of the procedure in Eq. (3.62), so

$$
\alpha_{mn}\sinh(\gamma_{mn}C) = \frac{4}{AB}\int_0^A \int_0^B \sin\left(\frac{m\pi x}{A}\right)\sin\left(\frac{n\pi y}{B}\right)V(x,y)dxdy. \tag{3.64}
$$

Now we have the complete solution to the problem of a specified potential on one surface with the other five surfaces grounded. The potential inside the box is given by Eq. (3.56), with the expansion coefficients α_{mn} given by Eq. (3.64). Since this solution satisfies Laplace's equation and a full set of Dirichlet boundary conditions, it is the unique solution to this problem.

The solution for this simplified problem of five (all but one) grounded surfaces can be used to generate the solution to the more general case of a non-zero potential being specified on all six faces. We could not easily solve this general problem all at once because we needed homogeneous boundary conditions on two of the three functions X, Y, Z to generate the eigenvalues and eigenfunctions.

The solution to the general case can be found from the solutions to the simpler grounded cases by solving the simplified problem six times, once for each surface not being grounded. We can label these six solutions as ϕ_n for $n = 1, 2, 3, 4, 5, 6$. Then the complete solution to the general problem is

$$
\phi = \sum_1^6 \phi_n. \tag{3.65}
$$

The sum of solutions still satisfies Laplace's equation. Since each solution satifies the boundary condition on one face with the other faces having zero potential, the sum of solutions will satisfy the complete boundary conditions for this problem. Therefor it is the unique solution. Care has to be taken in choosing the variables for each of the six cases. For instance, the case with $\phi(0, y, z)$ non-zero would involve the function $\sinh[\gamma_{mn}(A - x)]$.

It is interesting to note that the mathematics for steady state temperature distributions is the same as for electrostatics. This is because the heat flow equation for a steady state temperature distribution reduces to Laplace's equation. For instance, the temperature distribution inside a rectangular hockey arena with the floor kept at $0°C$ and the roof and four walls kept at $20°C$ would be given by

$$T(x, y, z) = \phi(x, y, z) - 20 \tag{3.66}$$

where $\phi(x, y, z)$ is the solution to the electrostatic problem solved above with the bottom surface at $V = 20$ and the other surfaces grounded.

For a temperature distribution, the Neumann boundary condition, correponding to specified heat currents at the surfaces is also of interest, but for electrostatics this boundary condition cannot easily be specified. For the Neumann case for the temperature distribution, the surface integral of the heat flows (corresponding to Gauss's law) would have to be zero in order to keep the temperature independent of time. This is the constraint we discussed earlier for the completely Neumann boundary condition.

As another example using Cartesian coordinates, we consider the case of an open channel bounding the region

$$x > 0, \qquad 0 < y < H. \tag{3.67}$$

as shown in Fig. 3.5. The bounding surfaces are assumed to have infinite extent in the z direction and, for simplicity, the boundary conditions have no z-dependence, so this is a two dimensional problem.

We first consider the boundary conditions

$$\phi(0, y) = V(y), \qquad \phi(x, 0) = \phi(x, H) = 0. \tag{3.68}$$

The solution for this problem is very much like the interior of the rectangular box done above, but in only two dimensions. The only change is that the implied x boundary condition that $\phi(x, y)$ remain finite as $x \to \infty$ requires that the x function be

$$X(x) = e^{-\gamma x}. \tag{3.69}$$

instead of a hyperbolic sinh. Thus the solution to this problem is

$$\phi(x, y) = \sum_n \alpha_n \sin\left(\frac{n\pi y}{H}\right) e^{-\frac{n\pi x}{H}}, \tag{3.70}$$

with

$$\alpha_n = \frac{2}{H} \int_0^H \sin\left(\frac{n\pi y}{H}\right) V(y) dy. \tag{3.71}$$

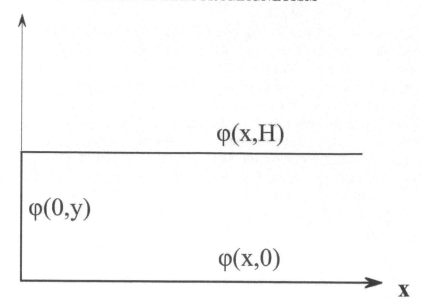

Figure 3.5: Open channel of height H.

3.3.3 Fourier sine integrals

A new kind of solution arises when we consider the boundary conditions

$$\phi(x, H) = f(x), \qquad \phi(0, y) = \phi(x, 0) = 0 \tag{3.72}$$

for the open channel. There is also an implicit boundary condition that $\phi(x, y)$ approaches zero fast enough as x becomes infinite, so that integrals over x will converge.

This problem can be considered the limit as L becomes infinite of an $L \times H$ enclosure. For finite L, the solution would be the two dimensional analogue of the potential inside a box

$$\phi(x, y) = \sum_n \alpha_n \sin\left(\frac{n\pi x}{L}\right) \sinh\left(\frac{n\pi y}{L}\right). \tag{3.73}$$

Applying the boundary condition at $y = H$ gives

$$\phi(x, H) = f(x) = \sum_n \alpha_n \sin\left(\frac{n\pi x}{L}\right) \sinh\left(\frac{n\pi H}{L}\right). \tag{3.74}$$

Equation (3.74) can be written as a Fourier sine series

$$f(x) = \sum_n \beta_n \sin\left(\frac{n\pi x}{L}\right), \tag{3.75}$$

with coefficients

$$\beta_n = \alpha_n \sinh\left(\frac{n\pi H}{L}\right) = \frac{2}{L} \int_0^L \sin\left(\frac{n\pi x}{L}\right) f(x)dx. \tag{3.76}$$

In order to take the limit as $L \to \infty$, we introduce a discrete variable k_n related to n by

$$k_n = \frac{n\pi}{L}. \tag{3.77}$$

Then Eq. (3.75) can be written as

$$f(x) = \sum_n \beta_n \sin(k_n x) \Delta n, \tag{3.78}$$

where we have introduced the quantity Δn, which just equals one. From Eq. (3.77), we see that

$$\Delta n = \frac{L}{\pi} \Delta k_n, \tag{3.79}$$

so Eq. (3.78) can be written

$$f(x) = \sum_n \frac{L}{\pi} \beta_n \sin(k_n x) \Delta k_n. \tag{3.80}$$

A new expansion coefficient

$$F(k_n) = \frac{L}{\pi} \beta_n \tag{3.81}$$

can be introduced, and then

$$f(x) = \sum_n F(k_n) \sin(k_n x) \Delta k_n. \tag{3.82}$$

Now, we can take the limit as $L \to \infty$, $\Delta k_n \to 0$ in Eq. (3.80). Then the sum in Eq. (3.80) approaches the usual definition of an integral, and we can write

$$f(x) = \int_0^\infty F(k) \sin(kx) dk, \tag{3.83}$$

with k now being a continuous variable. In this limit, equation (3.76) for the expansion coefficient becomes

$$F(k) = \frac{2}{\pi} \int_0^\infty \sin(kx) f(x) dx. \tag{3.84}$$

Equation (3.83) is called the **Fourier sine integral**, The expansion function $F(k)$ is called the **Fourier sine transform** of $f(x)$. In mathematics books, the factor $(2/\pi)$ in Eq. (3.84) is often treated in a different way, so that a factor of $\sqrt{(2/\pi)}$ appears symmetrically in each equation. We will keep the $(2/\pi)$ only in Eq. (3.84) so as to correspond more closely to the usual treatment of the discrete Fourier series expansion.

Applying the limit $L \to \infty$ to the Fourier sine expansion in Eq. (3.73) similarly leads to

$$\phi(x, y) = \int_0^\infty F(k) \sin(kx) \sinh(ky) dk, \tag{3.85}$$

with

$$F(k) = \frac{2}{\pi \sinh(kH)} \int_0^\infty \sin(kx) f(x) dx. \tag{3.86}$$

Equations (3.85) and (3.86) give the unique solution to the problem of the open channel shown in Fig. 3.5 with the top side kept at the potential f(x) and the other sides grounded.

As an illustration of the use of Fourier integrals, we solve the open channel problem for

$$f(x) = \phi(x, H) = V_0 e^{-\mu x} \tag{3.87}$$

on the upper surface. For this function, $F(k)$ is given by

$$F(k) = \frac{2V_0}{\pi \sinh(kH)} \int_0^\infty \sin(kx) e^{-\mu x} dx. \tag{3.88}$$

The integral in Eq. (3.88) is easily done:

$$\begin{aligned} I(k) &= \int_0^\infty \sin(kx) e^{-\mu x} dx \\ &= \text{Im} \int_0^\infty e^{-x(\mu - ik)} dx \\ &= \text{Im} \left(\frac{1}{\mu - ik} \right) \\ &= \frac{k}{k^2 + \mu^2}. \end{aligned} \tag{3.89}$$

Then

$$F(k) = \frac{2V_0}{\pi \sinh(kH)} \left(\frac{k}{k^2 + \mu^2} \right). \tag{3.90}$$

Substituting $F(k)$ into Eq. (3.85) gives

$$\phi(x, y) = \frac{2V_0}{\pi} \int_0^\infty \frac{k \sinh(ky) \sin(kx)}{(k^2 + \mu^2) \sinh(kH)} dk. \tag{3.91}$$

This final integral would have to be done numerically to get $\phi(x, y)$ anywhere inside the physical region. It is interesting to note that μ can be taken to be zero, corresponding to a constant potential V_0 on the top surface. The solution for ϕ is well behaved because the integrand in Eq. (3.91) approaces xy/H as μ approaches zero.

3.4 Surface Green's Function

An interesting thing happens if Eq. (3.62) for the expansion coefficients α_n is substituted into Eq. (3.60) for the Fourier sine series. We get

$$f(x) = \frac{2}{L} \sum_n \sin\left(\frac{n\pi x}{L}\right) \int_0^L \sin\left(\frac{n\pi x'}{L}\right) f(x') dx'. \tag{3.92}$$

Now we interchange the order of summation and integration. That is, we sum the series before integrating, which might be a perilous order to do things. The result is a rearrangement of Eq. (3.92)

$$f(x) = \int_0^L \left[\frac{2}{L} \sum_n \sin\left(\frac{n\pi x}{L}\right) \sin\left(\frac{n\pi x'}{L}\right) \right] f(x')dx'. \qquad (3.93)$$

Looking at Eq. (3.93) we see that the quantity in square brackets satisfies the definition of the Dirac delta function, and we have the result

$$\frac{2}{L} \sum_n \sin\left(\frac{n\pi x}{L}\right) \sin\left(\frac{n\pi x'}{L}\right) = \delta_L(x - x'). \qquad (3.94)$$

This is called the **Completeness Relation** for a set of functions. Its meaning is that the set of functions $\sin\left(\frac{n\pi x}{L}\right)$ are a complete set, sufficient to expand any function $f(x)$ that satisfies the integral relation in Eq. (3.93). We see now that the perilous step of summing before integrating has produced the delta function, so the series diverges for $x = x'$. However, when we realize that we will only use Eq. (3.94) under an integral, the danger is averted. We have added the subscript L to this delta function to remind ourselves that it is a periodic function of period L. Since we only intend to use it in the interval $(0, L)$, this periodicity does not affect our use of it.

Equation (3.94) is an example of a general result for a complete set of orthogonal functions $u_n(x)$ that have been **normalized** in an interval (a, b) so that

$$\int_a^b u_n^*(x)u_n(x)dx = 1. \qquad (3.95)$$

Such a set of normalized orthogonal functions is called a set of **orthonormal functions**. (However, be warned that physicists often still use the term "orthogonal functions" to describe such a set, with the normalization being understood.)

Note that in the general case of possibly complex functions the complex conjugate is taken of one of the pair of functions in the integral. The general completeness relation for such a set of orthonormal functions is

$$\sum_n u_n(x)u_n^*(x') = \delta_{ab}(x - x'). \qquad (3.96)$$

We have added the subscript ab to the delta function to emphasize that it is only guaranteed to act as the delta function in integrals over the interval (a,b).

The interesting mathematical question of exactly what class of functions $f(x)$ satisfy Eq. (3.93) and thus can be expanded in a Fourier sine series has not been completely solved in the mathematics literature. Sufficient conditions on the class of functions are known, but the full extent of necessary conditions has not yet been determined.

A set of conditions called the **Dirichlet conditions** have been found that include a large class of possible functions. Simply stated, the Dirichlet conditions are that any single valued function defined on an open interval (a, b) with a finite number of finite discontinuities, and a finite number of extrema, can be expanded in a Fourier series.

Another way to put it is that any single valued function you can draw in a finite length of time without lifting your pencil from the paper can be expanded in a Fourier series. This works because any "infinity" (infinite value, infinite discontinuity, infinite number of discontinuties or wiggles) in the function would require an infinite length of time to draw.

The wide range of functions permitted by the Dirichlet conditions has the physical significance that the potential specified on the bounding surfaces need not be continuous. A checkerboard pattern of potentials on the surfaces produces a well defined potential in the interior. Also, the Fourier sine series can be used to represent a function that does not vanish at the endpoints, even though every sine function in the sum does vanish at 0 and L. Mathematically, this is because the convergence of the Fourier series can be shown to be not uniform convergence. When a Taylor series converges, it does so with uniform convergence. Because of this, a Taylor series cannot represent a discontinuous function. This makes the Fourier series more useful in many physical situations.

We now apply the above considerations to the solution for the potential within a rectangular box. When the expansion coefficients α_{mn} given by Eq. (3.64) are substituted into Eq. (3.56) for the potential, the result is

$$\phi(x, y, z) = \frac{4}{AB} \sum_{mn} \sin\left(\frac{m\pi x}{A}\right) \sin\left(\frac{n\pi y}{B}\right) \frac{\sinh(\gamma_{mn} z)}{\sinh(\gamma_{mn} C)}$$

$$\int_0^A \int_0^B \sin\left(\frac{m\pi x'}{A}\right) \sin\left(\frac{n\pi y'}{B}\right) \phi(x', y', C) dx' dy'. \quad (3.97)$$

Interchanging the order of integration and summation now gives the result

$$\phi(x, y, z) = \int_0^A \int_0^B dx' dy' g(x, y, z; x', y', C) \phi(x', y', C). \quad (3.98)$$

The function $g(x, y, z; x', y', C)$ is given by a double sum

$$g(x, y, z; x', y', C) = \frac{4}{AB} \sum_{mn} \sin\left(\frac{m\pi x}{A}\right) \sin\left(\frac{n\pi y}{B}\right)$$

$$\sin\left(\frac{m\pi x'}{A}\right) \sin\left(\frac{n\pi y'}{B}\right) \frac{\sinh(\gamma_{mn} z)}{\sinh(\gamma_{mn} C)}. \quad (3.99)$$

We call the function $g(x, y, z; x', y', C)$ a **Surface Green's Function** (SGF), because it is related to more general Green's functions that we will discuss later on. The surface Green's function depends on six variables, three (x, y, z) inside the volume and three $(x', y', z' = C)$ on the surface. Integrating the product $g\phi$ over the surface variables x' and y', for z' fixed on the surface gives the potential ϕ inside the volume for any boundary value of ϕ on the surface.

We note from Eq. (3.99) that, on the surface $z = C$, the surface Green's function g is just the Dirac delta function $\delta(x - x', y - y')$, so the integration automatically gives the proper ϕ on the surface. An advantage of using the surface Green's function is that g does not depend on the value of ϕ on the surface, so once it has been calculated, it can be used for different surface values of ϕ without recalculation.

The surface Green's function can be used as a new method of finding the potential for Dirichlet boundary conditions. We call this the **Surface Green's Function Method**. The SGF method can be used when all but one of the surfaces have homogeneous boundary conditions $(\phi|_S = 0)$. It consists of two steps. The first step is simply to write down the solution in terms of the surface Green's function as

$$\phi(\mathbf{r}) = \int_S dS' g(\mathbf{r}, \mathbf{r}'_S)\phi(\mathbf{r}'_S), \qquad (3.100)$$

where the notation $\mathbf{r}'_S$ means that $\mathbf{r}'$ is restricted to the surface S.

Then the second step of the solution to the problem is finding the appropriate SGF. We construct the SGF by using its two defining properties:

I.   The SGF $g(\mathbf{r}, \mathbf{r}')$ must be a solution of Laplace's equation for the field variable $\mathbf{r}$, so

$$\nabla^2 g(\mathbf{r}, \mathbf{r}') = 0. \qquad (3.101)$$

II.   The SGF $g(\mathbf{r}, \mathbf{r}')$ must reduce to a two dimensional delta function when the field variable $\mathbf{r}$ is on the surface having the inhomogeneous boundary condition, so

$$g(\mathbf{r}_S, \mathbf{r}'_S) = \delta^{(2)}(\mathbf{r}_S - \mathbf{r}'_S). \qquad (3.102)$$

We can find an SGF satisfying these two conditions by first constructing a two dimensional delta function using the eigenfunctions of Laplace's equation for the homogeneous boundary conditions. In general, this will be

$$\delta(\mathbf{r} - \mathbf{r}') = \sum_n u_n(\mathbf{r})u_n^*(\mathbf{r}'), \qquad (3.103)$$

where the u_n represent a general set of complete orthonormal eigenfunctions and the subscript n stands for the complete set of subscripts. For the case of the rectangular box the two dimensional delta function is given by

$$\delta^{(2)}(\mathbf{r_S} - \mathbf{r}'_S) = \frac{4}{AB} \sum_{mn} \sin(\frac{m\pi x}{A}) \sin(\frac{m\pi x'}{A}) \sin(\frac{n\pi y}{B}) \sin(\frac{n\pi y'}{B}). \qquad (3.104)$$

The form given by Eq. (3.104) depends on only two of the three variables and is not a solution of Laplace's equation. The final step in finding the SGF is to construct a solution of Laplace's equation by multiplying by the appropriate function of the third variable. This function must also equal unity when the field variable **r** is on the surface. This last step gives the SGF

$$g(\mathbf{r}, \mathbf{r}') = \sum_n u_n^*(\mathbf{r}) u_n(\mathbf{r}') \frac{f_n(z)}{f_n(z_S)}. \tag{3.105}$$

We have written the third variable as z because so far we have only looked at Cartesian coordinates, but in a general problem the appropriate third variable would enter here. For the case of the rectangular box, the surface Green's function is that given by Eq. (3.99).

With the surface Green's function g constructed as above, the function given by integrating g with the surface potential as in Eq. (3.100) is guaranteed to be the unique solution of Laplace's equation satisfying the boundary conditions of the problem. The surface Green's function could be evaluated up to some order of the sum, and then be applied for any boundary condition. More often in the case of the double infinite sum, as in the rectangular box problem, the SGF method would just be used to write down the solution quickly. Then the double sum would be integrated term by term for a particular boundary condition. This use of the SGF method would be just like going through separation of variables, but would be a quicker, easier way of doing it.

The SGF method can also be applied to continuous solutions (integrals instead of sums) like the case of the open channel we did above using separation of variables. First we repeat the steps of Eqs. (3.92), (3.93), and (3.94) for the continuous case of the Fourier sine integral. Substituting the Fourier sine transform (3.84) into the Fourier sine integral (3.83) results in

$$f(x) = \int_0^\infty dk \sin(kx) \frac{2}{\pi} \int_0^\infty dx' \sin(kx') f(x'). \tag{3.106}$$

Then interchanging the orders of the two integration gives

$$f(x) = \int_0^\infty dx' f(x') \left[\frac{2}{\pi} \int_0^\infty dk \sin(kx) \sin(kx') \right]. \tag{3.107}$$

It can be seen that the integral in square brackets satisfies the definition of the Dirac delta function in the region $(0, \infty)$, so

$$\delta(x - x') = \frac{2}{\pi} \int_0^\infty \sin(kx) \sin(kx') dk. \tag{3.108}$$

This is an integral representation of the delta function using the complete set of orthogonal functions $\sin(kx)$. Equation (3.108) is the continuous version of the orthogonality relation equation (3.58) for discrete functions. At the same time, it is the continuous version of the completeness relation equation (3.94). We see that the dangerous operation of interchanging orders of integration has led to infinities, but when the second integral is performed everything works out satisfactorily.

When this interchange of integration order is applied to the solution of the open channel problem given by Eq. (3.85), the result can be written

$$\phi(x, y) = \int_0^\infty g(x, y; x', H)\phi(x', H)dx' \tag{3.109}$$

with the surface Green's function g(x,y;x',H) given by

$$g(x, y; x', H) = \frac{2}{\pi} \int_0^\infty \sin(kx) \sin(kx') \frac{\sinh(ky)}{\sinh(kH)} dk. \tag{3.110}$$

The steps of the SGF method for this problem would look like:

1. $\qquad \phi(x, y) \;=\; \displaystyle\int_0^\infty g(x, y; x', H)\phi(x', H) \hfill (3.111)$

2. $\qquad \delta(x - x') \;=\; \dfrac{2}{\pi} \displaystyle\int_0^\infty \sin(kx) \sin(kx')dk \hfill (3.112)$

3. $\; g(x, y; x', H) \;=\; \dfrac{2}{\pi} \displaystyle\int_0^\infty \sin(kx) \sin(kx') \dfrac{\sinh(ky)}{\sinh(kH)} dk. \hfill (3.113)$

3.5   Problems

1. A point charge q is a distance d from a grounded plane.

    (a) Integrate the surface charge to get the total charge induced on the plane.

    (b) Use Coulomb's law to find the force on the charge q by integration using the surface charge density on the plane.

2. Two semi-infinite grounded planes meet perpendicularly to form a square corner. A point charge q is placed a distance a from each plane.

    (a) Find the potential in the region between the planes.

    (b) Find the force on the point charge q.

    (c) Find the energy of this system. Evaluate the energy in eV if the particle is an electron and $a = 1\mathring{A}$.

3. A dipole **p** is a distance **d** away from an infinite grounded plane. The angle between **p** and **d** is $\theta = 60°$.

    (a) Find the energy of this system.

    (b) Find the force on the dipole.

    (c) Find the $\mathbf{p} \times \mathbf{E}$ and $\mathbf{r} \times \mathbf{F}$ torques on the dipole.

    (d) Describe in words (briefly) the motion of the dipole if it is released from rest.

4. A point charge q is a distance d $(>a)$ from the center of a neutral conducting sphere of radius a.

    (a) Find the induced dipole moment of the sphere. (Use the image charges.)

    (b) Find the dipole force on the point charge q.

5. A point charge q is a distance d $(>a)$ from the center of a grounded conducting sphere of radius a.

    (a) Find the surface charge density on the surface of the sphere.

    (b) Integrate the surface charge to find the total charge induced on the sphere.

    (c) Find the energy of this system. Evaluate the energy in eV if the particle is an electron and $d = 2\mathring{A}$ and $a = 1\mathring{A}$.

6. Find the Fourier sine series for the following functions defined on the interval (0,L):

    (a) $f(x) = 1$.

    (b) $f(x) = x$.

    (c) $f(x) = \cos(\pi x / L)$.

7. A function $f(x)$ defined in the interval $(0, L)$ can be expanded in a **Fourier cosine series** as

$$f(x) = \sum_n a_n \cos\left(\frac{n\pi x}{L}\right). \tag{3.114}$$

Find the coefficients a_n. (Note that there is now an a_0 term that must be treated separately.) The Fourier cosine series is useful for problems with the boundary condition $f'(0) = f'(L) = 0$.

8. A function $f(x)$ defined in the interval $(0, L)$ can be expanded in a **trigonometric Fourier series** as

$$f(x) = \sum_n \left[a_n \sin\left(\frac{\pi n x}{L}\right) + b_n \cos\left(\frac{\pi n x}{L}\right) \right]. \tag{3.115}$$

Find the coefficents a_n and b_n. The Fourier series will converge more quickly than the sine or cosine series for functions that do not satisfy homogeneous boundary conditions.

9. (a) A function $f(x)$ defined in the interval $(-L, L)$ can be expanded in an **exponential Fourier series** as

$$f(x) = \sum_{n=-\infty}^{+\infty} c_n e^{\frac{2\pi i n x}{L}}. \tag{3.116}$$

    Find the coefficients c_n.

    (b) The exponential Fourier series is equivalent to the trigonometric Fourier series. Find the coefficients a_n and b_n in terms of c_n.

10. A cavity in the shape of a regular polyhedron has differing uniform potentials on each face. Prove that the potential at the center of the cavity is the average of the potentials on the faces.

11. A cubical cavity of size $L \times L \times L$ has the top face at a uniform potential V_0, and the other five faces grounded.

    (a) Find the potential inside the cavity.

    (b) Find the potential at the center of the cavity.

    (c) Evaluate the potential (in terms of V_0) at the center of the cavity up to three significant figures.

12. Redo problem 10 for the boundary condition $\phi(x, y, L) = V_0 x/L$, with the other five faces grounded. Compare your numerical result for the potential at the middle of the cube with the result in problem 10. Explain the connection between these two results.

13. (a) The Fourier cosine sum can be be extended to an infinite interval, leading to the **Fourier cosine integral**

$$f(x) = \int_0^\infty F(k) \cos(kx) dk. \qquad (3.117)$$

   Derive the equation for $F(k)$ as an integral over $f(x)$.

   (b) The exponential Fourier sum can be extended to an infinite interval, leading to the **Fourier integral**

$$f(x) = \int_{-\infty}^\infty F(k) e^{ikx} dk. \qquad (3.118)$$

   The function $F(k)$ is called the **Fourier transform** of $f(x)$. Derive the equation for $F(k)$ as an integral over $f(x)$.

14. For an open channel with the boundary condition $\phi(0, y) = V_0$, with the other surfaces grounded, calculate the potential $\phi(H/2, H/2)$ to three significant figures.

15. Redo problem 14 for the boundary condition $\phi(0, y) = V_0 y/H$, with the other surfaces grounded. Explain the connection between this answer and that for problem 14.

16. For an open channel with the boundary condition $\phi(x, H) = V_0$, with the other surfaces grounded, calculate the potential $\phi(H/2, H/2)$ to three significant figures. Explain the connection between this answer and that for problem 14.

17. Find the surface Green's function $g(x, y; 0, y')$ for the $x = 0$ face of the open channel in Fig. 3.5.

Chapter 4

Spherical and Cylindrical Coordinates

4.1 General Orthogonal Coordinate Systems

A general coordinate system with three variables q_1, q_2, q_3 can be defined by relating the general coordinates to the Cartesian coordinates $x_1(=x)$, $x_2(=y)$, $x_3(=z)$. The defining equations can be taken as the three functions $x_i(q_1, q_2, q_3)$.

We would like to relate the unit vectors $\hat{\mathbf{q}}_i$ in the general system to the Cartesian unit vectors $\hat{\mathbf{n}}_1(=\hat{\mathbf{i}})$, $\hat{\mathbf{n}}_2(=\hat{\mathbf{j}})$, $\hat{\mathbf{n}}_3(=\hat{\mathbf{k}})$. The differential of the position vector $\mathbf{r}$ can be written as

$$
\begin{aligned}
\mathbf{dr} &= \sum_k \hat{\mathbf{n}}_k dx_k \\
&= \sum_k \hat{\mathbf{n}}_k \sum_i \frac{\partial x_k}{\partial q_i} dq_i \\
&= \sum_i \left[\sum_k \frac{\partial x_k}{\partial q_i} \hat{\mathbf{n}}_k \right] dq_i .
\end{aligned}
\tag{4.1}
$$

A unit vector $\hat{\mathbf{q}}_i$ in the general system is defined as that direction in which only the coordinate q_i changes. It can be seen from the last step in Eq. (4.1) that the quantity in square brackets is in the direction of $\hat{\mathbf{q}}_i$, but it may not have unit magnitude. So we write

$$
\mathbf{dr} = \sum_i h_i \hat{\mathbf{q}}_i dq_i
\tag{4.2}
$$

where

$$
h_i \hat{\mathbf{q}}_i = \sum_k \frac{\partial x_k}{\partial q_i} \hat{\mathbf{n}}_k
\tag{4.3}
$$

The quantity $h_i dq_i$ is the distance moved when the coordinate q_i changes by the differential amount dq_i, with the other general coordinates fixed.

We will restrict our considerations to orthogonal coordinate systems, defined by the condition

$$\hat{\mathbf{q}}_\mathbf{i} \cdot \hat{\mathbf{q}}_\mathbf{j} = \delta_{ij}. \tag{4.4}$$

This gives the orthogonality condition that

$$\sum_k \frac{\partial x_k}{\partial q_i} \frac{\partial x_k}{\partial q_j} = h_i h_j \delta_{ij}. \tag{4.5}$$

For $j = i$, Eq. (4.5) gives the magnitude of the metric coefficient h_i

$$h_i = \sqrt{\sum_k \left(\frac{\partial x_k}{\partial q_i}\right)^2}. \tag{4.6}$$

$$\hat{\mathbf{q}}_\mathbf{1} \times \hat{\mathbf{q}}_\mathbf{2} = \hat{\mathbf{q}}_\mathbf{3}, \quad \text{and cyclic} \tag{4.7}$$

We can now use the definitions of the vector differential operators to find their forms in a general othogonal coordinate system. The gradient is defined by Eq. (1.37)

$$d\phi(\mathbf{r}) = d\mathbf{r} \cdot \boldsymbol{\nabla}\phi. \tag{4.8}$$

This equation can be written in terms of the general coordinates q_i as

$$\sum_i \frac{\partial \phi}{\partial q_i} dq_i = \sum_i (\boldsymbol{\nabla}\phi)_i h_i dq_i. \tag{4.9}$$

From Eq. (4.9) and the fact that the displacements dq_i are arbitrary, we can see that

$$(\boldsymbol{\nabla}\phi)_i = \frac{1}{h_i} \frac{\partial \phi}{\partial q_i} \tag{4.10}$$

The definition of the divergence of a vector is given by Eq. (1.68)

$$\boldsymbol{\nabla} \cdot \mathbf{E} = \lim_{V \to 0} \frac{1}{V} \oint_S \mathbf{E} \cdot d\mathbf{A}. \tag{4.11}$$

The derivation following Eq. (1.68) in Section 1.5 for the form of the divergence in Cartesian coordinates can be extended to a general coordinate system. For general coordinates, the volume V to be divided by in Eq. (4.11) becomes

$$V = h_1 \Delta q_1 h_2 \Delta q_2 h_3 \Delta q_3, \tag{4.12}$$

while the surface integrals are also modified by the h_i's. So for general coordinates, Eq. (1.70) becomes

$$\begin{aligned} I+II &= \lim_{\Delta q_i \to 0} \frac{[(E_1 h_2 h_3)(q_1+\Delta q_1, q_2, q_3) - (E_1 h_2 h_3)(q_1, q_2, q_3)]\Delta q_2 \Delta q_3}{h_1 h_2 h_3 \Delta q_1 \Delta q_2 \Delta q_3} \\ &= \lim_{\Delta q_1 \to 0} \frac{[(E_1 h_2 h_3)(q_1 + \Delta q_1, q_2, q_3) - (E_1 h_2 h_3)(q_1, q_2, q_3)]}{h_1 h_2 h_3 \Delta q_1} \\ &= \frac{1}{h_1 h_2 h_3} \frac{\partial}{\partial q_1} (E_1 h_2 h_3). \end{aligned} \tag{4.13}$$

The integrals over faces III, IV, V, and VI are done similarly, so the total divergence for general coordinates can be written as

$$\mathbf{\nabla}\cdot\mathbf{E} = \frac{1}{h_1 h_2 h_3}\left[\frac{\partial}{\partial q_1}(E_1 h_2 h_3) + \text{cyclic}\right]. \tag{4.14}$$

The derivation of this general result can be used as a mnemonic for the form of the divergence in general coordinates. That is, the denominator $h_1 h_2 h_3$ can be remembered as the division by the volume element, while the factor $h_2 h_3$ multiplying E_1 can be recognized as the perpendicular surface element. Then the other two terms just come by cyclic substitution.

The Laplacian operator is the divergence of a gradient so Eqs. (4.10) and (4.14) can be combined to give

$$\mathbf{\nabla}^2\phi = \frac{1}{h_1 h_2 h_3}\left\{\frac{\partial}{\partial q_1}\left[\left(\frac{h_2 h_3}{h_1}\right)\frac{\partial\phi}{\partial q_1}\right] + \text{cyclic}\right\}. \tag{4.15}$$

This equation will be used in the following sections to solve Laplace's equation in general coordinate systems.

Although we will not use it in this text, we include below for completeness the expression for the curl in general coordinates:

$$(\mathbf{\nabla}\times\mathbf{E})_1 = \frac{1}{h_2 h_3}\left[\frac{\partial}{\partial q_2}(h_3 E_3) - \frac{\partial}{\partial q_3}(h_2 E_2)\right],\ \text{and cyclic.} \tag{4.16}$$

As a mnemonic, the $h_2 h_3$ in the denominator can be seen as the division by the area perpendicular to the 1 direction, and the factor $h_3 E_3$ can be seen as coming from the line integral of $\mathbf{E}\cdot\mathbf{dr}$ in the 3 direction in the definition of curl in Eq. (1.91).

4.2 Spherical Coordinates

Spherical coordinates, shown in Fig. 4.1, are defined by

$$\begin{aligned} q_1 &= r, & \text{the radial distance,} && (4.17)\\ q_2 &= \theta, & \text{the polar angle,} && (4.18)\\ q_3 &= \phi, & \text{the azimuthal angle.} && (4.19) \end{aligned}$$

They are related to the cartesian coordinates by

$$\begin{aligned} x &= r\sin\theta\cos\phi && (4.20)\\ y &= r\sin\theta\sin\phi && (4.21)\\ z &= r\cos\theta, && (4.22) \end{aligned}$$

as can be seen from Fig. 4.1.

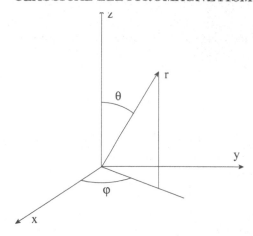

Figure 4.1: Spherical coordinates.

The spherical coordinate system is an orthogonal system, which can be verified by substitution of Eqs. (4.20) - (4.22) into the orthogonality condition of Eq. (4.5). The metric coefficients h_i can be calculated from Eq. (4.6):

$$h_1 = h_r \; = \; 1 \tag{4.23}$$
$$h_2 = h_\theta \; = \; r \tag{4.24}$$
$$h_3 = h_\phi \; = \; r\sin\theta. \tag{4.25}$$

The unit vectors could be related to the Cartesian unit vectors $\hat{\mathbf{i}}, \hat{\mathbf{j}}, \hat{\mathbf{k}}$ using Eq. (4.3). We will work strictly within the spherical coordinate system, and so will not need to use those relations.

With the h_i's known, we can substitute into Eq. (4.15) to get the Laplacian in spherical coordinates:

$$\boldsymbol{\nabla}^2\psi = \frac{[\partial_r(r^2\partial_r\psi)]}{r^2} + \frac{[\partial_\theta(\sin\theta\partial_\theta\psi)]}{r^2\sin\theta} + \frac{[\partial_\phi^2\psi]}{r^2\sin^2\theta}. \tag{4.26}$$

For spherical coordinates, we will designate the potential as $\psi(r,\theta,\phi)$ to avoid confusion between the potential and the ϕ coordinate.

4.2.1   Separation of variables in spherical coordinates

To apply separation of variables to Laplace's equation in spherical coordinates, we write the potential in factored form as

$$\psi(r,\theta,\phi) = R(r)\Theta(\theta)\Phi(\phi). \tag{4.27}$$

We rule out the trivial solution $\psi = 0$, so we can write Laplace's equation as

$$\frac{\boldsymbol{\nabla}^2\psi}{\psi} = \frac{[r^2R']'}{r^2R} + \frac{[\sin\theta\Theta']'}{r^2\sin\theta\,\Theta} + \frac{\Phi''}{r^2\sin^2\theta\,\Phi} = 0. \tag{4.28}$$

Since each of the functions R, Θ, Φ are functions of only one variable, the partial derivatives In Eq. (4.28) have all been replaced by ordinary derivatives with respect to the appropriate variable. In each term of Eq. (4.28), the two functions that are not differentiated have cancelled out of the quotient.

We make the simple step of rewriting Eq. (4.28) as

$$\frac{\sin^2\theta[r^2 R']'}{R} + \frac{\sin\theta[\sin\theta\Theta']'}{\Theta} = -\frac{\Phi''}{\Phi}. \tag{4.29}$$

Now we can see that the left hand side of Eq. (4.29) does not depend on the variable ϕ. Since the left hand side does not change when ϕ is varied, and the right hand side depends only on ϕ, the right hand side must equal a constant with respect to all three variables. Thus we can write

$$\Phi''(\phi) = \lambda_\phi \Phi(\phi). \tag{4.30}$$

This is an eigenvalue equation for the function Φ with eigenvalue λ_ϕ.

The boundary condition on $\Phi(\phi)$ follows from the fact that ϕ represents the angle of rotation about the z-axis. A rotation through an angle of 2π radians brings one back to the same point in space. The physical boundary condition we impose is that the potential $\psi(r,\theta,\phi)$ be single valued, and this requires the function $\Phi(\phi)$ to satisfy

$$\Phi(\phi + 2\pi) = \Phi(\phi). \tag{4.31}$$

This is called a **periodic boundary condition** with period 2π. This requires the solution of the differential equation (4.30) to be of the form

$$\Phi(\phi) = e^{im\phi}, \tag{4.32}$$

with the constant m being any positive or negative integer, or zero. The trigonometric functions $\sin(m\phi)$ and $\cos(m\phi)$ could also be used, but the standard notation in physics is to use the exponential form $e^{im\phi}$.

The ϕ eigenvalue is then given by

$$\lambda_\phi = -m^2, \tag{4.33}$$

and is always negative or zero. The left hand side of Eq. (4.29) must equal the constant m^2, so

$$\frac{\sin^2\theta[r^2 R']'}{R} + \frac{\sin\theta[\sin\theta\Theta']'}{\Theta} = m^2 \tag{4.34}$$

This equation can be rewritten as

$$\frac{[r^2 R']'}{R} = \frac{m^2}{\sin^2\theta} - \frac{[\sin\theta\Theta']'}{\Theta\sin\theta} = -\lambda_\theta. \tag{4.35}$$

We have used the same line of reasoning as was just used above to equate both sides of Eq. (4.35) to a constant $(-\lambda_\theta)$. Equation (4.35) can be written as an eigenvalue equation in the variable θ

$$[\sin\theta\,\Theta'(\theta)]' - m^2\Theta(\theta)/\sin\theta = \lambda_\theta \sin\theta\,\Theta(\theta). \tag{4.36}$$

4.2.2   Azimuthal symmetry, Legendre polynomials

We will first consider the solutions of this equation with $m = 0$ which would result if the boundary conditions on $\Psi(r, \theta, \phi)$ had no ϕ dependence (azimuthal symmetry). In that case, Eq. (4.36) becomes

$$[\sin\theta\,\Theta'(\theta)]' = \lambda_\theta \sin\theta\,\Theta(\theta). \tag{4.37}$$

This is the trigonometric form of **Legendre's equation**.

It is more convenient for solution to put it into algebraic form by the substitutions

$$x = \cos\theta, \qquad y(x) = \Theta(\theta). \tag{4.38}$$

Then, differentiation with respect to θ gets replaced by

$$
\begin{aligned}
\frac{d}{d\theta} &= \frac{dx}{d\theta}\frac{d}{dx} \\
&= -\sin\theta\frac{d}{dx} \\
&= -\sqrt{1-x^2}\frac{d}{dx},
\end{aligned} \tag{4.39}
$$

so Eq. (4.37) can be rewritten as

$$[(1-x^2)y'(x)]' = \lambda_\theta y(x). \tag{4.40}$$

This is also called Legendre's equation. Note that this differential equation has a singular behavior at $x = \pm 1$, where the derivative terms drop out. (Of course, the x and y in this form of Legendre's equation should not be confused with the Cartesian coordinates x and y.)

Equation (4.40) is a homogeneous differential equation with polynomial coefficients. The standard method of solving such a differential equation is the **method of Frobenius**, which develops a generalized power series solution. We write down a series for $y(x)$ of the form

$$y(x) = \sum_{n=0}^{\infty} a_n x^{s+n}, \quad a_0 \neq 0, \tag{4.41}$$

where the power s (called the **index** of the solution) and the coefficients a_n are to be determined by substitution of Eq. (4.41) into the differential equation.

We first form the terms of the differential equation by taking the appropriate derivatives

$$y''(x) = \sum_{n=0}^{\infty}(s+n)(s+n-1)a_n x^{s+n-2}, \tag{4.42}$$

$$(x^2 y'(x))' = \sum_{n=0}^{\infty}(s+n)(s+n+1)a_n x^{s+n}. \tag{4.43}$$

Next, the index n in Eq. (4.42) is adjusted to make all the exponents the same by making the substitution $n \to n + 2$. Then the expansion for y'' can be written as

$$y''(x) = \sum_{n=-2}^{\infty} (s+n+2)(s+n+1)a_{n+2}x^{s+n}. \qquad (4.44)$$

Note that the expansion for y'' now starts with $n = -2$, so the sum in Eq. (4.44) is just the same as that in Eq. (4.42).

We now combine the terms for the differential equation

$$x^s \sum_{n=-2}^{\infty} \{(s+n+2)(s+n+1)a_{n+2} - [(s+n)((s+n+1)+\lambda_\theta]a_n\}x^n = 0. \quad (4.45)$$

We have extended all the sums down to $n = -2$ with the understanding that any a_n is zero for negative n.

In order for the power series to equal zero, the coefficient of each power of x in Eq. (4.45) must vanish separately. We look first at the first coefficient in the sum with $n = -2$

$$s(s-1)a_0 = 0. \qquad (4.46)$$

This equation is called the **indicial equation**. It determines the index of the solution to be

$$s = 0 \quad \text{or} \quad s = 1. \qquad (4.47)$$

We look first at the case $s = 0$. Then the general term becomes

$$(n+2)(n+1)a_{n+2} - [(n+1)n + \lambda_\theta]a_n = 0. \qquad (4.48)$$

For $n = -1$, this reduces to $0 = 0$, so a_1 is arbitrary. For $n \geq 0$, Eq. (4.48) can be solved for a_{n+2} in terms of a_n

$$a_{n+2} = \frac{n(n+1)+\lambda_\theta}{(n+1)(n+2)}a_n. \qquad (4.49)$$

This recursion relation relates the coefficients of all the even powers of x to a_0, and all the odd powers to a_1. This means there are two independent solutions, an even function proportional to a_0, and an odd function proportional to a_1.

We now examine the convergence of the infinite series solution in the physical region defined by $-1 \leq x \leq +1$, since $x = \cos\theta$ where θ can vary from $0°$ to $180°$. We test the convergence of the solution with all even n. To facilitate this, we change the index by letting

$$m = n/2, \quad b_m = a_{2n}. \qquad (4.50)$$

Then the recursion relation becomes

$$b_{m+1} = \frac{m(m+\frac{1}{2}) + \frac{1}{4}\lambda_\theta}{(m+\frac{1}{2})(m+1)}b_m, \qquad (4.51)$$

which is a standard form of the recursion relation to test for convergence of the series.

For large m, we expand the recursion relation to first order in $1/m$ which gives for the ratio of adjacent terms in the series

$$\frac{b_{m+1}}{b_m}x^2 = (1 - \frac{1}{m})x^2. \tag{4.52}$$

This ratio approaches x^2 in the limit as $m \to \infty$. The simple ratio test for convergence requires the ratio of adjacent terms in an infinite series to be less than 1, and this is true here for all x less than 1 in magnitude. But for $x = \pm 1$, this limit equals one for this series, and the ratio test is indeterminate. Gauss's test for convergence [see Eq. (5.44a) of Arfken] indicates divergence of the series if the coefficient of the $1/m$ term equals 1 for large m, and that is the case for $x = \pm 1$.

We come to the conclusion that the infinite series would diverge in the physical region, and therefor must terminate for some finite n. We would reach the same conclusion for the odd series by similar reasoning. Looking at the recursion relation Eq. (4.49), we see that the series will terminate at $n = l$ if

$$\lambda_\theta = -l(l+1), \tag{4.53}$$

where l can be 0 or any positive integer. This means that the solution $y(x)$ will be a finite polynomial of order l that is even or odd depending on whether l is even or odd. The condition given in Eq. (15.55) is an eigenvalue condition on λ_θ.

The eigenfunction $y(x)$ is called a **Legendre polynomial**, and is designated by $P_l(x)$. The physical condition that has led to the θ eigenvalue is that the potential be finite throughout the physical region, especially at the endpoints (± 1) which are singular points of the differential equation.

We have not yet considered the other possibility that the index s equal 1 rather than 0, but it is easy to see that this would lead to the odd polynomial solution we have already obtained in the case of odd l. So we have found the complete set of eigenvalues and eigenfunctions for the θ eigenvalue problem.

The Legendre polynomials are given by the finite sums

$$P_l(x) = \sum_{n=0}^{l} a_n x^n, \tag{4.54}$$

with $a_1 = 0$ for even l, and $a_0 = 0$ for odd l. The coefficients are determined from the recursion relation

$$a_{n+2} = \frac{n(n+1) - l(l+1)}{(n+1)(n+2)}a_n. \tag{4.55}$$

At this point, many mathematics books go on to find a general form for the coefficients a_n. But that is unecessary, since everything we need to know is given by the recursion relation for the coefficients.

We list below a number of important properties of the Legendre polynomials:

Normalization: Legendre polynomials are proportional to the leading coefficient, either a_0 or a_1, which are arbitrary. These coefficients can be chosen so that

$$P_l(1) = 1. \tag{4.56}$$

This will prove to be a useful normalization for many applications of the Legendre polynomials.

Orthogonality: Legendre's equation can be written in operator form as

$$\mathcal{L}_L P_l(x) = \lambda_l P_l(x), \tag{4.57}$$

where the eigenvalue λ_l is given by $-l(l+1)$, and $\mathcal{L}_L$ is a differential operator defined by

$$\mathcal{L}_L f(x) = [(1-x^2)f'(x)]', \tag{4.58}$$

for any function $f(x)$. This operator is said to be **self-adjoint** (physicists also call this **Hermitian**) on the interval $-1 \leq x \leq +1$. In general, an operator $\mathcal{L}$ is self-adoint on a closed interval $[a, b]$ if it has the property that

$$\int_a^b \{f(x)^* \mathcal{L}g(x) - [\mathcal{L}f(x)]^* g(x)\} dx = 0 \tag{4.59}$$

for any two functions $f(x)$ and $g(x)$ in a particular class. [although $\mathcal{L}_L$ and the $P_l(x)$ are real, we consider possibly complex functions here for more generality.]

That the Legendre operator $\mathcal{L}_L$ is self-adjoint can be seen by substituting $\mathcal{L}_L$ into the integrand of Eq. (4.59)

$$\int_{-1}^{+1} \{f(x)^* [(1-x^2)g'(x)]' - g(x)[(1-x^2)f'(x)^*]'\} dx =$$
$$[f(x)^*(1-x^2)g'(x) - g(x)(1-x^2)f'(x)^*]_{-1}^{+1}. \tag{4.60}$$

We can see that the result in Eq. (4.60) will be 0 if the functions $f(x)$, $g(x)$ and their first derivatives are finite in the closed interval $[a, b]$. This class of functions includes the Legendre polynomials.

We can now use the self-adjoint property to show that the eigenvalues of a self-adjoint operator are real and that eigenfunctions with different eigenvalues are orthogonal. We form the integral

$$\int_a^b \{u_m(x)^* \mathcal{L}u_n(x) - [\mathcal{L}u_m(x)]^* u_n(x)\} dx$$
$$= (\lambda_n - \lambda_m^*) \int_a^b u_m(x)^* u_n(x) dx, \tag{4.61}$$

where we have used the fact that u_m and u_n are eigenfunctions of $\mathcal{L}$. But the integral on the left hand side of Eq. (4.61) equals 0 because $\mathcal{L}$ is a self-adjoint operator, so

$$(\lambda_n - \lambda_m^*) \int_a^b u_m(x)^* u_n(x) dx = 0. \qquad (4.62)$$

If we let $m = n$ in Eq. (4.62), we see that all the eigenvalues of a self-adjoint operator must be real. Then, if $\lambda_m \neq \lambda_n$, we see that the functions u_m and u_n are orthogonal on the interval [-1,+1].

For Legendre polynomials this orthogonality can be written

$$\int_{-1}^{+1} P_l(x) P_{l'}(x) dx = \frac{2}{2l+1} \delta_{ll'}. \qquad (4.63)$$

Since the normalization has already fixed by setting $P_l(1) = 1$, the orgonality integral cannot be set equal to 1 for $l = l'$. We will derive the normalization factor $2/(2l+1)$ below.

Generating function: If the function

$$g(x,t) = \frac{1}{\sqrt{1-2tx+t^2}} \qquad (4.64)$$

is expanded in powers of t, the result is

$$\frac{1}{\sqrt{1-2tx+t^2}} = \sum_n P_n(x) t^n, \quad |t| < 1. \qquad (4.65)$$

Note that the coefficients of each power of t in the expansion is the corresponding Legendre polynomial in the x variable. We will derive this property of the generating function later in this chapter.

The expansion of the generating function is very useful for developing properties of Legendre polynomials and for doing integrals involving Legendre polynomials. For example, we can use the generating function to derive the factor $2/(2l+1)$ in the Legendre polynomial normalization integral. We first expand the function $g^2(x,t)$, given by

$$g^2(x,t) = \frac{1}{1-2tx+t^2} = \sum_{l,l'} P_l(x) P_{l'}(x) t^{l+l'}, \quad |t| < 1. \qquad (4.66)$$

Next we integrate $g^2(x,t)$ with respect to x

$$\int_{-1}^{+1} \frac{dx}{1-2tx+t^2} = \frac{[\ln(1+t) - \ln(1-t)]}{t}$$

$$= \sum_{\text{odd } n} \frac{2t^{n-1}}{n}, \qquad (4.67)$$

where we have used the expansion

$$\ln(1-t) = -\sum_n \frac{t^n}{n}, \quad |t| < 1. \tag{4.68}$$

We also integrate the right hand side of Eq. (4.66) term by term, resulting in

$$\sum_{l,l'} t^{l+l'} \int_{-1}^{+1} P_l(x) P_{l'}(x) dx = \sum_l t^{2l} \int_{-1}^{+1} [P_l(x)]^2 dx, \tag{4.69}$$

where we have used the orthogonality of the P_l to set $l' = l$. To compare this sum on l with the sum on n in Eq. (4.67), we let $n = 2l + 1$. Then, the two sums will be equal if

$$\int_{-1}^{+1} [P_l(x)]^2 dx = \frac{2}{2l + 1}. \tag{4.70}$$

Recursion relation: A Legendre polynomial of a given l can be found in terms of the Legendre polynomials of lower l by the recursion relation

$$P_{l+1}(x) = \frac{(2l+1)x P_l(x) - l P_{l-1}(x)}{l+1}, \quad l = 1, 2, 3, ... \tag{4.71}$$

This recursion relation for the Legendre polynomials directly is more useful for practical calculations than the recursion relation (4.55) for the coefficients. For one thing, a Legendre polynomial of order l is rarely needed except in an expansion that requires all the lower order polynomials.

The recursion relation can be derived by acting on the generating function as follows

$$(1 - 2tx + t^2)\partial_t \frac{1}{\sqrt{1 - 2tx + t^2}} = \frac{x - t}{\sqrt{1 - 2tx + t^2}} \tag{4.72}$$

Expanding the generating function on each side of Eq. (4.72) results in

$$(1 - 2tx + t^2) \sum_n n P_n(x) t^{n-1} = (x - t) \sum_m P_m(x) t^m. \tag{4.73}$$

Now, changing the summation indices n and m appropriately (This is left as a problem.), leads to the Legendre polynomial recursion relation (4.71).

$\mathbf{P_0(x)}, \mathbf{P_1(x)}, \mathbf{P_2(x)}$: It is useful to commit to memory, the first three Legendre polynomials:

$$P_0(x) = 1, \tag{4.74}$$
$$P_1(x) = x, \tag{4.75}$$
$$P_2(x) = \frac{3x^2 - 1}{2}. \tag{4.76}$$

There is no need to memorize higher polynomials because they are more reliably obtained using the Legendre polynomial recursion relation (4.71).

4.2.3   Boundary value problems with azimuthal symmetry

Having found the eigenvalue $\lambda_\theta = -l(l+1)$, we can now put that into Eq. (4.35) for the radial function $R(r)$

$$[r^2 R'(r)]' = l(l+1)R(r). \tag{4.77}$$

Equation (4.77) not an eigenvalue equation, because the constant $l(l+1)$ has already been determined as the θ eigenvalue. It is an ordinary second order differential equation with two independent solutions

$$R(r) = r^l \quad \text{and} \quad R(r) = r^{-(l+1)}. \tag{4.78}$$

We now have the solution of Laplace's equation for a fixed value of l.

$$\psi_l(r,\theta) = \left[a_l r^l + \frac{b_l}{r^{l+1}}\right] P_l(\cos\theta), \tag{4.79}$$

where a_l and b_l are arbitrary constants. Since Laplace's equation is a homogeneous equation, any linear combination of solutions is also a solution. So the general solution of Laplace's equation in spherical coordinates with azimuthal symmetry is

$$\psi(r,\theta) = \sum_{l=0}^{\infty} \left[a_l r^l + \frac{b_l}{r^{l+1}}\right] P_l(\cos\theta). \tag{4.80}$$

The constants a_l and b_l are to be determined by the remaining boundary conditions for any particular problem.

Potential outside a sphere

A typical boundary value problem is to find the potential outside a sphere of radius R kept at a given potential $V(\theta)$ on its surface. This corresponds to the boundary condition

$$\psi(R,\theta) = V(\theta). \tag{4.81}$$

The first step is to write down the general solution for Laplace's equation outside the sphere

$$\psi(r,\theta) = \sum_l \frac{b_l}{r^{l+1}} P_l(\cos\theta), \quad r \geq R, \tag{4.82}$$

where we have left out the r^l terms because of the boundary condition that the potential go to zero at infinity.

Next, we evaluate the potential at the surface of the sphere, $r = R$

$$\psi(R,\theta) = V(\theta) = \sum_l \frac{b_l}{R^{l+1}} P_l(\cos\theta). \tag{4.83}$$

This is an expansion of the function $V(\theta)$ in the orthogonal set of Legendre polyomials, The coefficients $\frac{b_l}{R^{l+1}}$ can be found, just as we did for Fourier series, by using the orthogonality to project out a specific term.

We multiply each side of Eq. (4.83) by $P_n(\cos\theta)$ and integrate over the range $-1 \leq \cos\theta \leq +1$

$$\int_{-1}^{+1} V(\theta) P_n(\cos\theta) d(\cos\theta) = \int_{-1}^{+1} \sum_l \frac{b_l}{R^{l+1}} P_l(x) P_n(x) dx. \qquad (4.84)$$

Integrating the sum term by term, and using Eq. (4.63) for the orthogonality integral leads to the result

$$b_n = \frac{(2n+1)R^{n+1}}{2} \int_{-1}^{+1} V(\theta) P_n(\cos\theta) d(\cos\theta). \qquad (4.85)$$

This completes the solution of the problem with the potential outside the sphere given by Eq. (4.82).

If the sphere is hollow, the potential inside the sphere can be found in a similar manner, but omitting the $\frac{1}{r^{l+1}}$ terms corresponding to the implicit boundary condition that the potential be finite at the origin. The potential inside a hollow sphere is given by

$$\psi(r,\theta) = \sum_l \left(\frac{r^l}{R^{2l+1}} \right) b_l P_l(\cos\theta). \qquad (4.86)$$

Once the potential is known, the electric field $\mathbf{E}(r,\theta)$ can be found by taking the negative gradient of $\psi(r,\theta)$ in spherical coordinates. We can also find the surface charge density on a hollow sphere from Gauss's law by taking

$$\begin{aligned} \sigma(\theta) &= \frac{1}{4\pi}[E_r(R_+) - E_r(R_-)] \\ &= \frac{1}{4\pi}[\partial_r \psi(r,\theta)|_{r=R_-} - \partial_r \psi(r,\theta)|_{r=R_+}], \end{aligned} \qquad (4.87)$$

where $E_r(R_+)$ is the radial electric field just outside the sphere and $E_r(R_-)$ is the radial electric field just inside the sphere. Taking the radial derivatives results in

$$\sigma(\theta) = \frac{1}{4\pi} \sum_l \frac{(2l+1)}{R^{l+2}} b_l P_l(\cos\theta). \qquad (4.88)$$

As a simple example, we take the case of a constant potential V_0 on the upper hemisphere, and a constant potential $-V_0$ on the lower hemisphere of a hollow sphere of radius R. In this case, the coefficients b_l are given by

$$\begin{aligned} b_l &= (2l+1)V_0 R^{l+1} \int_0^1 P_l(x) dx, \quad \text{odd } l \\ &= 0, \quad \text{even } l, \end{aligned} \qquad (4.89)$$

where we have made use of the symmetry property

$$P_l(-x) = (-1)^l P_l(x). \tag{4.90}$$

We use the generating function to do the integral in Eq. (4.89)

$$\int_0^1 \frac{dx}{\sqrt{1 - 2tx + t^2}} = \frac{1 - t - \sqrt{1 + t^2}}{-t}$$

$$= 1 + \sum_{n=1}^{\infty} \binom{\frac{1}{2}}{n} t^{2n-1}. \tag{4.91}$$

In Eq. (4.91), we have used the binomial expansion

$$(1 + x)^a = \sum_{n=0}^{\infty} \binom{a}{n} x^n, \tag{4.92}$$

where the binomial coefficients are defined by the recursion relation

$$\binom{a}{0} = 1$$

$$\binom{a}{n+1} = \left(\frac{a-n}{n+1}\right) \binom{a}{n}, \tag{4.95a}$$

or directly by

$$\binom{a}{n} = \frac{a!}{n!(a-n)!}. \tag{4.95b}$$

Now to complete the evaluation of the integral in Eq. (4.89), we expand the generating function in the integral on the left hand side of Eq. (4.91), so

$$\sum_{l=0}^{\infty} t^l \int_0^1 P_l(x)dx = 1 + \sum_{n=1}^{\infty} \binom{\frac{1}{2}}{n} t^{2n-1}. \tag{4.96}$$

To equate coefficients of corresponding powers of t in the Taylor expansions, we let $n = (l+1)/2$ for odd l, with the result

$$\int_0^1 P_l(x)dx = \binom{\frac{1}{2}}{\frac{l+1}{2}}, \quad \text{odd } l. \tag{4.97}$$

We can also use Eq. (4.96) to see that the integral is zero for even l.

The result in Eq. (4.97) can be substituted into Eq. (4.89) to give

$$b_l = (2l+1)R^{l+1}V_0 \binom{\frac{1}{2}}{\frac{l+1}{2}}, \quad \text{odd } l$$

$$= 0, \quad \text{even } l. \tag{4.98}$$

Using this b_l in Eqs.(4.82) and (4.86) gives the potential for this problem:

$$\psi(r,\theta) = V_0 \sum_{\text{odd }l} (2l+1) \left(\frac{R}{r}\right)^{l+1} \left(\frac{\frac{1}{2}!}{\frac{l+1}{2}!}\right) P_l(\cos\theta), \quad r \geq R \qquad (4.99)$$

$$= V_0 \sum_{\text{odd }l} (2l+1) \left(\frac{r}{R}\right)^{l} \left(\frac{\frac{1}{2}!}{\frac{l+1}{2}!}\right) P_l(\cos\theta), \quad r \leq R. \qquad (4.100)$$

The first two explicit terms for the exterior potential are given by

$$\psi(r,\theta) = V_0 \left[\frac{3}{2}\left(\frac{R}{r}\right)^2 P_1(\cos\theta) - \frac{7}{8}\left(\frac{R}{r}\right)^4 P_3(\cos\theta)\right]. \qquad (4.101)$$

4.2.4 Multipole expansion

For any charge distribution of finite extent, we can always find some distance d so that none of the charge extends beyond a sphere of radius d. Then, for $r>d$, the potential of the charge distribution will satisfy Laplace's equation

$$\boldsymbol{\nabla}^2\psi(r,\theta,\phi) = 0, \quad r > d. \qquad (4.102)$$

If the charge distribution is axially symmetric, then the potential solving Laplace's equation will be given by

$$\psi(r,\theta) = \sum_l \frac{b_l}{r^{l+1}} P_l(\cos\theta), \quad r > d. \qquad (4.103)$$

This expansion is called the **multipole expansion** of the potential. The coefficients b_l are called the **multipole moments**, and are usually designated by p_l. The multipole expansion is generally an expansion in the ratio d/r, and is most useful for small d or large r where this ratio is small. The location of the origin for a multipole should be chosen, if possible, inside the charge distribution to make the distance d small.

The lth moment is called by the Latin name for 2^l, so

$l = 0$ is the monopole moment

$l = 1$ is the dipole moment

$l = 2$ is the quadrupole moment

$l = 3$ is the octopole moment.

For $l > 3$, say for $l = 4$, the moment is called the "two to the fourth moment". Note that the monopole, dipole, and quadrupole potentials given by l=0, 1, 2 in Eq. (4.103) are the same as the corresponding potentials derived in Chapter 1, using vector methods.

The multipole moments given as the b_l here are equal to the z-components of the vector (or dyadic) moments. But notice, we have extended the use of multipoles beyond just the monopole, dipole, and quadrupole moments introduced in the vector analysis.

One example of a multipole expansion is given by Eq. (4.99) for the potential outside a sphere. From that solution, we see that the multipole moments of the sphere are given by

$$p_l = (2l+1)V_0R^{l+1}\begin{pmatrix} \frac{1}{2} \\ \frac{l+1}{2} \end{pmatrix}, \quad \text{odd } l$$

$$= 0, \quad \text{even } l. \tag{4.104}$$

The multipole expansion of an axially symmetric potential has the remarkable property that the radial dependence is uniquely determined once the angular dependence is known, and vice-versa. In the split-sphere problem above we used this property to find the potential for all r, knowing its angular distribution at a smaller r. This property can also be used to find the potential outside an axially symmetric charge distribution, given the potential along the axis of symmetry.

Uniformly charged needle

As an example, we consider a uniformly charged needle of length L and charge q. We take the z-axis along the needle, and centered at the origin as shown in Fig. 4.2.

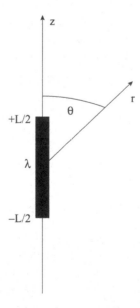

Figure 4.2: Uniformly charged needle of length L.

The potential on the axis of the needle for $|z| > L/2$ is easily found, by integrating Coulomb's law, to be

$$\psi(z) = \frac{q}{L}\left[\ln\left(|z| + \frac{L}{2}\right) - \ln\left(|z| - \frac{L}{2}\right)\right], \quad |z| > L/2. \tag{4.105}$$

To put this in the form of a multipole expansion, we realize that z is just the spherical coordinate r for $\theta = 0°$, so we can write

$$\psi(r, 0°) = \frac{q}{L}\left[\ln\left(1 + \frac{L}{2r}\right) - \ln\left(1 - \frac{L}{2r}\right)\right], \quad r > \frac{L}{2}. \tag{4.106}$$

Then we can expand the logarithms in Taylor series to get

$$\psi(r, 0°) = \frac{q}{L}\sum_{\text{odd } n}\frac{2}{n}\left(\frac{L}{2r}\right)^n, \quad r > \frac{L}{2}. \tag{4.107}$$

We make this sum look like the multipole expansion by changing the index from n to l with $n = l + 1$, so

$$\psi(r, 0°) = q\sum_{\text{even } l}\left[\frac{(L/2)^l}{(l+1)}\right]\left[\frac{1}{r^{l+1}}\right], \quad r > \frac{L}{2}. \tag{4.108}$$

The next step simply replaces the 1 in the numerator of Eq. (4.108) by $P_l(\cos 0°)$ leaving

$$\psi(r, 0°) = q\sum_{\text{even } l}\left[\frac{(L/2)^l}{(l+1)}\right]\left[\frac{P_l(\cos 0°)}{r^{l+1}}\right], \quad r > \frac{L}{2}. \tag{4.109}$$

This gives the potential at $0°$. To get the potential for all angles, we use the fact that the angular dependence of the $\frac{1}{r^{l+1}}$ term must be $P_l(\cos\theta)$, so we just have to put a line through the $0°$ making it θ. Then

$$\psi(r, \theta) = q\sum_{\text{even } l}\left[\frac{(L/2)^l}{(l+1)}\right]\left[\frac{P_l(\cos\theta)}{r^{l+1}}\right], \quad r > \frac{L}{2}. \tag{4.110}$$

This gives the potential of the needle at all angles for $r > L/2$.

The multipole moments of a uniformly charged needle can be read directly from Eq. (4.110) to be

$$p_l = \frac{q(L/2)^l}{(l+1)}. \tag{4.111}$$

For instance, the quadrupole moment of a uniformly charged needle is

$$p_2 = \frac{1}{12}qL^2. \tag{4.112}$$

This result agrees with the quadrupole moment obtained by integrating Eq. (2.94) over a uniformly charged needle. (*cf* problem 8a of chapter 2.)

We summarize the above procedure in three steps:

1. Solve for the potential on the axis of symmetry.

2. Expand the axial potential in a power series in $\frac{1}{r^{l+1}}$.

3. Multiply the lth term in the expansion by $P_l(\cos\theta)$.

Another problem that can be solved using the multipole expansion is to find the potential of a grounded sphere of radius R in a uniform electric field $\mathbf{E_0}$. The electric field will induce a surface charge distribution on the sphere. We describe the potential due to this charge distibution in terms of the multiple expansion and add it to the potential due to the uniform electric field (taken in the z direction).

$$\psi(r,\theta) = \sum_l \frac{p_l}{r^{l+1}} P_l(\cos\theta) - E_0 r\cos\theta, \quad r > R. \qquad (4.113)$$

We now evaluate the potential on the surface of the grounded sphere.

$$\psi(R,\theta) = 0 = \sum_l \frac{p_l}{R^{l+1}} P_l(\cos\theta) - E_0 R P_1(\cos\theta). \qquad (4.114)$$

Since the Legendre expansion is unique, the coefficients of each P_l must vanish separately, so

$$\begin{aligned} p_l &= 0, \quad l \neq 1, \\ p_1 &= E_0 R^3. \end{aligned} \qquad (4.115)$$

The potential is then given by

$$\psi(r,\theta) = \frac{E_0 R^3}{r^2}\cos\theta - E_0 r\cos\theta, \quad r > R. \qquad (4.116)$$

We see that the potential outside the sphere can be described as that due to the original electric field and an induced dipole on the sphere with dipole moment $p = E_0 R^3$.

Mutipole expansion for a point charge, derivation of the generating function for Legendre polynomials

A simple, but important, use of the multipole expansion is to apply it to a unit point charge located a vector distance $\mathbf{r'}$ from the origin. If we pick $\mathbf{r'}$ on the z-axis, then the potential on the z-axis is simply

$$\psi(r, 0°) = \frac{1}{r - r'}, \quad r > r'. \qquad (4.117)$$

We expand this in a power series (the geometric series)

$$\psi(r, 0°) = \sum_l \frac{r'^l}{r^{l+1}}, \quad r > r', \qquad (4.118)$$

which can be extended to all angles as

$$\psi(r, \theta) = \sum_l \frac{r'^l}{r^{l+1}} P_l(\cos\theta), \quad r > r'. \tag{4.119}$$

But, ψ is just the potential given by a unit charge at $\mathbf{r}'$, which is given by

$$\psi(\mathbf{r}) = \frac{1}{|\mathbf{r} - \mathbf{r}'|}. \tag{4.120}$$

Equating the two expressions for the potential due to a unit charge gives

$$\frac{1}{|\mathbf{r} - \mathbf{r}'|} = \sum_l \frac{r'^l}{r^{l+1}} P_l(\cos\theta), \tag{4.121}$$

for r' on the z-axis and less than r.

Since $\frac{1}{|\mathbf{r}-\mathbf{r}'|}$ is a vector expression symmetric in $\mathbf{r}$ and $\mathbf{r}'$, Eq. (4.121) can be generalized to the symmetric form

$$\frac{1}{|\mathbf{r} - \mathbf{r}'|} = \sum_l \frac{1}{r_>} \left(\frac{r_<}{r_>}\right)^l P_l(\cos\gamma), \tag{4.122}$$

where γ is the angle between $\mathbf{r}$ and $\mathbf{r}'$, and $\mathbf{r}$ and $\mathbf{r}'$ can have any orientation. The notation $r_<$ means the smaller of r and r', and $r_>$ means the larger.

A first use of Eq. (4.122) is to realize that it is just a representation of the generating function for Legendre polynomials. We can rewrite Eq. (4.121) in terms of the variables $t = r'/r$ and x, the cosine of the angle between $\mathbf{r}$ and $\mathbf{r}'$. It then becomes

$$\frac{1}{\sqrt{1 - 2xt + t^2}} = \sum_l t^l P_l(x), \quad t < 1. \tag{4.123}$$

which is just the definition of the generating function. This is the generating function derivation we promised earlier.

Point charge and grounded sphere

The multipole expansion of $\frac{1}{|\mathbf{r}-\mathbf{r}'|}$ can be used to find the potential of a point charge q outside a grounded sphere of radius a, at a distance d from the center. The point charge will induce a surface charge on the grounded sphere. The potential due to this induced charge can be expanded in multipoles for $r > a$.

Thus, for the region between the sphere and the point charge, we can write

$$\psi(r, \theta) = \sum_l \frac{p_l}{r^{l+1}} P_l(\cos\theta) + \sum_l \frac{qr^l P_l(\cos\theta)}{d^{l+1}}, \quad a < r < d. \tag{4.124}$$

The second sum in the potential is due to the point charge. Applying the boundary condition $\psi(a,\theta) = 0$, results in

$$p_l = -qa^l \left(\frac{a}{d}\right)^{l+1} \tag{4.125}$$

for the induced multipole moments of the sphere.

The potential outside the sphere is then

$$
\begin{aligned}
\psi(r,\theta) &= \sum_l \left[\frac{qr^l}{d^{l+1}} - \left(\frac{qa}{d}\right)\frac{(a^2/d)^l}{r^{l+1}}\right] P_l(\cos\theta), \quad a < r < d \\
&= \sum_l \left[\frac{qd^l}{r^{l+1}} - \left(\frac{qa}{d}\right)\frac{(a^2/d)^l}{r^{l+1}}\right] P_l(\cos\theta), \quad r > d.
\end{aligned} \tag{4.126}
$$

This can be seen to be the Legendre polynomial expansion of the potential that we found in Chapter 3 by using an image charge.

Multipole moment by integration

We can also find the general multipole expansion of the Coulomb law integral for an axially symmetric charge distribution of finite extent. We start with Coulomb's law

$$\psi(\mathbf{r}) = \int \frac{dq'}{|\mathbf{r} - \mathbf{r}'|}. \tag{4.127}$$

We first make the symmetry axis the z-axis, and find the potential on this axis by taking r on the z-axis. For large r (This means $r > \text{Max}\{r'\}$.), we expand the denominator in Eq. (4.127) and integrate the sum term by term, resulting in

$$\psi(r, 0°) = \sum_l \frac{1}{r^{l+1}} \int r'^l P_l(\cos\theta')dq'. \tag{4.128}$$

This is the potential for $\theta = 0°$, so we just multiply each term in the expansion by $P_l(\cos\theta)$ to get the potential at all angles

$$\psi(r, \cos\theta) = \sum_l \frac{P_l(\cos\theta)}{r^{l+1}} \int r'^l P_l(\cos\theta')dq'. \tag{4.129}$$

We see from Eq. (4.129) that the multipole moments of an axially symmetric charge distribution are given by an integral over the charge distribution.

$$p_l = \int r^l P_l(\cos\theta)dq. \tag{4.130}$$

This method is a third way to find the multipole moments of an axially symmetric charge distribution. The three methods are

1. Vector integration over the charge distribution, as in Eq. (2.41) or (2.96).

2. Identifying the coefficient of $\frac{1}{r^{l+1}}$ in the power series expansion of the potential on the symmetry axis, as in Sect. 2.1.4.

3. Integrating over the charge distribution weighted by $r^l P_l(\cos\theta)$, as in Eq. (4.130).

For the uniformly charged needle of length L and charge q, integration gives

$$
\begin{aligned}
p_l &= \frac{q}{L}\int_{-L/2}^{+L/2} z^l P_l(\cos\theta)dz \\
&= \frac{2q}{L}\int_0^{L/2} z^l dz = \frac{q(L/2)^l}{l+1}, \quad \text{even } l,
\end{aligned}
\tag{4.131}
$$

which is the same as given by Eq. (4.111). In Eq. (4.131), we have used the fact that $\cos\theta = +1$ for positive z and -1 for negative z. The multipole moments of the needle are zero for odd l.

4.2.5   Spherical harmonics

When there is ϕ dependence, the parameter m has to be included in the angular equation for $\Theta(\theta)$. Then the eigenvalue equation in the variable θ becomes

$$
[\sin\theta\,\Theta'(\theta)]' - \frac{m^2}{\sin\theta}\Theta(\theta) = \lambda_\theta \sin\theta\,\Theta(\theta).
\tag{4.132}
$$

This equation is called the **Associated Legendre Equation**. It can also be written in terms of the function $y(x) = \Theta(\theta)$, with $x = \cos\theta$,

$$
[(1-x^2)y'(x)]' - \frac{m^2 y(x)}{1-x^2} = \lambda_\theta y(x).
\tag{4.133}
$$

In either equation, m is already fixed by the solution to the ϕ equation, but the eigenvalue λ_θ is to be determined by the condition that $\Theta(\theta)$ or $y(x)$ be finite in the physical region.

The solution of the associated Legendre equation involves a good deal of complicated algebra, so we will just state the results without proof. A detailed derivation can be found in any good book on Mathematical Physics, for instance in Section 12.5 of [Arfken].

The solutions to Eq. (4.133) are **Associated Legendre Functions**, given by

$$
P_l^m(x) = (1-x^2)^{\frac{m}{2}}\left(\frac{d}{dx}\right)^m P_l(x), \quad 0 \le m \le l.
\tag{4.134}
$$

The eigenvalue is given by

$$
\lambda_\theta = -l(l+1).
\tag{4.135}
$$

This is the same eigenvalue as for $m = 0$, with no dependence on m. This is important because it means that all terms of the same l will have the same radial dependence.

Equation (4.134) gives $P_l^m(x)$ for non-negative m. Since the differential equation depends only on m^2 it is clear that the solution for negative m will have the same x dependence, but could differ by a constant factor. Several choices are made in the literature for this factor. We will use the definition most commonly used in mathematics texts that

$$P_l^{-m}(x) = (-1)^m \frac{(l-m)!}{(l+m)!} P_l^m(x). \tag{4.136}$$

Because of the restriction $-l \leq m \leq +l$, there are $2l + 1$ values of m for each l.

Legendre functions of the same m, but different l will be orthogonal, just as the Legendre polynomials were, because they are eigenfunctions with different eigenvalues of the same differential equation. There is no simple orthogonality for Legendre functions with different m because they are solutions of different differential equations. The orthogonality integral and normalization constant (for the same m) for the Legendre functions are

$$\int_{-1}^{+1} P_l^m(x) P_{l'}^m(x) dx = \left(\frac{2}{2l+1} \right) \frac{(l-m)!}{(l+m)!} \delta_{ll'}. \tag{4.137}$$

We now are in a position to define a complete solution to the angular part of Laplace's equation in terms of the two angular coordinates θ and ϕ. These are the **Spherical Harmonics** defined by

$$Y_l^m(\theta, \phi) = (-1)^m N_{lm} P_l^m(\cos\theta) e^{im\phi}, \tag{4.138}$$

with normalization given by

$$N_{lm} = \sqrt{\left(\frac{2l+1}{4\pi} \right) \frac{(l-m)!}{(l+m)!}}. \tag{4.139}$$

Although the normalization constant for the spherical harmonics appears complicated, it is the most useful for our purposes, as will be seen. The normalization constant and phase used above correspond to the **Condon-Shortley phase convention** that is consistently used throughout modern physics and in quantum mechanical applications. The slightest deviation from, or simplification of, this convention can lead (and has led) to disastrous consequences.

With this normalization, the orthogonality relation for the Y_l^m is

$$\int Y_l^{m^*}(\theta, \phi) Y_{l'}^{m'}(\theta, \phi) d\Omega = \delta_{ll'} \delta_{mm'}. \tag{4.140}$$

The integral is over all solid angle. The normalization constants were chosen so that the integral of $|Y_l^m|^2$ is unity, which will be convenient for expansions. In doing the integral, we should integrate first over ϕ so that the orthogonality of the $e^{im\phi}$ will make $m = m'$. Then the θ integration will lead to orthogonality with respect to l.

The spherical harmonics have useful symmetry properties that follow from their definition, and the definition of the associated Legendre functions. We list two of these:

1. Parity: The **Parity transformation,**

$$\theta \to \pi - \theta, \quad \phi \to \phi + \pi, \tag{4.141}$$

has the effect of taking the position vector **r** into its negative. Under the parity transformation, the Y_l^m satisfy

$$Y_l^m(\pi - \theta, \phi + \pi) = (-1)^l Y_l^m(\theta, \phi). \tag{4.142}$$

We see from this equation that the Y_l^m have parity $(-1)^l$.

2. Negative m: A simple form for Y_l^m of negative m is given by

$$Y_l^{-m}(\theta, \phi) = (-1)^m Y_l^{m*}(\theta, \phi). \tag{4.143}$$

The complete solution of Laplace's equation for a general angular distribution can be written using the spherical harmonics as

$$\psi(r, \theta, \phi) = \sum_{l,m} \left[a_{lm} r^l + \frac{b_{lm}}{r^{l+1}} \right] Y_l^m(\theta, \phi). \tag{4.144}$$

The sum on m is from $-l$ to $+l$ (all possible m), which will be true for all our m sums.

Potential outside a sphere

As an example of the use of the spherical harmonics, we find the potential outside a sphere of radius R with the potential on the sphere, $\psi(R, \theta, \phi)$, specified. Outside the sphere, the coefficients a_{lm} must be dropped, so the potential is given by

$$\psi(\theta, \phi) = \sum_{l,m} \frac{b_{lm}}{r^{l+1}} Y_l^m(\theta, \phi). \tag{4.145}$$

The coefficient b_{lm} given in terms of the boundary value $\psi(R, \theta, \phi)$ by using the orthogonality property of the Y_l^m as

$$b_{lm} = R^{l+1} \int Y_l^{m*}(\theta, \phi) \psi(R, \theta, \phi) d\Omega. \tag{4.146}$$

Note that the r dependence of the potential does not uniquely determine the angular distribution because there are $2l+1$ b_{lm} for each l. However, if the angular distribution corresponds to a given l, then the radial dependence must be $\frac{1}{r^{l+1}}$.

Multipole moments

The coefficients b_{lm} are related to the multipole moments of a general charge distribution. These moments, called the **multipole moments in the spherical basis**, are defined as

$$p_{lm} = \sqrt{\frac{2l+1}{4\pi}}\, b_{lm}. \qquad (4.147)$$

We have defined the multipole moments in this way so that p_{l0} corresponds to the moment p_l we had in the axially symmetric case.

The dipole $(l=1)$ moment has three components in either the spherical or vector basis. They are related by

$$
\begin{aligned}
p_x &= \frac{-1}{\sqrt{2}}(p_{1,1} - p_{1,-1}) \\
p_z &= p_{1,0} \\
p_y &= \frac{-i}{\sqrt{2}}(p_{1,1} + p_{1,-1}).
\end{aligned}
\qquad (4.148)
$$

The quadrupole moment $(l=2)$ in the spherical basis has five components. In the vector basis, the quadrupole moment is a symmetric dyadic, which has six independent elements. However the condition that it is traceless reduces the number of independent elements to five, as in the spherical basis. For higher moments, there are more complicated constraints that will always reduce the number of independent moments in the vector basis (if used) to the $2l+1$ of the spherical basis.

Rotation of axes

An important property of the potential $\psi(r, \theta, \phi)$ is that it is a scalar quantity. A defining property of a scalar is that it remains unchanged under a rotation of the coordinate system. This means that if there are two separate coordinate systems (r, θ_1, ϕ_1) and (r, θ_2, ϕ_2), connected by a rotation, then

$$\psi_1(r, \theta_1, \phi_1) = \psi_2(r, \theta_2, \phi_2). \qquad (4.149)$$

The subscripts on the potentials mean that $\psi_1(r, \theta_1, \phi_1)$ and $\psi_2(r, \theta_2, \phi_2)$ are different functions of their arguments. However, if ψ is a scalar function, the two functions are equal at the same point in space. It is clear that the potential has this scalar property because the Coulomb's law integral for the potential can be written with no reference to any coordinate system.

As an example of the scalar property of the potential, we consider the case of the sphere of radius R with the one half of the sphere kept at a potential $+V$ and the other half kept at $-V$. In one system of coordinates, the boundary condition would be written

$$\psi_1(R, \theta_1, \phi_1) \; = \; +V, \quad 0° < \theta_1 < 90°,$$
$$= \; -V, \quad 90° < \theta_1 < 180°. \tag{4.150}$$

If system 2 is rotated with respect to the first system, as in Fig. 4.3, then a typical point P in the figure has different coordinates in the second system. Because of this, the boundary condition changes to

$$\psi_2(R, \theta_2, \phi_2) \; = \; +V, \quad -90° < \phi_2 < +90°,$$
$$= \; -V, \quad 90° < \phi_2 < 270°. \tag{4.151}$$

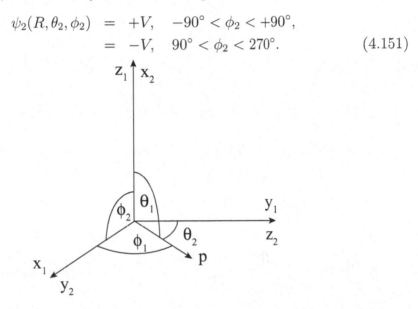

Figure 4.3: Rotated axes. The point P has angles $\theta_1 = 90°$, $\phi_1 = 45°$ in the x_1, y_1, z_1 system, and $\theta_2 = 45°$, $\phi_2 = 90°$ in the x_2, y_2, z_2 system. The potential is $+V$ for $0° < \theta_1 < 90°$, which corresponds to $-90° < \phi_2 < +90°$.

Although the two functions ψ_1 and ψ_2 are quite different, they correspond to the same physical boundary condition.

Addition theorem

We can use the scalar property of the potential to derive a useful relation involving the spherical harmonics. If we substitute b_{lm} given by Eq. (4.146) into Eq. (4.145), the result is

$$\psi(r, \theta, \phi) = \sum_{l,m} \frac{R^{l+1} Y_l^m(\theta, \phi)}{r^{l+1}} \int Y_l^{m*}(\theta', \phi') \psi(R, \theta', \phi') d\Omega'. \tag{4.152}$$

We now consider two sets of axes, x_1, y_1, z_1 and x_2, y_2, z_2, that are rotated with respect to one another as shown in Fig. 4.4.

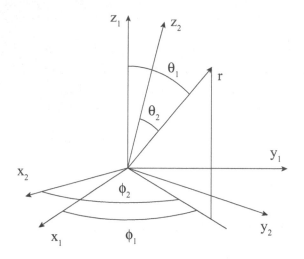

Figure 4.4: Rotated coordinate axes for the addition theorem.

We can write Eq. (4.152) for the potential in either coordinate system:

$$\psi_1(r, \theta_1, \phi_1) = \sum_{l,m} \frac{R^{l+1} Y_l^m(\theta_1, \phi_1)}{r^{l+1}} \int Y_l^{m*}(\theta_1', \phi_1') \psi_1(R, \theta_1', \phi_1') d\Omega_1' \qquad (4.153)$$

or

$$\psi_2(r, \theta_2, \phi_2) = \sum_{l,m} \frac{R^{l+1} Y_l^m(\theta_2, \phi_2)}{r^{l+1}} \int Y_l^{m*}(\theta_2', \phi_2') \psi_2(R, \theta_2', \phi_2') d\Omega_2'. \qquad (4.154)$$

We have not put subscripts on the radial variables r and R because they do not change in the rotation.

Since the potential ψ is a scalar, the two sums on the right hand sides of Eqs. (4.153) and (4.154) can be equated. Also, since each sum is a power series in $\frac{1}{r}$, we can equate the terms for each l separately. This leads to

$$\int d\Omega_1' \psi_1(R, \theta_1', \phi_1') \sum_m Y_l^m(\theta_1, \phi_1) Y_l^{m*}(\theta_1', \phi_1')$$

$$= \int d\Omega_2' \psi_2(R, \theta_2', \phi_2') \sum_m Y_l^m(\theta_2, \phi_2) Y_l^{m*}(\theta_2', \phi_2'). \qquad (4.155)$$

The functions $\psi_1(R, \theta_1', \phi_1')$ and $\psi_2(R, \theta_2', \phi_2')$ are equal and, since they are arbitrary functions of θ and ϕ, the other factors in the integrands must also be equal at each point of space. Then we have

$$\sum_m Y_l^m(\theta_1, \phi_1) Y_l^{m*}(\theta_1', \phi_1') = \sum_m Y_l^m(\theta_2, \phi_2) Y_l^{m*}(\theta_2', \phi_2'). \qquad (4.156)$$

This result is called the **addition theorem** for spherical harmonics. The four sets of angles in the addition theorem correspond to θ and ϕ for two different position vectors $\mathbf{r}$ and $\mathbf{r}'$, each in two different coordinate systems, connected by a rotation. Simply stated, the sum over all m in Eq. (4.156) is a scalar function under the rotation between system 1 and system 2.

The addition theorem can be put in a somewhat simpler form by letting the vector $\mathbf{r}'$ be on the z_2 axis so that $\theta'_2 = 0$. Then the sum on the right hand side of Eq. (4.156) collapses to only the $m = 0$ term, leaving

$$P_l(\cos\gamma) = \frac{4\pi}{2l+1} \sum_m Y_l^m(\theta,\phi) Y_l^{m*}(\theta',\phi'). \qquad (4.157)$$

The angle γ is the angle between $\mathbf{r}$ and $\mathbf{r}'$. The angles θ and ϕ are the coordinate angles for $\mathbf{r}$, while θ' and ϕ' are the coordinate for $\mathbf{r}'$. The relation between the angles is shown in Fig. 4.5.

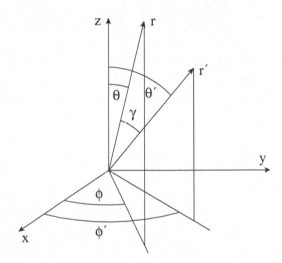

Figure 4.5: **Relation between the angles for the addition theorem.** γ **is the angle between two vectors r and r' which have spherical coordinates** (r,θ,ϕ) **and** (r',θ',ϕ'), **respectively.**

By this point, you may have noticed that the factor $4\pi/(2l+1)$ keeps cropping up in many different ways. A simple way to get it right in any particular equation is to check the equation for the $m = 0$ term, which can always be written purely in terms of P_l's where there should be no $4\pi/(2l+1)$.

Multipole moment by integration

The form of the addition theorem given in Eq. (4.157) can be used to extend the multipole expansion of $\frac{1}{|\mathbf{r}-\mathbf{r}'|}$ to a general (not necessarily symmetric) charge distribution. We substitute the addition theorem for $P_l(\cos\gamma)$ into Eq. (4.122), resulting in

$$\frac{1}{|\mathbf{r} - \mathbf{r}'|} = \sum_{l,m} \frac{1}{r_>} \left(\frac{r_<}{r_>}\right)^l \left(\frac{4\pi}{2l+1}\right) Y_l^m(\theta, \phi) Y_l^{m*}(\theta', \phi'). \qquad (4.158)$$

To write the multipole exansion for the potential of a general charge distribution, we just put the multipole expansion of Eq. (4.158) into the Coulomb law integral resulting in

$$\psi(r, \theta, \phi) = \sum_{l,m} \left(\frac{4\pi}{2l+1}\right) \frac{Y_l^m(\theta, \phi)}{r^{l+1}} \int r'^l Y_l^{m*}(\theta', \phi') dq'. \qquad (4.159)$$

Looking at Eq. (4.159), we see that we can identify the integral over the charge distribution with the multipole moments in the spherical basis by

$$p_{lm} = \sqrt{\frac{4\pi}{2l+1}} \int r^l Y_l^{m*}(\theta, \phi) dq. \qquad (4.160)$$

We now have two ways to find the multipole moments p_{lm} in the spherical basis. One is by integrating over the charge distribution as in Eq. (4.160). The other is to relate p_{lm} to the coefficient b_{lm} in the multipole expansion of the potential, as in Eq. (4.147).

The addition theorem can also be used to find the potential for an axially symmetric charge distribution in terms of axes that do not lie along the symmetry axis. For instance, it might important in a particular problem to know the potential of the uniformly charged needle when it is oriented as shown in Fig. 4.6, making angles θ' and ϕ' with respect to a fixed coordinate system.

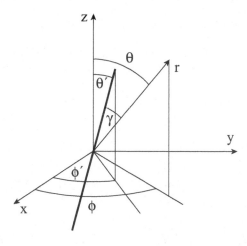

Figure 4.6: Charged needle at angles θ', ϕ', and point of observation $\mathbf{r}$ at angles θ, ϕ. γ is the angle between the needle and $\mathbf{r}$.

We can first write the potential in terms of the axial multipole moments of the needle

$$\psi(r, \gamma) = \sum_{l} \frac{p_l P_l(\cos \gamma)}{r^{l+1}}, \tag{4.161}$$

where γ is the angle between the field vector $\mathbf{r}$ and the axis of the needle. We then use the addition theorem to relate $P_l(\gamma)$ to the orientation of the needle, so

$$\psi(r, \theta, \phi) = \sum_{l,m} \left(\frac{p_l}{r^{l+1}} \right) \left(\frac{4\pi}{2l+1} \right) Y_l^m(\theta, \phi) Y_l^{m*}(\theta', \phi'). \tag{4.162}$$

4.3   Cylindrical Coordinates

Cylindrical coordinates, shown in Fig. 4.7, are defined by

$$\begin{aligned}
q_1 &= r, & &\text{the distance, from the symmetry axis,} & (4.163)\\
q_2 &= \theta, & &\text{the angle about the symmetry axis,} & (4.164)\\
q_3 &= z, & &\text{the distance along the symmetry axis.} & (4.165)
\end{aligned}$$

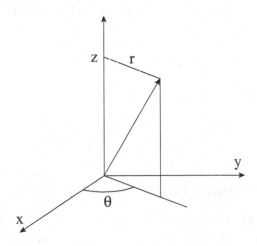

Figure 4.7: Cylindrical coordinates.

They are related to the cartesian coordinates by

$$\begin{aligned}
x &= r \cos \theta & (4.166)\\
y &= r \sin \theta & (4.167)\\
z &= z. & (4.168)
\end{aligned}$$

as can be seen from Fig. 4.7 . Note that the radial coordinate r and the angle θ in cylindrical coordinates are different than the r and θ used in spherical coordinates.

The cylindrical coordinate system is an orthogonal system, as can be verified by substitution of Eqs. (4.166) - (4.168) into the orthogonality condition of Eq. (4.5). The metric coefficients h_i can be calculated from Eq. (4.6):

$$h_1 = h_r = 1 \tag{4.169}$$
$$h_2 = h_\theta = r \tag{4.170}$$
$$h_3 = h_z = 1. \tag{4.171}$$

The unit vectors can be related to the Cartesian unit vectors $\hat{\mathbf{i}}, \hat{\mathbf{j}}, \hat{\mathbf{k}}$ using Eq. (4.3). We will work strictly within the cylindrical coordinate system, and so will not need to use those relations.

With the h_i's known, we can substitute into Eq. (4.15) to get the Laplacian in cylindrical coordinates:

$$\nabla^2 \psi = \frac{1}{r}[\partial_r(r\partial_r\psi)] + \frac{1}{r^2}\partial_\theta^2\psi + \partial_z^2\psi \tag{4.172}$$

4.3.1   Separation of Variables in Cylindical Coordinates

To apply separation of variables to Laplace's equation in cylindrical coordinates, we write the potential in factored form as

$$\psi(r,\theta,z) = R(r)\Theta(\theta)Z(z). \tag{4.173}$$

Then, Laplace's equation can be written

$$\frac{\nabla^2\psi}{\psi} = \frac{(rR')'}{rR} + \frac{\Theta''}{r^2\Theta} + \frac{Z''}{Z} = 0. \tag{4.174}$$

4.3.2   2 Dimensional cases (polar coordinates)

We first consider problems with infinite extent in the z direction having boundary conditions with no z dependence. Then the function $Z(z)$ will be a constant and the separated Laplace equation becomes

$$-\frac{r(rR')'}{R} = \frac{\Theta''}{\Theta} = \lambda_\theta. \tag{4.175}$$

We start first with the equation for $\Theta(\theta)$

$$\Theta(\theta)'' = \lambda_\theta\Theta(\theta). \tag{4.176}$$

The condition that the potential $\psi(r,\theta,z)$ be single valued requires the **periodic boundary condition** on $\Theta(\theta)$ that

$$\Theta(\theta+2\pi) = \Theta(\theta). \tag{4.177}$$

With this boundary condition, the differential equation for Θ has the periodic solution

$$\Theta(\theta) = \alpha \sin(n\theta) + \beta \cos(n\theta), \quad n = 0, 1, 2, \ldots \tag{4.178}$$

This restricts the eigenvalues λ_θ to

$$\lambda_\theta = -n^2, \tag{4.179}$$

with n any integer or zero.

The radial equation for $R(r)$ becomes

$$r(rR')' = n^2 R. \tag{4.180}$$

This has the simple solutions

$$R = r^{\pm n} \ , n > 0. \tag{4.181}$$

The case n=0 is somewhat different in that the two possible solutions are

$$R = 1 \quad \text{or} \quad R = \ln r. \tag{4.182}$$

We can now write the complete solution for $\psi(r, \theta)$ that satisfies the periodic boundary condition for θ.

$$\psi(r, \theta) = a_0 + b_0 \ln r + \sum_{n=1}^{\infty} (a_n r^n + b_n r^{-n})[\alpha_n \cos(n\theta) + \beta_n \sin(n\theta)]. \tag{4.183}$$

The coefficients a, b, α, β for each n are to be determined from the remaining boundary conditions.

Potential inside a cylinder

As a sample problem, we solve for the potential inside a hollow cylinder of radius R with the outer surface kept at a specified potential $V(\theta)$. The b_n terms are dropped from the sum to keep the potential finite at the origin, and the potential is

$$\psi(r, \theta) = \alpha_0 + \sum_{n=1}^{\infty} \left(\frac{r}{R}\right)^n [\alpha_n \cos(n\theta) + \beta_n \sin(n\theta)], \ r \leq R. \tag{4.184}$$

In Eq. (4.184), we have normalized the r dependence in the form $\frac{r}{R}$, so as to make the algebra below simpler. Applying the boundary condition at $r = R$ gives

$$\psi(R, \theta) = V(\theta) = \sum_{n=0}^{\infty} [\alpha_n \cos(n\theta) + \beta_n \sin(n\theta)]. \tag{4.185}$$

Fourier series

Equation (4.185) is a trigonometric Fourier series for the periodic function $V(\theta)$, which has the period 2π. It is a generalization of the Fourier sine series of period $2L$ given in Eq. (3.60). The sine and cosine functions in Eq. (4.185) have the same orthogonality properties as the sine functions in Eq. (3.58) with the periodicity 2π instead of $2L$. The complete set of orthogonality relations for the sine and cosine functions are

$$\int_0^{2\pi} \sin(nx)\sin(n'x)dx = \pi\delta_{nn'} \tag{4.186}$$

$$\int_0^{2\pi} \cos(nx)\cos(n'x)dx = \pi\delta_{nn'}, \, n > 0$$

$$= 2\pi, \, n = n' = 0 \tag{4.187}$$

$$\int_0^{2\pi} \sin(nx)\cos(n'x)dx = 0. \tag{4.188}$$

Note that these orthogonality integrals are over a full period of the sine and cosine functions. In Eq. (3.58) for the pure sine function expansion, the orthogonality integral could be taken over half a period because the sine functions have the symmetry property $\sin[n(\theta+\pi)] = (-1)^n\sin(n\theta)$.

Using the orthogonality relations of Eqs. (4.186)-(4.188), the expansion coefficients for $n > 0$ can be found as

$$\alpha_n = \frac{1}{\pi}\int_0^{2\pi}\cos(nx)V(x)dx. \tag{4.189}$$

$$\beta_n = \frac{1}{\pi}\int_0^{2\pi}\sin(nx)V(x)dx. \tag{4.190}$$

The case $n = 0$ for the cosine functions must be treated as a special case, with

$$\alpha_0 = \frac{1}{2\pi}\int_0^{2\pi}V(x)dx = \langle V\rangle, \tag{4.191}$$

where $\langle V\rangle$ is the average value of $V(\theta)$.

The solution for the potential outside the hollow cylinder is given by

$$\psi(r,\theta) = \langle V\rangle + b_0\ln(r/R) + \sum_{n=1}^{\infty}\left(\frac{R}{r}\right)^n[\alpha_n\cos(n\theta) + \beta_n\sin(n\theta)], \, r > R. \tag{4.192}$$

The coefficients α_n and β_n for the outside solution are the same as those in Eq. (4.184) for the inside solution, each being given by Eqs. (4.189) and (4.190).

The surface charge density on the cylinder is given by

$$\begin{aligned}
\sigma(\theta) &= -\frac{1}{4\pi}\hat{\mathbf{r}}\cdot\boldsymbol{\nabla}(\psi_{\text{out}} - \psi_{\text{in}}) \\
&= \frac{-b_0}{4\pi R} + \sum_{n=1}^{\infty}\left(\frac{2n}{4\pi R}\right)[\alpha_n\cos(n\theta) + \beta_n\sin(n\theta)].
\end{aligned} \tag{4.193}$$

Note that the potential outside the cylinder is not uniquely determined by $\psi(R,\theta)$ because the potential does not vanish at infinity. The average value of the surface charge density $\langle\sigma\rangle = -b_0/4\pi R$, which determines the asymptotic behavior of the potential, must also be specified.

Intersecting grounded planes

Another interesting two dimensional problem is to find the surface charge distribution near the intersection of two grounded conducting plates making an angle α with each other, as shown in Fig. 4.8.

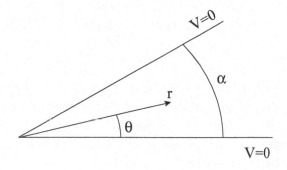

Figure 4.8: Grounded conducting planes with subtended angle α.

We assume that there is some remote charge distribution (beyond a distance d) producing the potential. The potential will satisfy Laplace's equation in the region defined by

$$0 \le r \le d, \quad 0 \le \theta \le \alpha. \tag{4.194}$$

The boundary condition for the potential at the grounded planes is

$$\psi(r,0) = 0 \quad \text{and} \quad \psi(r,\alpha) = 0. \tag{4.195}$$

With this boundary condition, the solution of the differential equation (4.176) for Θ is

$$\Theta(\theta) = \sin(m\pi\theta/\alpha), \tag{4.196}$$

with the eigenvalue

$$\lambda_\theta = -(m\pi/\alpha)^2, \tag{4.197}$$

where $m = 1, 2, \ldots$ is any positive integer. Then the radial equation is

$$r(rR')' = (m\pi/\alpha)^2 R \tag{4.198}$$

with the solution

$$R(r) = r^{\pm m\pi/\alpha}. \tag{4.199}$$

The regular solution for the potential near the corner is

$$\psi(r,\theta) = \sum_m a_m r^{m\pi/\alpha} \sin(m\theta/\alpha). \tag{4.200}$$

The constants a_m are generally non-zero, and would be determined from the specific features of the distant charge distribution.

The surface charge distribution on the conductor at $\theta = 0$ is given by

$$\begin{aligned}\sigma &= -\frac{1}{4\pi r}[\partial_\theta \psi(\theta)]|_{\theta=0} \\ &= \sum_m (m\pi a_m/4\alpha) r^{(\frac{m\pi}{\alpha}-1)}.F \end{aligned} \tag{4.201}$$

Since we are mainly concerned with the r dependence of the charge distribution as the edge of the plane is approached, we keep only the $m = 1$ term in the sum and can write

$$\sigma \sim r^{(\frac{\pi}{\alpha}-1)}. \tag{4.202}$$

The surface charge distribution will be the same at the upper plate at $\theta=\alpha$. We can see from Eq. (4.202) that σ goes to zero toward the intersection of the plates for any $\alpha < \pi$. For $\alpha > \pi$, σ increases towards the intersection. The angle $\alpha = 2\pi$ corresponds to the edge of a single conducting plate. For this case, the surface charge goes like

$$\sigma(\theta) \sim 1/\sqrt{r}, \tag{4.203}$$

becoming infinite near the edge.

4.3.3   3 Dimensional cases, Bessel functions

We return to Laplace's equation in all three cylindrical variables, and consider geometries within a cylinder. The requirement that the potential be single valued means that the $\lambda_\theta = -m^2$ ($m = 0, 1, 2, ...$), so Eq. (4.174) can be written

$$\frac{(rR')'}{rR} - \frac{m^2}{r^2} = -\frac{Z''}{Z} \tag{4.204}$$

The left hand side of Eq. (4.204) does not depend on z, so the right hand side must be a constant, which gives the z equation

$$Z'' = \lambda_z Z. \tag{4.205}$$

Equation (4.204) can be put into one of the standard forms of Bessel's equation. We write the z eigenvalue as

$$\lambda_z = k^2, \tag{4.206}$$

and then (4.204) becomes

$$(rR')' - \frac{m^2}{r}R = -k^2 rR. \tag{4.207}$$

Bessel functions

Equation (4.207) is called **Bessel's equation**. As with Legendre's equation, Bessel's equation as used here is an eigenvalue equation of the general form

$$\mathcal{L}_B R(r) = \lambda w(r) R(r), \tag{4.208}$$

with a self-adjoint operator $\mathcal{L}_B$ defined by

$$\mathcal{L}_B R(r) = [rR(r)']' - \frac{m^2}{r} R(r), \tag{4.209}$$

and eigenvalue $\lambda = -k^2$. A slight generalization is the appearance of the **weight function** $w(r) = r$ multiplying the eigenfunction in Eq. (4.207). The eigenvalue $-k^2$ will be determined by additional boundary conditions for any specific problem. Equations of the form of Bessel's equation or Legendre's equation are called **Sturm-Liouville** equations, and the problem of find the eigenfunction and eigenvalue are called Sturm-Liouville problems.

We can simplify Eq. (4.207) by changing the variable to

$$x = kr \quad \text{with} \quad y(x) = R(r). \tag{4.210}$$

This gives

$$x[xy(x)']' + (x^2 - m^2)y(x) = 0. \tag{4.211}$$

The eigenvalue seems to have dropped out of this form of Bessel's equation, but it is still included implicitly through Eq. (4.210) relating x and r.

We can solve Eq. (4.211) by the method of Frobenius, letting

$$y(x) = \sum_{n=0}^{\infty} a_n x^{(n+s)}. \tag{4.212}$$

The necessary derivatives of $y(x)$ are

$$x[xy(x)']' = \sum_{n=0}^{\infty} (n + s)^2 a_n x^{(n+s)}, \tag{4.213}$$

and

$$x^2 y(x) = \sum_{n'=0}^{\infty} a_{n'} x^{(n'+s+2)}. \tag{4.214}$$

We change the summation index in the sum for $x^2 y(x)$ to $n = n' + 2$ so that it can be rewritten as

$$x^2 y(x) = \sum_{n=2}^{\infty} a_{n-2} x^{(n+s)}. \tag{4.215}$$

The sum on n can be extended down to $n = 0$, with the understanding that $a_n = 0$ for negative n. Then the appropriate sums can be put into Eq. (4.211), resulting in

$$\sum_{n=0}^{\infty} \{[(n + s)^2 - m^2]a_n + a_{n-2}\} x^{(n+s)} = 0. \tag{4.216}$$

The coefficient of x to each power of $(n + s)$ must equal zero separately because of the uniqueness of a power series expansion. We look first at the $n = 0$ term for which

$$(s^2 - m^2)a_0 = 0. \tag{4.217}$$

This requires

$$s = +m \quad \text{or} \quad s = -m, \tag{4.218}$$

since a_0 cannot be zero. The next term, with $n = 1$ gives

$$[(s + 1)^2 - m^2]a_1 = 0. \tag{4.219}$$

This requires $a_1 = 0$ because s is already determined to be $\pm m$. For $n > 1$, we have the recursion relation

$$a_n = \frac{-a_{n-2}}{n(n + 2m)}. \tag{4.220}$$

Note that there will only be even terms in the sum since all odd terms are zero.

To look at the convergence of the infinite series, we change the summation index to $n' = \frac{n}{2}$ with $b_{n'} = a_n$. Then the power series for $y(x)$ becomes

$$y(x) = x^m \sum_{n'=0}^{\infty} b_{n'} x^{2n'}, \tag{4.221}$$

with the new recursion relation

$$b_{n'} = \frac{-b_{n'-1}}{4n'(n' + m)}. \tag{4.222}$$

The ratio test for convergence gives

$$R_{n'} = \frac{b_{n'} x^{2n'}}{b_{n'-1} x^{2(n'-1)}} = \frac{-x^2}{4n'(n' + m)}. \tag{4.223}$$

This ratio approaches zero as $n' \to \infty$ so the series is convergent for all finite x.

The solution we have just found is called a **Bessel function** $J_m(x)$, and is given by the power series

$$J_m(x) = x^m \sum_{n=0}^{\infty} b_n x^{2n}, \tag{4.224}$$

with

$$b_0 = \frac{1}{2^m m!} \tag{4.225}$$

and

$$b_n = \frac{-b_{n-1}}{4n(n + m)}. \tag{4.226}$$

The above normalization of b_0 is conventional for the definition of the Bessel function.

The Bessel function $J_m(x)$ is regular and finite at the origin, which is what we will need for potentials inside a cylinder. Note that the other choice of $s = -m$ for the index s does not lead to a second solution for integer m. This is because the recursion relation for B_m would become singular when $n = m$. There is another independent solution to Bessel's equation that is singular at the origin (called a **Neumann function**). We will discuss the Neumann functions in Chapter 11 in connection with coaxial wave guides.

We list, without proof, some useful properties of Bessel functions.

Recursion Relation:

$$J_{m+1}(x) + J_{m-1}(x) = \frac{2m}{x} J_m(x). \tag{4.227}$$

This recursion relation for Bessel functions is a better method to calculate the Bessel functions than the recursion relation for the power series coefficients. This is especially true if a number of Bessel functions is to be calculated. A convenient way to use the Bessel function recursion relation in a computer program is described in [Arfken] pp.576-7.

Derivative recursion relation:

$$J'_m(x) = \pm \left[\frac{m}{x} J_m(x) - J_{m\pm1} \right]. \tag{4.228}$$

This recursion relation is useful in relating derivatives of Bessel functions to Bessel functions.

Asymptotic forms:
At small x, the Bessel function is given by the first term in the power series expansion so

$$J(x) \sim \frac{x^m}{2^n n!} \quad \text{for small x.} \tag{4.229}$$

At large $x >> m^2/2$, the Bessel functions look like damped sine curves

$$J_m(x) \sim \sqrt{\frac{2}{\pi x}} \sin\left(x - \frac{m\pi}{2}\right) \quad \text{for large x.} \tag{4.230}$$

The first three Bessel functions are plotted in Fig. 4.9. Generally, the Bessel functions start like x^m, increase monotonically to a maximum near $x = m+1$, and then oscillate (but not periodically) with diminishing amplitude, finally reaching the asymptotic form of Eq. (4.230) for large x.

Solutions of Laplace's equation in cylindrical coordinates for each m that are finite at the origin can be written in terms of Bessel functions as

$$\psi_m(r, \theta, z) = J_m(kr)[\alpha_m \sin(m\theta) + \beta_m \cos(m\theta)]Z(z), \tag{4.231}$$

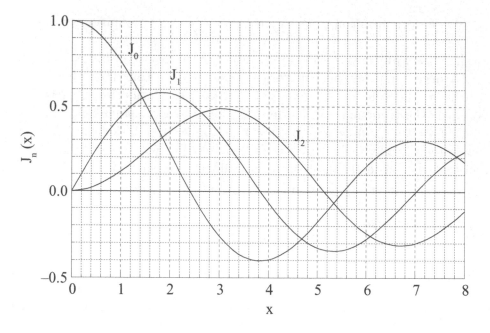

Figure 4.9: Bessel functions, $J_0(x)$, $J_1(x)$, $J_2(x)$.

where $Z(z)$ and values of k are still to be determined for specific cases.

Potential inside a cylinder

The first problem we consider is to find the potential inside a circular cylinder of radius R and length L having the curved surface and one end grounded, with a specified potential on the other end. We orient the cylinder with its axis along the z-axis and the grounded base at $z=0$. This makes the boundary conditions

$$\psi(r,\theta,0) = 0 \tag{4.232}$$
$$\psi(R,\theta,z) = 0 \tag{4.233}$$
$$\psi(r,\theta,L) = V(r,\theta), \tag{4.234}$$

where $V(r,\theta)$ is the specified potential on the top surface of the cylinder.

The boundary condition (4.233) that the potential vanish at $r = R$ requires that

$$J_m(kR) = 0. \tag{4.235}$$

This is an eigenvalue condition for the constant k. The k eigenvalues are given by

$$k_{m,n} = \frac{j_{m,n}}{R}, \tag{4.236}$$

where $j_{m,n}$ is the nth zero of J_m. It is convenient for this purpose to use tables of the zeroes of Bessel functions. The first few zeroes of J_0, J_1, and J_2 are given in Table 4.1. It can be seen in the table that the difference in succesive zeroes very quickly approaches π, although the asymptotic region has not yet been reached. The values of the derivative of J_m at its zero are also listed in Table 4.1. We will see shortly that these are useful for finding the potential. Table 4.2 lists the first few zeros of J'_m, which apply for the Neumann boundary condition that arises for circular wave guides and cavities. (See chapter 11.)

Table 4.1: Bessel function zeros $j_{m,n}$

n	$j_{0,n}$	$J'_0(j_{0,n})$	$j_{1,n}$	$J'_1(j_{1,n})$	$j_{2,n}$	$J'_2(j_{2,n})$
1	2.4048	-0.5191	3.8317	-0.4028	5.1356	-0.3397
2	5.5200	0.3403	7.0156	0.3001	8.4172	0.2714
3	8.6537	-0.2715	10.1735	-0.2497	11.6198	-0.2324
4	11.792	0.2325	13.3237	0.2184	14.7960	0.2065

Table 4.2: Bessel function derivative zeros $j'_{m,n}$

n	$j'_{0,n}$	$J_0(j'_{0,n})$	$j'_{1,n}$	$J_1(j'_{1,n})$	$j'_{2,n}$	$J_2(j'_{2,n})$
1	0	1	1.8412	0.5819	3.0542	0.4865
2	3.8317	-0.4028	5.3314	-0.3461	6.7061	-0.3135
3	7.0156	0.3001	8.5363	0.2733	9.9694	-0.2547
4	10.1734	-0.2497	11.7060	-0.2333	13.1704	-0.2209

Once k is known from the radial boundary condition, the Z differential equation (4.205) can be solved, giving

$$Z(z) = \sinh(k_{m,n}z). \tag{4.237}$$

We have picked the hyperbolic sine function to satisfy the homogeneous boundary condition that $\psi = 0$ at $z=0$. We now have all three functions, and the general solution for the potential inside the cylinder can be written as a sum of all possible solutions

$$\psi(r,\theta,z) = \sum_{n,m} [\alpha_{m,n}\cos(m\theta) + \beta_{m,n}\sin(m\theta)] J_m(k_{m,n}r)\sinh(k_{m,n}z). \tag{4.238}$$

The coefficients $\alpha_{m,n}$ and $\beta_{m,n}$ are determined by the remaining boundary condition at the top of the cylinder. Setting $z = L$ gives

$$\psi(r,\theta,L) = V(r,\theta) = \sum_{n,m} [\alpha_{m,n}\cos(m\theta) + \beta_{m,n}\sin(m\theta)] J_m(k_{m,n}r)\sinh(k_{m,n}L).$$

$$\tag{4.239}$$

Equation (4.239) is a double expansion. There is first a trigonometric Fourier expansion

$$V(r,\theta) = \sum_{m} [A_m(r)\cos(m\theta) + B_m(r)\sin(m\theta)]. \tag{4.240}$$

The functions $A_m(r)$ and $B_m(r)$ are Fourier coefficients given by integrals over $V(r, \theta)$ as in Eqs. (4.189)-(4.191). Then the $A_m(r)$ and $B_m(r)$ are expanded in Bessel function expansions as

$$
\begin{aligned}
A_m(r) &= \sum_n [\alpha_{m,n} J_m(k_{m,n} r) \sinh(k_{m,n} L)] \\
B_m(r) &= \sum_n [\beta_{m,n} J_m(k_{m,n} r) \sinh(k_{m,n} L)].
\end{aligned}
\tag{4.241}
$$

The $J_m(k_{m,n} r)$ in this sum with differing n are orthogonal because they are eigenfunctions (for Dirichlet boundary conditions) of Eq. (4.207) with different eigenvalues. This orthogonality takes the form

$$
\int_0^1 J_m(j_{m,n} x) J_m(j_{m,n'} x) x\, dx = N_{m,n} \delta_{nn'}.
\tag{4.242}
$$

Note that the weighting function $w(r) = r$ is included in the orthogonality integral. We have used the integration variable $x = r/R$ to make the integral dimensionless. The normalization constant is given by (see, for instance, [Arfken] for a derivation)

$$
N_{m,n} = \frac{1}{2} [J_m'(j_{m,n})]^2.
\tag{4.243}
$$

There is also an orthogonality relation for Neumann boundary conditions, for which $J_m'(j_{m,n}') = 0$. This orthogonality integral is

$$
\int_0^1 J_m(j_{m,n}' x) J_m(j_{m,n'}' x) x\, dx = N_{m,n}' \delta_{nn'},
\tag{4.244}
$$

with the normalization constant given by

$$
N_{m,n}' = \frac{1}{2} \left(1 - \frac{n^2}{j_{m,n}'^2} \right) [J_m(j_{m,n}')]^2.
\tag{4.245}
$$

Using the Dirichlet orthogonality relation of Eq. (4.242), the coefficients in the Bessel function expansion can be projected out as was done for the Legendre or trigonometric Fourier series. This gives

$$
\alpha_{m,n} \sinh(k_{m,n} L) = \frac{2}{[J_m'(j_{m,n})]^2} \int_0^1 J_m(j_{m,n} x) A_m(Rx) x\, dx,
\tag{4.246}
$$

and similarly for $\beta_{m,n}$. Equations (4.238), (4.241), and (4.246) constitute the solution for the potential inside the cylinder.

As a simple example, we consider the case when

$$
\psi(r, \theta, L) = V,
\tag{4.247}
$$

with V a constant potential. Then from the axial symmmetry, the Bessel index m must be zero and the potential inside the cylinder is given by

$$\psi(r,\theta,z) = \sum_n \alpha_n J_0(j_{0,n}r/R)\sinh(j_{0,n}z/R). \tag{4.248}$$

The coefficients α_n are given by

$$\alpha_n = \frac{2V}{\sinh(j_{0,n}L/R)[J_0'(j_{0,n})]^2}\int_0^1 J_0(j_{0,n}x)x\,dx. \tag{4.249}$$

This integral can be done by noticing from Bessel's equation (4.211) with $m=0$ that

$$xJ_0(x) = -[xJ_0(x)']'. \tag{4.250}$$

Then the integral in Eq. (4.249) becomes (letting $u = j_{0,n}x$)

$$\int_0^1 xJ_0(j_{0,n}x)dx = \frac{1}{j_{0,n}^2}\int_0^{j_{0,n}} uJ_0(u)du = -\frac{1}{j_{0,n}^2}\int_0^{j_{0,n}}[uJ_0(u)']'dx = \frac{-J_0'(j_{0,n})}{j_{0,n}}. \tag{4.251}$$

The potential inside the cylinder is then given by

$$\psi(r,\theta,z) = -2V\sum_n \frac{J_0(j_{0,n}r/R)\sinh(j_{0,n}z/R)}{j_{0,n}J_0'(j_{0,n})\sinh(j_{0,n}L/R)}. \tag{4.252}$$

Modified Bessel functions

The next case we consider has both ends of the cylinder grounded, and a given potential $U(z,\theta)$ on the curved surface. These boundary conditions are

$$\psi(r,\theta,0) = 0 \tag{4.253}$$
$$\psi(R,\theta,z) = U(\theta,z) \tag{4.254}$$
$$\psi(r,\theta,L) = 0, \tag{4.255}$$

To satisfy the zero boundary conditions at the ends of the cylinder, the solution of the differential equation (4.205) for $Z(z)$ must be

$$Z(z) = \sin(n\pi z/L), \tag{4.256}$$

with z eigenvalue

$$\lambda_z = k^2 = -(n\pi/L)^2, \tag{4.257}$$

where n is any integer. This means that

$$k = i(n\pi/L), \tag{4.258}$$

and is pure imaginary. All of Eqs. (4.211)-(4.230) still hold, only with the substitution $x \to ix$. This leads to Bessel functions of imaginary argument, called **modified Bessel functions**.

The conventional definition of the modified Bessel function that is regular at the origin is

$$I_m(x) = i^{-m} J_m(ix). \tag{4.259}$$

There is also an irregular modified Bessel function $K_m(x)$, but this does not enter for potentials that must be finite at the origin.

With the definition in Eq. (4.259), equations (4.224)-(4.227) for $J_m(x)$ get transformed into the following equations for $I_m(x)$:

$$I_m(x) = x^m \sum_{n=0}^{\infty} b_n x^{2n}, \tag{4.260}$$

with

$$b_0 = \frac{1}{2^m m!} \tag{4.261}$$

and

$$b_n = \frac{b_{n-1}}{4n(n+m)}, \tag{4.262}$$

$$I_{m-1}(x) - I_{m+1}(x) = \frac{2m}{x} I_m(x). \tag{4.263}$$

The first three $I_m(x)$ are plotted in Fig. 4.10.

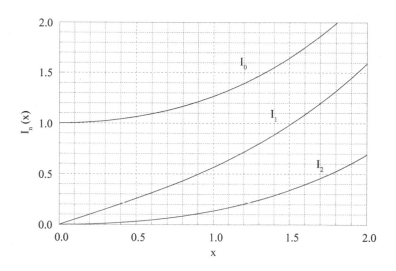

Figure 4.10: Modified Bessel functions, $I_0(x)$, $I_1(x)$, $I_2(x)$.

The solution for the potential inside the cylinder can be written using the $I_m(x)$ as

$$\psi(r, \theta, z) = \sum_{n,m} [\alpha_{m,n} \cos(m\theta) + \beta_{m,n} \sin(m\theta)] I_m(n\pi r/L) \sin(n\pi z/L). \tag{4.264}$$

The boundary condition on the curved surface of the cylinder gives

$$V(\theta, z) = \sum_{n,m} [\alpha_{m,n} \cos(m\theta) + \beta_{m,n} \sin(m\theta)] I_m(n\pi R/L) \sin(n\pi z/L). \quad (4.265)$$

This is a two dimensional Fourier transform. The coefficients $\alpha_{m,n}$ are given by

$$\alpha_{m,n} I_m(n\pi R/L) = \frac{2}{\pi L} \int_0^{2\pi} d\theta \cos(m\theta) \int_0^L dz \, \sin(n\pi z/L) V(\theta, z), \, m > 0. \quad (4.266)$$

$$= \frac{1}{\pi L} \int_0^{2\pi} d\theta \int_0^L dz \, \sin(n\pi z/L) V(\theta, z), \quad m = 0. \quad (4.267)$$

The $\beta_{m,n}$ are given by

$$\beta_{m,n} I_m(n\pi R/L) = \frac{2}{\pi L} \int_0^{2\pi} d\theta \sin(m\theta) \int_0^L dz \, \sin(n\pi z/L) V(\theta, z). \quad (4.268)$$

With these expressions for $\alpha_{m,n}$ and $\beta_{m,n}$, Eq. (4.264) gives the potential inside the cylinder.

4.4   Problems

1. (a) Write the potential of a dipole **p** in spherical coordinates.

   (b) Find **E** of the dipole for $r>0$ by taking the negative gradient in spherical coordinates. Show that this equals the result of writing the vector expression for **E** in spherical coordinates.

   (c) Calculate the curl and the divergence of **E** of the dipole in spherical coordinates for $r>0$. (Each should come out zero.)

2. Verify Eq. (4.63) explicitly for the cases $l, l' = 0, 1, 2$.

3. Show that the summation indices in Eq. (4.73) can be adjusted to derive the Legendre polynomial recursion relation (4.71).

4. The potential on a sphere of radius R is given by

$$\psi(R,\theta) \;=\; +V, \quad 0° \leq \theta < 90°$$
$$\psi(R,\theta) \;=\; -V, \quad 90° < \theta \leq 180°. \tag{4.269}$$

   Calculate the dipole moment of the sphere by integrating over the charge distribution. Check that this result agrees with the dipole term in the potential outside the sphere.

5. The potential on a sphere of radius R is given by

$$\psi(R,\theta) = V_0 \cos^2 \theta. \tag{4.270}$$

   (a) Find the potential outside the sphere.

   (b) Find the surface charge distribution on the sphere.

   (c) Integrate over the surface charge distribution to find the quadrupole moment of the sphere. Check that this result agrees with the $l = 2$ term in the potential outside the sphere.

6. The potential on a sphere of radius R is given by

$$\psi(R,\theta) = V_0 \sin^2 \theta. \tag{4.271}$$

   [Hint: Use the answers problem 5 to solve this problem.]

   (a) Find the potential outside the sphere.

   (b) Find the surface charge distribution on the sphere.

   (c) Integrate over the surface charge distribution to find the quadrupole moment of the sphere. Check that this result agrees with the $l = 2$ term in the potential outside the sphere.

7. This is a difficult and lengthy double credit problem.

   (a) Calculate the potential on the axis of a uniformly charged ring of charge Q and radius R.

   (b) Expand the potential to find the potential at all angles for $r>R$ and for $r<R$.

   (c) Identify the multipole moments of the ring from the expansion of the potential.

   (d) Calculate the multipole moments by integrating over the charge distribution of the ring and show that this result agrees with the previous answer. (Use the generating function to evaluate $P_l(0)$.

   (e) Integrate the potential on the axis of a ring to find the potential a distance z along the axis of a uniformly charged disk of charge Q and radius R.

   (f) Expand the potential to find the potential at all angles for $r>R$ and for $r<R$.

8. The potential on a sphere of radius R is given by

$$\begin{aligned} \psi_2(R, \theta_2, \phi_2) &= +V, & 90° < \phi_2 < 270° \\ &= -V, & -90° < \phi_2 < +90°. \end{aligned} \qquad (4.272)$$

   (a) Find the potential outside the sphere. You can leave your answer in terms of an integral over $d\theta$.

   (b) Calculate the dipole ($l = 1$) term in the potential explicitly, and identify the dipole moment for each m value.

   (c) Compare your answer to (b) with what you get by applying the addition theorem to the dipole moment in problem 4.

9. (a) Use Eqs. (4.83) and (3.33) to derive the completeness relation

$$\sum_l \left(\frac{2l+1}{2} \right) P_l(x) P_l(x') = \delta(x - x'), \quad -1 \le x \le +1 \qquad (4.273)$$

   for Legendre polynomials.

   (b) Use the completeness relation for Legendre polynomials to find the surface Green's function for the potential inside a sphere with a specified axially symmetric potential on its surface.

10. (a) Write the potential of a dipole **p** in cylindrical coordinates.

    (b) Find **E** of the dipole by taking the negative gradient in cylindrical coordinates. Show that this equals the result of writing the vector expression for **E** in cylindrical coordinates.

11. The potential on the surface of a hollow cylinder of radius R is given by $\phi(R,\theta) = V_0$ for $0° < \theta < 180°$, and $\phi(R,\theta) = -V_0$ for $0180° < \theta < 360°$.

   (a) Find the potential inside the cylinder.

   (b) Find the surface charge distribution on the cylinder.

12. (a) Find the potential at the center of a hollow cylinder of radius R and length $L = 2R$, having one end kept at a potential V, with the other end and the curved surface of the cylinder grounded.

   (b) Evaluate the potential at the center to three significant figures.

13. (a) Find the potential at the center of a hollow cylinder of radius R and length $L = 2R$, with both ends grounded and the curved surface kept at a potential V.

   (b) Evaluate the potential at the center to three significant figures.

   (c) What can you say about the numerical answers to problems 12 and 13?

Chapter 5

Green's Functions

5.1 Application of Green's Second Theorem

In this section we use Green's second theorem to find the solution to Poisson's equation for a general charge distribution with general boundary conditions. We start with Green's second theorem

$$\int_V [\phi \boldsymbol{\nabla}^2 \psi - \psi \boldsymbol{\nabla}^2 \phi] d\tau = \oint_S \mathbf{dA} \cdot [\phi \boldsymbol{\nabla} \psi - \psi \boldsymbol{\nabla} \phi]. \tag{5.1}$$

We take the function ϕ to be a solution of Poisson's equation

$$\boldsymbol{\nabla}^2 \phi = -4\pi\rho. \tag{5.2}$$

We pick the other function ψ, designated as a **Green's Function** $G(\mathbf{r}, \mathbf{r}')$, to be a function of two variables, satisfying

$$\boldsymbol{\nabla}'^2 G(\mathbf{r}, \mathbf{r}') = -4\pi\delta(\mathbf{r} - \mathbf{r}'). \tag{5.3}$$

(The Green's function is often defined without the -4π, but it is more convenient for electrostatics in Gaussian units to put the -4π here to make the Green's function equivalent to the potential of a unit charge.)

Then Green's second theorem can be written as

$$\int_V [\phi(\mathbf{r}') \boldsymbol{\nabla}'^2 G(\mathbf{r}, \mathbf{r}') - G(\mathbf{r}, \mathbf{r}') \boldsymbol{\nabla}'^2 \phi(\mathbf{r}')] d\tau' =$$
$$\oint_S \mathbf{dA}' \cdot [\phi(\mathbf{r}') \boldsymbol{\nabla}' G(\mathbf{r}, \mathbf{r}') - G(\mathbf{r}, \mathbf{r}') \boldsymbol{\nabla}' \phi(\mathbf{r}')], \tag{5.4}$$

where we integrate over $\mathbf{r}'$ keeping $\mathbf{r}$ fixed. Using Eqs. (5.2) and (5.3),in Eq. (5.4) results in

$$\phi(\mathbf{r}) = \int_V G(\mathbf{r}, \mathbf{r}') \rho(\mathbf{r}') d\tau' - \frac{1}{4\pi} \oint_S \mathbf{dA}' \cdot [\phi(\mathbf{r}') \boldsymbol{\nabla}' G(\mathbf{r}, \mathbf{r}') - G(\mathbf{r}, \mathbf{r}') \boldsymbol{\nabla}' \phi(\mathbf{r}')]. \tag{5.5}$$

5.2   Surface Boundary Conditions

We know from the uniqeness theorem that either ϕ or its normal derivative, but not both, must be given on every part of the surface for a unique solution to exist. If ϕ is given on any surface (Dirichlet boundary condition), then setting $G(\mathbf{r}, \mathbf{r}') = 0$ for $\mathbf{r}'$ on that surface will remove the surface integral over the unknown normal derivative of ϕ. If it is the normal derivative ϕ that is given on any surface (Neumann boundary condition), then setting the normal derivative with respect to $\mathbf{r}'$ of $G(\mathbf{r}, \mathbf{r}')$ equal to zero for $\mathbf{r}'$ on that surface will remove the surface integral over the unknown surface potential ϕ.

In either case, Eq. (5.5) will be the unique Green's function solution of Poisson's equation with the given boundary conditions. The general rule is that the Green's function satisfy the homogeneous boundary condition (Dirichlet or Neumann) corresponding to the inhomogeneous boundary condition satisfied by the potential. For the standard electrostatic problem, the potential ϕ is specified on the bounding surfaces for the volume integral in Eq. (5.5), which is the Dirichlet boundary condition.

5.3   Green's Function Solution of Poisson's Equation

The Green's function solution to Poisson's equation with the potential specified on all bounding surfaces can be written as

$$\phi(\mathbf{r}) = \int_V G(\mathbf{r}, \mathbf{r}')\rho(\mathbf{r}')d\tau' - \frac{1}{4\pi}\oint_S [d\mathbf{A}' \cdot \boldsymbol{\nabla}' G(\mathbf{r}, \mathbf{r}')]\phi(\mathbf{r}'). \qquad (5.6)$$

The surface integral can be written in terms of a surface Green's function, as

$$\phi(\mathbf{r}) = \int_V G(\mathbf{r}, \mathbf{r}')\rho(\mathbf{r}')d\tau' + \oint_S dA' g(\mathbf{r}, \mathbf{r}')\phi(\mathbf{r}'), \qquad (5.7)$$

where

$$g(\mathbf{r}, \mathbf{r}') = -\frac{1}{4\pi}[\hat{\mathbf{n}} \cdot \boldsymbol{\nabla}' G(\mathbf{r}, \mathbf{r}')], \qquad (5.8)$$

and $\hat{\mathbf{n}}$ is a unit vector normal to the surface. Equation (5.8) gives the surface Green's function, $g(\mathbf{r}, \mathbf{r}')$, introduced in Chapter 4 in terms of the three-dimensional Green's function, $G(\mathbf{r}, \mathbf{r}')$.

Using the Green's function, the solution to any electrostatic problem can be written down in two steps. The first step is immediate and straightforward. The solution is simply written down as given in terms of the Green's function by Eq (5.6). The second, more difficult, step is to find the Green's function by using its two defining properties:

I. $$\nabla'^2 G(\mathbf{r}, \mathbf{r}') = -4\pi\delta(\mathbf{r} - \mathbf{r}') \tag{5.9}$$

II. $$G(\mathbf{r}, \mathbf{r}'_\mathbf{S}) = 0, \tag{5.10}$$

where the notation $\mathbf{r}'_\mathbf{S}$ means that $\mathbf{r}'$ is on one of the bounding surfaces.

It is helpful to put the requirements for the Green's function into words:

> For the solution of Poisson's equation for an arbitrary charge distribution with the potential specified on all boundaries, the Green's function, $G(\mathbf{r}, \mathbf{r}')$, is the potential at $\mathbf{r}'$ due to a unit point charge at $\mathbf{r}$ with all surfaces acting as grounded conductors.

Note that, even if the original surfaces are not conductors, (They can't be conductors if the specified potential on them is not constant.) the Green's function is the point charge solution found as if all the surfaces were grounded conductors.

5.4 Symmetry of Green's Function

The Green's function is a symmetric function of its arguments. This can be shown by substituting two Green's functions, $G(\mathbf{r_1}, \mathbf{r}')$ and $G(\mathbf{r_2}, \mathbf{r}')$, with the same boundary conditions into Green's second theorem:

$$\int_V [G(\mathbf{r_1}, \mathbf{r}')\nabla'^2 G(\mathbf{r_2}, \mathbf{r}') - G(\mathbf{r_2}, \mathbf{r}')\nabla'^2 G(\mathbf{r_1}, \mathbf{r}')]d\tau' =$$
$$\oint_S d\mathbf{A}' \cdot [G(\mathbf{r_1}, \mathbf{r}')\nabla' G(\mathbf{r_2}, \mathbf{r}') - G(\mathbf{r_2}, \mathbf{r}')\nabla' G(\mathbf{r_1}, \mathbf{r}')]. \tag{5.11}$$

The surface integrals on the right hand of Eq. (5.11) are zero for either the Dirichlet or Neumann boundary conditions, and the volume integrals on the left hand side reduce to delta function integrals. This gives the result

$$G(\mathbf{r_1}, \mathbf{r_2}) = G(\mathbf{r_2}, \mathbf{r_1}). \tag{5.12}$$

Thus the Green's function is symmetric for either the Dirichlet or Neumann boundary conditions.

The symmetry of the Green's function means that the potential at a point $\mathbf{r_2}$ due to a point charge q at $\mathbf{r_1}$ equals the potential at $\mathbf{r_1}$ due to the same point charge q at $\mathbf{r_2}$, if all surfaces are grounded conductors. The symmetry of the Green's function is useful in solving for Green's functions, and in checking the final result for a Green's function. It also means you can afford to be a bit careless in deciding which variable, $\mathbf{r}$ or $\mathbf{r}'$, is which in using the Green's function.

5.5   Green's Reciprocity Theorem

A somewhat more general result, **Green's reciprocity theorem**, can be shown to hold for general charge distributions in a region bounded by conductors (not necessarily grounded). Green's reciprocity theorem states

$$\int [\rho_1\phi_2 - \rho_2\phi_1]d\tau = \sum_n [Q_2V_1 - Q_1V_2]_n, \qquad (5.13)$$

where Q and V represent the net charge and voltage on each conductor and the sum is over all conductors. The subscripts 1 and 2 refer to two different cases of charge distribution and resulting potential, with the same geometry for the conductors.

To prove the theorem, we start with Green's second theorem for the two different potentials

$$\int_V [\phi_1\nabla^2\phi_2 - \phi_2\nabla^2\phi_1]d\tau = \oint_S d\mathbf{A}\cdot[\phi_1\nabla\phi_2 - \phi_2\nabla\phi_1]. \qquad (5.14)$$

We now use Poisson's equation on the left hand side, and the fact that all surfaces on the right hand side are conductors to reduce Eq, (5.14) to Green's reciprocity theorem, Eq. (5.13). The surface integrals can all be done using the fact that the conductors are equipotentials, so the potential V can be taken out of the integral.

Green's reciprocity theorem is best illustrated by applying it to an example. We consider the configuration of a point charge q a distance d from the center of a neutral conducting sphere of radius a, with $a<d$. We want to find the potential of the conducting sphere.

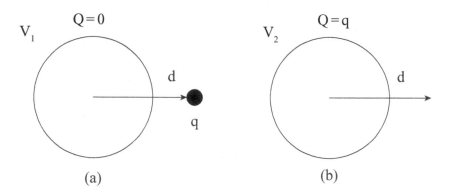

(a)   (b)

Figure 5.1: Two cases for Green's reciprocity theorem. (a) Point charge q outside a neutral sphere. (b) Sphere with charge q.

To apply Green's reciprocity theorem, we take the stated situation as case 1. Then

$$\rho_1(r) = q\delta(\mathbf{r} - \mathbf{d}), \qquad Q_1 = 0, \qquad (5.15)$$

as shown in Fig. 5.1a. The trick in applying the theorem is to choose a suitable case 2. We choose

$$\rho_2(\mathbf{r}) = 0, \qquad Q_2 = q, \tag{5.16}$$

as shown in Fig. 5.1b. We substitute these charge distributions into Eq. (5.13) to give

$$q\phi_2(\mathbf{d}) = qV_1. \tag{5.17}$$

We chose case 2 to be an easy case to solve, just an isolated conducting sphere of charge q. The potential a distance d from the center is

$$\phi_2(\mathbf{d}) = \frac{q}{d}. \tag{5.18}$$

Then Eq. (5.17) can be solved for V_1

$$V_1 = \frac{q}{d}, \tag{5.19}$$

which is the required potential on the sphere. This is the same answer we got by using images.

Note, that in using Green's reciprocity theorem, it is important to clearly distinguish between the two separate cases. It is best to do this with two separate pictures and corresponding sets of conditions, as we did above. Other examples of the use of Green's reciprocity theorem are given as problems.

5.6   Green's Functions for Specific Cases

The simplest Green's function is for the case where the potential is not specified on any surface, except for the requirement that it go to zero at infinity. This Green's function is given by

$$G(\mathbf{r}, \mathbf{r}') = \frac{1}{|\mathbf{r} - \mathbf{r}'|}. \tag{5.20}$$

The potential, given by Eq. (5.6), is just the Coulomb law integral we saw in Chapter 1.

We now find Green's functions for some of the examples considered earlier in the text.

Plane surface

In section 3.2, we solved for the potential due to a point charge a given distance from a grounded plane. The potential we found there is just the Green's function for the case where the potential is specified on a plane. That is, if the potential to be solved for is specified on the x-y plane, the Green's function will be the potential at a point x', y', z' due to a unit point charge at x, y, z and a negative unit charge at $x, y, -z$. The G for this is

$$G(x, y, z; x', y', z') \; = \; \cfrac{1}{\sqrt{(x' - x)^2 + (y' - y)^2 + (z' - z)^2}}$$
$$-\cfrac{1}{\sqrt{(x' - x)^2 + (y' - y)^2 + (z' + z)^2}}. \quad (5.21)$$

Note that the restriction $z, z' \geq 0$, which was required when solving the grounded conducting plane problem need not apply to the Green's function or its use. That is because the plane surface for which the Green's function is used is not generally a conducting plane.

The corresponding surface Green's function, $g(\mathbf{r}, \mathbf{r}')$, is given by Eq. (5.8) to be

$$g(x, y, z; x', y', 0) \; = \; -\frac{1}{4\pi}\partial_{z'} G(x, y, z; x', y', z')|_{z'=0}$$
$$= \; \frac{z}{2\pi[(x' - x)^2 + (y' - y)^2 + z^2]^{\frac{3}{2}}}. \quad (5.22)$$

A specific application of this surface Green's function is given as a problem.

Sphere

We also solved for the potential due to a point charge outside a grounded sphere. That potential serves as the Green's function for a potential that is specified on a sphere. Writing Eq. (3.25) for the potential due to the original charge and the image charge gives the Green's function

$$G(\mathbf{r}, \mathbf{r}') = \frac{1}{|\mathbf{r}' - \mathbf{r}|} - \frac{R/r}{|\mathbf{r}' - R^2\hat{\mathbf{r}}/r|} \quad (5.23)$$

when the potential is specified on a sphere of radius R.

5.7   Constructing Green's Functions

5.7.1   Construction of the Green's function from eigenfunctions

The two examples above found the Green's function by solving for the potential due to a point charge with the given surfaces taken as grounded conductors by using the method of images. Another way to find the Green's function is to directly construct it so that its Laplacian will give the delta function on the right hand side of Eq, (5.9). We demonstrate this method for solving Poisson's equation inside a rectangular cavity of dimensions $A \times B \times C$ whose walls are kept at specified potential distributions.

We start by finding the eigenfunctions of the Laplacian operator satisfying the equation

$$(\partial_x^2 + \partial_y^2 + \partial_z^2)u_\lambda(x,y,z) = \lambda u_\lambda(x,y,z), \tag{5.24}$$

with $u_\lambda = 0$ on all walls of the cavity. The eigenfunctions are easily found by the method of separation of variables to be

$$u_{lmn}(x,y,z) = \sqrt{\frac{8}{ABC}} \sin\left(\frac{l\pi x}{A}\right) \sin\left(\frac{m\pi y}{B}\right) \sin\left(\frac{n\pi z}{C}\right), \tag{5.25}$$

with the eigenvalues

$$\lambda_{lmn} = -\pi^2 \left[\left(\frac{l}{A}\right)^2 + \left(\frac{m}{B}\right)^2 + \left(\frac{n}{C}\right)^2 \right]. \tag{5.26}$$

We have normalized the eigenfunctions so that integrating $|u_{lmn}|^2$ over the full range of x, y, and z gives one. Such a set of normalized orthogonal functions are called a set of **orthonormal functions**. In section 3.4 we showed that the completeness sum over the normalized eigenfunctions of a self adjoint operator produced the delta function. That is the case here, so we have

$$\sum_{lmn} u_{lmn}^*(x',y',z')u_{lmn}(x,y,z) = \delta(\mathbf{r}' - \mathbf{r}). \tag{5.27}$$

If we act on the sum on the left hand side of Eq. (5.27) by the Laplacian operator, we get

$$\nabla'^2 \sum_{lmn} u_{lmn}^*(x',y',z')u_{lmn}(x,y,z) = \sum_{lmn} \lambda_{lmn} u_{lmn}^*(x',y',z')u_{lmn}(x,y,z). \tag{5.28}$$

This is no longer the delta function, because of the λ_{lmn} appearing in the sum. In order to construct the Green's function, we form the sum

$$G(x,y,z;x',y',z') = -4\pi \sum_{lmn} \frac{u_{lmn}^*(x',y',z')u_{lmn}(x,y,z)}{\lambda_{lmn}}. \tag{5.29}$$

Then, acting on G with the Laplacian produces the triple sum for the δ function in Eq. (5.27).

This gives the general construction of the Green's function as a triple sum over the eigenfunctions of the Laplacian operator. For the rectangular cavity, the u_{lmn} are given by Eq. (5.25) and λ_{lmn} by Eq. (5.26). Equation (5.29) would give the Green's function for any geometry if the appropriate eigenfunctions and eigenvalues were used.

5.7.2   Reduction to a one dimensional Green's function

Rectangular coordinates

The triple sum over eigenfunctions for the Green's function is easy to write down, but would be cumbersome to actually use. We can derive a somewhat more useful double sum formula for the Green's function. We demonstrate this in Cartesian coordinates for the potential inside a rectangular cavity. We start by forming a two dimensional delta function from eigenfunctions of the Laplacian in two of the three variables

$$\sum_{lm} u_{lm}^*(x', y') u_{lm}(x, y) = \delta(x' - x)\delta(y' - y). \tag{5.30}$$

The eigenfunctions $u_{lm}(x, y)$ satisfy

$$(\partial_x^2 + \partial_y^2) u_{lm}(x, y) = -\gamma_{lm}^2 u_{lm}(x, y), \tag{5.31}$$

and also satisfy the homogeneous boundary conditions in the x and y variables. The eigenfunctions for the rectangular cavity of dimensions $A \times B \times C$ are given by

$$u_{lm}(x, y) = \sqrt{\frac{4}{AB}} \sin\left(\frac{l\pi x}{A}\right) \sin\left(\frac{m\pi y}{B}\right), \tag{5.32}$$

and the corresponding eigenvalues are

$$\gamma_{lm}^2 = \pi^2 \left[\left(\frac{l}{A}\right)^2 + \left(\frac{m}{B}\right)^2\right]. \tag{5.33}$$

To construct a Green's function in the three variables x, y, z, we introduce a function $f_{lm}(z, z')$ inside a double sum over the eigenfunctions, writing

$$G(x, y, z; x', y', z') = -4\pi \sum_{lm} u_{lm}^*(x', y') u_{lm}(x, y) f_{lm}(z, z'). \tag{5.34}$$

This double sum, including the function $f_{lm}(z, z')$, will satisfy the differential equation

$$(\partial_{x'}^2 + \partial_{y'}^2 + \partial_{z'}^2) \sum_{lm} u_{lm}^*(x', y') u_{lm}(x, y) f_{lm}(z, z') = \delta(x' - x)\delta(y' - y)\delta(z' - z),$$
$$\tag{5.35}$$

provided that $f_{lm}(z, z')$ satisfies the differential equation

$$\partial_{z'}^2 f_{lm}(z, z') - \gamma_{lm}^2 f_{lm}(z, z') = \delta(z' - z). \tag{5.36}$$

The function $f_{lm}(z, z')$ must also satisfy the homogeneous boundary conditions

$$f_{lm}(z, 0) = f_{lm}(z, C) = 0. \tag{5.37}$$

Equations (5.36) and (5.37) are the defining properties of a one dimensional Green's function, so we have reduced the original three dimensional problem to a one dimensional Green's function problem.

We can solve for $f(z, z')$ (We temporarily drop the subscript lm.) by using the fact that the delta function on right hand side of Eq. (5.36) is zero except at one point. Then $f(z, z')$ satisfies the homogeneous differential equation

$$\partial_{z'}^2 f(z, z') - \gamma^2 f(z, z') = 0, \quad z' \neq z. \tag{5.38}$$

We break the interval [0,C] into two regions as shown in Fig. 5.2.

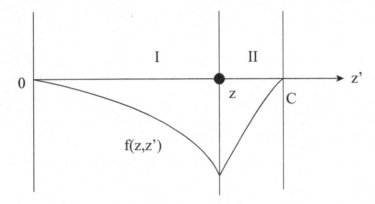

Figure 5.2: Two regions for Green's function solution.

$$\text{Region I}: \quad 0 \leq z' < z, \qquad f(z, z') = \sinh[\gamma z'] \tag{5.39}$$
$$\text{Region II}: \quad z < z' \leq C, \qquad f(z, z') = \sinh[(\gamma(z' - C)]. \tag{5.40}$$

The solutions to the differential equation (5.38), appropriate to the boundary condition in each region, are given in Eqs. (5.39) and (5.40).

The function $f(z, z')$ must be continuous at the point $z = z'$. Otherwise its first derivative would not be finite at that point. We still have the freedom of multiplying each function $f(z, z')$ by an arbitrary function of z, without affecting the differential equation. We can use this freedom now to make the resulting $f(z, z')$ continuous at $z' = z$.

$$\text{Region I}: \quad 0 \leq z' < z, \quad f(z, z') = N \sinh[\gamma(z - C)] \sinh[\gamma z'] \tag{5.41}$$
$$\text{Region II}: \quad z < z' \leq C, \quad f(z, z') = N \sinh[\gamma z] \sinh[\gamma(z' - C)]. \tag{5.42}$$

The trick was to multiply each function of z' by the function of z that held for the other region. This was a simple way of guaranteeing that the resulting $f(z, z')$ would be continuous at $z' = z$. Notice that this also made $f(z, z')$ symmetric in the variables z and z', which we showed was always true for a Green's function. In fact, this could have been used as another simple way to find the right function $f(z, z')$.

The last step in making $f(z, z')$ a Green's function is to make its second partial derivative with respect to z' a delta function at $z' = z$. Notice that we have included a constant N in our equation for $f(z, z')$. We now show how to pick this constant so that we get the delta function.

We first demonstrate that one representation of the delta function is the first derivative of the unit step function. The **unit step function** is defined by

$$\theta(x) = 0, \quad x < 0; \qquad \theta(x) = 1, \quad x > 0. \tag{5.43}$$

We can then represent the delta function as

$$\delta(x) = \frac{d\theta(x)}{dx}. \tag{5.44}$$

As with any representation of the delta function, this representation does not make sense by itself, being either zero or undefined (infinite). But when put under an integral it gives

$$
\begin{aligned}
\int_a^b \frac{d\theta(x)}{dx} dx &= \theta(b) - \theta(a) \\
&= 0 \quad \text{if } x \neq 0 \text{ between a and b} \\
&= 1 \quad \text{if } a < 0 < b.
\end{aligned}
\tag{5.45}
$$

This result is a defining property of the delta function.

Now we can look at the derivatives of $f(z, z')$. The first derivative with respect to z' is discontinuous at $z' = z$, and has the behavior

$$\partial_z' f(z, z')|_{z-} = \gamma N \sinh[\gamma(z - C)] \cosh[\gamma z] \tag{5.46}$$

$$\partial_z' f(z, z')|_{z+} = \gamma N \sinh[\gamma z] \cosh[\gamma(z - C)] \tag{5.47}$$

The notation $z-$ and $z+$ stands for z' just below and just above z. In order for the next derivative with respect to z' to give the delta function, the discontinuity in $\partial_z' f(z, z')$ at $z' = z$ should equal one so that it will behave like the unit step function. The finite and smooth variation of $f(z, z')$ away from $z' = z$ does not affect the singular behavior at $z' = z$.

Setting the discontinuity equal to one results in

$$\partial_z' f(z, z')|_{z+} - \partial_z' f(z, z')|_{z-} = \gamma N \sinh(\gamma C) = 1. \tag{5.48}$$

Note that the z dependence of the discontinuity at $z' = z$ has cancelled out in the hyperbolic trigonometry. This is a general feature, and is necessary since the constant N could not depend on z. Solving for N gives

$$N = \frac{1}{\gamma \sinh(\gamma C)}. \tag{5.49}$$

The final one dimensional Green's function $f(z, z')$ is shown in Fig. 5.2.

We can now write down the full three variable Green's function as

$$G(x, y, z; x', y', z') =$$

$$\frac{-16\pi}{AB} \sum_{lm} \frac{\sin\left(\frac{l\pi x}{A}\right) \sin\left(\frac{l\pi x'}{A}\right) \sin\left(\frac{m\pi y}{B}\right) \sin\left(\frac{m\pi y'}{B}\right) \sinh[\gamma(z_> - C)] \sinh[\gamma z_<]}{\gamma \sinh(\gamma C)}, \quad (5.50)$$

with

$$\gamma = \pi \sqrt{\left(\frac{l}{A}\right)^2 + \left(\frac{m}{B}\right)^2}. \quad (5.51)$$

The notation $z_>$ means the larger of z and z', and $z_<$ means the smaller. Equation (5.50) is thus really two equations, one for $z' < z$ and one for $z' > z$.

The double sum Green's function given by Eq. (5.50) looks quite different from the triple sum Green's function given by Eq. (5.29), but, by the uniqueness theorem, they must represent the same function. They can be shown to be equal by expanding the function $f_{lm}(z, z')$ in a Fourier sine series. Also, two additional equivalent forms of the double sum Green's function can be found by suitably interchanging the x, y, z variables.

The surface Green's function at the surface $z = C$ is given by the normal derivative of $G(x, y, z; x', y', z')$ at $z = C$

$$g(\mathbf{r}, \mathbf{r}'_\mathbf{S}) = -\frac{1}{4\pi} \partial_{z'} G(x, y, z, x', y', z')|_{z'=C}$$

$$= \frac{4}{AB} \sum_{lm} \frac{\sin\left(\frac{l\pi x}{A}\right) \sin\left(\frac{l\pi x'}{A}\right) \sin\left(\frac{m\pi y}{B}\right) \sin\left(\frac{m\pi y'}{B}\right) \sinh(\gamma z)}{\sinh(\gamma C)}. \quad (5.52)$$

The surface Green's function found this way is the same as that constructed in Eq. (3.99) using separation of variables. Surface Green's functions for the other surfaces of the box can be found by suitable interchange of the variables in Eq. (5.52).

Spherical coordinates

We next solve for the Green's function for a potential specified on a sphere of radius R. For spherical coordinates, the spherical harmonics $Y_l^m(\theta, \phi)$ are eigenfunctions of the angular part of the Laplacian, so

$$\left[\frac{\partial_\theta(\sin\theta \partial_\theta)}{\sin\theta} + \frac{\partial_\phi^2}{sin^2\theta}\right] Y_l^m(\theta, \phi) = -l(l+1)Y_l^m(\theta, \phi). \quad (5.53)$$

The Y_l^m eigenfunctions satisfy the completeness relation

$$\sum_{lm} Y_l^{m*}(\theta, \phi) Y_l^m(\theta', \phi') = \delta(\cos\theta - \cos\theta')\delta(\phi - \phi'). \quad (5.54)$$

We can form a double sum Green's function by introducing a radial function $f_l(r, r')$. Then

$$G(r, \theta, \phi; r', \theta', \phi') = -4\pi \sum_{lm} f_l(r, r') Y_l^{m*}(\theta, \phi) Y_l^m(\theta', \phi'). \quad (5.55)$$

The function f_l satisfies the differential equation

$$\frac{1}{r'^2}\partial_{r'}[r'^2\partial_{r'}f_l(r,r')] - l(l+1)f_l(r,r') = \delta(r-r') \tag{5.56}$$

and the boundary conditions that $f_l(r,R)) = 0$, and that $f_l(r,r')$ be finite for $r' = 0$ and go to zero as $r' \to \infty$.

We solve Eq. (5.56) in the same way as we did the Cartesian function by requiring that $f_l(r,r')$ satisfy the homogeneous equation except at $r' = r$, where it must be continuous but have a unit discontinuity in its first derivative. For $0 \le r, r' < R$ (inside the sphere), this results in

$$\text{Region I:}\quad 0 \le r' < r, \quad f_l(r,r') = \frac{-r'^l}{2l+1}\left[\frac{1}{r^{l+1}} - \frac{r^l}{R^{2l+1}}\right] \tag{5.57}$$

$$\text{Region II:}\quad r < r' \le R, \quad f_l(r,r') = \frac{-r^l}{2l+1}\left[\frac{1}{r'^{l+1}} - \frac{r'^l}{R^{2l+1}}\right]. \tag{5.58}$$

The Green's function can then be written as

$$G(r,\theta,\phi;r',\theta',\phi') =$$
$$\sum_{lm}\frac{4\pi}{2l+1}\left[\frac{r_<^l}{r_>^{l+1}} - \frac{r_>^l r_<^l}{R^{2l+1}}\right]Y_l^{m*}(\theta,\phi)Y_l^m(\theta',\phi'), \quad r,r' \le R. \tag{5.59}$$

For $r, r' \ge R$ (outside the sphere), a similar derivation gives

$$G(r,\theta,\phi;r',\theta',\phi') =$$
$$-\sum_{lm}\frac{4\pi}{2l+1}\left[\frac{r_<^l}{r_>^{l+1}} - \frac{R^{2l+1}}{(r_>r_<)^{l+1}}\right]Y_l^{m*}(\theta,\phi)Y_l^m(\theta',\phi'), \quad r,r' \ge R. \tag{5.60}$$

Comparing Eqs. (5.59) and (5.60) with the vector form of the Green's function for a sphere given by Eq. (5.23), we see that they are the spherical harmonic expansions of the vector form.

We can find the surface Green's function for a spherical surface by taking the radial derivative of the Green's function, resulting in

$$g(r,\theta,\phi;R,\theta',\phi') = -\frac{1}{4\pi}\partial_{r'}G(r,\theta\phi;R,\theta',\phi')|_{r'=R}$$
$$= \frac{1}{R^2}\sum_{lm}\left(\frac{r}{R}\right)^l Y_l^{m*}(\theta,\phi)Y_l^m(\theta',\phi'), \quad r \le R.$$
$$= \frac{1}{R^2}\sum_{lm}\left(\frac{R}{r}\right)^{l+1} Y_l^{m*}(\theta,\phi)Y_l^m(\theta',\phi'), \quad r \ge R. \tag{5.61}$$

5.8   Problems

1. A point charge q is located between two parallel grounded conducting planes, as shown below Use Green's reciprocity theorem to find the net charge induced on each plane in terms of q and the distances a and b.

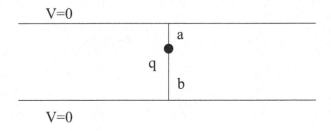

2. A point charge q is located between two connected grounded conducting planes, as shown below.
   Use Green's reciprocity theorem to find the net charge induced on each plane in terms of q and the angles α and β.

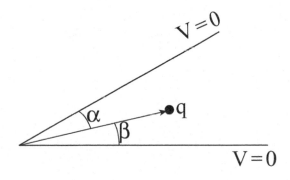

3. An infinite plane (Take it as the x-y plane, and use cylindrical coordinates r, θ, z.) has the potential distribution

$$V(r, \theta) = V_0, \quad r < a$$
$$V(r, \theta) = 0, \quad r > a. \tag{5.62}$$

   (a) Use the surface Green's function for a plane to find the potential $\psi(r, z)$ above the plane. (Don't evaluate the integral.)

   (b) Evaluate the integral for the case $r = 0$, and find the potential on the z axis.

4. Expand the function $\sinh[\gamma(z_> - C)]\sinh[\gamma z_<]$ in Eq. (5.50) in a Fourier sine series, and show that the double sum Green's function equals the triple sum Green's function of Eq. (5.29).

5. A cubical cavity of size $L \times L \times L$ has the top face at a uniform potential V_0, and the other five faces grounded.

   (a) Use the triple sum Green's function of Eq. (5.29) to find the surface Green's function for the top face.

   (b) Use the surface Green's function to find the potential inside the cavity.

   (c) Find the potential at the center of the cavity.

6. The potential on the surface of a sphere of radius R is given by

$$V(\theta) = V_0 \sin^2(\theta). \tag{5.63}$$

Use the surface Green's function of Eq. (5.61) to find the potential outside the sphere.

Chapter 6

Electrostatics in Matter

6.1 Polarization

Up to now, we have not considered the effects of matter other than as a medium to hold or conduct charge. However the molecules that make up matter can also affect electrostatics in another way, through polarization of the matter. This can occur when existing electric fields either polarize the molecules, or align molecules that already have electric dipole moments. Such polarizable materials that are also insulators are called **dielectrics**. We will discuss mechanisms for producing the polarization later in this chapter. For now, let us see what effect the polarization of the matter has on the electric potential.

The polarization can be described in terms of a polarization density vector, **P**, such that a small bit of matter of volume ΔV will have an electric dipole moment

$$\mathbf{p} = \mathbf{P}\Delta V. \tag{6.1}$$

In general, the polarization density will be a function of position, $\mathbf{P}(\mathbf{r})$. When there is a polarization density, the derivation of the potential of a continuous charge distribution leading to Eq. (1.34) has to be modified to take the dipole field due to the polarization into account.

The electric dipole potential due to the polarization in a collection of volume elements ΔV_n will be given by

$$\phi_{\mathrm{P}}(\mathbf{r}) = \sum_n \left[\frac{(\mathbf{r} - \mathbf{r_n}) \cdot \mathbf{P}(\mathbf{r_n})}{|\mathbf{r} - \mathbf{r_n}|^3} \right] \Delta V_n. \tag{6.2}$$

In the limit that the ΔV_n approach zero, the sum becomes the integral

$$\phi_{\mathrm{P}}(\mathbf{r}) = \int_V d\tau' \left[\frac{(\mathbf{r} - \mathbf{r}') \cdot \mathbf{P}(\mathbf{r}')}{|\mathbf{r} - \mathbf{r}'|^3} \right] = \int_V d\tau' \mathbf{P}(\mathbf{r}') \cdot \mathbf{\nabla}' \left[\frac{1}{|\mathbf{r} - \mathbf{r}'|} \right], \tag{6.3}$$

We can use the divergence trick to write

$$\phi_{\mathrm{P}}(\mathbf{r}) = - \int_V d\tau' \frac{\mathbf{\nabla}' \cdot \mathbf{P}(\mathbf{r}')}{|\mathbf{r} - \mathbf{r}'|} + \oint_S \frac{d\mathbf{A}' \cdot \mathbf{P}(\mathbf{r}')}{|\mathbf{r} - \mathbf{r}'|}. \tag{6.4}$$

The surface integral is over the surface S enclosing the volume V.

The full electric potential is given by the sum of $\phi_{\rho\sigma}$ due to volume and surface charge distributions and ϕ_P due to the polarization, so

$$\phi(\mathbf{r}) = \phi_{\rho\sigma}(\mathbf{r}) + \phi_P(\mathbf{r})$$

$$= \int_V d\tau' \frac{[\rho(\mathbf{r}') - \boldsymbol{\nabla}'\cdot\mathbf{P}(\mathbf{r}')]}{|\mathbf{r} - \mathbf{r}'|}$$

$$+ \oint_S \frac{[\sigma dA' + \mathbf{P}(\mathbf{r}')\cdot d\mathbf{A}']}{|\mathbf{r} - \mathbf{r}'|}. \tag{6.5}$$

The corresponding electric field is given by

$$\mathbf{E}(\mathbf{r}) = -\boldsymbol{\nabla}\phi$$

$$= \int_V d\tau' \frac{[\rho(\mathbf{r}') - \boldsymbol{\nabla}'\cdot\mathbf{P}(\mathbf{r}')](\mathbf{r} - \mathbf{r}')}{|\mathbf{r} - \mathbf{r}'|^3}$$

$$+ \oint_S \frac{[\sigma dA' + \mathbf{P}(\mathbf{r}')\cdot d\mathbf{A}'](\mathbf{r} - \mathbf{r}')}{|\mathbf{r} - \mathbf{r}'|^3}. \tag{6.6}$$

Equations (6.5) and (6.6) show that the negative divergence of the polarization acts like an additional volume charge density, which we designate as

$$\rho_b(\mathbf{r}) = -\boldsymbol{\nabla}\cdot\mathbf{P}(\mathbf{r}), \tag{6.7}$$

and the normal component of the polarization at the surface of a dielectric acts like an additional surface charge density, which we designate as

$$\sigma_b(\mathbf{r}) = \hat{\mathbf{n}}\cdot\mathbf{P}(\mathbf{r}). \tag{6.8}$$

The subscript b refers to the fact that these effective charge densities are due to the **bound charge** in each molecule that has been displaced microscopically, producing the polarization. The usual charge densities $\rho(\mathbf{r})$ and $\sigma(\mathbf{r})$ are sometimes called ρ_{free} and σ_{free} because they are due to charges that can move about within the material, or at least are not bound to one particular molecule.

Although the charges designated as ρ_b and σ_b are not free to move about the material, they are real charge distributions. This can be seen in the exaggerated picture of polarized molecules in Fig. 6.1, which shows a polarization increasing toward the right. It can be seen on the figure that there is an excess of positive charge on the right face of the material, and an excess of negative charge in the body of the dielectric.

That $-\boldsymbol{\nabla}\cdot\mathbf{P}$ and $\hat{\mathbf{n}}\cdot\mathbf{P}$ correspond to bound charge densities can be seen by calculating the force on a neutral dialectric. This force is given by

$$\mathbf{F} = \int d\tau(\mathbf{P}\cdot\boldsymbol{\nabla})\mathbf{E}$$

$$= \int d\tau[\boldsymbol{\nabla}\cdot(\mathbf{P}\mathbf{E}) - \mathbf{E}(\boldsymbol{\nabla}\cdot\mathbf{P})]$$

$$= \oint (d\mathbf{S}\cdot\mathbf{P})\mathbf{E} - \int d\tau\mathbf{E}(\boldsymbol{\nabla}\cdot\mathbf{P}), \tag{6.9}$$

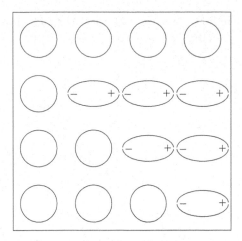

Figure 6.1: A dielectric with polarization increasing from left to right. There is excess positive charge on the right-hand surface, and excess negative charge in the body of the dielectric even though each charge is bound to its molecule.

which confirms the role of $-\boldsymbol{\nabla}\cdot\mathbf{P}$ and $\hat{\mathbf{n}}\cdot\mathbf{P}$ as bound charge densities.

6.2   The Displacement Vector D.

It can be seen from Eqs. (6.7) and (6.8) that the divergence of $\mathbf{E}$ will now be given by

$$\boldsymbol{\nabla}\cdot\mathbf{E} = 4\pi\rho - 4\pi\boldsymbol{\nabla}\cdot\mathbf{P}, \tag{6.10}$$

and the discontinuity in $\mathbf{E}$ across a surface between two regions by

$$\hat{\mathbf{n}}\cdot(\mathbf{E_2} - \mathbf{E_1}) = 4\pi[\sigma + \hat{\mathbf{n}}\cdot(\mathbf{P_1} - \mathbf{P_2})], \tag{6.11}$$

where $\hat{\mathbf{n}}$ is the normal unit vector from region 1 to region 2.

As a mathematical convenience, we define a vector $\mathbf{D}$, given by

$$\mathbf{D(r)} = \mathbf{E(r)} + 4\pi\mathbf{P(r)}. \tag{6.12}$$

$\mathbf{D}$ is called the **Displacement** Vector. Its introduction simplifies the divergence equation and the discontinuity relation, which can now be written as

$$\boldsymbol{\nabla}\cdot\mathbf{D(r)} = 4\pi\rho(\mathbf{r}) \tag{6.13}$$

and

$$\hat{\mathbf{n}}\cdot(\mathbf{D_2} - \mathbf{D_1}) = 4\pi\sigma. \tag{6.14}$$

It looks as if the polarization has disappeared from the equations, but it has just been hidden away in the definition of $\mathbf{D}$. Nothing useful can be done with Eqs. (6.13) and (6.14) unless there is some relation that specifies the polarization $\mathbf{P}$. Such a relation is called a **constitutive relation**. The simplest such relation is when $\mathbf{P}$ is proportional to $\mathbf{E}$

$$\mathbf{P}(\mathbf{r}) = \chi_e \, \mathbf{E}(\mathbf{r}). \tag{6.15}$$

The proportionality constant χ_e is called the **electric susceptibility**. The equation relating $\mathbf{D}$ to $\mathbf{E}$ is

$$\mathbf{D}(\mathbf{r}) = \epsilon \, \mathbf{E}(\mathbf{r}). \tag{6.16}$$

The constant

$$\epsilon = (1 + 4\pi\chi_e) \tag{6.17}$$

is called the **permittivity**. Note that in these Gaussian units, the vacuum's susceptibility is zero and its permittivity is one.

In SI units, Eqs. (6.10)-(6.17) are quite different. We briefly list some of them here:

$$\begin{align}
\mathbf{\nabla \cdot E}(\mathbf{r}) &= \frac{1}{\epsilon_0}[\rho(\mathbf{r}) - \mathbf{\nabla \cdot P}(\mathbf{r})], \quad \mathbf{SI} \tag{6.18}\\
\mathbf{D}(\mathbf{r}) &= \epsilon_0 \, \mathbf{E}(\mathbf{r}) + \mathbf{P}(\mathbf{r}), \quad \mathbf{SI} \tag{6.19}\\
\mathbf{\nabla \cdot D}(\mathbf{r}) &= \rho(\mathbf{r}), \quad \mathbf{SI} \tag{6.20}\\
\mathbf{P}(\mathbf{r}) &= \epsilon_0 \, \chi_e \, \mathbf{E}(\mathbf{r}), \quad \mathbf{SI} \tag{6.21}\\
\mathbf{D}(\mathbf{r}) &= \epsilon \, \mathbf{E}(\mathbf{r}), \quad \mathbf{SI} \tag{6.22}\\
\epsilon &= \epsilon_0(1 + \chi_e), \quad \mathbf{SI}. \tag{6.23}
\end{align}$$

In SI units, another constant

$$\kappa = \frac{\epsilon}{\epsilon_0} = 1 + \chi_e \quad \mathbf{SI} \tag{6.24}$$

is introduced. κ is called the **dielectric constant**. Numerically, the dielectric constant κ in the SI system and the permittivity ϵ in the Gaussian system are the same dimensionless constant.

So far, we have just discussed the simple case where the permittivity ϵ is a constant number multiplying $\mathbf{E}$. There are more general possibilities for the permittivity, which we list here:

1. If the permittivity is just a constant number, the material is called a **simple material**.

2. The permittivity can depend on position as $\epsilon(\mathbf{r})$. This is called an **inhomogeneous** material.

3. The permittivity can depend on E as $\epsilon(E)$, so D is no longer proportional to E. This is called a **non-linear** material.

4. The dependence of D on E can be so non-linear, or not even single valued, that a permittivity cannot be usefully defined. This is called a **ferroelectric** material, in analogy with ferromagnetism.

5. The permittivity may be a dyadic $[\epsilon]$, so

$$D = [\epsilon] \cdot E. \qquad (6.25)$$

Then **D** and **E** will generally not be in the same direction. This is called a **non-isotropic** material. Under rotations, $[\epsilon]$ transforms as a second rank tensor (See Chapter 14.), so it is usually called the **permittivity tensor** or the **dielectric tensor**.

6. The permittivity may depend on the frequency as $\epsilon(\omega)$. This is called a **dispersive** material.

The most general permittivity would be written as $[\epsilon(r, E, \omega)]$. A simple material could also be called a homogeneous, linear, isotropic, non-dispersive material. We will discuss some of these properties of the permittivity in more detail later in the text.

Although the introduction of **D** can simplify calculations, it is only **E** that has a physical significance because **D** arises from only part of the charge distribution. In polarizable matter, the potential ϕ is still related to **E** by the relations in Section 1.3. But, it is only for the case of a simple material that the potential satisfies Poisson's equation

$$\nabla^2 \phi(\mathbf{r}) = -\frac{4\pi}{\epsilon} \rho(\mathbf{r}). \qquad (6.26)$$

For a general permittivity, the differential equation for the potential is more complicated

$$\nabla \cdot \{[\epsilon(r, E(r))] \cdot \nabla \phi(\mathbf{r})\} = -4\pi \rho(\mathbf{r}). \qquad (6.27)$$

6.3 Uniqeness Theorem with Polarization

The basic differential equations of electrostatics in a polarizable material are

$$\nabla \cdot \mathbf{D} = 4\pi \rho \qquad (6.28)$$
$$\nabla \times \mathbf{E} = 0. \qquad (6.29)$$

There also must be a constitutive relation between **D** and **E**, and boundary conditions on **D** and **E**. We now prove that the solutions to the **D** and **E** equations are unique for certain classes of constitutive relation and appropriate boundary conditions. If uniqeness holds, Eqs. (6.28) and (6.29) can be considered an independent starting point for the electrostatics of polarizable matter.

The proof of uniqueness follows the general procedure used in Section 2.1.1. We start by assuming two different sets of solutions, $\mathbf{D_1}\,\mathbf{E_1}$ and $\mathbf{D_2}\,\mathbf{E_2}$. Their differences

$$\mathcal{D} \;=\; \mathbf{D_1} - \mathbf{D_2} \tag{6.30}$$

$$\mathcal{E} \;=\; \mathbf{E_1} - \mathbf{E_2} \tag{6.31}$$

will satisfy the homogeneous equations

$$\boldsymbol{\nabla}\!\cdot\!\mathcal{D} \;=\; 0 \tag{6.32}$$

$$\boldsymbol{\nabla}\!\times\!\mathcal{E} \;=\; 0. \tag{6.33}$$

Since the curls of $\mathbf{E_1}$, $\mathbf{E_2}$, $\mathcal{E}$ all vanish, we can write

$$\mathbf{E_1} = -\boldsymbol{\nabla}\phi_1, \quad \mathbf{E_2} = -\boldsymbol{\nabla}\phi_2, \quad \mathcal{E} = -\boldsymbol{\nabla}\psi, \tag{6.34}$$

$$\psi = \phi_1 - \phi_2. \tag{6.35}$$

We apply the divergence theorem to the combination $\psi\mathcal{D}$, leading to

$$\int_V \boldsymbol{\nabla}\!\cdot\!(\psi\mathcal{D})d\tau \;=\; \int_V (\mathcal{D}\!\cdot\!\boldsymbol{\nabla}\psi + \psi\boldsymbol{\nabla}\!\cdot\!\mathcal{D})d\tau \tag{6.36}$$

$$\oint_S \psi\mathcal{D}\!\cdot\!\mathbf{dA} \;=\; -\int_V \mathcal{D}\!\cdot\!\mathcal{E}\,d\tau. \tag{6.37}$$

In deriving Eq. (6.37), we have used the divergence theorem on the left hand side of Eq. (6.36), and used the fact that $\boldsymbol{\nabla}\!\cdot\!\mathcal{D}=0$ on the right hand side. The surface integral in Eq. (6.37) will vanish if either ψ or $\hat{\mathbf{n}}\!\cdot\!\mathcal{D}$ ($\hat{\mathbf{n}}$ is the normal to the surface area.) vanish on each bounding surface. So, suitable boundary conditions for unique solutions are

$$\textbf{Dirichlet BC}: \quad \phi \quad \text{given on the surface} \tag{6.38}$$

$$\textbf{Neumann BC}: \quad \hat{\mathbf{n}}\!\cdot\!\mathbf{D} \quad \text{given on the surface.} \tag{6.39}$$

For either of these boundary conditions, the volume integral of $\mathcal{D}\!\cdot\!\mathcal{E}$ will vanish. But this does not necessarily imply that $\mathcal{D}$ and $\mathcal{E}$ must both vanish. In the case of a non-isotropic material, the permittivity dyadic $[\epsilon]$ must be 'positive definite'. This means that, taken as a matrix, its three eigenvalues must all be real and positive. Then, in the system in which $[\epsilon]$ is diagonal, it does follow that the vanishing of the volume integral requires that both $\mathcal{D}$ and $\mathcal{E}$ must be zero everywhere.

This completes the proof of the uniqueness theorem in polarizable material. If the permittivity is not positive definite, then the uniqueness theorem does not hold. The uniqeness theorem also does not generally apply for ferroelectrics, where $\mathbf{D}$ and $\mathbf{E}$ can be unconnected.

6.4 Boundary Value Problems with Polarization.

6.4.1 Boundary conditions on D, E, and ϕ

The differential equations (6.28) and (6.29) are equivalent, using the divergence theorem and Stokes' theorem, respectively, to Gauss's law for **D**

$$\oint_S \mathbf{D} \cdot \mathbf{dA} = 4\pi Q_{\text{enclosed}}, \tag{6.40}$$

and to **E** being a conservative field

$$\oint_C \mathbf{dr} \cdot \mathbf{E} = 0. \tag{6.41}$$

These integral relations can be used to specify the boundary conditions on **D** and **E** at the interface between two different materials where the permittivity can change discontinuously. Such an interface is shown in Fig. 6.2, with fields $\mathbf{D_1}$ and $\mathbf{E_1}$ in region 1 with permittivity ϵ_1, and fields $\mathbf{D_2}$ and $\mathbf{E_2}$ in region 2 with permittivity ϵ_2. The unit vector $\hat{\mathbf{n}}$ is perpendicular to the interface in the direction from region 1 to region 2. In general, there can be a surface charge σ on the interface.

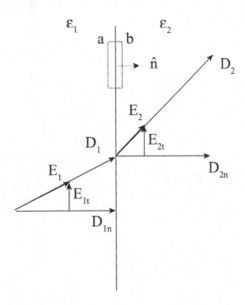

Figure 6.2: Interface between two different materials. The case shown has $\sigma = 0$, so D_n is continuous across the boundary.

Applying Gauss's law for **D** to a pillbox enclosing an area ΔA on the interface results in

$$\hat{\mathbf{n}} \cdot (\mathbf{D_2} - \mathbf{D_1}) \Delta A = 4\pi\sigma \Delta A,$$

$$\hat{n}\cdot(\mathbf{D_2} - \mathbf{D_1}) \;=\; 4\pi\sigma. \tag{6.42}$$

If there is no free surface charge, this means that D_n (for $D_{\text{normal}} = \hat{n}\cdot\mathbf{D}$) is continuous across the interface.

Applying Eq. (6.41) to the contour shown on Fig. (6.2) (dashed line) gives

$$\hat{n}\times(\mathbf{E_1} - \mathbf{E_2}) = 0. \tag{6.43}$$

This means that $\mathbf{E_t}$ (for $\mathbf{E}_{\text{tangential}} = \hat{n}\times\mathbf{E}$) is continuous across the interface. We can also get a condition on the potential at the interface by integrating along one of the short legs of the rectangular contour. This gives

$$\phi_2 - \phi_1 = -\int_a^b \mathbf{E}\cdot d\mathbf{r} = 0, \tag{6.44}$$

in the limit as the distance between a and b goes to zero. This means the the potential ϕ is continuous across the interface. Either the continuity of $\mathbf{E_t}$ or of ϕ across the interface may be used, depending on which is more convenient.

6.4.2   Needle or Lamina.

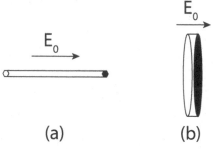

Figure 6.3: Needle (a) and lamina (b) in a uniform electric field.

For a dielectric needle of permittivity ϵ, aligned parallel to an electric field $\mathbf{E_0}$ as shown in Fig. 6.3a, the boundary condition that determines the fields inside the needle is that on E_t, since the end effects will be negligible. Therefore, $\mathbf{E}$ inside the needle will be the same as the external field, while $\mathbf{D}$ will equal $\epsilon\mathbf{E_0}$.

$$\text{Needle}: \qquad \mathbf{E} = \mathbf{E_0}, \quad \mathbf{D} = \epsilon\mathbf{E_0}. \tag{6.45}$$

The fields inside a dielectric lamina placed perpendicular to $\mathbf{E_0}$, as in Fig. 6.3b, will be determined by the boundary condition on D_n. Therefore, $\mathbf{D}$ inside will equal the external $\mathbf{E}$ field, and $\mathbf{E}$ inside the lamina will equal $\mathbf{E_0}/\epsilon$:

$$\text{Lamina}: \qquad \mathbf{E} = \mathbf{E_0}/\epsilon, \quad \mathbf{D} = \mathbf{E_0}. \tag{6.46}$$

These results can be directly applied to a long thin, needle-like cavity or a large flat, disc-like cavity in a dielectric. The only difference for these cavities would be that the ϵ in Eqs. (6.45) and (6.46) would be replaced by $1/\epsilon$.

6.4.3 Capacitance

A parallel plate capacitor is shown in Fig. 6.4. It is composed of two large flat conducting plates of Area A, a short distance d apart, with $A \gg d^2$. A dielectric with permittivity ϵ fills the volume between the plates. A charge $+Q$ is on one plate and $-Q$ on the other. Because of the geometry, the field lines $\mathbf{E}$ and $\mathbf{D}$ are parallel lines perpendicular to the plates, and the charge is evenly distributed on the plates. We are neglecting end effects.

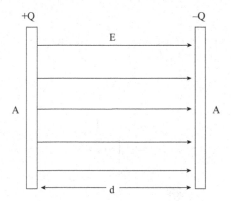

Figure 6.4: Parallel plate capacitor.

From Gauss's law at the surface of either plate

$$D = 4\pi\sigma = 4\pi Q/A. \tag{6.47}$$

Since the lines of $\mathbf{E}$ are straight lines going from one plate to the other, the potential difference on the plates is

$$V = Ed = Dd/\epsilon = \frac{4\pi d}{\epsilon A}Q. \tag{6.48}$$

This shows that the potential difference V is proportional to Q. The constant of proportionality is called the **capacitance** C. For a parallel plate capacitor, the capacitance is given by

$$C = \frac{Q}{V} = \frac{\epsilon A}{4\pi d}. \tag{6.49}$$

For any configuration of two conductors, the ratio $\frac{Q}{V}$ is constant, and defines the capacitance of the system. Calculating the capacitance of more complicated systems is given as problems. In Gaussian units, the capacitance has the unit of cm, although the unit **statfarad** is also used.

The corresponding equation for the capacitance of a parallel plate capacitor in SI units is

$$C = \frac{Q}{V} = \frac{\epsilon A}{d} \quad \textbf{SI}. \tag{6.50}$$

In SI units, ϵ has complicated units, and the new unit **farad** is introduced for capacitance. The farad can be related to the esu unit for capacitance by

$$1\,\text{farad} = \frac{1\,\text{coulomb}}{1\,\text{volt}} = \frac{3 \times 10^9 \text{statcoulomb}}{\frac{1}{300}\text{statvolt}} = 9 \times 10^{11}\,\text{statfarad}. \quad (6.51)$$

A farad turns out to be a huge unit, so laboratory capacitances are typically given in μf (microfarads).

6.4.4 Images

We consider the case of a point charge q, a distance d from a large, simple dielectric with a flat face, as in Fig. 6.5.

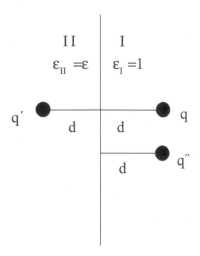

Figure 6.5: A point charge q a distance d from a flat dielectric with permittivity ϵ.

We want to find the potential and fields for region I to the right of the dielectric with $\epsilon_I = 1$, and region II inside the dielectric where the permittivity is $\epsilon_{II} = \epsilon$.

To find the potential in region I, we replace the polarized dielectric by a single image charge q'. As an educated guess, we place the image charge a distance d to the left. Then, the surface of the dielectric will be equidistant from the original charge and the image charge. To find the potential in region II, we consider all space to have to have permittivity ϵ, and place an image charge q'' at the location of the original charge.

Then, we can write for the potential in each region

$$\text{I}: \quad \phi_I(\mathbf{r}) = \frac{q}{|\mathbf{r}-\mathbf{d}|} + \frac{q'}{|\mathbf{r}+\mathbf{d}|} \quad (6.52)$$

$$\text{II}: \quad \phi_{II}(\mathbf{r}) = \frac{q''}{\epsilon|\mathbf{r}-\mathbf{d}|}. \quad (6.53)$$

These simple forms for the potential are only possible for a simple material, where ϵ is a constant number. We assume such a material for this problem and the following boundary value problems in this section.

The continuity of ϕ at the interface where all the denominators are equal requires

$$q + q' = \frac{q''}{\epsilon}. \tag{6.54}$$

The continuity of D_n at the interface means that $\epsilon \hat{\mathbf{d}} \cdot \boldsymbol{\nabla} \phi$ is continuous. This gives

$$q - q' = q''. \tag{6.55}$$

Because of the location of the image charges, the boundary conditions hold for all points on the surface of the dielectric.

The solutions to Eqs. (6.54) and (6.55) are

$$q' = \frac{1 - \epsilon}{1 + \epsilon} q \tag{6.56}$$

$$q'' = \frac{2\epsilon}{1 + \epsilon} q. \tag{6.57}$$

The potential will be given in each region by Eqs. (6.52) and (6.53) with q' and q'' given by Eqs. (6.57) and (6.57).

There is a force on the point charge equal to the Coulomb force between it and the image charge q':

$$\mathbf{F} = \frac{q q' \hat{\mathbf{d}}}{4 d^2} = \left[\frac{1 - \epsilon}{1 + \epsilon} \right] \frac{q^2 \hat{\mathbf{d}}}{4 d^2}. \tag{6.58}$$

This force will be attractive for the usual case of $\epsilon > 1$. By Newton's 3rd Law, there is an equal and opposite force on the dielectric. This induced force is the familiar force that is often our first experience with static electricity. The teacher combed her hair with a hard rubber comb, and then used the comb to make torn bits of paper dance, without even touching them. You may have wondered at the time (I did.) how the paper could be picked up even though it had no charge. This is how. Although uncharged, it had become polarized.

An interesting feature of Eq. (6.57) for the image charge behind the surface, and Eq. (6.58) for the force, is that taking the limit $\epsilon \to \infty$ reproduces the results we found in Chapter 2 for conductors. It is easy to see why this happens. If ϵ becomes infinite for finite $\mathbf{D}$, then $\mathbf{E}$ inside the material will approach zero. Although this is not the physical reason why $\mathbf{E} = \mathbf{0}$ inside conductors, it can be used as a mathematical trick to reduce any dielectric result to the equivalent result for a conductor. This is not the easiest way to solve problems with conductors, but it can serve as a quick check on the solution to any dielectric problem.

6.4.5   Dielectric sphere in a uniform electric field.

We next consider a simple dielectric sphere of radius R placed into a uniform electric field E_0, as shown in Fig. 6.6.

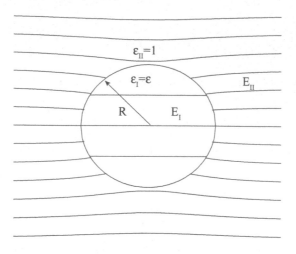

Figure 6.6: Lines of force for a dielectric sphere in a uniform electric field.

We take the field in the z-direction, so the potential (before the introduction of the sphere) is given by

$$\phi_0 = -E_0 r \cos\theta. \tag{6.59}$$

With the sphere in place, the potential satisfies Laplace's equation in two separate regions,

$$\text{I}: \quad r < R$$
$$\text{II}: \quad r > R.$$

It can be expanded in Legendre polynomials in each region:

$$\phi_{\text{I}}(r,\theta) = \sum_l a_l r^l P_l(\cos\theta) - E_0 r P_1(\cos\theta), \quad r < R \tag{6.60}$$

$$\phi_{\text{II}}(r,\theta) = \sum_l \frac{b_l}{r^{l+1}} P_l(\cos\theta) - E_0 r P_1(\cos\theta), \quad r > R \tag{6.61}$$

We have written the potential due to the external field and that due to the polarized sphere separately.

The continuity of ϕ at $r = R$ requires

$$b_l = R^{(2l+1)} a_l. \tag{6.62}$$

The boundary condition that D be continuous at $r = R$, requires

$$\epsilon \partial_r \phi_I(r,\theta)|_{r=R} = \partial_r \phi_{II}(r,\theta)|_{r=R}. \tag{6.63}$$

This leads to

$$\sum_l \epsilon l a_l R^{(l-1)} P_l(\cos\theta) - \epsilon E_0 P_1(\cos\theta) =$$
$$-\sum_l (l+1) a_l R^{(l-1)} P_l(\cos\theta) - E_0 P_1(\cos\theta). \tag{6.64}$$

The expansion in the P_l is unique, so Eq. (6.64) must hold for each l separately. This requires that all the coefficients a_l must be zero except for a_1, which is given by setting $l = 1$ in the sum. This gives

$$a_1 = \left[\frac{\epsilon-1}{\epsilon+2}\right] E_0. \tag{6.65}$$

Substituting this into Eqs. (6.60) and (6.61), the potential is given by

$$\phi_\mathrm{I}(r,\theta) = -\frac{3E_0 r \cos\theta}{\epsilon+2}, \quad r < R \tag{6.66}$$
$$\phi_\mathrm{II}(r,\theta) = \left[\frac{\epsilon-1}{\epsilon+2}\right]\frac{E_0 R^3}{r^2}\cos\theta - E_0 r\cos\theta, \quad r > R. \tag{6.67}$$

The electric field inside the dielectric sphere is uniform, and is given by

$$\mathbf{E}_\mathrm{I}(\mathbf{r}) = \frac{3}{\epsilon+2}\mathbf{E}_0, \quad r < R. \tag{6.68}$$

Outside the sphere, the potential and electric field due to the polarized sphere is just that of a dipole

$$\mathbf{p} = \frac{\epsilon-1}{\epsilon+2}\mathbf{E}_0 R^3. \tag{6.69}$$

The electric field outside the sphere is the original uniform field plus the field due to the induced dipole

$$\mathbf{E}_\mathrm{II}(\mathbf{r}) = \frac{3(\mathbf{p}\cdot\hat{\mathbf{r}})\hat{\mathbf{r}} - \mathbf{p}}{r^3} + \mathbf{E}_0, \quad r > R. \tag{6.70}$$

Lines of force for these electric fields are shown in Fig. (6.6).

Note that the above results for a dielectric sphere in a uniform electric field will also hold for the nonlinear case where ϵ is a function of E. This is because $\mathbf{E}$ turns out to be a constant vector inside the sphere, so the potential ϕ will still satisfy Laplace's equation.

6.4.6 Dielectric sphere and point charge.

A point charge q is a distance d from the center of a dielectric sphere of radius R, as shown in Fig. 6.7.

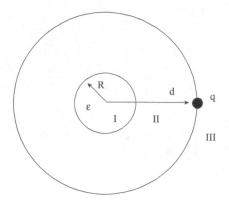

Figure 6.7: A point charge a distance d outside a dielectric sphere of radius R.

The potential for this case satisfies Laplace's equation in three separate regions:

$$
\begin{aligned}
\text{I}: &\quad r < R \\
\text{II}: &\quad R < r < d \\
\text{III}: &\quad r > d.
\end{aligned}
$$

We expand the potential in Legendre polynomials separately for each region.

$$
\phi_I(r,\theta) \;=\; \sum_l a_l r^l P_l(\cos\theta) + \sum_l \frac{qr^l}{d^{l+1}} P_l(\cos\theta), \quad r < R \tag{6.71}
$$

$$
\phi_{II}(r,\theta) \;=\; \sum_l \frac{b_l}{r^{l+1}} P_l(\cos\theta) + \sum_l \frac{qr^l}{d^{l+1}} P_l(\cos\theta), \quad R < r < d \tag{6.72}
$$

$$
\phi_{III}(r,\theta) \;=\; \sum_l \frac{b_l}{r^{l+1}} P_l(\cos\theta) + \sum_l \frac{qd^l}{r^{l+1}} P_l(\cos\theta), \quad r > d. \tag{6.73}
$$

We have written the Legendre sums separately for the potential due to the sphere and that due to the point charge. Note that we have expanded the potential due to the point charge in Legendre polynomials as in Eq. (4.122).

We can solve for the coefficients a_l and b_l by using the boundary conditions on ϕ and $\mathbf{D}$ at the surface of the sphere, $r = R$. The continuity of ϕ gives

$$
a_l R^l = \frac{b_l}{R^{l+1}}, \tag{6.74}
$$

so

$$
b_l = a_l R^{2l+1}. \tag{6.75}
$$

We have used the uniqueness of the Legendre polynomial expansion to equate the terms for each l separately. The continuity of D_n at $r = R$ means that

$$
\epsilon \partial_r \phi_I(r,\theta)|_{r=R} = \partial_r \phi_{II}(r,\theta)|_{r=R}. \tag{6.76}
$$

This gives

$$\epsilon l a_l R^{l-1} + \frac{\epsilon l q R^{l-1}}{d^{l+1}} = \frac{-(l+1)b_l}{R^{l+2}} + \frac{lqR^{l-1}}{d^{l+1}}. \tag{6.77}$$

Equations (6.75) and (6.77) can be solved for a_l, giving

$$a_l = \frac{(1-\epsilon)lq}{(\epsilon l + l + 1)d^{l+1}}. \tag{6.78}$$

The potential is given by Eqs. (6.71)-(6.73), with a_l given by Eq. (6.78) and b_l given by Eq. (6.75).

There will be a force on the point charge due to the electric field of the polarized sphere, given by

$$\begin{aligned} F_q &= -q\partial_r\phi_{\text{sphere}}(r,0)|_{r=d} \\ &= \sum_l \frac{l(l+1)(1-\epsilon)q^2}{(\epsilon l + l + 1)d^2}\left(\frac{R}{d}\right)^{2l+1}. \end{aligned} \tag{6.79}$$

This is an attractive force for $\epsilon > 1$. There will be an equal force attracting the sphere to the point charge. This is the force you observe when you bring a charged rubber rod close to a hanging neutral pith ball in demonstrating static electricity.

At large distances, the $l = 1$ dipole term will dominate, and the force will be

$$F_q = \left(\frac{1-\epsilon}{\epsilon+2}\right)\frac{2q^2R^3}{d^5}. \tag{6.80}$$

This is just the force between a point charge and an induced dipole

$$\mathbf{p} = \left(\frac{1-\epsilon}{\epsilon+2}\right)\frac{qR^3}{d^2}\hat{\mathbf{d}}. \tag{6.81}$$

The force between a point charge and a dipole ordinarily falls like $\frac{1}{d^3}$, but we see here that the force between a point charge and an induced dipole falls like $\frac{1}{d^5}$. This is because the induced dipole moment has a $\frac{1}{d^2}$ dependence.

6.5 Induced Dipole-dipole force, the Van der Waals Force

The force in Eq. (6.80) between a point charge and the induced dipole moment of a dielectric sphere could have been found somewhat more easily from the earlier result for a sphere in a uniform electric field, by noticing that the electric field of the point charge at the sphere is $E = \frac{q}{d^2}$. Then the induced dipole moment of the sphere given by Eq. (6.81) is the same as that given by Eq. (6.69) for a dielectric sphere in this external electric field. Since we are just interested in the dipole moment of the sphere, we can neglect the variation of the external field, and can consider it as a uniform field in the region of the sphere.

We can use this reasoning to find the dipole-dipole force between a dipole p_0 and the induced dipole moment of a neutral dielectric sphere. The electric field at the sphere, due to the dipole a distance $\mathbf{d}$ from the center of the sphere, is

$$\mathbf{E_0} = \frac{3(\mathbf{p_0}\cdot\hat{\mathbf{d}})\hat{\mathbf{d}} - \mathbf{p_0}}{d^3}. \tag{6.82}$$

This will induce a dipole moment $\mathbf{p}$ in the sphere, given by

$$\begin{aligned}
\mathbf{p} &= \left(\frac{\epsilon-1}{\epsilon+2}\right)\mathbf{E_0}R^3 \\
&= \left(\frac{\epsilon-1}{\epsilon+2}\right)\frac{R^3}{d^3}[3(\mathbf{p_0}\cdot\hat{\mathbf{d}})\hat{\mathbf{d}} - \mathbf{p_0}].
\end{aligned} \tag{6.83}$$

Substituting the dipole moments $\mathbf{p}$ and $\mathbf{p_0}$ into Eq. (2.62) for the force between two dipoles gives (after some algebra)

$$\mathbf{F} = -3\left(\frac{\epsilon-1}{\epsilon+2}\right)\frac{R^3}{d^7}[4(\mathbf{p_0}\cdot\hat{\mathbf{d}})^2\hat{\mathbf{d}} + p_0^2\hat{\mathbf{d}} - (\mathbf{p_0}\cdot\hat{\mathbf{d}})\mathbf{p_0}]. \tag{6.84}$$

This is the force on a neutral dipole $\mathbf{p_0}$ a distance $\mathbf{d}$ from a neutral dielectric sphere of radius R (for $d>>R$). The force is non-central for a general orientation, but angular momentum is still conserved when the spin torque on the dipole is included. (See problem 2.5.) Averaging the force over all directions of $\mathbf{p_0}$ gives the average force for unpolarized dipoles

$$\mathbf{F} = -6p_0^2\left(\frac{\epsilon-1}{\epsilon+2}\right)\frac{R^3}{d^7}\hat{\mathbf{d}}. \tag{6.85}$$

The force given in Eq. (6.85) falls off with distance like d^{-7}. This is the characteristic spatial behavior of the **Van der Waals force** between neutral, but polarizable molecules. (An understanding of the actual details of the physical force requires quantum mechanics, but this classical treatment shows the general features of the force.) The d^{-7} comes from two sources. The dipole-dipole force goes like d^{-4}, while the induced dipole is proportional to the electric field due to the first dipole, which goes like d^{-3}.

6.6   Molecular Polarizability

6.6.1   Microscopic electric field

In order to investigate the polarizability of a molecule, we have to determine the local electric field that acts on the molecule. This is not the same as the field we have been calculating by averaging over a small volume of matter ΔV, as in Eq. (1.15) or Eq. (6.2). That average was purposely chosen to smooth out the sharp variation in the actual electric field from molecule to molecule. The resulting field (that we have been using up to now) is called the **macroscopic electric field (E_M)**. This field would act on a charge probe that may be small, but would still be composed of many molecules. The field that acts on an individual molecule must be a **microscopic electric field (E_μ)** that includes the separate contribution of each individual charge.

It would be too hard to include all of the huge number of charges individually, so we break the problem of finding the microscopic field into two parts. We can write

$$E_\mu = E_{near} + E_{far}, \tag{6.86}$$

where E_{near} comes from those molecules within a radius R, and E_{far} comes from everything beyond the radius R. (See Fig. 6.8.)

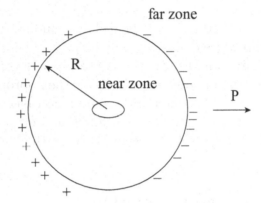

Figure 6.8: Near and far zones for microscopic electric field. The polarization in the far zone produces surface charges at the radius R.

The radius R is chosen to include a large number of molecules, but to be small enough that the macroscopic field E_M can be assumed constant within the sphere of radius R.

For most materials the symmetry of the charges in the near region is such that the field E_{near} vanishes. For instance, the fields due to a spherically symmetric collection of similarly polarized dipoles about the molecule vanishes. This is because

$$\oint d\Omega E_P(r) = \oint d\Omega \frac{[3(p \cdot \hat{r})\hat{r} - p]}{r^3} = 0. \tag{6.87}$$

This is also true for most isotropic materials, and we will assume that this near field is zero for the following derivation.

The remaining electric field at the molecule is due to everything beyond the radius R. The only change in this field is the fact that there is now a surface charge σ_b on the surface at radius R. This surface charge density is given by

$$\sigma_b = -\hat{r} \cdot P. \tag{6.88}$$

The $-\hat{\mathbf{r}}$ in Eq. (6.88) is the normal vector out of the dielectric at the surface of the sphere. The electric field due to this bound surface charge should be added to the original macroscopic field $\mathbf{E_M}$, so the microscopic electric field at the molecule at the center of the sphere is given by

$$
\begin{aligned}
\boldsymbol{E}_\mu(0) &= \mathbf{E_M} - \oint \frac{\sigma(\mathbf{r})\hat{\mathbf{r}}\,dA}{R^2} \\
&= \mathbf{E_M} + \oint d\Omega\,\hat{\mathbf{r}}\hat{\mathbf{r}}\cdot\mathbf{P} \\
&= \mathbf{E_M} + \frac{4\pi}{3}\mathbf{P}.
\end{aligned}
\tag{6.89}
$$

In doing the angular integral, we used the fact that the radius R is small enough that the polarization $\mathbf{P}$ can be assumed constant on the surface.

We see that the actual microscopic field acting on a single molecule is stronger by $\frac{4\pi}{3}\mathbf{P}$ than the macroscopic field $\mathbf{E_M}$. Note that the microscopic field $\mathbf{E}_\mu$ is not the same as the electric field that would exist inside a cavity of radius R in the material. (See problem 6.9)

Although the microscopic field is the field that acts on a single molecule fixed in the material, the macroscopic field $\mathbf{E_M}$ is the appropriate field in most other cases. For instance, the force on a conduction electron (see Sec. 6.7) should be given by $-e\mathbf{E_M}$. Although this force acts on a point charge, the conduction electron moves through the material so the averaged field, given by $\mathbf{E_M}$, should be used.

6.6.2   Clausius-Mossotti relation

We can now relate the polarization $\mathbf{P}$ to the polarizability of the molecules in the dielectric. The **polarizability** α is defined so that the dipole moment of the molecule is given by

$$
\mathbf{p} = \alpha\boldsymbol{E}_\mu.
\tag{6.90}
$$

We recall that the polarization vector corresponds to the dipole moment per unit volume, so

$$
\mathbf{P} = n\,\mathbf{p},
\tag{6.91}
$$

where n is the number of molecules per unit volume (number density). Then we can write

$$
\begin{aligned}
\mathbf{P} &= n\,\alpha\mathbf{E}_\mu \\
\mathbf{P} &= n\,\alpha\left(\mathbf{E_M} + \frac{4\pi}{3}\mathbf{P}\right) \\
\mathbf{P} &= \frac{n\,\alpha\mathbf{E_M}}{1 - \frac{4\pi}{3}n\,\alpha}.
\end{aligned}
\tag{6.92}
$$

Using the relation

$$
\mathbf{P} = \chi_e\,\mathbf{E_M},
\tag{6.93}
$$

results in

$$\chi_e = \frac{n\,\alpha}{1 - \frac{4\pi}{3}n\,\alpha}, \tag{6.94}$$

giving the susceptibility in terms of the molecular polarizability. For a material with several different kinds of molecule, this result for a single type of molecule would be summed over each molecule using the number density n_i corresponding to molecules of type i.

We can solve Eq. (6.94) for α, giving

$$\alpha = \frac{\chi_e}{n(1 + \frac{4\pi}{3}\chi_e)}, \tag{6.95}$$

for the molecular polarizability in terms of the susceptibility. Equation (6.95) is usually written in terms of the permittivity as

$$\alpha = \frac{3(\epsilon - 1)}{4\pi n(\epsilon + 2)}. \tag{6.96}$$

In this form, it is called the **Clausius-Mossotti Relation** giving the molecular polarization in terms of the number density and the permittivity.

6.6.3 Models for molecular polarization

There are two common cases for which an electric field produces polarization in a dielectric. The electric field can either produce a polarizing distortion of a molecule, or the field can orient polar molecules (molecules that have permanent electric dipole moments) preferentially along the direction of the field. We will look at each of these mechanisms.

Molecules are held together by internal forces between the nuclei and electrons in the molecule. These force are themselves electrostatic, but we will consider them here as phenomenological forces that hold the system in equilibrium. An imposed electric field will tend to displace the charges within the molecule, resulting in an induced electric dipole moment. For small displacements, the restoring forces within the molecule will be linear elastic forces. For a simple case of a single bound electron of charge $-e$, Newton's force equation for equilibrium is

$$-e\mathbf{E} = m\omega_0^2\mathbf{r}. \tag{6.97}$$

In Eq. (6.97), we have written the restoring force constant in terms of the mass m of the electron and its natural angular frequency of oscillation ω_0. Note that in this section the electric field $\mathbf{E}$ represents the microscopic field that acts on each molecule. (For the remainder of the book, the symbol $\mathbf{E}$ will refere to the macroscopic electric field.) The electric dipole moment induced by the displacement in $\mathbf{r}$ is

$$\mathbf{p} = -e\mathbf{r} = \frac{e^2}{m\omega_0^2}\mathbf{E}. \tag{6.98}$$

For a more complex molecule, m would be the effective mass of a normal mode of oscillation having angular frequency ω_0.

From Eq. (6.98) we see that the polarizability of this molecule is given by

$$\alpha = \frac{e^2}{m\omega_0^2}. \tag{6.99}$$

The electric susceptibility or permittivity can then be given using Eq. (6.94) or Eq. (6.96), respectively.

Polar molecules, in the absence of an external electric field, will tend to be randomly oriented, so the system will have no net electric dipole moment. In an imposed electric field $\mathbf{E}$, the molecules will tend to align themselves along the direction of the field. The extent of this alignment will be determined by the Boltzmann distribution of molecules of varying energy

$$\mathcal{P}(U) = Ne^{-\frac{U}{kT}}. \tag{6.100}$$

The distribution $\mathcal{P}(U)dU$ is the probability of finding a molecule with energy between U and $U + dU$. The coefficent N is a normalization constant chosen to make the total probability unity, k is Boltzmann's constant, and T is the Kelvin temperature.

The energy of polar molecules in the electric field is just that of electric dipoles p_0

$$U = -\mathbf{p_0} \cdot \mathbf{E}. \tag{6.101}$$

The average dipole moment of this distribution, averaged of all angles of $\mathbf{p_0}$, using the Boltzmann energy distribution, is

$$\langle \mathbf{p} \rangle = N \oint \mathbf{p_0} \, \exp\left[\frac{\mathbf{p_0} \cdot \mathbf{E}}{kT}\right] d\Omega_{\mathbf{p_0}} \tag{6.102}$$

The magnitude of the energy U is usually small compare to kT, so the exponential can be expanded to first order, simplifying the integrals with the result

$$\langle \mathbf{p} \rangle = \frac{p_0^2}{3kT}\mathbf{E}. \tag{6.103}$$

This average polarization can be used to define an effective molecular polarizability

$$\alpha = \frac{p_0^2}{3kT}. \tag{6.104}$$

Thus polar molecules will have a temperature dependent permittivity resulting from the temperature dependence of the polarizability.

We see that these simple models for molecular polarizability predict a permittivity that depends on either the natural frequency of oscillation of the molecules or on the temperature of the material. Some materials have both of these characteristics.

6.7 Electrostatic energy in dielectrics

We can determine the electrostatic energy in the presence of dielectrics by calculating the work required to assemble a collection of charges and dielectrics. We follow the derivation in Sec. 2.2, but now with dielectrics included.

Before introducing any charge, we assemble the dielectrics (without their charge) in the final configuration. Then we introduce the charge (including any free charge on dielectrics) as in Sec. 1.8. This leads to the same energy as in Eq. (2.14)

$$U = \int d\tau \sum_n \phi'_n \delta\rho'_n \to \int d\tau \int_0^{\rho(\mathbf{r})} \phi' d\rho'. \tag{6.105}$$

In order to proceed further, we have to assume that the dielectric is linear. Then, any fractional increase in ϕ' is equal to the fractional increase in ρ', so

$$\frac{\delta\phi'}{\phi'} = \frac{\delta\rho'}{\rho'}, \tag{6.106}$$

and

$$\rho'\delta\phi' = \phi'\delta\rho'. \tag{6.107}$$

Then, it follows that

$$\delta(\rho'\phi') = \rho'\delta\phi' + \phi'\delta\rho' = 2\phi'\delta\rho', \tag{6.108}$$

as in Sec. 1.8. This leads to the same final result as in Sec. 1.8

$$U = \frac{1}{2}\int d\tau \int_0^{\rho(\mathbf{r})\phi(\mathbf{r})} d(\rho'\phi') = \frac{1}{2}\int \rho\phi d\tau \tag{6.109}$$

for linear dielectrics. For non-linear dielectrics, Eq. (6.105) would have to be integrated to find U.

We can use Eq. (6.109) to find the energy stored in a capacitor. All the charge is on the conducting surfaces so

$$U = \frac{1}{2}QV. \tag{6.110}$$

Using the definition of capacitance, this energy can be written as

$$U = \frac{1}{2}CV^2 = \frac{Q^2}{2C}. \tag{6.111}$$

The electrostatic energy can also be found in terms of $\mathbf{D}$ and $\mathbf{E}$. We start with Eq. (6.105) and introduce $\mathbf{D}$ and $\mathbf{E}$

$$
\begin{aligned}
U &= \int d\tau \int_0^{\rho(\mathbf{r})} \phi' d\rho' \\
&= \frac{1}{4\pi}\int d\tau \int_0^{\mathbf{D}(\mathbf{r})} \phi' \boldsymbol{\nabla}\cdot(d\mathbf{D}') \\
&= \frac{1}{4\pi}\int d\tau \int_0^{\mathbf{D}(\mathbf{r})} [\boldsymbol{\nabla}\cdot(\phi' d\mathbf{D}') - (\boldsymbol{\nabla}\phi')\cdot d\mathbf{D}'] \\
&= \frac{1}{4\pi}\int d\tau \int_0^{\mathbf{D}(\mathbf{r})} \mathbf{E}'\cdot d\mathbf{D}'. \tag{6.112}
\end{aligned}
$$

The total divergence in Eq. (6.112) has been converted to a surface integral at infinity that vanishes for static fields.

For non-linear dielectrics, Eq. (6.112) must be used to find U. For a linear dielectric, we can write

$$\mathbf{E}' \cdot d\mathbf{D}' = (d\mathbf{E}') \cdot \mathbf{D}' = \frac{1}{2} d(\mathbf{E}' \cdot \mathbf{D}'). \tag{6.113}$$

Then

$$U = \frac{1}{8\pi} \int d\tau \int_0^{\mathbf{E}(\mathbf{r}) \cdot \mathbf{D}(\mathbf{r})} d(\mathbf{E}' \cdot \mathbf{D}') = \frac{1}{8\pi} \int d\tau \, \mathbf{E} \cdot \mathbf{D}. \tag{6.114}$$

Either Eq. (6.114) or Eq. (6.109) can be used to find the electrostatic energy.

6.8   Forces on dielectrics

We can find the force on a dielectric by considering a virtual infinitesimal displacement of a coordinate ξ of the dielectric. Then

$$\mathbf{F} = -\boldsymbol{\nabla}_\xi U|_\rho, \tag{6.115}$$

where $\boldsymbol{\nabla}_\xi U|_\rho$ is the gradient with respect to the coordinate ξ with all charge distributions (except any free charge fixed in the dielectric) held constant. This is necessary so that no work is done on these charge distributions during the displacement of the dielectric.

We now consider some examples of finding the force on a dielectric using Eq. (6.115). For a point charge q a distance x outside a large flat dielectric (Section 6.4.4), the energy is

$$U = \frac{1}{2} \int \rho(\mathbf{r}) \phi(\mathbf{r}) d\tau = \left[\frac{1-\epsilon}{1+\epsilon}\right] \frac{q^2}{4x}. \tag{6.116}$$

In Eq. (6.116), we have used the potential due to the image charge given by Eq. (6.57), and have used the charge density

$$\rho(\mathbf{r}) = q\delta(\mathbf{r} - \mathbf{x}). \tag{6.117}$$

We have been careful to include only the free charge in ρ, so the original charge q is the only charge that is included. Now, taking the gradient with respect to the variable x results in

$$\mathbf{F} = -\frac{d}{dx} \left\{ \left[\frac{1-\epsilon}{1+\epsilon}\right] \frac{q^2}{4x} \right\} = \left[\frac{1-\epsilon}{1+\epsilon}\right] \frac{q^2}{4x^2}, \tag{6.118}$$

which is the same force as given in Eq. (6.58).

Next we consider the parallel plate capacitor shown in Fig. 6.9, which has a dielectric extending a distance x into the space between the plates.

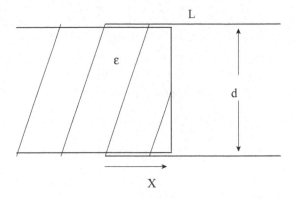

Figure 6.9: Dielectric being drawn into a parallel plate capacitor.

The capacitance is given by (See Problem 6.10.)

$$C(x) = \frac{L}{4\pi d}[L + (\epsilon - 1)x].$$ (6.119)

We write the energy as

$$U(x) = \frac{Q^2}{2C(x)}.$$ (6.120)

We use this form so that the charge Q can be held fixed. Taking the derivative of U with respect to x gives the force

$$F = \frac{(\epsilon - 1)L}{4\pi d} \left[\frac{Q}{C}\right]^2 = \frac{(\epsilon - 1)LV^2}{4\pi d}.$$ (6.121)

For $\epsilon > 1$, this force will be to the right in Fig. 6.9, tending to draw the dielectric into the capacitor.

The force on a dielectric can also be calculated by taking the gradient of the energy while holding the voltage on all conductors constant. But, in this case the sign of the derivative is positive, so

$$\mathbf{F} = +\boldsymbol{\nabla}_\xi U|_V.$$ (6.122)

This change of sign can be seen by the following argument. We consider a conductor with original voltage V and original charge Q. (For a set of conductors, the energies in the following steps would be summed over all the conductors.) We accomplish the infinitesimal displacement of the dielectric with fixed voltage in two steps:

In **step 1**, the dielectric is displaced holding all charges fixed. This leads to a change in energy we will call δU_1. During step 1, the voltage will have changed by an amount δV_1, and the change in energy is given by

$$\delta U_1 = Q\delta V_1.$$ (6.123)

In **step 2**, we hold the dielectric fixed, but bring the voltage back to its original value. This can be accomplished by attaching a battery of the appropriate voltage to the conductor. Then an infinitesimal current will flow, bringing the conductor back to its original voltage. There will thus be a change in voltage of

$$\delta V_2 = -\delta V_1. \tag{6.124}$$

The infinitesimal current will also produce a change in charge of δQ_2. The procedure in step 2 will do an amount of work producing a change in energy δU_2. This change in energy is given by

$$\delta U_2 = V\delta Q_2 + Q\delta V_2. \tag{6.125}$$

Now, if the dielectric is linear we can write

$$Q\delta V_2 = V\delta Q_2. \tag{6.126}$$

This means that

$$\delta U_2 = 2Q\delta V_2 = -2Q\delta V_1 = -2\delta U_1. \tag{6.127}$$

Then the total change in energy with no overall change in voltage will be

$$\delta U|_V = \delta U_1 + \delta U_2 = -\delta U|_Q. \tag{6.128}$$

Thus the infinitesimal change in energy at constant voltage is just the negative of the change in energy at constant charge, which explains the change of sign in Eq. (6.122).

Although the change in energy has the opposite sign, the force will be in the same direction in either case. This can be seen by considering the case treated above of a dielectric being pulled into a capacitor. At constant charge, the force is given by

$$F = -\partial_x U|_Q = -\partial_x \left[\frac{Q^2}{2C(x)} \right] = + \left[\frac{Q^2}{2C^2} \right] \partial_x C(x) = \frac{1}{2} V^2 \partial_x C(x). \tag{6.129}$$

At constant voltage, the force is

$$F = +\partial_x U|_V = \partial_x \left[\frac{1}{2} C(x) V^2 \right] = \frac{1}{2} V^2 \partial_x C(x), \tag{6.130}$$

so the force is the same in either case. As a mnemonic for the direction of the force, we can think of the dielectric tending to move so as to decrease the energy at constant charge. But at constant voltage, batteries are connected that want to do work, so the dielectric will tend to move so as to increase the energy. As a general rule, the polarization force on a dielectric will tend to move it into a region of stronger electric field.

One troubling question remains for the force on the dielectric in the case of a parallel plate capacitor. How could there be a horizontal force on the dielectric in Fig. (6.9) if all the lines of **E** are vertical?

The answer lies in our neglect of fringing fields in the parallel plate capacitor. The assumption that the lines of **E** are simply parallel lines, fully contained within the capacitor (as shown in Fig. 6.3 when $d << L$) is a good approximation for calculating the capacitance and energy. This approximation is still good when using Eq. (6.115) or Eq. (6.122) to find the force on the dielectric. But a more exact description of the electric field, including fringing fields, is necessary when using a direct force equation like Eq. (1.21).

The fringe electric field outside the capacitor is illustrated in Fig. 6.10.

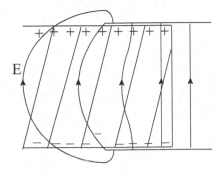

Figure 6.10: Fringe electric field at the edge of a capacitor. A force to the right is exerted on the bound surface charge on the dielectric.

It can be seen that there is a horizontal component to the original field $\mathbf{E_0}$ that produces an force to the right on the bound surface charge. Although this use of the direct force equation demonstrates that there is a force pulling the dielectric into the capacitor, the force on the dielectric is more easily calculated using either of the energy equations (6.115) or (6.122).

6.9 Steady state currents

So far, we have discussed dielectrics with no conductivity, so no current will flow. Now we consider the case where electric charge can flow through the material as an electric current.

6.9.1 Current density and continuity equation

The rate at which charge flows through a differential cross-section area **dA** is given in terms of a **current density j** by

$$d\left(\frac{\Delta q}{\Delta t}\right) = \mathbf{j} \cdot \mathbf{dA}. \tag{6.131}$$

The rate at which charge passes a finite area $\mathbf{A}$, called the **current** I through that area, is given by an integral over the area

$$I = \frac{\Delta q}{\Delta t} = \int_A \mathbf{j} \cdot d\mathbf{A}. \tag{6.132}$$

This for, instance, would be the current that flows in a wire.

If we assume **conservation of charge**, then the rate at which charge is leaving any volume must equal the rate of decrease of the charge inside the volume. For an infinitesimal volume ΔV bounded by a surface ΔS, as shown in Fig. 6.11, this gives the relation

$$-(\partial_t \rho)\Delta V = \oint_{\Delta S} \mathbf{j} \cdot d\mathbf{A}. \tag{6.133}$$

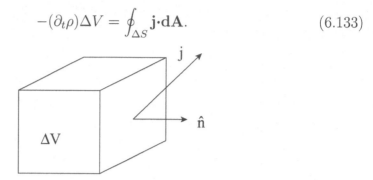

Figure 6.11: Current leaving a volume ΔV.

Dividing by the infinitesimal volume ΔV and taking the limit $\Delta V \to 0$, the right hand side becomes our definition of the divergence of $\mathbf{j}$, so

$$\boldsymbol{\nabla} \cdot \mathbf{j} = -\partial_t \rho. \tag{6.134}$$

This is called the **continuity equation** for the current density. From the derivation of the continuity equation, it is clear that it depends on conservation of electric charge. Even though charges are moving, a current is called a **steady state current** if all time derivatives are zero, including $\partial_t \rho$. In that case, the continuity equation becomes

$$\nabla \cdot \mathbf{j} = 0. \tag{6.135}$$

6.9.2   Ohm's law

The movement of charge that constitutes the current is generally caused by an electric field. In many materials the current density is proportional to the electric field, so

$$\mathbf{j} = \sigma \mathbf{E}. \tag{6.136}$$

The constant of proportionality, σ, is called the **conductivity**. Equation (6.136) is called **Ohm's law** for the current density. The **resistivity**

$$\rho = \frac{1}{\sigma} \tag{6.137}$$

is also used, and then Ohm's law is given by

$$\mathbf{j} = \mathbf{E}/\rho. \tag{6.138}$$

It has become standard notation to use σ for both the conductivity and for the surface charge density, and to use ρ for both the resistivity and for the volume charge density. It is usually clear from the context how to interpret each symbol.

The conductivity (or resistivity) can have all the complicating properties that were described for the permittivity in Section 6.2. Most of the time, we will treat **simple ohmic materials** for which Eqs. (6.136) and (6.138) hold.

The line integral of the electric field along a wire carrying a current I equals the voltage difference V between the endpoints of the integral

$$V = \int_a^b \mathbf{E} \cdot \mathbf{dr}. \tag{6.139}$$

If the material of the wire satisfies Ohm's law, then the voltage difference will be given by

$$V = \int_a^b \rho \mathbf{j} \cdot \mathbf{dr}. \tag{6.140}$$

Although the resistivity is usually a constant for any wire, we have kept ρ under the integral to include cases where it might vary along the length of the wire.

The ratio of V given by Eq. (6.140) to I given by Eq. (6.132) will be a constant for any given wire, since the magnitude of $\mathbf{j}$ will cancel in the ratio. Then we can write

$$V = IR, \tag{6.141}$$

with the constant R (the **resistance** of the wire) given by

$$R = \frac{\int \rho \mathbf{j} \cdot \mathbf{dr}}{\int \mathbf{j} \cdot \mathbf{dA}}. \tag{6.142}$$

Since the divergence of $\mathbf{j}$ vanishes for a steady current, the integral of $\mathbf{j} \cdot \mathbf{dA}$ can be taken over any cross section of the wire, even for a wire of varying cross section. For a length L of a wire of constant ρ and cross-sectional area $\mathbf{A}$, the resistance is

$$R = \frac{\rho L}{A}. \tag{6.143}$$

Equations (6.131)-(6.143) are the same in either the Gaussian or SI systems, but the units are different. The SI unit for resistance is the **ohm** designated by the symbol Ω. In Gaussian units, the conductivity has the unit sec^{-1} and the resistivity has the unit sec, so the resistance has the unit sec/cm, although the term **statohm** is sometimes used. The ohm is related to the esu unit by

$$1\,\text{ohm} = \frac{1\,\text{volt}}{1\,\text{ampere}} = \frac{\frac{1}{300}\text{statvolt}}{3 \times 10^9\,\text{statampere}} = \frac{1}{9 \times 10^{11}}\,\text{statohm}. \tag{6.144}$$

6.9.3   Relaxation constant

Before continuing with steady currents, we will use the time dependent continuity equation (6.134) to investigate the transient approach to the steady state. If we substitute Ohm's law of Eq. (6.136) into the continuity equation for an ohmic material with permittivity ϵ, we get

$$\partial_t \rho = -\sigma \nabla \cdot \mathbf{E} = -\frac{4\pi\sigma}{\epsilon}\rho. \tag{6.145}$$

This differential equation has the solution

$$\rho(\mathbf{r}, t) = e^{-\frac{4\pi\sigma}{\epsilon}t}\rho(\mathbf{r}, 0). \tag{6.146}$$

There is thus an exponential decay of the original charge distribution until no charge is left.

For an extended material, the charge distribution retains its shape but decays to zero charge. Where does the charge go, and why does no charge appear beyond the original distribution? An electric current does not require an excess of charge. The material beyond the original charge distribution remains neutral, but the negative charges are moving with respect to the positive charges, producing the current. For a material of infinite extent, the charge in the current just keeps moving out to infinity, and there is never an imbalance of charge.

For the case of a blob of charge originally inside a conductor of finite extent, the original blob will retain its shape but decrease exponentially to zero. At the same time, a charge distribution will grow on the outer surface of the conductor. The charge on the outer surface will eventually arrange itself so as not to produce any electric field in the body of the conductor, since this must be the steady state condition. At no time will there be any excess of charge inside the conductor, other than the original charge distribution which decays to zero.

The decay equation (6.146) can be written in terms of a **relaxation constant** τ as

$$\rho(\mathbf{r}, t) = e^{-t/\tau}\rho(\mathbf{r}, 0), \tag{6.147}$$

with τ given by

$$\tau = \frac{\epsilon}{4\pi\sigma}. \tag{6.148}$$

A conductor is deemed a 'good' conductor or a 'bad' conductor, depending on the size of the ratio σ/ϵ. Whether the conductor is a good or a bad conductor, the steady state will be the same inside the conductor. The only difference will be in how fast the steady state is reached.

There is a large range of relaxation times. A good conductor like copper has a relaxation time of about 10^{-19} seconds, while rubber has a relaxation time of about 3 hours. Note, however, that even for a good insulator (bad conductor) like rubber, charge will eventually leak off.

6.9.4   Effective resistance

If the original charge distribution is not replenished, then the steady state condition will have no charge, current, or electric field inside the conductor, as discussed in Section 2.1. However, if there are regions of the conductor kept at different potentials by external means such as batteries, then a steady state current will flow.

A simple case is a conducting sphere of radius a kept at a voltage V that is imbedded in a medium having permittivity ϵ and conductivity σ. There will be an electric field inside the medium given by

$$\mathbf{E} = \frac{aV\hat{\mathbf{r}}}{r^2}. \tag{6.149}$$

Therefore, the current inside the medium will be

$$\mathbf{j} = \sigma\mathbf{E} = \frac{\sigma aV\hat{\mathbf{r}}}{r^2}. \tag{6.150}$$

The total current leaving the sphere is

$$I = \oint \mathbf{j}\cdot d\mathbf{A} = 4\pi\sigma aV. \tag{6.151}$$

Consequently, the effective resistance for this configuration is

$$R = \frac{V}{I} = \frac{1}{4\pi\sigma a}. \tag{6.152}$$

There is a simple relation connecting the effective resistance between two electrodes and the capacitance of the two electrodes for an extended material having constant σ and ϵ. (By **electrode**, we mean a conductor of any shape, kept at a constant potential.) The effective resistance is given by the ratio of integrals

$$R = \frac{V}{I} = \left|\frac{\int \mathbf{E}\cdot d\mathbf{r}}{\oint \mathbf{j}\cdot d\mathbf{A}}\right|. \tag{6.153}$$

In Eq. (6.153), the line integral in the numerator is taken over any path between the electrodes, and the surface integral in the denominator is taken over any closed surface enclosing either electrode. We take the absolute value of the ratio so that we will not have to worry about signs. Applying Ohm's law $\mathbf{j} = \sigma\mathbf{E}$ to the denominator of Eq. (6.153) gives

$$R = \frac{V}{I} = \left|\frac{\int \mathbf{E}\cdot d\mathbf{r}}{\sigma \oint \mathbf{E}\cdot d\mathbf{A}}\right|. \tag{6.154}$$

The capacitance of the two electrodes is given the ratio Q/V, where Q is the magnitude of the charge on each electrode given, from Gauss's law by

$$Q = \left|\frac{\epsilon}{4\pi} \oint \mathbf{E}\cdot d\mathbf{A}\right|. \tag{6.155}$$

This gives the capacitance as the ratio of integrals

$$C = \frac{Q}{V} = \left| \frac{\epsilon}{4\pi} \frac{\oint \mathbf{E} \cdot \mathbf{dA}}{\int \mathbf{E} \cdot \mathbf{dr}} \right|. \tag{6.156}$$

Comparing Eq. (6.156) for the capacitance C with Eq. (6.154) for the resistance R, we see that the integrals cancel in the product RC, so

$$RC = \frac{\epsilon}{4\pi\sigma} = \tau \tag{6.157}$$

for any configuration. We see that the product RC just equals the relaxation time for the material. This result also holds for the example given above of a single electrode in an extended conductor if the second electrode is taken to be an infinite surface at infinity.

6.10 Problems

1. Find the bound surface charge density for the dielectric (with permittivity ϵ)

   (a) lamina in Eq. (6.46).

   (b) in a parallel plate capacitor with charge Q and plate area A.

2. (a) Find the capacitance of a spherical capacitor with inner radius A and outer radius B filled with a dielectric of permittivity ϵ.

   (b) Find the capacitance if the dielectric fills only the lower half of the capacitor. (Hint: Use the uniqueness theorem to make a guess about the shape of the lines of **E**.)

3. (a) Find the capacitance per unit length of a coaxial cylindrical capacitor with inner radius A and outer radius B filled with a dielectric of permittivity ϵ.

   (b) Find the capacitance per unit length if the dielectric fills only the lower half of the capacitor.

4. (a) Find the bound surface charge density σ_b on the flat face of a dielectric with a point charge q a distance d from the face.

   (b) Integrate the bound surface charge density to find the total bound charge on the flat face.

   (c) Find the force exerted on the dielectric by the point charge by integrating over the bound surface charge density.

5. A dielectric sphere is placed in a uniform electric field $\mathbf{E_0}$. Find

   (a) the bound surface charge density σ_b on the sphere.

   (b) the induced dipole moment of the sphere by integrating over the bound surface charge density.

6. Find the electric field inside a dielectric shell (of permittivity ϵ) with inner radius A and outer radius B that is placed in a uniform electric field $\mathbf{E_0}$.
   (Hint: Expand the potential in three regions, using only $l = 1$.)

7. Derive Eqs. (6.84) and (6.85).

8. For each of the following equations, show that letting $\epsilon \to \infty$ results in the corresponding equation for a conductor.

   (a) Eq. (6.58).

(b) Eq. (6.69).

(c) Eq. (6.80).

9. Complete the derivation of Eq. (6.103).

10. Verify Eq. (6.119) for the capacitance of the capacitor in Fig. 6.9.

11. (a) Use the result in Eq. (6.68) to find the electric field E_{cavity} inside a spherical cavity of radius R in a dielectric with permittivity ϵ that originally had a uniform electric field E_0.

    (b) Compare the result of (a) to the microscopic field E_μ given by Eq. (6.89). Show that $E_\mu > E_{\text{cavity}}$ for any value of ϵ. Explain in words why this inequality holds.

12. For the current density

$$\mathbf{j} = j_0 \mathbf{k} \cos\left(\frac{\pi r}{2R}\right) \tag{6.158}$$

in a wire of radius R in cylindrical coordinates (r, θ, z), find

(a) $\boldsymbol{\nabla}\cdot\mathbf{j}$.

(b) $\boldsymbol{\nabla}\times\mathbf{j}$.

13. Find the resistance and capacitance for the following electrode configurations and check whether $RC = 4\pi\epsilon/\sigma$ for each case.

    (a) A single electrode of radius a in an extended material.

    (b) The configuration given in Problem 2 (a).

    (c) The configuration given in Problem 2 (b).

Chapter 7

Magnetostatics

Magnetism was one of the earliest physical phenomena that fascinated man. The mystery and magic of the magnetic lodestone predates history. For many of us, a toy magnet or compass was our first encounter with 'science' and 'experiment'. Einstein recalled in a memoir his first profound experience with a toy compass given to him by his father when he was four years old. Because magnetism was so familiar, without understanding, its early history was shrouded in mysticism and a bit of confusion, even into the nineteenth century (not to mention magnetic healing bracelets, still popular today).

7.1 Magnetic Forces Between Electric Currents

The modern scientific study of magnetism started in 1820 when Hans Christian Oersted observed (reportedly by accident during a lecture) the **Oersted effect**. This was the surprising deflection of a magnetic needle by an electric current. This was the first example of a connection between the ancient phenomena of magnetism and more recent observations of electricity. It could be called the birth of **electromagnetism**.

The discovery of the Oersted effect led to extensive experimental and theoretical studies of the magnetic effects of electric currents by a large number of early physicists. The swift pace and overlap of important results has led to a bit of confusion in attributions and the naming of the laws of magnetism. Different modern books still give conflicting names to these laws. We will give our understanding of what is appropriate, but will also point out other usages.

In 1823, after three years of experimental work on the forces between current loops, and remarkable theoretical insight, André-Marie Ampère proposed what was then called **Ampere's law**. This states that the magnetic force on a differential current element $I\mathbf{dl}$ a distance $\mathbf{r}$ from a differential current element $I'\mathbf{dl'}$ is given by

$$\mathbf{d^2F}_{II'} = kII'\frac{[\mathbf{dl}\times(\mathbf{dl'}\times\mathbf{r})]}{r^3}.$$

(7.1)

Ampere's law can be recognized as the magnetic equivalent of Coulomb's law, and will be the starting point for our study of magnetostatics. First we must note, as was known to Ampere, that Eq. (7.1) can not be used as it stands since static currents can only exist in closed current loops. In actual usage, Ampere's law must be integrated over two closed loops to give

$$\mathbf{F}_{cc'} = kII' \oint_c \oint_{c'} \frac{\mathbf{dr} \times [\mathbf{dr'} \times (\mathbf{r} - \mathbf{r'})]}{|\mathbf{r} - \mathbf{r'}|^3} \tag{7.2}$$

for the force on current loop c due to loop c'. The differential vectors $\mathbf{dr}$ and $\mathbf{dr'}$ are taken tangent to each respective current loop. The choice of the constant k is determined by the choice of units. We will discuss the various choices that have been used for k after some further development of the force law.

A second thing to note about Ampere's law is that Eq. (7.1) does not satisfy Newton's third law. However, that is OK because Eq. (7.1) is a mathematical law that only has physical meaning after the double integration over closed loops in Eq. (7.2). Even this does not appear to satisfy Newton's third law, but we now show that, in fact, it does.

We apply bac-cab to the numerator in Eq. (7.2) to give

$$\mathbf{F}_{cc'} = kII' \left[\oint_{c'} \mathbf{dr'} \oint_c \frac{\mathbf{dr} \cdot (\mathbf{r} - \mathbf{r'})}{|\mathbf{r} - \mathbf{r'}|^3} - \oint_c \oint_{c'} \frac{(\mathbf{dr} \cdot \mathbf{dr'})(\mathbf{r} - \mathbf{r'})}{|\mathbf{r} - \mathbf{r'}|^3} \right]. \tag{7.3}$$

The first integral over loop c vanishes because it is an integral over a closed contour of a perfect differential. (In fact it is just like the integral of $\mathbf{E} \cdot \mathbf{dr}$ for a point charge.) This leaves

$$\mathbf{F}_{cc'} = -kII' \oint_c \oint_{c'} \frac{(\mathbf{dr} \cdot \mathbf{dr'})(\mathbf{r} - \mathbf{r'})}{|\mathbf{r} - \mathbf{r'}|^3} \tag{7.4}$$

for the force on current loop c due to loop c'. This form of the force law satisfies Newton's third law

$$\mathbf{F}_{c'c} = -\mathbf{F}_{cc'}. \tag{7.5}$$

Equation (7.4) for the force between current loops is called **Neumann's formula**, after Franz Neumann who first derived it.

As an example of the use of the Neumann formula, we use it to find the force per unit length between two long straight parallel wires a distance $\mathbf{d}$ apart, carrying currents I and I' respectively. By "long", we mean that the length of each wire is very much greater than the distance between them.

The two wires and the variables for the integration are shown in Fig. 7.1. In order to close the circuit for each wire, there would have to be large semicircular wires connecting the ends of the straight parts of each wire. We assume that these are far enough away from the point of measurement to give negligible force.

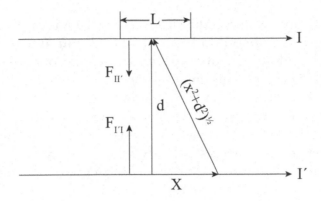

Figure 7.1: Force between two parallel wires.

We choose the x-axis to be along the positive direction of current in the wires. By symmetry, the horizontal force on either wire will vanish. The integral for the vertical force per unit length L of the upper wire can be written as

$$
\begin{aligned}
\frac{\mathbf{F}_{II'}}{L} &= -kII' \int_{-\infty}^{\infty} \frac{\mathbf{d}\,dx}{(x^2 + d^2)^{\frac{3}{2}}} \\
&= \frac{-2kII'\hat{\mathbf{d}}}{d}.
\end{aligned}
\tag{7.6}
$$

This derivation for the force on parallel wires was made easy by the simple geometry and the use of symmetry. But we should point out that Neumann's formula is difficult to use for most other cases, and is not generally a practical means of calculating forces.

Equation (7.6) can also be considered an analogue of Coulomb's law, although, because of the different geometry, the force falls off as $1/r$. The currents I and I' in Eq. (7.6) can be positive (corresponding to right in Fig. 7.1) or negative (for left in the figure). The force is seen to be attractive for 'like' (parallel) currents and repulsive for 'unlike' (anti-parallel) currents. This is just the opposite of Coulomb's law.

7.2   Units of Electricity and Magnetism

Equation (7.6) has been used to define the constant k, and to define and measure the units of electric current. The simplest, and first, choice was to choose $k = 1$ and use Eq. (7.6) to define the unit of current. This led to what is called the **emu system** (for "electromagnetic units") in which the unit of current is the **abampere**, defined in words by **The force per unit length between two long parallel wires 1 centimeter apart, each carrying a current of 1 abampere is 2 dynes per centimeter.** Note that emu units are simplest in the cgs system. This definition of the abampere then was used to define the emu unit **abcoulomb** of charge as **A current of 1 abampere corresponds to a flow of 1 abcoulomb of charge per second.**

The original choices of emu units for magnetic phenomena and esu units for electric phenomena were simplest for each individual case, but left them unconnected. There were two units of charge, the emu abcoulomb and the esu statcoulomb. Both were units of electric charge, but were applicable to either magnetic or electric phenomena separately.

An important step towards correlating electricity and magnetism was the determination of the ratio of these two charge units in 1856 by Wilhelm Weber and Friedrich Kohlrausch. They found

$$\frac{1 \text{ abcoulomb}}{1 \text{ statcoulomb}} = 3.1 \times 10^{10} \text{cm/sec.} \tag{7.7}$$

Knowing this ratio, the same set of units could be used for both electric and magnetic phenomena by giving all charges and currents in the same esu system, but dividing the charges and currents in magnetic equations by the ratio in Eq. (7.7), which is generally designated as c.

This system of units, in coordination with cgs units, is called the **Gaussian system**, and is the system of units used in this text. The Gaussian system can be considered as the esu system with the constant k appearing in the equation for the force between currents taken equal to $1/c^2$. The introduction of the Gaussian system was an important step in the progress toward the unification of electricity and magnetism.

A general rule for the Gaussian system is to divide any current or charge producing or receiving a magnetic force by c. Another way to do this, somewhat in the spirit of Special Relativity, is to use c as a conversion factor. Whenever the time and space units of the left and right hand sides of an equation don't agree, the appropriate factors of c should be introduced to make the units consistent.

In the Gaussian system, the force between two current loops is given by

$$\mathbf{F}_{cc'} = \frac{II'}{c^2} \oint_c \oint_{c'} \frac{d\mathbf{r} \times [d\mathbf{r}' \times (\mathbf{r} - \mathbf{r}')]}{|\mathbf{r} - \mathbf{r}'|^3}, \tag{7.8}$$

and the force per unit length between parallel wires by

$$\frac{F}{L} = \frac{2II'}{c^2 d}. \tag{7.9}$$

Note, in each of the above equations, the division by c^2 on the right hand side gives the force in units of charge^2/distance^2.

Of course, we notice immediately the close approximation of the ratio abcoulomb/statcoulomb, which appears in the Gaussian system as the constant

c, with the speed of light. This remarkable coincidence did not escape early investigators either, and led some to conjecture that light and electromagnetism were related.

Michael Faraday had observed the rotation of the plane of polarization of light by a magnetic field in 1845, which was the first indication of a connection between magnetism and light. Then, in 1857, Gustav Kirchoff showed that an electromagnetic pulse would propagate along a wire at the speed of light. The culmination was Maxwell's derivation in 1864 (in a paper titled "A Dynamical Theory of the Electromagnetic Field") of electromagnetic waves in vacuum that propagated at the speed of light. Maxwell's derivation was not universally accepted until it was experimentally confirmed by Heinrich Hertz in 1888. Hertz's experimental verification of Maxwell's theory completed the unification of electricity and magnetism into one **electromagnetic field**.

In the SI system, the unit of current is predetermined as the **ampere**. The ampere is closely connected to the abampere, with[1].

$$1 \, \text{ampere} = 0.1 \, \text{abampere}. \tag{7.10}$$

This change in the current unit was probably made because someone felt the abampere was not a convenient size. With this definition of the ampere, the constant k in Eqs. (7.1-7.6) comes out to be exactly 10^{-7}. Because the SI system uses rationalized units, this is written as

$$k = \frac{\mu_0}{4\pi} = 10^{-7}. \quad \textbf{SI} \tag{7.11}$$

Then the Gaussian equations (7.8) and (7.9) become

$$\mathbf{F}_{cc'} = \frac{\mu_0 I I'}{4\pi} \oint_c \oint_{c'} \frac{d\mathbf{r} \times [d\mathbf{r}' \times (\mathbf{r} - \mathbf{r}')]}{|\mathbf{r} - \mathbf{r}'|^3}, \quad \textbf{SI} \tag{7.12}$$

for the force between current loops, and

$$\frac{F}{L} = \frac{\mu_0 I I'}{2\pi}. \quad \textbf{SI} \tag{7.13}$$

for the force per unit length between long straight wires.

The constant μ_0 is often called the "permeability of free space", but, as was the case with ϵ_0, it has nothing to do with a property of free space. μ_0 has complicated units, but it is not necessary to write them down if care is taken to have all other quantities in any equation in the appropriate SI units. The

---

[1]In 2019, the SI unit of current, the ampere, was given a new definition, as a current of one coulomb/second, with the coulomb being defined by, "One coulomb equals $1.602176634 \times 10^{-19} e$", where e is the charge on the electron, as determined by several experimental measurements.

appearance of μ_0 and ϵ_0 in separate equations makes the connection with the speed of light less transparent, although their product must satisfy

$$\frac{1}{\epsilon_0\mu_0} = c^2 \quad \textbf{SI} \tag{7.14}$$

to be consistent with Gaussian units.

The fact that $\mu_0/4\pi$ equals a simple power of 10 suggests that it does not have physical significance, but is just the result of a mismatch of units[2]. Had the ampere been chosen to equal the abampere, and the SI system been written in cgs units, $\mu_0/4\pi$ would equal 1. Then, the SI system would just be the emu implementation of the Gaussian system.

7.3   The Magnetic Field B

As we did for electrostatic forces, we can break up the double integration in Eq. (7.2) into two steps. We first define the **magnetic field B** due to a current I' in a loop c' by the integral

$$\mathbf{B}(\mathbf{r}) = \frac{I'}{c}\oint_{c'} \frac{d\mathbf{r}'\times(\mathbf{r}-\mathbf{r}')}{|\mathbf{r}-\mathbf{r}'|^3}, \tag{7.15}$$

and then calculate the force on any other current loop c with current I by the integral

$$\mathbf{F} = \frac{I}{c}\oint_c d\mathbf{r}\times\mathbf{B}(\mathbf{r}). \tag{7.16}$$

The combination of these two steps reproduces the Ampere force law of Eq. (7.8). As with the electric case, if we assume linearity the total magnetic field would be the vector sum over all loops c'.

In SI units, Eq. (7.15) for the magnetic field is written as

$$\mathbf{B}(\mathbf{r}) = \frac{\mu_0 I'}{4\pi}\oint_{c'} \frac{d\mathbf{r}'\times(\mathbf{r}-\mathbf{r}')}{|\mathbf{r}-\mathbf{r}'|^3} \quad \textbf{SI} \tag{7.17}$$

and Eq. (7.16) for the magnetic force on a current loop by

$$\mathbf{F} = I\oint_c d\mathbf{r}\times\mathbf{B}(\mathbf{r}). \quad \textbf{SI} \tag{7.18}$$

Comparing the SI equations with their Gaussian counterparts, we can formulate a simple rule for converting a magnetism equation from Gaussian to SI form. In Gaussian equations for producing **B** from currents, make the transformation

$$\frac{1}{c} \rightarrow \frac{\mu_0}{4\pi} \quad \text{(for Gaussian} \rightarrow \text{SI).} \tag{7.19}$$

---

[2] $\mu_0/4\pi$ was redefined in 2019, so that it is now an experimental quantity in SI units, but still equals 10^{-7} to nine significant figures.

In Gaussian equations that give the force on currents, just erase the $\frac{1}{c}$.

Equation (7.15) (or 7.17) is called the **law of Biot-Savart**, although that designation has also sometimes been given to the differential form

$$\mathbf{dB} = \frac{I\mathbf{dl}\times\mathbf{r}}{cr^3}, \tag{7.20}$$

which gives the magnetic field a distance $\mathbf{r}$ from a differential current element $I\mathbf{dl}$. Actually Eq. (7.15) was first given by Laplace in an analysis of the extensive experiments of Biot and Savart. The distinction between Ampere's force law and the Biot-Savart law is that Ampere's law is for the forces he had measured between current loops, while Biot and Savart had measured the magnetic field **B** of current loops by using the deflection of a compass needle.

You may have noticed that we have used the term **magnetic field** for **B**, although most authors still use that term for the **H** field discussed later in this chapter. The most common name for the **B** field in the literature has been "magnetic induction", while the SI system seems to encourage the use of the term "magnetic flux density". The phrase "magnetic induction" was used by Maxwell, and also by earlier authors in their studies of the effects of permanent magnets. But the use of that term now obscures the strong physical connection between the **B** and **E** fields that is the result of the advances by Maxwell, Einstein, and all the Quantum Electrodynamics of the 20th century.

It is important throughout a consistent development of electromagnetism to treat the **E** and **B** fields in a similar manner to ease their inevitable merger, in Special Relativity, into the same tensor quantity. For this reason we will call **B** what it is... the **magnetic field**. In cases where there might be ambiguity, we will simply refer to them as the **B field** and the **H field**.

7.4 Applications of the Biot-Savart Law

We now use the Biot-Savart law to find the magnetic field a distance **d** away from a long straight wire carrying a current $\mathbf{I'}$. We represent the current as a vector here to include its direction along the wire. We use the variables shown in Fig. 7.1 in Eq. (7.15) to get the field a distance **d** from a wire:

$$\begin{aligned}\mathbf{B} &= \frac{1}{c}\int_{-\infty}^{\infty}\frac{\mathbf{I'}\times\mathbf{d}dx}{(x^2+d^2)^{\frac{3}{2}}} \\ &= \frac{2\mathbf{I'}\times\hat{\mathbf{d}}}{cd}.\end{aligned} \tag{7.21}$$

Sighting along the wire, the lines of **B** will be circles with the positive direction of **B** given by the cross product $\mathbf{I'}\times\hat{\mathbf{d}}$. This can be conveniently described by a right hand rule. **If the thumb of your right hand is placed**

along the positive direction of the current, then as you close your fingers they will curve in the positive direction of B.

The force per unit length on a second parallel wire as in Fig. 7.1 will be given, using Eq. (7.16), by

$$\frac{\mathbf{F}_{II'}}{L} = \frac{2\mathbf{I}\times(\mathbf{I'}\times\hat{\mathbf{d}})}{c^2 d} = -\frac{2II'\hat{\mathbf{d}}}{c^2 d}. \tag{7.22}$$

This is the same as given by direct application of Neumann's formula.

We can modify the above derivation to calculate the magnetic field due to the current in a straight segment of wire as shown in Fig. 7.2. The only change is in the limits of integration, which are now x_1 and x_2. The variable x and the angle θ is positive if to the right of the line of **d** in the figure, and negative if to the left.

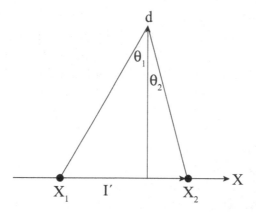

Figure 7.2: Magnetic field of a current segment.

The field is given by

$$
\begin{aligned}
\mathbf{B} &= \frac{1}{c}\int_{x_1}^{x_2} \frac{\mathbf{I'}\times\mathbf{d}\,dx}{(x^2+d^2)^{\frac{3}{2}}} \\
&= \frac{\mathbf{I'}\times\hat{\mathbf{d}}}{cd}\int_{\theta_1}^{\theta_2}\cos\theta\,d\theta \\
&= \frac{\mathbf{I'}\times\hat{\mathbf{d}}}{cd}[\sin\theta_2 - \sin\theta_1].
\end{aligned} \tag{7.23}
$$

with the angles θ_1 and θ_2 as shown in Fig. 7.2. Using Eq. (7.23), the magnetic field of a wire segment of length L, lying along the x axis from $x = -L/2$ to $x = +L/2$, is given by

$$\mathbf{B}(x,y) = \frac{\mathbf{I}\times\hat{\mathbf{j}}}{cy}\left[\frac{(L/2-x)}{\sqrt{(L/2-x)^2+y^2}} + \frac{(L/2+x)}{\sqrt{(L/2+x)^2+y^2}}\right]. \tag{7.24}$$

Equation (7.24) cannot be used by itself because the current segment is not closed, but is useful if several straight segments are connected to form a closed circuit.

Another useful application of the Biot-Savart law is to calculate the magnetic field on the axis of a circular current loop. As shown in Fig. 7.3, we consider a loop of radius a carrying a current I. From the axial symmetry, the horizontal component of the field at the axis cancels.

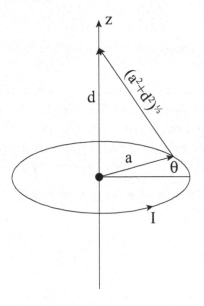

Figure 7.3: Magnetic field of a current loop.

The remaining field on the axis a distance $\mathbf{d}$ above the loop is given by

$$\begin{aligned}
\mathbf{B} &= \frac{I}{c} \oint \frac{[\mathbf{dr} \times (\mathbf{r} - \mathbf{r}')]}{|\mathbf{r} - \mathbf{r}'|^3} \\
&= \frac{I}{c} \int_0^{2\pi} \frac{a^2 d\theta \hat{\mathbf{d}}}{(a^2 + d^2)^{\frac{3}{2}}} \\
&= \frac{2\pi I a^2 \hat{\mathbf{d}}}{c(a^2 + d^2)^{\frac{3}{2}}}.
\end{aligned} \tag{7.25}$$

There is a convenient right hand rule for the direction of the field through the loop. **If you curl your fingers along the wire in the direction of the current, your thumb will extend in the direction of the field through the loop.** You could also use the right hand rule for a straight wire by placing your thumb along the wire in the direction of the current. Then your fingers will curl in the direction of the field through the loop. These two right hand rules are illustrated in Fig. 7.4.

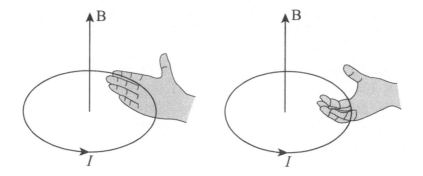

Figure 7.4: Two right-hand rules for the field of a current loop.

The result for the field on the axis of a current loop can be used to find the field on the axis of a solenoid, which is a tightly wound series of connected current loops. Such a solenoid is shown in Fig. (7.5). The current distribution is characterized by n, the number of turns per unit length, which for a solenoid of length L with N total turns is given by $n = N/L$.

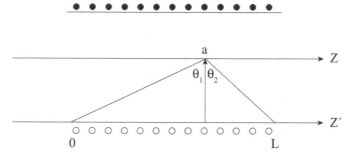

Figure 7.5: Solenoid of length L and radius a.

We can consider the solenoid to be a set of current loops, each with current $nIdz'$ in a differential length dz' of the solenoid. The total field of the solenoid is given by the integral over the length of the solenoid of the field on the axis of a loop with this current:

$$\mathbf{B} = \frac{2\pi n I a^2}{c} \int_{-\frac{L}{2}}^{\frac{L}{2}} \frac{dz'}{[a^2 + (z' - z)^2]^{\frac{3}{2}}}. \tag{7.26}$$

Making the substitution

$$z - z' = a \tan \theta \tag{7.27}$$

reduces this integral to

$$
\begin{aligned}
\mathbf{B} &= \frac{2\pi n I}{c} \int_{\theta_1}^{\theta_2} \cos \theta \, d\theta \\
&= \frac{2\pi n I}{c} [\sin \theta_2 - \sin \theta_1]. \tag{7.28}
\end{aligned}
$$

The angles θ_1 and θ_2, as shown in Fig. 7.5, are positive if to the right of vertical at the point of measurement (z), and negative if to the left.

From Eq. (7.28), we see that the field near the middle of a long solenoid is given by

$$\mathbf{B} = \frac{4\pi nI}{c}, \tag{7.29}$$

while it is one half of that at either end. The magnitude of $\mathbf{B}$ along the axis of the solenoid is plotted in Fig. 7.4 for the case of a solenoid with a current of 1 ampere (3×10^9 statamperes) in a solenoid with a length of 10 cm and radius of 1 cm, having 1,000 turns. For comparison, the dashed line is the axial field of a single coil with the same central field.

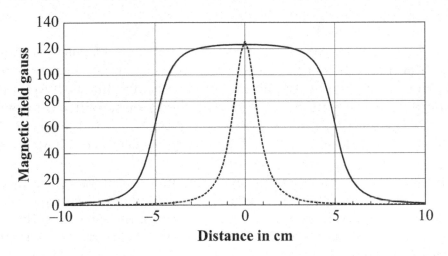

Figure 7.6: Magnetic field (solid curve) on the axis of a solenoid of length 10 cm and radius 1 cm, with a current 1 ampere in 1,000 turns. The dashed line is the field for a single coil with the same central field.

7.5   Magnetic Effects on Charged Particles

We can define effective currents of moving charged particles by the association

$$I\mathbf{dl} \rightarrow q\mathbf{v}, \tag{7.30}$$

since either term gives the amount of charge passing through a surface per unit time. This connection, along with Eq. (7.16) for the magnetic force on a current element, gives a magnetic force on a moving charge

$$\mathbf{F}_{\text{magnetic}} = \frac{q}{c}\mathbf{v} \times \mathbf{B}. \tag{7.31}$$

Since the particle has an electric charge, the total force on it will be the sum of the electric and magnetic forces:

$$\mathbf{F} = q\mathbf{E} + \frac{q}{c}\mathbf{v} \times \mathbf{B}. \tag{7.32}$$

Equation (7.32), for the total electromagnetic force on a moving charge is called the **Lorentz force**. It is unusual for having an absolute, rather than a relative, velocity appearing in the force equation. This causes no problem here because the system in which the velocity is defined is taken to be whichever system the **E** and **B** fields were calculated in.

Before the introduction of special relativity, the problem of transforming the Lorentz force equation to another system moving with constant velocity with respect to the original system was unsolvable, although several attempts were made. It turns out, as a fortunate result in special relativity (see Chapter 14), that the Lorentz force equation is exactly the same equation in any moving system (By "moving system", we will generally mean one moving with constant velocity with respect to the original system.), even though each of the quantities **F**, **E**, **v**, **B** change in complicated ways.

The Lorentz force can be used to define the units of magnetic field **B**. In Gaussian units, this is the **gauss**, defined for a purely magnetic force as **A particle with one statcoulomb of charge moving with velocity v perpendicular to a magnetic field of one gauss, will experience a force of v/c dynes.** Because the velocity appears as v/c in the Lorentz force in Gaussian units, the electric field **E** and magnetic field **B** can have the same units.

This fact has usually been obscured, even in the Gaussian system, by using the unit gauss only for **B**, while giving **E** in units of either dynes per statcoulomb, or statvolts per centimeter. The use of the same unit, the gauss, for both **E** and **B** has been advocated by Melvin Schwartz in his textbook *Principles of Electrodynamics*. We second his proposal here. The numerical value of E in the Gaussian system is the same whether the unit is called "dyne per statcoulomb", "statvolt per centimeter", or "gauss".

In SI units, the Lorentz force equation is written without the divisor c:

$$\mathbf{F} = q\mathbf{E} + q\mathbf{v} \times \mathbf{B}. \quad \textbf{SI} \tag{7.33}$$

The SI unit for magnetic field is the **tesla**, defined in terms of Eq. (7.33) as **A particle with one Coulomb of charge moving with a velocity of one meter per second perpendicular to a magnetic field of one tesla, will experience a force of one Newton.**

One tesla equals 10^4 gauss. The tesla is a rather large unit for typical laboratory use (The earth's magnetic field is about half a gauss.), although magnetic fields of up 8.3 tesla have been achieved at the LHC with superconducting electromagnets carrying 11,000 amperes.

Because the divisor c does not appear in the SI equation for the Lorentz force, the SI definition of the tesla is a bit simpler, but then **E** and **B** cannot have the same units. The usual SI unit for electric field is the volt per meter, with the conversion

$$1 \, \text{gauss} = 1 \, \text{statvolt/cm} = 3 \times 10^4 \, \text{V/m}. \tag{7.34}$$

A purely magnetic force, as given by Eq. (7.31), is perpendicular to the velocity, and does no work, so the magnitude of the velocity remains constant. Thus a charged particle injected at right angles into a magnetic field will move in a circle. If the particle is injected with a component of velocity along the field, that component of velocity will remain constant, so the ensuing path will be a helix with its axis along the magnetic field. The radius R of the circle (or helix) is determined by

$$F = \frac{qvB}{c} = \frac{mv_\perp^2}{R}, \tag{7.35}$$

where $v_\perp$ is the component of the velocity perpendicular to **B**. Solving for R:

$$R = \frac{mvc}{qB}. \tag{7.36}$$

This was the equation that J. J. Thomson used to measure e/m for the electron (thus 'discovering' the electron) by measuring the bending radius of cathode rays of known velocity in a constant magnetic field. Luckily for Thomson, his electrons were moving with a velocity much less than c. As we will see in Chapter 14, Eq. (7.36) for the radius gets changed in special relativity.

The angular velocity of electrons in circular motion is given by

$$\omega = \frac{v}{R} = \frac{qB}{mc}. \tag{7.37}$$

Note that ω is independent of the velocity of the electrons and the radius R. That is the principle behind the **cyclotron** particle accelerator. A schematic drawing of a cyclotron is shown in Fig. 7.7.

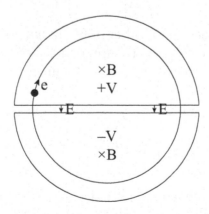

Figure 7.7: The two D's of a cyclotron.

The cyclotron consists of two hollow semicircular conductors called D's, with a small separation between them. There is a magnetic field into the paper that keeps an electron in a circular path within the D's. An electric field $\mathbf{E}$ between the D's accelerates the electron each time it passes between the D's.

The D's are kept at opposite voltages $V(t) = V_0 \cos(\omega t)$. Each time an electron passes between the D's its energy is increased by $2eV_0$. Since the frequency ω is independent of the velocity of the electron, the frequency can be kept constant and the electron will experience a boost in energy at each passage between the D's. The cyclotron can accelerate charged particles in this way as long as their velocity is small compared to c, so that Eq. (7.37) holds. The angular frequency ω in Eq. (7.37) is called the **cyclotron frequency**.

7.6   Magnetic Effects of Current Densities

7.6.1   Volume current density j

The results for currents in wires can be extended to volume current densities $\mathbf{j}$ by the connection

$$I\mathbf{dl} \rightarrow \mathbf{j}d\tau, \tag{7.38}$$

which follows directly from Eq. (6.132) for the current in a wire where $\mathbf{j}$ is directed along the length of the wire. Then, the Biot-Savart law becomes

$$\mathbf{B}(\mathbf{r}) = \frac{1}{c} \int \frac{\mathbf{j}(\mathbf{r}')\times(\mathbf{r} - \mathbf{r}')}{|\mathbf{r} - \mathbf{r}'|^3} d\tau', \tag{7.39}$$

and the magnetic force on the currents in a volume V is given by

$$\mathbf{F} = \frac{1}{c} \int_V \mathbf{j}(\mathbf{r})\times\mathbf{B}(\mathbf{r})d\tau. \tag{7.40}$$

Although the current density $\mathbf{j}$ (and currents in wires) are actually composed of particles in motion, we will see in chapter 14 that all our magnetic equations involving steady state currents turn out to be unchanged by special relativity.

Equations (7.39) and (7.40) can be combined to give the force on a current distribution $\mathbf{j_b}$ due to another current distributiion $\mathbf{j_a}$:

$$\mathbf{F_{ba}} = \frac{1}{c^2} \int d\tau_b \int d\tau_a \frac{\mathbf{j_b}(\mathbf{r_b})\times[\mathbf{j_a}(\mathbf{r_a})\times(\mathbf{r_b} - \mathbf{r_a})]}{|\mathbf{r_b} - \mathbf{r_a}|^3}. \tag{7.41}$$

As we did in deriving Neumann's formula for the force between two current loops, the force law in Eq. (7.41) can be put into the more symmetric form. (See problem 7.10.)

$$\mathbf{F_{ba}} = -\frac{1}{c^2} \int d\tau_b \int d\tau_a \frac{[\mathbf{j_b}(\mathbf{r_b})\cdot\mathbf{j_a}(\mathbf{r_a})](\mathbf{r_b} - \mathbf{r_a})}{|\mathbf{r_b} - \mathbf{r_a}|^3}. \tag{7.42}$$

7.6.2 Surface current density K

Electric current can also occur as a surface current density **K**, constrained to flow on a surface S. The surface current density is defined so that the current flowing past a differential line element **dL** on the surface is given by

$$dI = \mathbf{K}\cdot(\mathbf{dL}\times\hat{\mathbf{n}}), \tag{7.43}$$

where $\hat{\mathbf{n}}$ is a unit vector normal to the surface, as shown on Fig. 7.8.

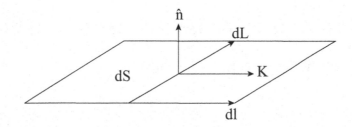

Figure 7.8: Surface current K.

The magnetic effects of a surface current follow from the connection

$$I\mathbf{dl} \rightarrow \mathbf{K}dS, \tag{7.44}$$

where the differential surface dS is given by

$$dS = |\mathbf{dl}\times\mathbf{dL}|. \tag{7.45}$$

The magnetic field due to a surface current density is given by

$$\mathbf{B}(\mathbf{r}) = \frac{1}{c}\int\frac{\mathbf{K}(\mathbf{r}')\times(\mathbf{r}-\mathbf{r}')}{|\mathbf{r}-\mathbf{r}'|^3}dS. \tag{7.46}$$

Also, the magnetic force on the currents on a surface S is given by

$$\mathbf{F} = \frac{1}{c}\int_S \mathbf{K}(\mathbf{r})\times\mathbf{B}(\mathbf{r})dS'. \tag{7.47}$$

We see that a differential current element **dI** can take any of the forms

$$\mathbf{dI} \leftrightarrow I\mathbf{dl} \leftrightarrow \mathbf{j}d\tau \leftrightarrow \mathbf{K}dS, \tag{7.48}$$

which can be used interchangeably where appropriate.

7.6.3   Magnetic effects of moving charges.

Extending magnetic effects to moving charges by the association $I\mathbf{dl} \to q\mathbf{v}$ gives

$$\mathbf{B} = \frac{q\mathbf{v} \times \mathbf{r}}{cr^3} \qquad (7.49)$$

for the magnetic field a distance $\mathbf{r}$ from a charge q moving with velocity $\mathbf{v}$. This result takes us out of the realm of magnetostatics, because the magnetic field has a time derivative, even for constant velocities, and there are relativistic corrections of the order $(v/c)^2$ to Eq. (7.49) that we will consider in Chapter 15.

There are problems raised by this equation. To write down an equation for the force between two moving charges, a suggestion from using the magnetic field of Eq. (7.49) in the Lorentz force equation would be

$$F_{qq'} = \frac{qq'(\mathbf{r} - \mathbf{r}')}{|\mathbf{r} - \mathbf{r}'|^3} + \frac{qq'\mathbf{v} \times [\mathbf{v}' \times (\mathbf{r} - \mathbf{r}')]}{c^2|\mathbf{r} - \mathbf{r}'|^3}. \qquad (7.50)$$

This force equation, when it was first derived, was a 'first' in physics. It was the first force between particles to depend on the absolute velocities of the two particles, rather than their relative velocity. This causes conceptual as well as practical problems.

A major problem is how to determine the appropriate reference system for the velocities. This seems necessary because the magnetic term on the right hand side changes with reference system. Theories were proposed by Gauss, Weber, Helmholtz, and others who considered corrections of order $(v/c)^2$ to Eq. (7.50) to compensate for the deficiencies of the magnetic term.

The various suggested corrections were inconsistent with each other, and each suggestion corrected one fault but introduced another. Fortunately for each theory, particle velocities high enough to test the theories were unobtainable in the nineteenth century.

Those brilliant, but disjoint, theories are critically discussed in the last chapter of Maxwell's book, *A Treatise on Electricity and Magnetism*. Their contradictory failures convinced Maxwell at that time of the necessary existence of a real, material medium (the **aether**). The last sentence in Maxwell's treatise reads "Hence all these theories lead to the conception of a medium in which the propagation takes place, and if we admit this medium as an hypothesis, I think it ought to occupy a prominent place in our investigations, and that we ought to endeavor to construct a mental representation of all the details of its action, and this has been my constant aim in this treatise." Unfortunately, Maxwell died in 1879 at the age of 49, shortly after writing those words.

Perhaps, if Maxwell had lived a bit longer, the internal inconsistencies of classical nineteenth century electromagnetism, with or without an aether, might have led him to a "Maxwell's Theory of Relativity" to resolve the problems, of which Eq. (7.50) is one example. That not happening, Einstein's Special Theory of Relativity did provide a consistent, and correct, resolution of all the pre-quantum conceptual and practical problems with electromagnetism.

Another problem is that Newton's third law is not satisfied by the force law in Eq. (7.50). The violation also holds in special relativity, where it shows up as a lack of conservation of momentum by two moving charges. The resolution is that the momentum of the electromagnetic field, which we will discuss in Chapter 9, also has to be considered. There we will see that it is the sum of both mechanical and electromagnetic field momenta that is conserved.

7.7 Differential Form of Magnetostatics

As you recall, we started our discussion of electrostatics with Coulomb's force law, and then gave, in Chapter 2, an alternate development of electrostatics from differential equations for $\mathbf{E}$. Here, we make a similar alternate development of magnetostatics, starting with two differential equations for $\mathbf{B}$.

We can convert the integral for $\mathbf{B}$ in Eq. (7.39) into the two differential equations. First, we calculate the divergence of $\mathbf{B}$:

$$
\begin{aligned}
\boldsymbol{\nabla}\cdot\mathbf{B}(\mathbf{r}) &= \frac{1}{c}\int \boldsymbol{\nabla}\cdot\left[\frac{\mathbf{j}(\mathbf{r}')\times(\mathbf{r}-\mathbf{r}')}{|\mathbf{r}-\mathbf{r}'|^3}\right]d\tau' \\
&= -\frac{1}{c}\int \mathbf{j}(\mathbf{r}')\cdot\boldsymbol{\nabla}\times\left[\frac{(\mathbf{r}-\mathbf{r}')}{|\mathbf{r}-\mathbf{r}'|^3}\right]d\tau' \\
&= 0,
\end{aligned}
\tag{7.51}
$$

where the **curl** in the last integral vanishes. In reducing the integral in Eq. (7.51), we have used the fact that the operator $\boldsymbol{\nabla}$ does not act on $\mathbf{j}(\mathbf{r}')$, and then applied the cyclic property of the triple scalar product.

Next, we take the **curl** of $\mathbf{B}$:

$$
\begin{aligned}
\boldsymbol{\nabla}\times\mathbf{B}(\mathbf{r}) &= \frac{1}{c}\int \boldsymbol{\nabla}\times\left[\frac{\mathbf{j}(\mathbf{r}')\times(\mathbf{r}-\mathbf{r}')}{|\mathbf{r}-\mathbf{r}'|^3}\right]d\tau' \\
&= \frac{1}{c}\int \mathbf{j}(\mathbf{r}')\boldsymbol{\nabla}\cdot\left[\frac{(\mathbf{r}-\mathbf{r}')}{|\mathbf{r}-\mathbf{r}'|^3}\right]d\tau' - \frac{1}{c}\int \mathbf{j}(\mathbf{r}')\cdot\boldsymbol{\nabla}\left[\frac{(\mathbf{r}-\mathbf{r}')}{|\mathbf{r}-\mathbf{r}'|^3}\right]d\tau'. \tag{7.52}
\end{aligned}
$$

In Eq. (7.52), we have used bac-cab, and the fact that the $\boldsymbol{\nabla}$ operator does not does not act on $\mathbf{j}(\mathbf{r}')$. Now we show that the last integral in Eq. (7.52) vanishes:

$$
\int \mathbf{j}(\mathbf{r}')\cdot\boldsymbol{\nabla}\left[\frac{(\mathbf{r}-\mathbf{r}')}{|\mathbf{r}-\mathbf{r}'|^3}\right]d\tau' = -\int \mathbf{j}(\mathbf{r}')\cdot\boldsymbol{\nabla}'\left[\frac{(\mathbf{r}-\mathbf{r}')}{|\mathbf{r}-\mathbf{r}'|^3}\right]d\tau'
$$

$$= -\int \left\{ \nabla' \cdot \left[\frac{\mathbf{j}(\mathbf{r}')(\mathbf{r} - \mathbf{r}')}{|\mathbf{r} - \mathbf{r}'|^3} \right] - \frac{(\mathbf{r} - \mathbf{r}')\nabla' \cdot \mathbf{j}(\mathbf{r}')}{|\mathbf{r} - \mathbf{r}'|^3} \right\} d\tau'.$$

(7.53)

Each of the two integrals in the last line of Eq. (7.53) vanishes. We can use the divergence theorem on the first integral to get a surface integral over a sphere of large enough radius that $\mathbf{j}$ vanishes on the surface. The second integral includes $\nabla \cdot \mathbf{j}$ which vanishes for currents with no time depedence. Going back to Eq. (7.52), we are left with the first integral in the last line, which gives

$$\begin{aligned} \nabla \times \mathbf{B}(\mathbf{r}) &= \frac{1}{c} \int \mathbf{j}(\mathbf{r}') \nabla \cdot \left[\frac{(\mathbf{r} - \mathbf{r}')}{|\mathbf{r} - \mathbf{r}'|^3} \right] d\tau' \\ &= \frac{4\pi}{c} \int \mathbf{j}(\mathbf{r}')\delta(\mathbf{r} - \mathbf{r}')d\tau' \\ &= \frac{4\pi}{c} \mathbf{j}(\mathbf{r}). \end{aligned}$$

(7.54)

So we see that the magnetostatic field $\mathbf{B}$ has no scalar sources

$$\nabla \cdot \mathbf{B} = 0,$$

(7.55)

but does have a vector source

$$\nabla \times \mathbf{B} = \frac{4\pi}{c} \mathbf{j}.$$

(7.56)

This is just the opposite of the electrostatic field $\mathbf{E}$, which had a scalar source ($4\pi\rho$), but no vector source. The two differential equations (7.55) and (7.56) can be taken as the starting point for magnetostatics. In the next section, we solve them for the magnetic field $\mathbf{B}$ by introducing the magnetic vector potential $\mathbf{A}$.

7.8   The Vector Potential A

From the fact that the divergence of $\mathbf{B}$ vanishes, we can infer that $\mathbf{B}$ can be written as the curl of another vector

$$\mathbf{B} = \nabla \times \mathbf{A}.$$

(7.57)

The vector $\mathbf{A}$ is called the **vector potential**. Putting Eq. (7.57) into the curl of $\mathbf{B}$ gives

$$\begin{aligned} \nabla \times \mathbf{B} &= \nabla \times (\nabla \times \mathbf{A}) \\ &= \nabla(\nabla \cdot \mathbf{A}) - \nabla^2 \mathbf{A}. \end{aligned}$$

(7.58)

We show below how we can make the first term in this equation vanish, and then show how to treat the second term.

7.8.1   Gauge transformation

To simplify Eq. (7.58), we now show that we can always choose to make the divergence of the vector potential vanish. In traditional classical magnetostatics, the vector potential is considered just a mathematical construct with no physical significance, its only purpose being to facilitate finding the physical magnetic field **B**. Since **B** equals the curl of **A**, adding a gradient to **A** will have no physical effect. Such a change in **A**

$$\mathbf{A} \to \mathbf{A}' = \mathbf{A} + \boldsymbol{\nabla}\psi \qquad (7.59)$$

is called a **gauge transformation**. Another gauge transformation, adding a constant to the scalar potential ϕ, leaves the electric field **E** unchanged, but this is so trivial and obvious that we did not mention when we introduced ϕ.

Actually, in quantum mechanics, and in some alternate approaches to electromagnetism discussed in Chapter 16, the vector potential does have physical significance. In these cases, the physics is still unchanged by a gauge transformation. In fact, the gauge transformation takes on even more importance in these cases. For now, we will use the gauge transformation to show how we can always set the divergence of **A** to zero.

Let us suppose a case where the divergence of **A** were not zero, but

$$\boldsymbol{\nabla}\cdot\mathbf{A} = \chi, \qquad (7.60)$$

where χ is some scalar field. We can make a gauge transformation on **A**

$$\mathbf{A} \to \mathbf{A}' = \mathbf{A} + \boldsymbol{\nabla}\psi. \qquad (7.61)$$

Then, taking the divergence of **A**' gives

$$\boldsymbol{\nabla}\cdot\mathbf{A}' = \boldsymbol{\nabla}\cdot\mathbf{A} + \nabla^2\psi = \chi + \nabla^2\psi. \qquad (7.62)$$

We now can set $\boldsymbol{\nabla}\cdot\mathbf{A}' = 0$ if ψ is the solution of Poisson's equation

$$\nabla^2\psi = -\chi \qquad (7.63)$$

over all space. The solution to this is similar to what we found for the scalar potential in electrostatics:

$$\psi(\mathbf{r}) = \frac{1}{4\pi} \int \frac{\chi(\mathbf{r}')d\tau'}{|\mathbf{r}-\mathbf{r}'|}. \qquad (7.64)$$

Thus, we can always find a ψ to make the divergence of the new vector potential **A**' vanish.

7.8.2   Poisson's equation for A

Now we can go back to putting a vector potential with zero divergence into Eq. (7.58) for the curl of **B**. The result is

$$\nabla^2 \mathbf{A} = -\frac{4\pi \mathbf{j}}{c},$$

(7.65)

which is Poisson's equation for **A** with a vector source **j**.

This is our first encounter with the Laplacian of a vector. The Laplacian acting on a vector is relatively simple in cartesan coordinates:

$$\nabla^2 A_i = (\partial_x^2 + \partial_y^2 + \partial_z^2) A_i,$$

(7.66)

but this simple form holds only in Cartesian coordinates, and the Laplacian of a vector is quite complicated in other coordinate systems. It is usually best to avoid the use of a coordinate system when ∇^2 acts on a vector. The Lapacian of a vector could be found from the two right hand sides of Eq. (7.58)

$$\nabla^2 \mathbf{A} = \nabla(\nabla \cdot \mathbf{A}) - \nabla \times (\nabla \times \mathbf{A}),$$

(7.67)

but that would just take us backwards in our derivation.

We show below, using the facr that the Laplacian is the divergence of the gradient (even when it's the gradient of vector), that the vector potential,

$$\mathbf{A}(\mathbf{r}) = \frac{1}{c} \int \frac{\mathbf{j}(\mathbf{r}')d\tau'}{|\mathbf{r} - \mathbf{r}'|},$$

(7.68)

is a solution of Poisson's equation.

The Laplacian of **A** is

$$
\begin{aligned}
\nabla^2 \mathbf{A}(\mathbf{r}) &= \nabla \cdot (\nabla \mathbf{A}) = \frac{1}{c} \nabla \cdot \int d\tau \, \nabla \left[\frac{\mathbf{j}(\mathbf{r}')}{|\mathbf{r} - \mathbf{r}'|} \right] \\
&= -\frac{1}{c} \nabla \cdot \int d\tau' \left[\frac{(\mathbf{r} - \mathbf{r}')\mathbf{j}(\mathbf{r}')}{|\mathbf{r} - \mathbf{r}'|^3} \right] \\
&= -\frac{1}{c} \int d\tau' \nabla \cdot \left[\frac{(\mathbf{r} - \mathbf{r}')\mathbf{j}(\mathbf{r}')}{|\mathbf{r} - \mathbf{r}'|^3} \right] \\
&= -\frac{1}{c} \int d\tau' \left\{ \nabla \cdot \left[\frac{(\mathbf{r} - \mathbf{r}')}{|\mathbf{r} - \mathbf{r}'|^3} \right] \right\} \mathbf{j}(\mathbf{r}') \\
&= -\frac{4\pi}{c} \int d\tau' \delta(\mathbf{r} - \mathbf{r}')\mathbf{j}(\mathbf{r}') \\
&= -\frac{4\pi}{c} \mathbf{j}(\mathbf{r}),
\end{aligned}
$$

(7.69)

which confirms that Eq. (7.68) is a solution of Eq. (7.65).

We can find **B** by taking the curl of **A**, which gives

$$\begin{aligned}
\mathbf{B(r)} &= \boldsymbol{\nabla} \times \mathbf{A} \\
&= \frac{-1}{c} \int \mathbf{j}(\mathbf{r'}) \times \boldsymbol{\nabla} \left[\frac{1}{|\mathbf{r} - \mathbf{r'}|} \right] d\tau' \\
&= \frac{1}{c} \int \frac{\mathbf{j}(\mathbf{r'}) \times (\mathbf{r} - \mathbf{r'}) d\tau'}{|\mathbf{r} - \mathbf{r'}|^3}.
\end{aligned} \tag{7.70}$$

This is just the Biot-Savart law, Eq. (7.39), for a current density. This demonstrates that the Biot-Savart integral for **B**, and the divergence and curl differential equations for **B** are equivalent starting points for magnetostatics.

The vector potential due to a current loop can be written as

$$\mathbf{A(r)} = \frac{I}{c} \oint \frac{d\mathbf{r'}}{|\mathbf{r} - \mathbf{r'}|}, \tag{7.71}$$

or for a surface current as

$$\mathbf{A(r)} = \frac{1}{c} \int \frac{\mathbf{K} dS'}{|\mathbf{r} - \mathbf{r'}|}. \tag{7.72}$$

In each case, taking the curl of the integral for **A** gives the Biot-Savart law for that current distribution.

7.9 Ampere's Circuital Law

Applying Stokes' theorem to Eq. (7.54) for the curl of **B**, we can derive a useful integral relation:

$$\begin{aligned}
\int_S d\mathbf{S} \cdot (\boldsymbol{\nabla} \times \mathbf{B}) &= \frac{4\pi}{c} \int_S d\mathbf{S} \cdot \mathbf{j} \\
\oint_c d\mathbf{r} \cdot \mathbf{B} &= \frac{4\pi}{c} I_S,
\end{aligned} \tag{7.73}$$

where I_S is the net current passing through the surface S. Equation (7.73) was originally called **Ampere's Circuital Law.** Stated in words, it says **The line integral of the tangential component of B around any closed loop equals $\frac{4\pi}{c}$ times the net current through the loop.**

Following the sign convention used for Stokes' theorem, the positive direction for current flow through the loop is given by a right hand rule. **If you curl the fingers of your right hand in the direction of integration around the loop, your extended thumb will be in the positive direction for current through the loop.**

The recent trend has been to call Eq. (7.73) simply **Ampere's law,** and to give no name to his law for the force between current elements, given in Eq.

(7.1) . To avoid confusion, we will follow this practice and refer to Eq. (7.73) as Ampere's law for the remainder of this text.

We can use Ampere's law to determine the discontinuity $\mathbf{B_2} - \mathbf{B_1}$ on either side of a surface current density $\mathbf{K}$ as shown on Fig. 7.9.

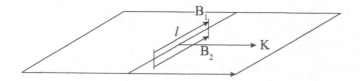

Figure 7.9: Discontinuity in B due to a surface current K.

We take for the path of integration, a rectangle with two sides of length l on either side of the current sheet, and two infinitesimal legs closing the rectangle. The current within the integration path is Kl, so Ampere's law gives

$$(B_{2t} - B_{1t})l = \frac{4\pi}{c}Kl. \tag{7.74}$$

We divide out the common factor l, and also put the discontinuity into vector form giving

$$\hat{\mathbf{n}} \times (\mathbf{B_2} - \mathbf{B_1}) = \frac{4\pi}{c}\mathbf{K}, \tag{7.75}$$

where $\hat{\mathbf{n}}$ is the normal unit vector from side 1 to side 2 of the current sheet.

Ampere's law is the magnetic equivalent of Gauss's law in electrostatics. Like Gauss's law, Ampere's law can be used to give simple derivations of the magnetic field when symmetry can be used to make the line integral trivial.

To find the magnetic field a distance $\mathbf{r}$ away from a long straight wire carrying current $\mathbf{I}$, we pick an Amperian loop that is a circle of radius r about the wire, as shown in Fig. 7.10.

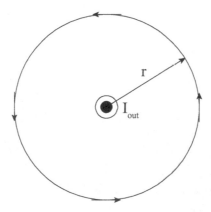

Figure 7.10: Amperian circle for a long straight current out of the page.

By symmetry, there will be no magnetic field in the direction of the long wire. There will also be no field in the direction of **r**. To state this, we have to use a bit more than just Ampere's law by recalling that all previous expressions for the magnetic field involve **r** in a cross product. This means that **B** will be in the $\hat{\mathbf{I}} \times \hat{\mathbf{r}}$ direction, as given by the right hand rule for a long wire.

By the axial symmetry of the configuration, the magnitude of **B** will be constant and its direction will always be along the circular path of integration. Then B can be taken out of the line integral in Ampere's law, leaving

$$2\pi r B = \frac{4\pi}{c} I. \tag{7.76}$$

Solving for B gives

$$B = \frac{2I}{cr}, \tag{7.77}$$

as we found from a somewhat more complicated integration using the law of Biot-Savart.

This use of Ampere's law shows that the radius of the wire does not enter, so Eq. (7.77) holds for any wire as long as its radius is less than the r for which we want to know B. It also holds for a varying current distribution in the wire, as long as this variation preserves the axial symmetry. In this case, B depends only on the net current, and not on the current distribution.

Ampere's law can also be used to find the magnetic field inside a long straight wire if the current distribution is known. Then

$$2\pi r B(r) = \frac{8\pi^2}{c} \int_0^r j(r) r \, dr. \tag{7.78}$$

For a uniform current distribution in a wire of radius R,

$$j(r) = \frac{I}{\pi R^2}, \tag{7.79}$$

so Eq. (7.78) gives

$$B(r) = \frac{2Ir}{cR^2}. \tag{7.80}$$

Ampere's law can also be used to find the magnetic field near the middle of a long solenoid. We consider a long solenoid of length L, carrying a current I in N turns. We choose a rectangular Amperian path, as shown on Fig. 7.11, with a short length l, and the far end of the rectangle far enough away for the magnetic field to be negligible. Near the middle of the solenoid, the field will be horizontal in the figure, so the contribution to Ampere's integral will vanish on the two long vertical legs.

This leaves for the Ampere law integral

$$\frac{4\pi n l I}{c} = \oint \mathbf{B} \cdot \mathbf{dr} = \int B \, dz = Bl. \tag{7.81}$$

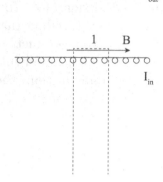

Figure 7.11: Three sides of an Amperian rectangle (dashed line) for a solenoid.

Solving for B, we get

$$B = \frac{4\pi n I}{c},$$ (7.82)

in agreement with Eq. (7.29) from the Biot-Savart law.

An important configuration is that of a **torus**, which can be considered as a long solenoid bent into a circle, as shown in Fig. 7.12. Using an Amperian loop that is a circle inside the torus, we see that the magnetic field inside is the same as it would be near the middle of the solenoid, and is given by Eq. (7.82). The torus is a useful configuration because the field inside is fairly uniform throughout with no end effects, and with zero magnetic field outside the torus.

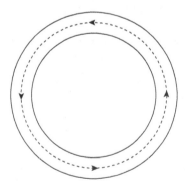

Figure 7.12: Amperian circle (dashed line) for a torus. The arrowheads represent the direction of B.

7.10   Magnetic Scalar Potential

The curl of the magnetic field, **B**, does not vanish in the presence of currents, so **B** cannot be given as the gradient of a scalar over all space. But it does turn out to be useful to define a magnetic scalar potential, whose gradient gives the magnetic field in those regions that have no current. The magnetic scalar potential, ϕ_m, would be related to the magnetic field **B** by the relation

$$\mathbf{B} = -\boldsymbol{\nabla}\phi_m, \tag{7.83}$$

just as the electric potential is related to the electric field.

We can find the magnetic scalar potential of a current loop by using the definition of the gradient, and the law of Biot-Savart for the magnetic field:

$$\begin{aligned} d\phi_m &= \mathbf{dr}\cdot\boldsymbol{\nabla}\phi_m = -\mathbf{dr}\cdot\mathbf{B} \\ &= -\mathbf{dr}\cdot\oint\frac{I\mathbf{dr}'\times(\mathbf{r}-\mathbf{r}')}{c|\mathbf{r}-\mathbf{r}'|^3}. \end{aligned} \tag{7.84}$$

Then, ϕ_m will be given by the double integral

$$\begin{aligned} \phi_m(\mathbf{r}) &= -\frac{I}{c}\int\oint\frac{\mathbf{dr}\cdot[\mathbf{dr}'\times(\mathbf{r}-\mathbf{r}')]}{|\mathbf{r}-\mathbf{r}'|^3} \\ &= -\frac{I}{c}\int\oint\frac{(\mathbf{dr}\times\mathbf{dr}')\cdot(\mathbf{r}-\mathbf{r}')]}{|\mathbf{r}-\mathbf{r}'|^3}, \end{aligned} \tag{7.85}$$

where the integration over **r** is from one side of the current loop to the other.

We can recognize the combination $\mathbf{dr}\times\mathbf{dr}'$ as the area[3] $\mathbf{dS}'$ of an infinitesimal parallelogram inside the current loop. This means that we can write Eq. (7.85) as

$$\begin{aligned} \phi_m(\mathbf{r}) &= \frac{I}{c}\int\frac{\mathbf{dS}'\cdot(\mathbf{r}'-\mathbf{r})]}{|\mathbf{r}-\mathbf{r}'|^3} \\ &= \frac{I}{c}\int d\Omega = \frac{I}{c}\Omega, \end{aligned} \tag{7.86}$$

where Ω is the solid angle subtended by the current loop as seen from the point of observation **r**.

Although Eq. (7.86) is a neat expression for the magnetic scalar potential of a current loop, its application is limited to current loops, and even then it can be quite difficult to evaluate. An alternate expression which is more generally applicable and easier to calculate is to derive the magnetic scalar potential as the integral

$$\phi_m(\mathbf{r}) = \int_{\mathbf{r}}^{\infty}\mathbf{B}\cdot\mathbf{dr}, \tag{7.87}$$

just as we had found the electric potential from the electric field.

It is necessary to use Eq. (7.87) in cases where the magnetic field is not due to a simple current loop. In fact, an important to use of the magnetic scalar

---

[3]Since **A** now represents the vector potential, we will use **dS** for the differential vector area.

potential is in finding the magnetic field in the presence of magnetic material where there is no current loop. We will see in chapter 8 that using the magnetic scalar potential makes magnetostatics in the presence of magnetic material just the same as electrostatics was in the presence of polarizability material. The magnetic scalar potential is also required for a simple understanding of the **B** and **H** fields of permanent magnets.

7.10.1   Magnetic field of a current loop

As an example of the utility of the magnetic scalar potential, we use it to find the field off the axis of a circular current loop of radius a carrying a current I, as shown in Fig. 7.13.

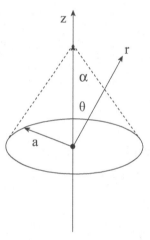

Figure 7.13: Circular current loop. The dashed lines show the extent of the solid angle subtended by the loop from a point z on the axis of the loop. The vector r is the extension of z at an angle θ off the axis.

First, we find the potential on the axis of the loop, a distance z above the plane of the loop. The potential is given in terms of the solid angle subtended by the loop as

$$
\begin{aligned}
\phi_m(z) &= \frac{I}{c}\Omega \\
&= \frac{I}{c}2\pi \int_0^\alpha \sin\theta d\theta \\
&= \frac{2\pi I}{c}(1 - \cos\alpha) \\
&= \frac{2\pi I}{c}\left[1 - \frac{z}{\sqrt{z^2 + a^2}}\right].
\end{aligned}
\tag{7.88}
$$

The magnetic field on the axis is the negative gradient of ϕ_m:

$$
\begin{aligned}
\mathbf{B} &= -\frac{2\pi I \hat{\mathbf{k}}}{c} \partial_z \left[1 - \frac{z}{\sqrt{z^2 + a^2}} \right] \\
&= \frac{2\pi I \hat{\mathbf{k}} a^2}{c(z^2 + a^2)^{\frac{3}{2}}},
\end{aligned}
\tag{7.89}
$$

which is the same as given by Eq. (7.25) using the Biot-Savart law. We also could have found ϕ by integrating the magnetic field along the z axis.

The magnetic scalar potential can be extended off the z- axis using Legendre polynomials in the same way as we used them for the electric scalar potential in Sec. 4.2.3. First, we expand the potential in powers of (a/z)

$$
\begin{aligned}
\phi_m &= \frac{2\pi I}{c} \left\{ 1 - \left[1 + \left(\frac{a}{z} \right)^2 \right]^{-\frac{1}{2}} \right\} \\
&= \frac{2\pi I}{c} \left[1 - \sum_{n=0}^{\infty} \binom{-\frac{1}{2}}{n} \left(\frac{a}{z} \right)^{2n} \right] \\
&= \frac{-2\pi I}{c} \sum_{n=1}^{\infty} \binom{-\frac{1}{2}}{n} \left(\frac{a}{z} \right)^{2n}.
\end{aligned}
\tag{7.90}
$$

We change the summation index from n to

$$
l = 2n - 1,
\tag{7.91}
$$

and also change the variable from z to the spherical coordinates r and $\theta = 0°$. We can now write the sum in the form of a standard Legendre polynomial expansion:

$$
\phi_m(r, 0°) = \frac{-2\pi I}{c} \sum_{\text{odd } l}^{\infty} \binom{-\frac{1}{2}}{\frac{l+1}{2}} \left(\frac{a}{r} \right)^{l+1} P_l(\cos 0°).
\tag{7.92}
$$

The inclusion of $P_l(\cos 0°)$ can be made because this term equals one. We now just just replace the angle of $0°$ by the general angle θ, so

$$
\phi_m(r, \theta) = \frac{-2\pi I}{c} \sum_{\text{odd } l}^{\infty} \binom{-\frac{1}{2}}{\frac{l+1}{2}} \left(\frac{a}{r} \right)^{l+1} P_l(\cos \theta)
\tag{7.93}
$$

gives the potential at any angle off the axis.

The magnetic field $\mathbf{B}$ off the axis is given as the negative gradient of ϕ_m in spherical coordinates

$$
\begin{aligned}
B_r(r, \theta) &= -\partial_r \phi_m(r, \theta) \\
&= -\frac{2\pi I}{c} \sum_{\text{odd } l}^{\infty} (l+1) \binom{-\frac{1}{2}}{\frac{l+1}{2}} \frac{a^{l+1} P_l(\cos \theta)}{r^{l+2}},
\end{aligned}
\tag{7.94}
$$

$$
\begin{aligned}
B_\theta(r, \theta) &= -\frac{1}{r} \partial_\theta \phi_m(r, \theta) \\
&= -\frac{2\pi I}{c} \sum_{\text{odd } l}^{\infty} \binom{-\frac{1}{2}}{\frac{l+1}{2}} \frac{a^{l+1} P_l^1(\cos \theta)}{r^{l+2}}.
\end{aligned}
\tag{7.95}
$$

The associated Legendre function $P_l^1(\cos\theta)$ in Eq. (7.95) arises from the derivative

$$\partial_\theta P_l(\cos\theta) = -\sin\theta\,\partial_{(\cos\theta)}P_l(\cos\theta) = -P_l^1(\cos\theta). \qquad (7.96)$$

Since the curl of **B** does not vanish everywhere, the magnetic scalar potential ϕ_m must be treated with some caution. The magnetic scalar potential on the axis above the current loop is positive. This follows from the right hand rule for determining the sign of the current.

However, below the plane of the current loop, the current is in the negative direction, while the cosine of the angle α in Fig. 7.13 is given by

$$\cos\alpha = \frac{-a}{\sqrt{z^2+a^2}}. \qquad (7.97)$$

This results in

$$\phi_m = \frac{-2\pi I}{c}\left[1 + \frac{z}{\sqrt{z^2+a^2}}\right] \qquad (7.98)$$

on the negative z axis, so the sign of ϕ_m is negative below the plane.

Comparison of Eq. (7.98) with Eq. (7.88) for ϕ_m on the positive z axis shows that there is a discontinuity of magnitude $4\pi I/c$ at $z=0$ on the axis of the loop. Looking at Fig. 7.13, we can see that this discontinuity in ϕ_m will be the same for crossing any part of the plane of the loop inside the loop. This is because the solid angle just above the loop is $+2\pi$, and is -2π just below the loop. The scalar potential ϕ_m will be continuous when passing through the plane of the loop outside the loop, because, in this case, the solid angle is zero both just above and just below the plane.

The result is that ϕ_m derived from Eq. (7.86) will have a discontinuity of magnitude $4\pi I/c$ when passing through a current loop. Although the discontinuity in ϕ_m was demonstrated here for a plane circular loop, the same discontinuity will hold for a loop of irregular shape (even non-planar). Note that, although the magnetic scalar potential ϕ_m is discontinous, its negative gradient, the magnetic field **B**, which is the physical quantity, will be continuous everywhere.

Another method to calculate the magnetic scalar potential is by using Eq. (7.87)

$$\begin{aligned}
\phi_m(z) &= \int_z^\infty \mathbf{B}(\mathbf{r}')\cdot d\mathbf{r}' \\
&= \int_z^\infty \frac{2\pi I a^2 dz'}{c(a^2+z'^2)^{\frac{3}{2}}} \\
&= \frac{2\pi I}{c}\left[1 - \frac{z}{\sqrt{z^2+a^2}}\right].
\end{aligned} \qquad (7.99)$$

For positive z, this is the same potential as derived from Eq. (7.86), but the result in Eq. (7.99) has no discontinuity at the plane of the loop. So this form for ϕ_m is not exactly the same as the form using the solid angle.

While this form has no discontinuity crossing inside the plane of the current loop, it does have its own peculiar behavior. Because of Ampere's law

$$\oint \mathbf{B}(\mathbf{r}') \cdot d\mathbf{r}' = \frac{4\pi I}{c}, \qquad (7.100)$$

the magnetic scalar potential derived from Eq. (7.87) will be multiple valued, increasing by $(4\pi I/c)$ each time the current loop is passed through.

There are thus two different representations of the magnetic scalar potential of a current loop, one discontinuous inside the plane of loop, and the other continuous but not single valued. Since each of these give the same continuous and single valued magnetic field $\mathbf{B}$, either one may be used. In the absence of magnetic monopoles, the magnetic scalar potential is just a mathematical object used for convenience in finding the physical magnetic field.

However, if a magnetic monopole (the magnetic equivalent of an electric charge) existed, then the integral in Eq. (7.87) would have the physical significance of the energy change per unit magnetic charge as the charge was moved. In fact, the energy of a magnetic charge g would not be a single valued function of position, but would increase by $(4\pi I g/c)$ each time the magnetic charge passed through a current loop. This would not lead to perpetual motion because the energy given to the magnetic charge would come from the battery or generator keeping the current in the loop constant while the magnetic charge moved.

7.11 Magnetic Dipole Moment.

7.11.1 Magnetic multipole expansion

The expansion in Eq. (7.93) can be recognized as a multipole expansion of the magnetic scalar potential ϕ_m. The magnetic multipole moments are the coefficients of the factor $\frac{P_l(\cos\theta)}{r^{l+1}}$ for each l. For the circular current loop, these are given by

$$m_l = \frac{-2\pi I a^{l+1}}{c} \begin{pmatrix} -\frac{1}{2} \\ \frac{l+1}{2} \end{pmatrix}, \quad l \text{ odd}, \qquad (7.101)$$

and are zero for even l. The most important multipole is the dipole ($l = 1$), which dominates at large r. For the circular loop, it is given by

$$m_1 = \frac{I\pi a^2}{c}. \qquad (7.102)$$

It is useful to introduce the concept of an elementary magnetic dipole $\boldsymbol{\mu}$, defined by taking the limit of Eq. (7.102 as $a \to 0$, with the product $I a^2$ fixed:

$$\boldsymbol{\mu} = \lim_{a \to 0} \frac{I\pi a^2}{c}. \qquad (7.103)$$

Unfortunately, the symbol μ is used for both the magnetic moment and for the permeability. Fortunately, the context usually makes clear which physical quantity it stands for in any particular case. Also, we will generally use the bold face $\boldsymbol{\mu}$ to designate the vector magnetic moment.

7.11.2 Magnetic dipole scalar potential of a current loop

From the multipole expansion in Eq. (7.93), we see that the dipole part of the magnetic scalar potential is given by

$$\phi_{\mathrm{md}} = \frac{\mu\cos\theta}{r^2} = \frac{\boldsymbol{\mu}\cdot\mathbf{r}}{r^3}, \quad r \neq 0, \tag{7.104}$$

and the dipole magnetic field by

$$\mathbf{B_{md}} = -\boldsymbol{\nabla}\left(\frac{\boldsymbol{\mu}\cdot\mathbf{r}}{r^3}\right) = \frac{3(\boldsymbol{\mu}\cdot\hat{\mathbf{r}})\hat{\mathbf{r}} - \boldsymbol{\mu}}{r^3}, \quad r \neq 0. \tag{7.105}$$

Equations (7.104) and (7.105) are seen to be the magnetic analogues of the equations for the potential and electric field of an electric dipole. The restriction $r \neq 0$ is necessary because of the singularity at the origin, which will be treated in the next section using the vector potential. This singularity cannot be treated by the magnetic scalar potential, which is not valid in the any region where there is current.

We can also derive the magnetic dipole moment of a general (not necessarily circular or planar) current loop directly from Eq. (7.86) for the scalar potential by expanding to lowest order in powers of (r'/r) [Note: In this and other expansions for the dipole moment, we will usually keep only enough terms in the expansion to give the dipole contribution.]

$$\begin{aligned}\phi_{md}(\mathbf{r}) &= \left[\frac{I}{c}\int_S \frac{(\mathbf{r}-\mathbf{r}')\cdot d\mathbf{S}'}{|\mathbf{r}-\mathbf{r}'|^3}\right]\\ &= \frac{I}{c}\int_S d\mathbf{S}'\cdot\left[\frac{\mathbf{r}}{r^3}+...\right]\\ &= \frac{I\mathbf{r}}{cr^3}\cdot\int_S d\mathbf{S}'\\ &= \frac{I\mathbf{S}\cdot\mathbf{r}}{cr^3}.\end{aligned} \tag{7.106}$$

This result shows that the magnetic dipole moment of a current loop is given by

$$\boldsymbol{\mu} = \frac{I\mathbf{S}}{c}, \tag{7.107}$$

where $\mathbf{S}$ is the projected vector area of the loop. The magnetic moment is given by Eq. (7.107) for any shape loop, including non-planar loops. The result in Eq. (7.102) is a special case, for a circular loop, of this more general formula.

7.11.3 Magnetic dipole vector potential of a current loop

The magnetic dipole moment and its magnetic field can also be treated using the vector potential $\mathbf{A}$ by expanding the integral for $\mathbf{A}$ in Eq. (7.71) in powers of (r'/r). This gives

$$
\begin{aligned}
\mathbf{A}(\mathbf{r}) &= \frac{I}{c} \oint \frac{d\mathbf{r}'}{|\mathbf{r} - \mathbf{r}'|} \\
&= \frac{I}{cr} \oint d\mathbf{r}' + \frac{I}{cr^3} \oint (\mathbf{r}\cdot\mathbf{r}')d\mathbf{r}' + ...
\end{aligned}
\tag{7.108}
$$

Each integral in Eq. (7.108) can be evaluated by using the integral identity in Eq. (1.124), which we repeat here:

$$
\int_S d\mathbf{S}\times[\nabla\psi] = \oint_C d\mathbf{r}\psi,
\tag{7.109}
$$

where ψ is an arbitrary scalar function. The first integral in Eq. (7.108) vanishes because the function ψ there is just a constant.

The second integral gives the vector potential, $\mathbf{A}_{md}$, of a magnetic dipole. For this integral, the function ψ in Eq. (7.109) is given by $(\mathbf{r}\cdot\mathbf{r}')$, and we get

$$
\begin{aligned}
\mathbf{A}(\mathbf{r})_{md} &= \frac{I}{cr^3} \oint_c (\mathbf{r}\cdot\mathbf{r}')d\mathbf{r}' \\
&= \frac{I}{cr^3} \int_S d\mathbf{S}'\times[\nabla'(\mathbf{r}\cdot\mathbf{r}')] \\
&= \frac{I}{cr^3} \int_S d\mathbf{S}'\times\mathbf{r} \\
&= \frac{I\mathbf{S}}{c}\times\left(\frac{\mathbf{r}}{r^3}\right).
\end{aligned}
\tag{7.110}
$$

We recognize $\frac{I\mathbf{S}}{c}$ as the magnetic dipole moment of the loop, so the vector potential of a magnetic dipole is given by

$$
\mathbf{A}_{md} = \frac{\boldsymbol{\mu}\times\mathbf{r}}{r^3}.
\tag{7.111}
$$

The magnetic dipole field $\mathbf{B}_{md}$ is the curl of $\mathbf{A}_{md}$:

$$
\begin{aligned}
\mathbf{B}_{md} &= \nabla\times\left(\frac{\boldsymbol{\mu}\times\mathbf{r}}{r^3}\right) \\
&= \boldsymbol{\mu}\nabla\cdot\left(\frac{\mathbf{r}}{r^3}\right) - \boldsymbol{\mu}\cdot\nabla\left(\frac{\mathbf{r}}{r^3}\right) \\
&= \boldsymbol{\mu}\nabla\cdot\left(\frac{\mathbf{r}}{r^3}\right) + \boldsymbol{\mu}\times\left[\nabla\times\left(\frac{\mathbf{r}}{r^3}\right)\right] - \nabla\left(\frac{\boldsymbol{\mu}\cdot\mathbf{r}}{r^3}\right) \\
&= \boldsymbol{\mu}\nabla\cdot\left(\frac{\mathbf{r}}{r^3}\right) - \nabla\left(\frac{\boldsymbol{\mu}\cdot\mathbf{r}}{r^3}\right).
\end{aligned}
\tag{7.112}
$$

The first term in Eq. (14.161) leads to a delta function, while the second term is just the negative gradient that gives the electric field due to an electric dipole. Thus, we can write

$$\mathbf{B_{md}} = 4\pi\boldsymbol{\mu}\delta(\mathbf{r}) + \mathbf{B_{ed}} \tag{7.113}$$

$$= \frac{3(\mathbf{p}\cdot\hat{\mathbf{r}})\hat{\mathbf{r}} - \mathbf{p}}{r^3} - 4\pi\hat{\mathbf{r}}(\hat{\mathbf{r}}\cdot\mathbf{p})\delta(\mathbf{r}) + 4\pi\boldsymbol{\mu}\delta(\mathbf{r}), \tag{7.114}$$

where $\mathbf{B_{md}}$ is the magnetic field due to a magnetic dipole $\boldsymbol{\mu}$, and $\mathbf{B_{ed}}$ is the field we would get if we used the electric dipole formula for the magnetic field.

The delta function singularity in Eq. (7.114) is only effective for an infinitesimal current loop or an elementary dipole. For a finite size loop, the expansion in powers of r'/r will break down as the origin is approached, and there will be no singularity.

For $r \neq 0$, we see that the magnetic dipole field has the same form as the field of an electric dipole, and as given by Eq. (7.105) using the magnetic scalar potential. However, the singular behavior at the origin is different for the two cases. From Eq. (2.74), the singular part of the electric dipole field, averaged over angle, is given by

$$\mathbf{E}(\mathbf{r})_{\mathbf{sing}} = -\frac{4\pi}{3}\mathbf{p}\delta(\mathbf{r}). \tag{7.115}$$

Adding the delta function in Eq. (7.114) gives, for the averaged singular part of the magnetic dipole field of a current loop,

$$\mathbf{B}(\mathbf{r})_{\mathbf{sing}} = +\frac{8\pi}{3}\boldsymbol{\mu}\delta(\mathbf{r}). \tag{7.116}$$

We see that the singular part of the magnetic dipole field has the opposite sign from that of the electric dipole.

7.11.4   Magnetic dipole moment of a current density

We can find the magnetic dipole moment of a confined current density by expanding the integral for the vector potential in Eq. (7.71) in powers of (r'/r):

$$\mathbf{A}(\mathbf{r}) = \frac{1}{c}\int\frac{\mathbf{j}(\mathbf{r}')d\tau'}{|\mathbf{r} - \mathbf{r}'|}$$

$$= \frac{1}{cr}\int\mathbf{j}(\mathbf{r}')d\tau' + \frac{\mathbf{r}}{cr^3}\cdot\int\mathbf{r}'\mathbf{j}(\mathbf{r}')d\tau' + ... \tag{7.117}$$

We show that the first integral in Eq. (7.117) is zero by introducing the divergence trick as follows:

$$\int\mathbf{j}d\tau = \int(\mathbf{j}\cdot\boldsymbol{\nabla})\mathbf{r}d\tau$$

$$= \int\boldsymbol{\nabla}\cdot(\mathbf{j}\mathbf{r})d\tau - \int\mathbf{r}\boldsymbol{\nabla}\cdot\mathbf{j}d\tau \tag{7.118}$$

The first integral can be converted into a surface integral by the divergence theorem, with surface taken outside the current distribution so that the integral vanishes. The second integral vanishes because $\nabla\cdot\mathbf{j} = 0$ for steady currents.

We evaluate the second integral in Eq. (7.117) by breaking it into symmetric and antisymmetric parts:

$$\int \mathbf{jr}d\tau = \frac{1}{2}\int(\mathbf{rj}+\mathbf{jr})d\tau + \frac{1}{2}\int(\mathbf{rj}-\mathbf{jr})d\tau. \qquad (7.119)$$

The symmetric integral vanishes by again introducing a divergence as follows:

$$\int(\mathbf{rj}+\mathbf{jr})d\tau = \int[\mathbf{r}(\mathbf{j}\cdot\nabla\mathbf{r})+(\mathbf{j}\cdot\nabla\mathbf{r})\mathbf{r}]d\tau$$
$$= \int\nabla\cdot(\mathbf{jrr})d\tau - \int(\nabla\cdot\mathbf{j})\mathbf{rr}d\tau. \qquad (7.120)$$

Both these integrals vanish, the first by going to a surface integral outside the current distribution, and the second because $\nabla\cdot\mathbf{j} = 0$.

Putting only the antisymmetric integral into Eq. (7.117) gives, for the magnetic dipole part of the vector potential

$$\mathbf{A(r)_{md}} = \frac{1}{2cr^3}\int\mathbf{r}\cdot[\mathbf{r'j(r')}-\mathbf{j(r')r'}]d\tau'. \qquad (7.121)$$

The above integrand can be recognized as bac-cab in the expansion of the triple cross product, so we get

$$\mathbf{A(r)_{md}} = \left[\frac{1}{2c}\int\mathbf{r'}\times\mathbf{j(r')}d\tau'\right]\times\frac{\mathbf{r}}{r^3}. \qquad (7.122)$$

Comparison with Eq. (7.111) shows that the magnetic dipole moment of a current distribution is given by

$$\boldsymbol{\mu} = \frac{1}{2c}\int\mathbf{r}\times\mathbf{j}d\tau. \qquad (7.123)$$

The magnetic moment of a current loop can be inferred from Eq. (7.123) by the substitution $\mathbf{j}d\tau\to I\mathbf{dl}$ in Eq. (7.123) to put it into the form:

$$\boldsymbol{\mu} = \frac{I}{2c}\oint\mathbf{r}\times\mathbf{dr}. \qquad (7.124)$$

This form for the magnetic moment of a current loop looks quite different than that given by Eq. (7.107) using the magnetic scalar potential, but they can be shown to be equivalent. (See Problem 7.6)

7.11.5 Gyromagnetic ratio

A charge distribution $\rho_e(\mathbf{r})$ that is moving with a velocity distribution $\mathbf{v}_e(\mathbf{r})$ has an effective current density

$$\mathbf{j} = \rho_e \mathbf{v}_e. \tag{7.125}$$

This follows because $\rho \mathbf{v} \cdot \mathbf{dS}$ is the rate at which charge will pass through a differential area $\mathbf{dS}$. Then, the integral for the magnetic moment of a moving charge distribution is given by

$$\boldsymbol{\mu} = \frac{1}{2c} \int \mathbf{r} \times (\rho_e \mathbf{v}_e) d\tau. \tag{7.126}$$

This integral is very much like the integral for the angular momentum of a mass distribution $\rho_m(\mathbf{r})$ moving with a velocity distribution $\mathbf{v}_m(\mathbf{r})$:

$$\mathbf{L} = \int \mathbf{r} \times (\rho_m \mathbf{v}_m) d\tau. \tag{7.127}$$

If the effective charged current distribution $(\rho_e \mathbf{v}_e)$ is proportional to the effective mass current distribution $(\rho_m \mathbf{v}_m)$, then the magnetic moment of the distribution will be proportional to the angular momentum, with

$$\boldsymbol{\mu} = \frac{q}{2Mc} \mathbf{L}, \tag{7.128}$$

where q and M are the total charge and mass of the distributions.

The ratio of the magnetic moment to the angular momentum of any object is called the **gyromagnetic ratio** G:

$$\boldsymbol{\mu} = G\mathbf{L}. \tag{7.129}$$

For a collection of electrons in orbital motion (as in an atom), the gyromagnetic ratio is given by

$$G = \frac{-e}{2mc}, \tag{7.130}$$

where $-e/m$ is the charge to mass ratio of an electron. We know from quantum mechanics that the orbital angular of electrons in an atom are given in integral units of $\hbar$ (Planck's constant divided by 2π). This makes the **Bohr magneton**, defined by

$$\mu_B = \frac{e\hbar}{2mc}, \tag{7.131}$$

a natural unit for atomic magnetic moments. With quantum mechanics, Eq. (7.128) would be written as

$$\boldsymbol{\mu} = -\mu_B \mathbf{L}, \tag{7.132}$$

with L now being a positive integer representing the orbital angular momentum of the electrons.

Another constant, simply called the **g factor**, is defined as the ratio of the actual magnetic moment to the magnetic moment that would be given by Eq. (7.132) for electrons of total angular momentum $\hbar J$. If $g = 1$, the magnetic moment is said to be 'normal'. An 'anomalous magnetic moment' is one for which g does not equal 1.

7.11.6 The Zeeman effect

When atoms are placed in a magnetic field B, the energy is changed by $\Delta U = -\boldsymbol{\mu}\cdot\mathbf{B}$. (See the next section.) This leads to the **Zeeman effect**, a splitting of spectral lines in a magnetic field, given by

$$\Delta U = g\mu_B B L_z \tag{7.133}$$

where L_z is the component of $\mathbf{L}$ in the direction of the magnetic field. When the g factor equals one, a single spectral line will break up into equally spaced lines as given by Eq. (7.133). This is the **Normal Zeeman effect**.

In what is called the **anomalous Zeeman effect**, the g factor differs from one, and is different for each L_z value, so the line splittings are not equally spaced. This was a puzzle for the semi-classical explanation given above. The anomalous Zeeman effect is explained in quantum mechanics by introducing an intrinsic angular momentum called the **spin** of the electron, with magnitude $s=\frac{\hbar}{2}$. At the same time, the electron has an anomalous g factor of $g=2$, (predicted by relativistic quantum mechanics), so for electrons, $\boldsymbol{\mu}=-2\mu_B\mathbf{s}$.

The total angular momentum of an atom with total spin $\mathbf{S}$ (the vector sum of all the electron spins) is given by

$$\mathbf{J} = \mathbf{L} + \mathbf{S}, \tag{7.134}$$

while the magnetic moment of the atom is given by

$$\boldsymbol{\mu} = -\mu_B(\mathbf{L} + 2\mathbf{S}). \tag{7.135}$$

Because of this mismatch between the total angular momentum $\mathbf{J}$ and the magnetic moment $\boldsymbol{\mu}$, the line splitting in a magnetic field is given by

$$\Delta U = -g_L\mu_B B J_z \tag{7.136}$$

where g_L (called the **Landé g factor**, named for Alfred Landé who first derived it from quantum addition of the angular momenta) now depends on J, J_z, L, and S. This gives an anomalous non-linear spacing of energy levels. The normal Zeeman effect, with equal spacing, occurs only for atoms with total spin $S = 0$.

7.11.7　Intrinsic magnetic moments

In addition to the two forms of magnetic dipole moment derived above (from classical current loops or current densities), intrinsic magnetic moments can arise from quantum mechanics. These magnetic moments are
• intrinsic magnetic moments of leptons and quarks, and the magnetic moments of baryons and mesons composed of quarks. These magnetic moments can usually be considered to be point magnetic moments for use in classical electromagnetism.
• magnetic moments of quantized atomic or nuclear orbits. Although these magnetic moments arise from orbital currents, they behave differently than purely classical magnetic moments because they are quantized. They can also usually be considered to be point magnetic moments.
• permanent magnets (See Chapter 8). These generally arise from quantum mechanical cooperative effects that align the spins of atomic electrons.

In each of these cases, the vector potential is given by $\mathbf{A} = \boldsymbol{\mu} \times \mathbf{r}/r^3$, and the magnetic field by $\mathbf{B} = \boldsymbol{\nabla} \times \mathbf{A}$. These intrinsic magnetic moments cannot be derived in classical electromagnetism and must be considered as given objects, in the same way that point charges are given and cannot be derived classically. Many attempts (especially by H. A. Lorentz) tried to derive a point charge classically as the limit of its radius approaching zero, but failed.

7.12　Magnetic dipole force, torque, and energy.

7.12.1　Magnetic dipole force on a current loop

The force on a current loop is given by Eq. (7.16):

$$\mathbf{F} = \frac{I}{c} \oint d\mathbf{r} \times \mathbf{B}(\mathbf{r}). \tag{7.137}$$

The line integral can be transformed into a surface integral using the vector identity

$$\oint d\mathbf{r} \times \mathbf{V} = \int (\boldsymbol{\nabla} V) \cdot d\mathbf{S} - \int (\boldsymbol{\nabla} \cdot \mathbf{V}) d\mathbf{S}. \tag{7.138}$$

To derive this identity, we start by dotting the left-hand side of Eq. (7.138)) by an arbitrary constant vector $\mathbf{k}$:

$$\begin{aligned}
\mathbf{k} \cdot \oint d\mathbf{r} \times \mathbf{V} &= \oint \mathbf{k} \cdot (d\mathbf{r} \times \mathbf{V}) \\
&= \oint d\mathbf{r} \cdot (\mathbf{V} \times \mathbf{k}) \\
&= \int [\boldsymbol{\nabla} \times (\mathbf{V} \times \mathbf{k})] \cdot d\mathbf{S}
\end{aligned}$$

$$= \int [(\mathbf{k} \cdot \boldsymbol{\nabla})\mathbf{V} - \mathbf{k}(\boldsymbol{\nabla} \cdot \mathbf{V})] \cdot d\mathbf{S}$$

$$= \mathbf{k} \cdot \int (\boldsymbol{\nabla}\mathbf{V}) \cdot d\mathbf{S} - \mathbf{k} \cdot \int (\boldsymbol{\nabla} \cdot \mathbf{V}) d\mathbf{S}. \qquad (7.139)$$

Since $\mathbf{k}$ is an arbitrary constant vector, this completes the derivation of the identity.

We apply this identity to the force equation (7.137)),

$$\begin{aligned} \mathbf{F} &= \frac{I}{c} \oint d\mathbf{r} \times \mathbf{B}(\mathbf{r}) \\ &= \frac{I}{c} \int (\boldsymbol{\nabla}\mathbf{B}) \cdot d\mathbf{S} - \frac{I}{c} \int (\boldsymbol{\nabla} \cdot \mathbf{B}) d\mathbf{S} \\ &= \frac{I}{c} \int (\boldsymbol{\nabla}\mathbf{B}) \cdot d\mathbf{S}. \end{aligned} \qquad (7.140)$$

In the limit of the area of the current loop approaching zero, $\boldsymbol{\nabla}B$ can be taken as constant in the surface integral, so

$$\begin{aligned} \mathbf{F_{md}} &= (\boldsymbol{\nabla}\mathbf{B}) \cdot \left[\frac{I}{c} \int d\mathbf{S} \right] \\ &= \boldsymbol{\nabla}(\boldsymbol{\mu} \cdot \mathbf{B}). \end{aligned} \qquad (7.141)$$

This gives the force on a magnetic dipole $\boldsymbol{\mu}$ in a magnetic field $\mathbf{B}$, evaluated at the position of the dipole.

If the current producing the magnetic field does not overlap with the current distribution of the magnetic dipole (so $\boldsymbol{\nabla} \times \mathbf{B} = 0$), we can use bac-cab and the fact that $\boldsymbol{\mu}$ is a constant vector to get another form for the force on the dipole:

$$\begin{aligned} \mathbf{F_{md}} &= \boldsymbol{\nabla}(\boldsymbol{\mu} \cdot \mathbf{B}) \\ &= \boldsymbol{\mu} \times (\boldsymbol{\nabla} \times \mathbf{B}) + (\boldsymbol{\mu} \cdot \boldsymbol{\nabla})\mathbf{B} \\ &= (\boldsymbol{\mu} \cdot \boldsymbol{\nabla})\mathbf{B} \end{aligned} \qquad (7.142)$$

The two forms for the force on a magnetic dipole in an external magnetic field are the same as those on an electric dipole in an external electric field. The magnetic dipole force given in Eq. (7.142) only holds if the external currents don't overlap the dipole, but always holds for an electric dipole.

7.12.2 Magnetic dipole energy

For intrinsic magnetic moments, orbital magnetic moments of charged particles in quantized orbits, and magnetic moments of permanently magnetized material, the energy of the magnetic moment in a magnetic field is derived in quantum mechanics and has to be treated as a given in classical electromagnetism. This energy is

$$U = -\boldsymbol{\mu} \cdot \mathbf{B}. \qquad (7.143)$$

(This magnetic energy term was used in the previous section to explain the Zeeman effect). This equation for the energy of a magnetic dipole is the same form as that for the energy of an electric dipole. In each case, if energy is conserved, the force on the dipole is given by

$$\mathbf{F} = -\boldsymbol{\nabla} U = \boldsymbol{\nabla}(\boldsymbol{\mu}\cdot\mathbf{B}). \tag{7.144}$$

This is the same energy as we derived for the force on the magnetic moment of a current loop.

It is important to note that the minus sign in Eq. (7.143) applies only to magnetic moments (like those above) that are maintained without the influence of any external interaction. For magnetic moments due to electric currents sustained by an external battery or other electrical energy source, the sign would be positive. This is because the external energy source provides twice the energy change (with the opposite sign) that would occur for an intrinsic magnetic moment, and changes the sign of the original energy change. (We will discuss this further in Chapter 9.)

The effect is similar to what we saw in Chapter 6 for the change in electrostatic energy due to the movement of a dielectric. In that case we showed that an energy change at constant voltage was equal and opposite in sign to an energy change at constant charge. For the magnetic case, it is the change in energy with current held fixed by an external source that is opposite in sign to the energy change of an intrinsic magnetic moment. It does turn out, for the fixed current case, that the force equals the positive gradient of the magnetic energy, and is still given by Eq. (7.144). That is, the sign of the force does not depend on the nature of the magnetic moment, only the magnetic energy does.

7.12.3   Magnetic dipole torque

We can derive the torque on a magnetic dipole in three different ways. First, we can derive the torque from the magnetic dipole energy by the same steps that we used for the torque on an electric dipole, leading to Eq. (2.56). This gives

$$\boldsymbol{\tau} = \boldsymbol{\mu}\times\mathbf{B}. \tag{7.145}$$

We can also derive the torque on a magnetic dipole directly from the force law of Eq. (7.16). For a current loop, we get

$$\begin{aligned}\boldsymbol{\tau} &= \frac{I}{c} \oint \mathbf{r}\times(\mathbf{dr}\times\mathbf{B}) \\ &= \frac{I}{c} \oint \mathbf{dr}(\mathbf{r}\cdot\mathbf{B}) - \frac{I}{c} \oint \mathbf{B}(\mathbf{r}\cdot\mathbf{dr}).\end{aligned} \tag{7.146}$$

For the magnetic dipole torque, we need keep only the constant $\mathbf{B}$ term in an expansion of $\mathbf{B}$ in powers of r. Then, the second integral in Eq. (7.146) vanishes because it is the line integral of $\mathbf{dr}$ dot a gradient. We use the identity of Eq. (1.124) in the remaining integral to get

$$
\begin{aligned}
\frac{I}{c}\oint (\mathbf{r}\cdot\mathbf{B})\mathbf{dr} &= \frac{I}{c}\int_S \mathbf{dS}\times\boldsymbol{\nabla}(\mathbf{r}\cdot\mathbf{B}) \\
&= \frac{I\mathbf{S}}{c}\times\mathbf{B} \\
&= \boldsymbol{\mu}\times\mathbf{B}.
\end{aligned}
\tag{7.147}
$$

Finally, for the magnetic dipole torque on a current density, we use the magnetic force law to get

$$
\begin{aligned}
\boldsymbol{\tau} &= \frac{1}{c}\int \mathbf{r}\times(\mathbf{j}\times\mathbf{B})d\tau \\
&= \frac{1}{c}\int \mathbf{j}(\mathbf{r}\cdot\mathbf{B})d\tau - \frac{1}{c}\int \mathbf{B}(\mathbf{j}\cdot\mathbf{r})d\tau.
\end{aligned}
\tag{7.148}
$$

The second integral in Eq. (7.148) vanishes. This is because we can write it as

$$
\begin{aligned}
\int (\mathbf{j}\cdot\mathbf{r})d\tau &= \frac{1}{2}\int \mathbf{j}\cdot\boldsymbol{\nabla}(r^2)d\tau \\
&= \frac{1}{2}\int [\boldsymbol{\nabla}\cdot(\mathbf{j}r^2) - r^2\boldsymbol{\nabla}\cdot\mathbf{j}]d\tau \\
&= 0.
\end{aligned}
\tag{7.149}
$$

The first integral in Eq. (7.148) is the same as the integral in Eq. (7.119). The same method of integration leads to

$$
\boldsymbol{\tau} = \boldsymbol{\mu}\times\mathbf{B},
\tag{7.150}
$$

so all three of the above methods lead to the same equation for the torque on a magnetic dipole.

7.12.4 Fermi-Breit interaction between magnetic dipoles

For $\mathbf{r}\neq 0$, the interaction energy between two magnetic dipoles has the same form as that given for two electric dipoles by Eq. (2.61):

$$
U_{\boldsymbol{\mu}\boldsymbol{\mu}'} = \frac{\boldsymbol{\mu}\cdot\boldsymbol{\mu}' - 3(\boldsymbol{\mu}\cdot\hat{\mathbf{r}})(\boldsymbol{\mu}'\cdot\hat{\mathbf{r}})}{r^3}, \quad r\neq 0.
\tag{7.151}
$$

However, the singular energy for $\mathbf{r} = 0$ is different for the two cases. The singular energy for two electric dipoles (averaged over angle) is given by Eq. (2.74):

$$
\langle U_{\mathbf{p}\mathbf{p}'}\rangle = \frac{4\pi}{3}(\mathbf{p}\cdot\mathbf{p}')\delta(\mathbf{r}-\mathbf{r}'),
\tag{7.152}
$$

while the singular energy for two magnetic dipoles is given, using Eq. (7.116), by

$$\langle U_{\mu\mu'}\rangle = -\frac{8\pi}{3}(\boldsymbol{\mu}\cdot\boldsymbol{\mu}')\delta(\mathbf{r}-\mathbf{r}'). \tag{7.153}$$

The singular interaction between two magnetic dipoles plays an important role in atomic and elementary particle physics. In quantum mechanics, the singular term enters for wave functions that do not vanish at the origin. This interaction between the magnetic dipole moments of a bound electron and the nucleus produces hyperfine splitting of atomic energy levels. The magnetic dipole-dipole hyperfine interaction between bound quarks is one component of the small mass differences between the neutron and proton, and between other baryons in isotopic multiplets in the quark model.

In classical terms, the question arises as to how to treat the intrinsic magnetic moments of the electron, nucleus, and quarks. These moments do not necessarily arise from orbital motion, and so might not be caused by simple current loops. Another possibility could be that these intrinsic moments are caused by North and South magnetic poles, as in the elementary electric dipole shown in Fig. 1.18. The difference between these two models is in the sign of the hyperfine interaction, which is positive for the magnetic pole model, and negative for the current loop model. Experimentally, the sign of the hyperfine interaction for both atoms and baryons is found to be negative, so a classical picture of the intrinsic magnetic moments would that of an infinitesimal current loop.

Actually, it is not correct to use the classical model to describe intrinsic magnetic moments. Intrinsic moments of elementary particles arise from relativistic quantum mechanics, and have nothing to do with current loops. This is an example of an incorrect classical model (current loops) that happens to give the correct answer. Although intrinsic moments of elementary particles are not current loops, Eqs. (7.151) and (7.116) do give their interaction energy correctly.

The singular dipole-dipole interaction of Eq. (7.153) is called the **Fermi-Breit interaction**, after Enrico Fermi who suggested it as a reasonable outcome of a one-body relativistic wave equation (the Dirac equation), and Gregory Breit who derived the interaction from a two-body relativistic wave equation (the Breit equation).

7.13 Problems

1. Two parallel straight wires, each of length L, are a distance d apart. Each wire carries a current I.

   (a) Find the force between the wires. (Neglect the force on the wires required to close the circuit.) [Ans: $F = \frac{2I^2}{cd}(\sqrt{L^2 + d^2} - d)$.

   (b) Find the ratio d/L required for the force between the wires to be only 1% lower than the force that would be given using Eq. (7.22).

2. A square loop with sides of length a carrying a current I is a distance a from a long straight wire also carrying current I, as shown below.

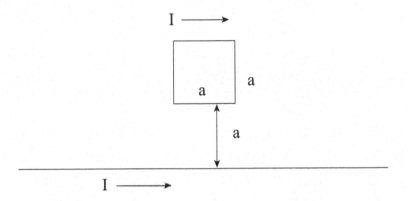

   (a) Find the force on the square loop.

   (b) Find the force on the long straight wire due to the current square loop using Eq. (7.23).

3. (a) Find the magnetic field at a distance $\mathbf{d}$ along the axis of a square current loop composed of four current elements, each of length $2a$, carrying a current I.

   (b) Compare your answer to part (a) with the field for a circular loop of radius a carrying the same current I.

      i. Which loop has the larger field at the center of the loop? Explain why this is reasonable.

      ii. Which loop has the larger field for large d?
          Can you explain this?

4. Two circular current loops of radius R, each carrying a current I (in the same direction), are a distance L apart on the same axis, as shown on below.

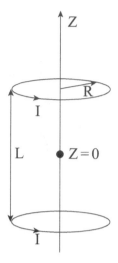

(a) Find the magnetic field on the z-axis between the loops.

(b) Show that the first derivative $\frac{dB}{dz}$ vanishes at $z=0$.

(c) Find the ratio of L/R for which the second derivative $\frac{d^2B}{dz^2}$ vanishes at $z=0$. (This combination of loops are called **Helmholtz coils**, and are used to provide a relatively constant field between the coils.)

5. (a) What magnetic field (in gauss) is required to bend electrons with kinetic energy 200 eV into a circular path with radius 5 cm?

   (b) What is the velocity and the angular velocity (cyclotron frequency) of the electrons?

6. A charged particle is moving upward (tending toward positive z) in approximate helical motion in a magnetic field whose z component is increasing in the z direction.

   (a) Show that the magnetic field must have a component in the negative r direction in cylindrical coordinates.

   (b) Show that there will be a downward component of force on the particle, tending to slow and reverse its upward motion.

   This result is the basis for "magnetic mirror" attempts to confine highly ionized gases (plasmas), and for the reflection near the Earth's poles of electrons in the Van Allen belt.

7. An electron is initially at rest in a uniform electric field E in the negative y direction and a uniform magnetic field B in the negative z direction.

(a) Solve the equations of motion given by the Lorentz force, and show that

$$x(t) = \frac{cE}{\omega B}(\omega t - \sin \omega t) \qquad (7.154)$$

$$y(t) = \frac{cE}{\omega B}(1 - \cos \omega t), \qquad (7.155)$$

where $\omega = (eB/mc)$.

(b) Show that $x(t)$ and $y(t)$ satisfy the constraint equation

$$(x - \omega R t)^2 + (y - R)^2 = R^2, \qquad (7.156)$$

with $R = (cE/\omega B)$. The functions $x(t)$ and $y(t)$ are the parametric equations for a cycloid, which is the locus of a point of the rim of a wheel rolling along the x axis.

(c) Sketch the path of the electron in the x-y plane for several cycles of the motion.

8. Two charged particles are moving with constant velocities, as shown below.

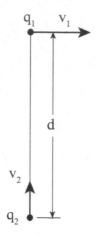

(a) Find the force on each particle.

(b) Find the net unbalanced force on the system of both electrons.

9. For each of the following cases, integrate the **B** field along the z axis from $-\infty$ to $+\infty$ to check agreement with Ampere's law:

(a) a circular current loop.

(b) a square current loop.

(c) a solenoid.

10. (a) Derive Eq. (7.42), the symmetric form of the force between two current densities, from Eq. (7.41).

(b) Show directly that the divergence of the vector potential **A** given by the integral in Eq. (7.68) vanishes.

11. A long straight wire of radius R carrying a current I has a circular portion of radius a cut out at a distance d from the center, as shown below. Find the magnetic field inside the cut-out portion. (Hint: Use a trick to reduce this to a problem of two long straight wires.)

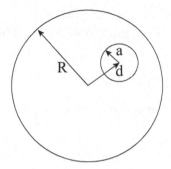

12. A long straight wire of radius R has a current distribution $j(r) = j_0 \cos\left(\frac{\pi r}{2R}\right)$.

    (a) Find the total current in the wire.

    (b) Find the magnetic field outside the wire.

    (c) Find the magnetic field at a radius $r < R$.

13. (a) Find the magnetic scalar potential on the axis of the square current loop in problem 3.

    (b) Use the magnetic scalar potential to find the magnetic field on the axis of the square loop.

14. Find the magnetic moment of the square current loop in problems 3 and 13

    (a) from its area.

    (b) by expanding the magnetic scalar potential for large distance along the axis of the loop.

15. Two circular current loops of radius R, each carrying a current I (in the same direction), are a distance $L = 4R$ apart on the same axis, Find the force between the two coils

    (a) by applying the magnetic dipole field of the bottom coil to the upper coil.

    (b) by finding the dipole-dipole force between the two coils. Use the same formula as for the electric dipole-dipole force in Chapter 2.

16. Show that $\oint_c \mathbf{r} \times d\mathbf{r} = 2\mathbf{S}$, where $\mathbf{S}$ is the projected area of the loop c. This shows that the magnetic scalar potential and the vector potential lead to the same magnetic moment (Hint: Dot the integral with a constant vector $\mathbf{k}$, and apply Stokes' theorem.)

17. A sphere rotating with angular velocity ω has a uniform mass density ρ_m and a uniform surface charge density σ. The sphere has a total mass M and a net charge Q.

    (a) Find its magnetic moment.

    (b) Find its gyromagnetic ratio and its g factor.

18. A uniformly charged disk of radius R carrying a charge Q is rotating with angular velocity ω about its axis. Find its magnetic moment. (Integrate over a series of current loops.)

19. The probability distribution of the electron in the hydrogen atom is given by $P(r) = e^{-2r/a}/\pi a^3$, where a $(=.53\,\text{Å})$ is called the "Bohr radius" of hydrogen. Calcuate the Fermi-Breit hyperfine interaction energy between the magnetic moment of the electron and that of the proton, which has a magnetic moment of $\mu_p = 2.79\ \mu_N$.

20. (a) Calculate the Bohr magneton in units of eV/gauss.
        (Answer: $\mu_B = 5.79 \times 10^{-9}$ eV/gauss.)

    (b) The **nuclear magneton** μ_N uses the proton mass, instead of the electron mass in defining the magneton. Calculate the nuclear magneton in units of eV/gauss.
        (Answer: $\mu_N = 3.15 \times 10^{-12}$ eV/gauss.)

Chapter 8

Magnetization and Ferromagnetism

8.1 Magnetic Field Including Magnetization

In magnetic materials, the molecules can have magnetic dipole moments due to circulating currents or the intrinsic magnetic moments of valence electrons that can become aligned due to an applied magnetic field. In either case, the Biot-Savart law for finding the total magnetic field has to be modified to include the dipole field due to the molecular magnetic moments. The procedure is very much like the modification of Coulomb's law in section 6.1.

The polarization of the molecules can be characterized by a magnetic dipole moment per unit volume, called the **magnetization M**. In terms of the magnetization, a small bit of matter of volume ΔV will have a magnetic dipole moment

$$\boldsymbol{\mu} = \mathbf{M}\Delta V. \tag{8.1}$$

In general, the magnetization will be a function of position $\mathbf{M}(\mathbf{r})$.

When there is magnetization, the derivation of the vector potential of a current density has to be modified to take the dipole field due to the magnetization into account. Using Eq. (8.1), the vector potential due to the magnetization in a collection of volume elements ΔV_n is given by

$$\mathbf{A_M}(\mathbf{r}) = \sum_n \frac{\mathbf{M}(\mathbf{r_n}) \times (\mathbf{r} - \mathbf{r_n})\Delta V_n}{|\mathbf{r} - \mathbf{r_n}|^3}. \tag{8.2}$$

In the limit that the ΔV_n approach zero, the sum becomes the integral

$$\mathbf{A_M}(\mathbf{r}) = \int \frac{\mathbf{M}(\mathbf{r}') \times (\mathbf{r} - \mathbf{r}')d\tau'}{|\mathbf{r} - \mathbf{r}'|^3}. \tag{8.3}$$

This integral can be modified by using the curl theorem of Eq. (1.120):

$$\mathbf{A_M}(\mathbf{r}) \;=\; \int \frac{\mathbf{M}(\mathbf{r}') \times (\mathbf{r} - \mathbf{r}')d\tau'}{|\mathbf{r} - \mathbf{r}'|^3}$$

$$= \int \left[\mathbf{M}(\mathbf{r}') \times \mathbf{\nabla}' \frac{1}{|\mathbf{r} - \mathbf{r}'|} \right] d\tau'$$

$$= \int \left\{ -\mathbf{\nabla}' \times \left[\frac{\mathbf{M}(\mathbf{r}')}{|\mathbf{r} - \mathbf{r}'|} \right] + \frac{\mathbf{\nabla}' \times \mathbf{M}(\mathbf{r}')}{|\mathbf{r} - \mathbf{r}'|} \right\} d\tau'$$

$$= \int \frac{\mathbf{M}(\mathbf{r}') \times d\mathbf{S}'}{|\mathbf{r} - \mathbf{r}'|} + \int \frac{\mathbf{\nabla}' \times \mathbf{M}(\mathbf{r}') d\tau'}{|\mathbf{r} - \mathbf{r}'|}. \tag{8.4}$$

The full vector potential is given by the sum of a contribution, $\mathbf{A}_{jK}$, due to applied volume and surface current densities, and the magnetization contribution $\mathbf{A}_{\mathbf{M}}$ given above. The sum of these two contributions is

$$\begin{aligned} \mathbf{A}(\mathbf{r}) &= \mathbf{A}_{jK}(\mathbf{r}) + \mathbf{A}_{\mathbf{M}}(\mathbf{r}) \\ &= \frac{1}{c} \int d\tau' \frac{[\mathbf{j}(\mathbf{r}') + c\mathbf{\nabla}' \times \mathbf{M}(\mathbf{r}')]}{|\mathbf{r} - \mathbf{r}'|} \\ &\quad + \frac{1}{c} \int \frac{\mathbf{K}(\mathbf{r}') dS' + c\mathbf{M}(\mathbf{r}') \times d\mathbf{S}'}{|\mathbf{r} - \mathbf{r}'|}. \end{aligned} \tag{8.5}$$

The full magnetic field is given by the curl of $\mathbf{A}$:

$$\begin{aligned} \mathbf{B}(\mathbf{r}) &= \frac{1}{c} \int d\tau' [\mathbf{j}(\mathbf{r}') + c\mathbf{\nabla}' \times \mathbf{M}(\mathbf{r}')] \times \frac{(\mathbf{r} - \mathbf{r}')}{|\mathbf{r} - \mathbf{r}'|^3} \\ &\quad + \frac{1}{c} \int [\mathbf{K}(\mathbf{r}') dS' + c\mathbf{M}(\mathbf{r}') \times d\mathbf{S}'] \times \frac{(\mathbf{r} - \mathbf{r}')}{|\mathbf{r} - \mathbf{r}'|^3}. \end{aligned} \tag{8.6}$$

Equation (8.6) shows that the curl of the magnetization acts like an additional volume current density, given by

$$\mathbf{j_b} = c\mathbf{\nabla} \times \mathbf{M}, \tag{8.7}$$

and the tangential component of $\mathbf{M}$ at a surface acts like a surface curent density, given by

$$\mathbf{K_b} = c\mathbf{M} \times \hat{\mathbf{n}}. \tag{8.8}$$

The subscript $\mathbf{b}$ refers to the fact that these effective current densities are due to the motion and magnetic moments of electrons that are bound in each molecule, producing the magnetization. The usual current densities $\mathbf{j}$ and $\mathbf{K}$ are sometimes called $\mathbf{j_{free}}$ and $\mathbf{K_{free}}$ because they are due to electrons that move through the material. Whenever we write $\mathbf{j}$ or $\mathbf{K}$ with no subscript, we will mean the free current densities.

8.2   The H field, Susceptibility and Permeability

The equation for the curl of $\mathbf{B}$ is modified by the additional bound current term in the numerator of Eq. (8.6), so

$$\mathbf{\nabla} \times \mathbf{B} = \frac{4\pi}{c}(\mathbf{j} + \mathbf{j_b})$$

$$= \frac{4\pi}{c}\mathbf{j} + 4\pi\boldsymbol{\nabla}\times\mathbf{M}. \qquad (8.9)$$

This will change Ampere's law for **B** to

$$\oint [\mathbf{B}(\mathbf{r}) - 4\pi\mathbf{M}(\mathbf{r})]\cdot\mathbf{dr} = \frac{4\pi}{c}I. \qquad (8.10)$$

As a mathematical convenience, we define a vector field **H**, given by

$$\mathbf{H} = \mathbf{B} - 4\pi\mathbf{M}. \qquad (8.11)$$

The introduction of the field **H** simplifies the curl equation, which can now be written as

$$\boldsymbol{\nabla}\times\mathbf{H} = \frac{4\pi}{c}\mathbf{j}, \qquad (8.12)$$

and Ampere's law for **H** is

$$\oint \mathbf{H}(\mathbf{r})\cdot\mathbf{dr} = \frac{4\pi I}{c}. \qquad (8.13)$$

It looks as if the magnetization has disappeared from the equations, but it has just been hidden away in the definition of **H**. Nothing useful can be done with Eq. (8.12) unless there is a constituitive relation that specifies the magnetization **M**. The simplest such relation is when **M** is proportional to the **B** field that causes the magnetization.

At this point, the historical development of the theory of magnetostatics departed from the logical development of electrostatics by writing the proportionality between the magnetization and the magnetic field in terms of the vector **H** rather than the physical field **B**, so the constituitive relation for the magnetization is written as

$$\mathbf{M}(\mathbf{r}) = \chi_m\,\mathbf{H}(\mathbf{r}). \qquad (8.14)$$

The proportionality constant χ_m is called the **magnetic susceptibility**. Even though this constitutive relation gives **M** as a function of **H**, it is important to remember that it is really **B** rather than **H** that acts to produce the magnetization.

B is directly related to **H** by the equation

$$\mathbf{B}(\mathbf{r}) = (1 + 4\pi\chi_m)\mathbf{H}(\mathbf{r}) = \mu\,\mathbf{H}(\mathbf{r}). \qquad (8.15)$$

The constant

$$\mu = (1 + 4\pi\chi_m) \qquad (8.16)$$

is called the **permeability**. Note that in these Gaussian units the vacuum susceptibility is zero and its permeability is one

In SI units, Eqs (8.9)-(8.16) are quite different. We briefly list them here:

$$\nabla \times \mathbf{B} = \mu_0[\mathbf{j} + \nabla \times \mathbf{M}], \quad \mathbf{SI} \tag{8.17}$$

$$\mathbf{H} = \frac{\mathbf{B}}{\mu_0} - \mathbf{M}, \quad \mathbf{SI} \tag{8.18}$$

$$\nabla \times \mathbf{H} = \mathbf{j}, \quad \mathbf{SI} \tag{8.19}$$

$$\mathbf{M} = \chi_m \mathbf{H}, \quad \mathbf{SI} \tag{8.20}$$

$$\mathbf{B} = \mu \mathbf{H}, \quad \mathbf{SI} \tag{8.21}$$

$$\mu = \mu_0(1 + \chi_m), \quad \mathbf{SI}. \tag{8.22}$$

In SI units, another constant

$$\mu_r = \frac{\mu}{\mu_0} = 1 + \chi_m \quad \mathbf{SI} \tag{8.23}$$

is introduced. μ_r is called the **relative permeability**. Then, in SI units, the actual permeability is written

$$\mu = \mu_r \mu_0, \tag{8.24}$$

using three contants instead of one. Luckily, the relative permeability μ_r in the SI system and the permeability μ in the Gaussian system are the same dimensionless numerical constant.

The SI unit for H is ampere-turns per meter, corresponding to the SI form for the magnetic field near the middle of a solenoid:

$$H = \frac{NI}{L}. \tag{8.25}$$

In the Gaussian system, **B** and **H** can have the same units (gauss), but for historical reasons and to honor another great physicist, the **H** field has been given the unit Oersted. Then, in free space, **B** and **H** would have the same numerical value, but different units. We will not follow that practice in this text, using gauss as the unit for all four electromagnetic fields: **B, H, E, D**.

The names given to the electromagnetic fields are different in SI units, and in Gaussian units. In both SI and Gaussian units, E is called the 'electric field' and D is called the 'displacement field'. However, B and H have different names in the two unit systems. In SI units, the H field is called the 'magnetic field'. This is probably because it is the H field that is directedly produced by electrical engineers with electric currents. B then is given the name 'magnetic flux density' and the unit Webers per square meter.

In Gaussian units, it is B that is called the magnetic field. This preserves the strong connection between B and E, which are contained in the same tensor in special relativity. There could be confusion due to these different naming schemes, but fortunately the standard practice throughout the world is to refer to them simply as the E, D, B, and H fields.

Just as was the case for the permittivity in electrostatics, the permeability can have a more complicated character than just being a simple number multiplying **H**. These properties, which parallel those of the permittivity, are listed below:

1. If the permeability is just a constant number, the material is called a **simple** magnetic material.

2. The permeability can depend on position as $\mu(\mathbf{r})$. This is called an **inhomogeneous** magnetic material.

3. The permeability can depend on H as $\mu(H)$, so B is no longer proportional to H. This is called a **non-linear** magnetic material.

4. The dependence of B on H can be so non-linear, or not even single valued, that a permeability cannot be usefully defined. This is called a **ferromagnetic** material.

5. The permeability may be a dyadic $[\boldsymbol{\mu}]$ so

$$\mathbf{B} = [\boldsymbol{\mu}]\cdot\mathbf{H}. \tag{8.26}$$

Then **B** and **H** will generally not be in the same direction. This is called a **non-isotropic** magnetic material. Under rotations, $[\boldsymbol{\mu}]$ transforms as a second rank tensor (See Chapter 14.), so it is usually called the **permeability tensor**.

6. The permeability may depend on the frequency as $\mu(\omega)$. This is called a **dispersive** magnetic material.

The most general permeability would be written as $[\boldsymbol{\mu}(\mathbf{r}, \mathbf{H}, \omega)]$. A simple magnetic material could also be called a homogeneous, linear, isotropic, non-dispersive magnetic material.

If the susceptiblility is positive and, consequently, the permeability greater than one, the material is said to be **paramagnetic**. This arises if the molecules of the material have permanent magnetic dipole moments that become oriented along an applied magnetic field, producing a positive magnetization, analogous to that given by Eq. (6.104) for the corresponding electric polarization.

If the molecules of a material have no permanent magnetic moment, introducing a magnetic field will induce a current in each molecule that will produce a magnetic moment anti-parallel to the field. This is because of Lenz's law discussed in the next chapter. This leads to a negative susceptibility, and a permeability less than one. Such a material is called a **diamagnetic**.

Note that for magnetization, the two different effects discussed above have opposite signs for the susceptibility, while for electric polarization discussed in subsection 6.6.3 both effects give the same sign, so the electric susceptibilty is generally positive and the permittivity greater than one.

You may have noticed that we have used the term 'diamagnetic' for the case with permeability less than one, but the term 'dielectric' referred to permittivity greater than one. The reason for this disparity in terminology is that the electrostatic relation $\mathbf{D} = \epsilon\mathbf{E}$ gives the mathematical construct $\mathbf{D}$ in terms of the physical field $\mathbf{E}$, while the magnetostatic relation $\mathbf{B} = \mu\mathbf{H}$ gives the physical field $\mathbf{B}$ in terms of the mathematical construct $\mathbf{H}$.

8.3 Comparison of Magnetostatics and Electrostatics

We now consider the equations for $\mathbf{B}$ and $\mathbf{H}$ in regions where there are no free currents. Then

$$\boldsymbol{\nabla}\cdot\mathbf{B} = 0 \tag{8.27}$$
$$\boldsymbol{\nabla}\times\mathbf{H} = 0. \tag{8.28}$$

Just as for the electrostatic case in Section 6.4.1, these differential equations lead to boundary conditions on $\mathbf{B}$ and $\mathbf{H}$ at the interface of two materials:

$$\mathbf{B}_{\text{normal}} \text{ is continuous}, \quad \mathbf{H}_{\text{tangential}} \text{ is continuous}. \tag{8.29}$$

We can also define, in analogy with electrostatics, a magnetic scalar potential related to the $\mathbf{H}$ field by

$$\mathbf{H} = -\boldsymbol{\nabla}\phi_m. \tag{8.30}$$

Note that these magnetostatic equations and boundary conditions are the same as the electrostatic relations satisfied by $\mathbf{D}$ and $\mathbf{E}$. Thus, with the substitutions

$$\mathbf{D} \to \mathbf{B}, \quad \mathbf{E} \to \mathbf{H}, \tag{8.31}$$

the electrostatic equations become the magnetostatic equations. This means we have already solved many magnetostatic problems if we just write in $\mathbf{B}$ and $\mathbf{H}$ in place of $\mathbf{D}$ and $\mathbf{E}$.

We list below a number of electrostatic results that also follow for simple magnetic materials, citing the corresponding electrostatic equation.

For a needle aligned with a magnetic field $\mathbf{B_0}$ [Eq. (6.45)]

$$\mathbf{H} = \mathbf{B_0}, \quad \mathbf{B} = \mu\mathbf{B_0}. \tag{8.32}$$

For a lamina aligned perpendicular to a magnetic field $\mathbf{B_0}$ [Eq. (6.46)]

$$\mathbf{H} = \mathbf{B_0}/\mu, \quad \mathbf{B} = \mathbf{B_0}. \tag{8.33}$$

For a sphere of radius R placed into a uniform magnetic field $\mathbf{B_0}$, the magnetic field inside the sphere is [Eq. (6.68)]

$$\mathbf{B(r)} = \frac{3\mu}{\mu + 2}\mathbf{B_0}, \quad r < R. \tag{8.34}$$

The magnetic field outside the sphere is the original uniform field plus a field due to a magnetic dipole $\boldsymbol{\mu}$ induced in the sphere [Eq. (6.70)]:

$$\mathbf{B(r)} = \frac{3(\boldsymbol{\mu}\cdot\hat{\mathbf{r}})\hat{\mathbf{r}} - \boldsymbol{\mu}}{r^3} + \mathbf{B_0}, \quad r > R. \tag{8.35}$$

The induced magnetic moment is given by [Eq. (6.69)]

$$\boldsymbol{\mu} = \frac{\mu - 1}{\mu + 2}\mathbf{B_0}R^3. \tag{8.36}$$

(Be careful not to confuse the magnetic moment $\boldsymbol{\mu}$ with the permeability μ.) Field lines for $\mathbf{H}$ are those shown in Fig. (6.6) for $\mathbf{E}$. Field lines for $\mathbf{B}$ would be the same outside the sphere, but the $\mathbf{B}$ field would be stronger by a factor μ inside.

8.4   Ferromagnetism

The molecules of **Ferromagnetic** materials like iron, or iron based alloys, have large magnetic moments that can become permanently aligned due to cooperative interactions between the moments. A good understanding of this process requires quantum statistical mechanics. We will just assume that the cooperative alignment occurs and leads to a large magnetization $\mathbf{M}$, much larger than the $\mathbf{H}$ field produced by free currents.

There are two classes of ferromagnetism, so called **soft ferromagnetism** and **hard ferromagnetism**. But don't take these terminologies too literally. All ferromagnets are hard if you get hit by them. Soft ferromagnets can acquire a large magnetization, with a susceptibility that is a highly non-linear function $\chi(\mathbf{H})$. When the currents producing the $\mathbf{H}$ field are turned off, the magnetization $\mathbf{M}$ of a soft ferromagnet will return to zero. For this reason, soft ferromagnets are used in electromagnets so that the piece of iron you pick up will drop when the current is turned off. Hard ferromagnets have no simple relation between $\mathbf{M}$ and $\mathbf{H}$, so the concept of a susceptibility or a permeability becomes meaningless. The first rule in treating hard ferromagnets is **don't try to use** χ_m or μ. The magnetization in a hard ferromagnet is often opposite to the direction of $\mathbf{H}$, and will usually remain large when the currents producing $\mathbf{H}$ go to zero. Because of this last property, hard ferromagnets can be used as permanent magnets.

CLASSICAL ELECTROMAGNETISM

8.5   Hysteresis

Figure 8.1 shows the relation between **B** and **H** for a ferromagnetic material placed inside a toroidal solenoid. The scales for **B** and **H** are different because **B** can become very much larger than **H**.

It is best to have the magnetic material in the shape of a torus, with the current windings wrapped continuously around it. This is to avoid end effects, which we shall see can be very important. (A long rod inside a much longer solenoid can be used if the fields are measured far from the ends of the rod.) Inside the torus or solenoid, **H** will be given by

$$H = \frac{4\pi n I}{c},\qquad(8.37)$$

and can be varied at will by changing the current I.

Since there are no free currents at the surface of the magnetic material, the tangential component of **H** will be continuous at the surface of the material, and will be the same inside the magnetic material as in the space outside. Thus the value of **H** inside the material will also be given by Eq. (8.37).

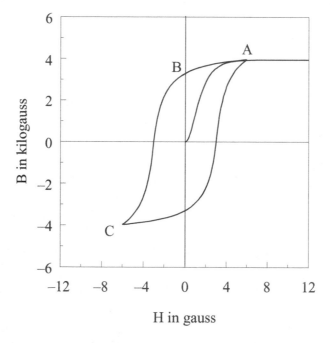

Figure 8.1: Hysteresis curve for **B** *vs* **H**.

For a linear paramagnetic or diamagnetic material, the graph of **B** *vs* **H** would be a straight line, with slope greater than one (paramagnetic) or slightly less than one (diamagnetic). The curves for these cases would look almost horizontal on Fig. 8.1 because **B** is plotted in kilogauss to keep **B** and **H** on the same plot.

For a ferromagnetic material, the curve of **B** *vs* **H** on Fig. 8.1 shoots up steeply until **B** is thousands of times bigger than **H**. (The steepness of this rise is not so obvious in Fig. 8.1 because of the difference in scales for **B** and **H**). The slope of the ferromagnetic curve then decreases to unity in what is called **saturation**. This occurs when the cooperative alignment of the molecular magnetic moments has reached its maximum. **M** is constant beyond that point on the curve.

As the **H** field is decreased for a soft ferromagnet, the **B** field will retrace the same curve as for its increase, going to zero as **H** does. For a hard ferromagnet, as **H** decreases to zero from any point A, the magnetic field, **B**, follows the curved path shown on the figure as AB. Note that **B** is quite large, even for **H**=0. This constitutes **permanent magnetization**. That is, the magnetization, given by $\mathbf{M} = (\mathbf{B} - \mathbf{H})/4\pi$, will also be large when **H** goes back to zero.

As **H** becomes negative, representing an **H** in the opposite direction, **B** continues on the smooth curve BC, approaching saturation in the other direction. Note that for the first part of this curve, **B** and **H** are in opposite directions. Now, if **H** is returned to zero and then to its maximum value in the original direction, **B** follows the curve CA. This leads to a closed path if $\mathbf{H_C}$ is chosen as the negative of $\mathbf{H_A}$. The resulting curve ABCA is called a **hysteresis loop**.

We have to emphasize that the standard hysteresis loop for any material is only valid for the toroidal geometry we have described above. If a different experimental geometry is used either for the coils or the material inside, corrections may have to be made for end effects, or the resulting curve of **B** *vs* **H** could be quite different. The end effects will be small for a long rod near the middle of a longer solenoid, so that can be used as a substitute for a torus.

The language we used in the above description made it sound like **H** were a physical field producing **B**, but that is not the actual case. It is the **B** field that produces the magnetic torques on the molecules that result in the magnetization. In fact, we could interchange the axes on Fig. 8.1, which then could be considered a plot of the combination $\mathbf{B} - 4\pi\mathbf{M}$ as a function of **B** as the independent variable.

To further illustrate this point of view, we show in Fig. 8.2 a plot of $4\pi\mathbf{M}$ *vs* **B** for a ferromagnetic material. We see that the increase of $4\pi\mathbf{M}$ as **B** increases is almost linear, lying just below the (dashed) line $4\pi\mathbf{M} = \mathbf{B}$. Eventually **M** saturates at a value $\mathbf{M_{sat}}$.

8.6 Permanent Magnetism

As **B** decreases to zero in a soft ferromagnet, **M** decreases along the same path as its increase. For a hard ferromagnet, **M** tends to stay locked near its maximum value because of the cooperative effect of the polarized molecules

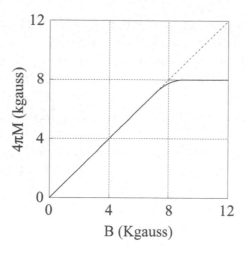

Figure 8.2: Magnetization curve for $4\pi\mathbf{M}$ *vs* $\mathbf{B}$.

on each other. Then, it is really $\mathbf{B}$ that becomes a function of $\mathbf{M}$, and of the geometry, through Eq. (8.6):

$$
\begin{aligned}
\mathbf{B}(\mathbf{r}) &= \frac{1}{c}\int d\tau'[\mathbf{j}(\mathbf{r}') + c\boldsymbol{\nabla}'\times\mathbf{M}(\mathbf{r}')]\times\left(\frac{\mathbf{r}-\mathbf{r}'}{|\mathbf{r}-\mathbf{r}'|^3}\right)\\
&\quad+\frac{1}{c}\int[\mathbf{K}(\mathbf{r}')dS' + c\mathbf{M}(\mathbf{r}')\times d\mathbf{S}']\times\left(\frac{\mathbf{r}-\mathbf{r}'}{|\mathbf{r}-\mathbf{r}'|^3}\right)\\
&= \mathbf{B_{jK}} + \mathbf{B_M}.
\end{aligned}
\tag{8.38}
$$

Equation (8.38) shows that $\mathbf{B}$ has two kinds of sources: external volume and surface currents leading to $\mathbf{B_{jK}}$, and effective magnetization currents leading to $\mathbf{B_M}$.

As the external currents, $\mathbf{j}$ and $\mathbf{K}$ are reduced to zero in a hard ferromagnet, $\mathbf{M}$ tends to stay constant while $\mathbf{B}$ decreases. Thus, as the external currents are turned off, the curve for $\mathbf{M}$ *vs* $\mathbf{B}$ will be an almost horizontal line, with $\mathbf{B}$ decreasing until it reaches the value given by the integrals over $\mathbf{M}$ in Eq. (8.38). The final value of $\mathbf{B}$ will be determined by the geometry, and will be independent of the magnetic properties of the medium other than the value of the magnetization $\mathbf{M}$.

For simplicity we will consider the case of an 'ideal' ferromagnet for which $\mathbf{M}$ is uniform inside the material, so $\mathbf{B}$ is given only by the last surface integral in Eq. (8.6):

$$
\mathbf{B}(\mathbf{r}) = \frac{1}{c}\int c\mathbf{M}(\mathbf{r}')\times d\mathbf{S}'\frac{(\mathbf{r}-\mathbf{r}')}{|\mathbf{r}-\mathbf{r}'|^3}.
\tag{8.39}
$$

This shows that the tangential component of the magnetization, M_t, on the surface at the side of the ferromagnet is equivalent to a surface current I_M, given by

$$
\frac{NI_M}{L} = cM_t.
\tag{8.40}
$$

The resulting **B** for no external current depends on the geometry of the surface. For a torus when the external current is turned off, **B** is given by

$$\mathbf{B} = \frac{4\pi N I_M}{cL} = 4\pi\mathbf{M} \tag{8.41}$$

throughout the torus.

A long thin magnetized bar acts like a long solenoid with current I_M, and then **B** is the solenoid field given by Eq. (7.28). Near the middle of the long thin magnet, **B** is nearly constant and equals $4\pi\mathbf{M}$. But **B** decreases to $2\pi\mathbf{M}$ at either end of the magnet, just as it does for the solenoid.

The value of **H** in the permanent magnet has nothing to do with permeability. It is determined by the shape of the magnet. It is also not appropriate to use this **H** in a hysteresis curve, which is valid only for the toroidal geometry or some other configuration with negligible end effects.

8.7 The Use of the H Field for a Permanent Magnet

Although the **H** field has no physical meaning in a permanent magnet, it can be quite useful in relating the **B** field to the magnetization **M**. We start with the original definition of **H**:

$$\mathbf{H} = \mathbf{B} - 4\pi\mathbf{M}. \tag{8.42}$$

For a permanent magnet, **M** depends on neither **B** nor **H**.

In the absence of free currents, this definition leads to the following equations for **H**:

$$\boldsymbol{\nabla}\cdot\mathbf{H} = -4\pi\boldsymbol{\nabla}\cdot\mathbf{M} \tag{8.43}$$
$$\boldsymbol{\nabla}\times\mathbf{H} = 0. \tag{8.44}$$

These are the same two differential equations satisfied by **E** for an effective charge distribution ρ_m, given by

$$\rho_m = -\boldsymbol{\nabla}\cdot\mathbf{M}. \tag{8.45}$$

Consequently, we can take all the results of chapters 1-5 for electrostatics and apply them directly to permanent magnets, using the effective magnetic charge density ρ_m. We also can adapt the integral result for **E**, given by Eq. (6.5) in Chapter 6 to the case of a finite chunk of magnetized material:

$$\mathbf{H}(\mathbf{r}) = -\int d\tau' \frac{(\mathbf{r}-\mathbf{r}')\boldsymbol{\nabla}'\cdot\mathbf{M}(\mathbf{r}')}{|\mathbf{r}-\mathbf{r}'|^3} + \oint \frac{d\mathbf{A}'\cdot\mathbf{M}(\mathbf{r}')(\mathbf{r}-\mathbf{r}')}{|\mathbf{r}-\mathbf{r}'|^3}. \tag{8.46}$$

This can be written as

$$\mathbf{H}(\mathbf{r}) = \int d\tau' \frac{(\mathbf{r} - \mathbf{r}')\rho_m(\mathbf{r}')}{|\mathbf{r} - \mathbf{r}'|^3} + \oint \frac{dA'(\mathbf{r} - \mathbf{r}')\sigma_m(\mathbf{r}')}{|\mathbf{r} - \mathbf{r}'|^3}. \qquad (8.47)$$

This means that we can treat a permanently magnetized piece of material as if it had an effective magnetic charge density ρ_m given by Eq. (8.45) and an effective magnetic surface charge distribution σ_m given by

$$\sigma_m = \hat{\mathbf{n}} \cdot \mathbf{M}. \qquad (8.48)$$

The magnetic material has no other effect, and can be disregarded except for these charge distributions.

Then, any magnetostatic problem with a permanent magnet can be solved by just writing down the answer for the corresponding electrostatic problem with these charge distributions. We do want to emphasize that, although they are extremely useful, ρ_m and σ_m are **not** real magnetic charge distributions, but just mathematical constructs that act like charge distributions.

8.8   Bar Magnet

We consider a long thin hard ferromagnet of length L, cross sectional area A (with $L^2 >> A$), and a constant permanent magnetization $\mathbf{M}$ directed along the length of the cylinder, as shown in Fig. 8.3. This constitutes a **bar magnet**.

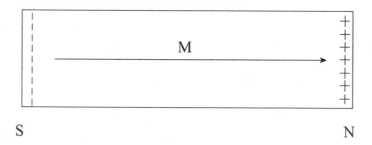

Figure 8.3: Bar magnet with magnetization $\mathbf{M}$ and effective magnetic surface charges at each end.

There will be an effective surface charge distribution

$$\sigma_m = +M \qquad (8.49)$$

at the right face of the bar magnet in Fig. 8.3, and

$$\sigma_m = -M \qquad (8.50)$$

at the left face. Thus, the bar magnet behaves like a long cylinder with a constant magnetic surface charge at either end, and a total magnetic charge at each end,

$$g = +MA, \quad \text{right end},$$
$$g = -MA, \quad \text{left end}. \tag{8.51}$$

The magnitude of the effective magnetic charge at either end is called the **pole strength** of the magnet. Pole strength has the units of gauss-cm^2. We will use $(+)$ and $(-)$ to designate the sign of the magnetic charge, in analogy with electrostatics. Historically, the designations N (for North) and S (for South) have been used, because the N pole of a compass needle points close to the geographic North pole of the Earth. (Note that this means that the N pole of a compass needle is really a South magnetic pole.)

We investigate the magnetic field of a bar magnet at various distances from the magnet, as indicated on Fig. 8.4. Some of these results were already obtained in Section 8.6 using **B**, but they will be more transparent using the magnetostatics of the **H** field.

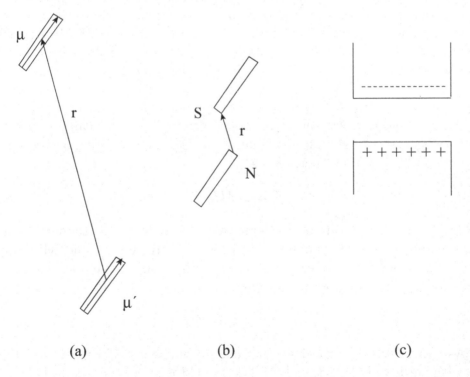

        (a)   (b)   (c)

Figure 8.4: Bar magnets at several distances: (a) Dipoles at a large distance. (b) Magnetic poles at moderate distance. (c) Sheets of magnetic charge at close distance.

(a) At large distances, $r \gg L$, the bar magnet will act like a magnetic dipole of magnetic moment

$$\boldsymbol{\mu} = \int \mathbf{M} d\tau = \mathbf{M} AL. \qquad (8.52)$$

Here, we have used the definition of $\mathbf{M}$ as the dipole moment density in the material. The magnetic moment can also be calculated as it would be in electrostatics as

$$\boldsymbol{\mu} = g\mathbf{L} = \mathbf{M} AL. \qquad (8.53)$$

The magnetic field of the dipole is

$$\mathbf{B} = \frac{3(\boldsymbol{\mu}\cdot\hat{\mathbf{r}})\hat{\mathbf{r}} - \boldsymbol{\mu}}{r^3}, \quad r \gg L. \qquad (8.54)$$

The force between two bar magnets at a large distance apart, as in Fig. 8.4a, will be the dipole-dipole force given by Eq. (2.62).

(b) For r small compared to L, but with $r^2 \gg A$, and much closer to one end than the other, the magnetic charge at the close end can be approximated by a point magnetic pole, $g = MA$. Then, the magnetic field in this region will be just that of a point charge. This is the origin of the early notion that a bar magnet had point charge like poles at each end. It also explains why breaking a bar magnet in half produces two new poles at the broken ends. The force between two bar magnets oriented as in Fig. 8.4b will be just the Coulomb-like force between the two close magnetic poles.

(c) The force between two bar magnets placed end to end as in Fig. 8.4c, will be like the force between two charged plates. This force can be calculated as the magnetic field $\mathbf{B} = 2\pi\mathbf{M}$ at the end of one magnet acting on the surface charge $\sigma_m = M'$ of the other, so

$$F = 2\pi M M' A. \qquad (8.55)$$

This is the magnitude of the force required to pull two magnets apart or to push them together. The force will be attractive for $\mathbf{M}$ and $\mathbf{M}'$ parallel, and repulsive if they are anti-parallel. Stated in terms of the pole strengths of each magnet, the force between them is

$$F = 2\pi g g'/A. \qquad (8.56)$$

As either end is approached, that end looks like a uniformly charged lamina with surface charge $\sigma_m = M$. The $\mathbf{H}$ field is the same as the $\mathbf{E}$ field for a uniformly charged disk. Very close to the end of the bar magnet, the end begins to look like an infinite plane with surface charge $\sigma_m = M$. This is shown in an enlarged picture of the right end of a bar magnet in Fig 8.5.

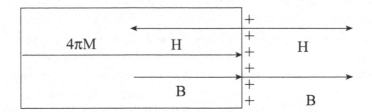

Figure 8.5: Magnetization M, and B and H fields, near the end of a bar magnet. H is discontinuous like the E field of a sheet of charge, while B is continuous at the end of the magnet.

The **H** field is given by $H = 2\pi M$, directed away from the face of the end. As shown on the figure, **H** has the same magnitude just inside and just outside the end face, but is in opposite directions. **B** just outside the face is equal to **H**. Just inside, $\mathbf{B} = 4\pi\mathbf{M} + \mathbf{H} = 2\pi\mathbf{M} = -\mathbf{H}$, so **B** and **H** are in opposite directions just inside the face. Note that **B** is continuous through the face as required by $\mathbf{\nabla \cdot B} = 0$. These relations between **B**, **H**, and **M** depend only on the geometry at the end of the magnet, and not on the nature of the material, as long as it is magnetized .

Close to a bar magnet, but not too close to either end, a bar magnet looks like two point charges, $+g$ and $-g$. This means that at the middle of a bar magnet,

$$\mathbf{H} = -\frac{8g\hat{\mathbf{L}}}{L^2} = -\frac{8\mathbf{M}A}{L^2}. \tag{8.57}$$

For a long thin magnet **H** will be much smaller than **M**, so **B** will approximately equal $4\pi\mathbf{M}$.

We have seen that either the **B** field of bound currents or the **H** field of effective magnetic charges can be used to study a permanent magnet with a given magnetization. Use of the **H** field was simpler, but all the results should always be equivalent. We give as a problem to demonstrate that the **B** and **H** fields outside a bar magnet are the same for either method, and that **B** and **H** differ by $4\pi\mathbf{M}$ inside.

We saw above that the **H** field in a bar magnet is in the opposite direction of **M**. Because of this, it is sometimes called a "demagnetizing field". But it is not appropriate to put this **H** field into a hysteresis curve like that in Fig. 8.1, which only applies when there are no end effects. There is no physical field or force in the magnet opposing the magnetization. The **B** field is weaker near the ends, and this does make the magnet more prone to lose some magnetization if it is struck or heated.

For this reason, it helps to store bar magnets side by side in pairs, with the N end of one next to the S end of the other. Then **B** will approach its maximum value more quickly inside the magnet. We will see in the next section that this can also be accomplished by storing the bar magnet on a soft ferromagnet, like a steel surface.

The demagnetizing effect of **B** being weaker near the end of a bar magnet can be eliminated by bending a bar magnet into the shape of the letter U, forming a **horseshoe magnet**. Then placing a soft ferromagnetic piece (called a **keeper**) across the two poles of the magnet eliminates most of the end effects and **B** is at its maximum value thoughout the magnet. The horseshoe shape also effectively doubles the lifting capacity of the magnet, because now two poles act when the magnet touches a flat surface.

8.9   Magnetic Images

As with electrostatics, magnetic image charges can be used for some ferromagnetic cases. Because the polarization image charges in Section 6.4.4 depended on a linear relation between **D** and **E**, they cannot be used reliably for ferromagnets with their large non-linearity. However the permeability for soft ferromagnetic materials is usually so large that they can be treated like conductors in electrostatics, as we showed in Section 6.4.4. This means that the image charges for conductors in Section 3.2 can be applied to soft ferromagnets.

We list below a number of electrostatic image results that also follow for ferromagnets, citing the corresponding electrostatic equation.

1. The force of attraction on a flat surface of a soft ferromagnet by a point magnetic charge g (as approximated by the near end of a magnet) a distance **d** away is [Eq. (3.20)]

$$\mathbf{F} = -\frac{g^2 \hat{\mathbf{d}}}{4d^2}. \tag{8.58}$$

This force would be a reasonable approximation for the force on any large object when the end of the magnet is close enough for the near surface of the object to be relatively flat, but far enough away that the end of the magnet is point-like.

2. When the end of the magnet gets close to or touches a flat surface of the object, then the end of the magnet acts like a large charged plane. This will induce an image charge in the flat surface that is just the negative of the magnet's charge. The force on the object will be equal the force between two identical touching magnets, and will be given by Eq. (8.55) or (8.56).

For this case, the magnetic field between the two planes will be constant, so a magnetic image charge can be used even for non-linear, permeability. We can extend Eq. (6.57) for images in dielectrics to this magnetic case as

$$g' = \left(\frac{1-\mu}{1+\mu}\right) g \tag{8.59}$$

Then, the force of attraction on a flat object of permeability μ by a touching magnet with magnetization M is

$$F = 2\pi M^2 A \left(\frac{\mu - 1}{\mu + 1} \right). \tag{8.60}$$

3. If a magnet is laid flat on a ferromagnetic surface as shown in Fig. 8.6,

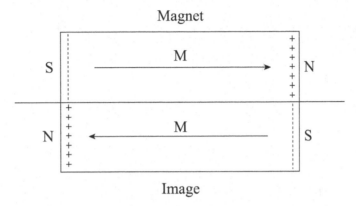

Figure 8.6: Bar magnet on ferromagnetic surfrace.

a negative image magnet will be induced in the surface. The force required to pick up the magnet will be that due to the force between the two sheets of charge at each end of the magnet and its image. Also, the **H** field in the magnet will be reduced by the negative image charge, so **B** inside the magnet will more quickly approach its maximum value. This will reduce the demagnetizing effect of a smaller **B** near the end of the magnet, and help preserve the magnetization of a bar magnet placed on a ferromagnetic surface.

4. When one pole of a magnet is close to a soft ferromagnet that has approximately spherical shape, we can consider this to be like a point charge outside a neutral conducting sphere. The magnetic force on a sphere of radius a a distance **d** from a magnetic pole (one end of a magnet) of strength g will be [Eq. (3.34)]

$$\mathbf{F} = -\frac{g^2 a \mathbf{d}}{(d^2 - a^2)^2} + \frac{g^2 a \mathbf{d}}{d^4}. \tag{8.61}$$

The force on a small sphere ($a<<d$) can be approximated by

$$\mathbf{F} = -\frac{2g^2 a^3 \hat{\mathbf{d}}}{d^5}. \tag{8.62}$$

8.10   Problems

1. A disk of permeability $\mu(>> 1)$ has a radius a and a thickness t. Its axis makes an angle θ with a constant magnetic field $\mathbf{B_0}$.

   (a) Find the induced magnetic moment of the disk.

   (b) Show that the torque on the disk is $\frac{1}{8}\mu a^2 t B_0^2 \sin(2\theta)$.

2. A sphere of radius R has constant magnetization $\mathbf{M}$.

   (a) Find the bound current density in the sphere.

   (b) Find the bound surface current density on the sphere.

   (c) Use the bound surface current density to calculate the magnetic moment of the sphere.

3. A sphere of radius R has constant magnetization $\mathbf{M}$.

   (a) Find the effective magnetic charge density in the sphere.

   (b) Find the effective magnetic surface charge density on the sphere.

   (c) Use the magnetic surface charge density to find the magnetic moment of the sphere.

4. A sphere of radius R has the magnetization $\mathbf{M} = a\hat{\mathbf{k}}(\hat{\mathbf{k}}\cdot\mathbf{r})^2$.

   (a) Find the bound magnetic charge density in the sphere.

   (b) Find the bound magnetic surface charge density on the sphere.

   (c) Find the magnetic moment of the sphere by integrating over the bound densities.

5. A sphere of radius R has the magnetization $\mathbf{M} = a\hat{\mathbf{k}}(\hat{\mathbf{k}}\cdot\mathbf{r})^2$.

   (a) Find the bound current density in the sphere.

   (b) Find the bound surface current density on the sphere.

   (c) Find the magnetic moment of the sphere by integrating over the bound densities.

6. A sphere of radius R has constant magnetization $\mathbf{M}$.

   (a) Find the magnetic scalar potential ϕ_m outside and inside the sphere. (Use the Legendre polynomial expansion, keeping only $l = 1$.)

   (b) Use ϕ_m to find $\mathbf{H}$, and $\mathbf{B}$ inside the sphere.

7. A bar magnet of length L with a circular cross-section of radius a has a constant magnetization $\mathbf{M}$.

(a) Find $H_z(z)$ on the axis of the bar magnet by treating it as two sheets of magnetic charge located at $z = +L/2$ and $z = -L/2$.

(b) Show that for $z > L/2$ Eq. (7.28) for $B_z(z)$ of a solenoid gives the same result as found in part (a).

(c) Show that, for $-L/2 < z < +L/2$, B_z and H_z differ by $4\pi M$.

8. What minimum magnetization is required in each magnet for one uniformly magnetized iron sphere of radius $R = 5$ cm to be able to pick up an identical magnetized sphere?

9. (a) What minimum magnetization is required in each magnet for one iron bar magnet of length $L = 10$ cm to be able to pick up an identical magnet?

(b) What minimum magnetization would be required if the second iron bar were not a magnet, but a soft ferromagnet with $\mu = 4$?

(c) What would be the easiest way to distinguish between the second bar magnet in (a) and the soft ferromagnet in (b)?

Chapter 9

Time varying fields, Maxwell's equations

9.1 Faraday's Law.

So far we have discussed only static fields or fields for which time derivatives could be neglected. With time variation, a number of new and unexpected phenomena arise that change the nature of electromagnetism. Historically, the first of these was Faraday's law, discovered in 1831 by Michael Faraday.

What Faraday observed was that a changing magnetic field can induce currents in a closed electric circuit. The change in the magnetic field can be caused by a change in the current producing the magnetic field, or by moving the circuit producing the field, or by moving a magnet producing the field.

To put these effects in consistent quantitative terms, we introduce the concepts of **magnetic flux** and **electromotive force**. The magnetic flux Φ through a closed loop is defined by

$$\Phi = \int \mathbf{B} \cdot \mathbf{dS}, \tag{9.1}$$

where the integration is over a surface enclosed by the loop. Because the divergence of $\mathbf{B}$ is zero, the integration surface can be any surface bounded by the loop. (Use the divergence theorem to show this.) The electromotive force (EMF), $\mathcal{E}$, in a closed loop is defined as the work done in moving a unit test charge around the loop. This equals the line integral of the tangential component of $\mathbf{E}$ around the loop:

$$\mathcal{E} = \oint \mathbf{E} \cdot \mathbf{dr}. \tag{9.2}$$

The observations made by Faraday can be summarized in **Faraday's law**, which states that

$$\mathcal{E} = -\frac{1}{c}\frac{d\Phi}{dt}. \tag{9.3}$$

251

In words, **The EMF around a closed loop equals the rate of change of the magnetic flux through the loop divided by c.** The minus sign in the law is appropriate for the sign conventions we have used for the positive direction of the line integration and for the surface integration. That is, if the fingers of your right hand are curled in the direction of the line integration, your extended thumb will point in the positive direction for the surface integration.

The direction of the EMF is also given by **Lenz's law**, which states that **the direction of current induced in a closed circuit by any change in the flux through the circuit will induce a secondary magnetic field in a direction to oppose the change in total flux through the circuit.** That is to say that the induced current tries to resist any change in magnetic flux through a closed circuit. Although Lenz's law is stated in terms of the current induced in a circuit, this is directly related to the EMF through Ohm's law $\mathcal{E} = IR$. However, we note that no actual wire is necessary for Faraday's law, which holds for the EMF even for a mathematical contour.

We originally stated Faraday's law for a stationary loop, with the flux through it changing due to changing sources. But the flux through a loop can also change due to motion of the loop. This can be through rotation or distortion of the loop, or by translation of the loop through a non-uniform magnetic field. In each case, there will be a change in flux through the loop and a consequent EMF given by Eq. (9.3).

The motional EMF of Faraday's law can be derived from the Lorentz force law. We start with the action of the Lorentz force on a conduction electron in a wire moving with velocity $\mathbf{v}$ in a magnetic field:

$$\mathbf{F} = -\frac{e}{c}\mathbf{v}\times\mathbf{B}. \qquad (9.4)$$

In the rest frame of the moving wire, this force on a charge $-e$ is the same as if there were an electric field given by

$$\mathbf{E} = \frac{\mathbf{v}}{c}\times\mathbf{B}. \qquad (9.5)$$

Although we used the mechanism of a moving wire to find the $\mathbf{E}$ field in a moving system, the field exists whether or not there actually is a wire, so Eq. (9.5) can be considered a relation between the two fields. Equation (9.5) relating $\mathbf{E}$ in a moving system to $\mathbf{B}$ in a rest system has corrections of order v^2/c^2 arising in special relativity. It turns out, however, that this is the only equation in this section that is modified by special relativity. Although quantities on each side of some of the other equations have complicated transformations when going to a moving system, each side transforms in the same way so that the equality is preserved.

Using Eq. (9.5) for $\mathbf{E}$, the motional EMF in the loop is given by

$$\mathcal{E} = \frac{1}{c}\oint(\mathbf{v}\times\mathbf{B})\cdot d\mathbf{r}.) \qquad (9.6)$$

We first consider the case of a loop moving with velocity $\mathbf{v}$ in an inhomogenous magnetic field $\mathbf{B}$. We can apply Stokes' theorem to Eq. (9.6) to get

$$
\begin{aligned}
\mathcal{E} &= \frac{1}{c}\int [\boldsymbol{\nabla}\times(\mathbf{v}\times\mathbf{B})]\cdot\mathbf{dS} \\
&= \frac{1}{c}\int [\mathbf{v}(\boldsymbol{\nabla}\cdot\mathbf{B}) - (\mathbf{v}\cdot\boldsymbol{\nabla})\mathbf{B}]\cdot\mathbf{dS} \\
&= -\frac{1}{c}\int [(\mathbf{v}\cdot\boldsymbol{\nabla})\mathbf{B}]\cdot\mathbf{dS}.
\end{aligned}
\tag{9.7}
$$

The EMF in Eq. (9.7) Is due to the translational motion of the loop. An additional EMF is produced by the rate of change, $\partial_t\mathbf{B}$, of the magnetic field. Including this, the EMF is given by

$$
\mathcal{E} = -\frac{1}{c}\int [(\mathbf{v}\cdot\boldsymbol{\nabla})\mathbf{B} + \partial_t\mathbf{B}]\cdot\mathbf{dS}
\tag{9.8}
$$

The total time derivative in a system that is moving with respect to a fixed field $\mathbf{B}(\mathbf{r}, t)$ is given by

$$
\frac{d\mathbf{B}}{dt} = (\mathbf{v}\cdot\boldsymbol{\nabla})\mathbf{B} + \partial_t\mathbf{B},
\tag{9.9}
$$

so Eq. (9.7) can be written as

$$
\mathcal{E} = -\frac{1}{c}\int \left(\frac{d\mathbf{B}}{dt}\right)\cdot\mathbf{dS}.
\tag{9.10}
$$

There is also a motional EMF if the loop rotates or is distorted. For this effect, we write the EMF as

$$
\mathcal{E} = -\frac{1}{c}\oint \mathbf{B}\cdot(\mathbf{v}\times\mathbf{dr}) = -\frac{1}{c}\mathbf{B}\cdot\oint \mathbf{v}\times\mathbf{dr}.
\tag{9.11}
$$

We have taken $\mathbf{B}$ out of the integral because it can be considered constant here, since its space variation is fully included in Eq. (9.8). The velocity $\mathbf{v}$ in Eq. (9.11) is now a function of $\mathbf{r}$, varying according to the displacement of the integration contour.

The integral $\oint \mathbf{v}\times\mathbf{dr}$ can be reduced to

$$
\oint \mathbf{v}\times\mathbf{dr} = \frac{d\mathbf{S}_l}{dt},
\tag{9.12}
$$

where $\mathbf{S}_l$ is the surface area enclosed by the loop. (Although the shape of the surface is arbitrary, its vector area $\mathbf{S}_l = \int \mathbf{dS}$ depends only on the bounding loop.) This can be seen in Fig. 9.1 which shows that the infinitesimal surface area given by $(\mathbf{v}\Delta t)\times\boldsymbol{\Delta}\mathbf{r}$ is the change in the surface area of the loop in time Δt due to the loop segment $\boldsymbol{\Delta}\mathbf{r}$.

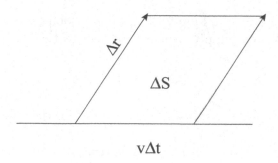

Figure 9.1: The change in surface area ΔS is the area of the parallelogram given by $\mathbf{v}\Delta t \times \Delta \mathbf{r}$.

Then, taking the limits as Δt and $\Delta \mathbf{r}$ go to zero, and completing the integral around the loop leads to Eq. (9.12). This gives the EMF due to loop distortion or rotation as

$$\mathcal{E} = -\frac{1}{c}\int \mathbf{B}\cdot\frac{d\mathbf{S}_l}{dt}. \tag{9.13}$$

Combining Eqs. (9.10) and (9.13) gives

$$\mathcal{E} = -\frac{1}{c}\frac{d}{dt}\int \mathbf{B}\cdot d\mathbf{S} = -\frac{1}{c}\frac{d\Phi}{dt}. \tag{9.14}$$

for the EMF in a closed loop due to its distortion or its motion in a time dependent magnetic field.

The connection with Faraday's law, which we originally stated for a stationary loop in a time varying magnetic field, comes through applying the physical principle of Galilean invariance, which states that physical laws should not depend on which coordinate system is used. Here, that means that the EMF in the loop should be the same in the rest frame of the loop as in the system in which the loop is moving.

In fact, the final form for the motional EMF in Eq. (9.14) makes no mention of reference frame and is valid in any frame. Equation (9.14) applies to all modes of change in the flux, and is the full statement of Faraday's law. This shows that even though Faraday's law was first discovered experimentally, it could have been suggested by applying the principle of Galilean invariance to the motional EMF in electromagnetism.

Using the definitions of EMF and magnetic flux, Faraday's law can be stated as an integral relation between $\mathbf{E}$ and $\mathbf{B}$:

$$\oint \mathbf{E}\cdot d\mathbf{r} = -\frac{1}{c}\frac{d}{dt}\int \mathbf{B}\cdot d\mathbf{S}. \tag{9.15}$$

To derive the differential form of this relation, we apply Eq. (9.14) to an arbitrary mathematical loop that is fixed in space. Then the total time derivative acts only on the explicit time dependence of $\mathbf{B}$, so we can write

$$\oint \mathbf{E}\cdot d\mathbf{r} = -\frac{1}{c}\int (\partial_t \mathbf{B})\cdot d\mathbf{S}. \tag{9.16}$$

We use Stokes' theorem on the left hand integral to get

$$\int (\nabla \times \mathbf{E}) \cdot \mathbf{dS} = -\frac{1}{c} \int (\partial_t \mathbf{B}) \cdot \mathbf{dS}. \tag{9.17}$$

Since the loop is arbitrary, Eq. (9.17) must hold at every point, so

$$\nabla \times \mathbf{E} = -\frac{1}{c} \partial_t \mathbf{B}. \tag{9.18}$$

9.2 Inductance.

One effect of Faraday's law is that a changing current in one circuit will induce an EMF in a second circuit. The magnetic field of the first circuit is proportional to its current (for linear materials), so the flux through the second circuit will be proportional to the current in the first circuit. The constant of proportionality is called the **mutual inductance** M_{21} of the two circuits, so

$$\Phi_2 = cM_{21}I_1. \tag{9.19}$$

The constant c is inserted here so that it will drop out of equations for electric circuits. The EMF is $-\frac{1}{c}\frac{d\Phi}{dt}$, and is given by

$$\mathcal{E}_2 = -M_{21}\frac{dI_1}{dt}. \tag{9.20}$$

The flux through circuit 2 is given by

$$\Phi_2 = \int \mathbf{dS_2} \cdot \mathbf{B_1}, \tag{9.21}$$

where $\mathbf{B_1}$ is the magnetic field due to circuit 1. This flux can be written in terms of the magnetic vector potential $\mathbf{A_1}$ as

$$\Phi_2 = \int \mathbf{dS_2} \cdot (\nabla \times \mathbf{A_1}) = \oint_2 \mathbf{dr_2} \cdot \mathbf{A_1}, \tag{9.22}$$

where we have used Stokes theorem. The subscript 2 on the integral is to indicate that the integral is taken around circuit 2. $\mathbf{A_1}$ is given by Eq. (7.71) as an integral of the current in circuit 1, so we can write

$$\Phi_2 = \oint_2 \mathbf{dr_2} \cdot \frac{I_1}{c} \oint_1 \frac{\mathbf{dr_1}}{|\mathbf{r_2} - \mathbf{r_1}|}, \tag{9.23}$$

and the mutual inductance between the two circuits is given by the double integral

$$M_{21} = \frac{1}{c^2} \oint_2 \oint_1 \frac{\mathbf{dr_2} \cdot \mathbf{dr_1}}{|\mathbf{r_2} - \mathbf{r_1}|}. \tag{9.24}$$

This form for the mutual inductance shows **reciprocity**, defined as

$$M_{21} = M_{12}. \tag{9.25}$$

A circuit can also have an EMF due to changing flux from current in the same circuit. The proportionality constant for this is the **self inductance** L, with

$$\Phi = cLI, \tag{9.26}$$

and

$$\mathcal{E} = -L\frac{dI}{dt}. \tag{9.27}$$

The sign in Eq. (9.27) follows from Lenz's law, while the sign in Eq. (9.20) depends on the sign convention used for each of the two circuits involved.

The Gaussian unit for inductance can be seen from Eq. (9.24) to be $\sec^2/\text{cm}$. The SI unit is the henry, with the conversion

$$1 \sec^2/\text{cm} = 9 \times 10^{11} \text{ henry}. \tag{9.28}$$

Equations (9.20) and (9.27) are the same in either SI or Gaussian units. That means they can be used directly for SI units with L and M in henrys.

The self inductance in an electric circuit produces a persistence of currents in the circuit if the original source of the EMF (either from a changing magnetic flux as in a generator, or due to a battery) is turned off. From Eq. (9.27) and Ohm's law, we get

$$IR = -L\frac{dI}{dt}. \tag{9.29}$$

This is a simple differential equation for $I(t)$, with the solution

$$I(t) = I(0)e^{-\frac{R}{L}t}. \tag{9.30}$$

This means that the current in any circuit will decay like

$$I(t) = I(0)e^{-\frac{t}{\tau}} \tag{9.31}$$

with a time constant given by

$$\tau = \frac{L}{R}. \tag{9.32}$$

9.3   Displacement Current, Maxwell's Equations.

With the inclusion of time variation and Faraday's law, the differential equations of electromagnetism would look like:

$$\nabla \cdot \mathbf{D} \;=\; 4\pi\rho \tag{9.33}$$

$$\mathbf{\nabla} \times \mathbf{E} = -\frac{1}{c}\partial_t \mathbf{B} \qquad (9.34)$$

$$\mathbf{\nabla} \cdot \mathbf{B} = 0 \qquad (9.35)$$

$$\mathbf{\nabla} \times \mathbf{H} = \frac{4\pi \mathbf{j}}{c}. \qquad (9.36)$$

But there is a glaring inconsistency in these equations. Do you see it? Maxwell did, and saw the need for an important addition to one of the equations.

We know the divergence of a curl is zero. Taking the divergence of $\mathbf{\nabla} \times \mathbf{E}$ in the second equation does give zero since $\mathbf{\nabla} \cdot \mathbf{B}=0$. However, for the fourth equation we get

$$\mathbf{\nabla} \cdot (\mathbf{\nabla} \times \mathbf{H}) = \frac{4\pi}{c}\mathbf{\nabla} \cdot \mathbf{j} = -\frac{4\pi}{c}\partial_t \rho, \qquad (9.37)$$

which follows from the continuity equation (6.134). Since the divergence of a curl must equal zero, Eq. (9.37) is inconsistent for a time dependent charge distribution.

The resolution of this inconsistency can be found by looking at Eq. (9.33). where $\partial_t \rho$ is related to $\mathbf{D}$ by

$$-\frac{4\pi}{c}\partial_t \rho = -\frac{1}{c}\partial_t \mathbf{\nabla} \cdot \mathbf{D}. \qquad (9.38)$$

This shows that adding a term $\frac{1}{c}\partial_t \mathbf{D}$ to Eq. (9.36) for curl $\mathbf{H}$ will correct the inconsistency. Doing so gives

$$\mathbf{\nabla} \times \mathbf{H} = \frac{4\pi \mathbf{j}}{c} + \frac{1}{c}\partial_t \mathbf{D}, \qquad (9.39)$$

and now each side has zero divergence.

The addition of this term to the curl $\mathbf{H}$ equation may seem obvious to us, over a hundred years after the fact, but for Maxwell in 1864 it was a giant leap that constitutes one of the pivotal advances in the history of physics. Maxwell called the term $\frac{1}{4\pi}\partial_t \mathbf{D}$ the **displacement current density** because he related it to the microscopic displacement of charges within molecules. The displacement current also exists in vacuum, and Maxwell considered that this was due to the microscopic displacement of charges within a material "aether" that he hypothesized permeated the vacuum.

With the addition of the displacement current density, the four equations (9.33), (9.34), (9.35), and (9.39), called **Maxwell's equations**, are the basic equations for all further development of electromagnetism.

9.4 Electromagnetic Energy

We can find the energy content of electric and magnetic fields (**electromagnetic energy**) by using Maxwell's equations to calculate the rate of energy input into the fields. The rate at which an electric field gives energy to a charge q moving with velocity $\mathbf{v}$ is $q\mathbf{v}\cdot\mathbf{E}$. The magnetic field does not enter here because its force on the charge is always perpendicular to the velocity, and does no work.

The current density $\mathbf{j}$ is composed of many of these charges, such that

$$\mathbf{j}\Delta V = \sum_i q_i \mathbf{v_i}, \tag{9.40}$$

where the sum is over all the charges in the volume ΔV. Although there are many charges within the volume ΔV, it can be small enough that the macroscopic electric field is constant throughout the volume. Then the rate at which the electric field puts energy put into matter within ΔV is given by

$$\left[\frac{dU}{dt}\right]_{\text{matter}} = \sum_i q_i \mathbf{v_i}\cdot\mathbf{E} = \mathbf{j}\cdot\mathbf{E}\Delta V. \tag{9.41}$$

If we sum over all these volumes in the limit as $\Delta V \to 0$, we get the integral relation

$$\left[\frac{dU}{dt}\right]_{\text{matter}} = \int_V \mathbf{j}\cdot\mathbf{E}d\tau. \tag{9.42}$$

This energy input into matter generally results in ohmic heating of the matter.

We can replace the current $\mathbf{j}$ in Eq. (9.42) by using the curl $\mathbf{H}$ equation, and performing some vector manipulation using Maxwell's equations:

$$\begin{aligned}
\int_V \mathbf{j}\cdot\mathbf{E}d\tau &= \frac{1}{4\pi}\int_V [c\mathbf{E}\cdot(\boldsymbol{\nabla}\times\mathbf{H}) - \mathbf{E}\cdot\partial_t\mathbf{D}]d\tau \\
&= \frac{1}{4\pi}\int_V [c\boldsymbol{\nabla}\cdot(\mathbf{H}\times\mathbf{E}) + c\mathbf{H}\cdot(\boldsymbol{\nabla}\times\mathbf{E}) - \mathbf{E}\cdot\partial_t\mathbf{D}]d\tau \\
&= -\frac{c}{4\pi}\oint_S d\mathbf{S}\cdot(\mathbf{E}\times\mathbf{H}) - \frac{1}{4\pi}\int_V [\mathbf{H}\cdot\partial_t\mathbf{B} + \mathbf{E}\cdot\partial_t\mathbf{D}]d\tau. \quad (9.43)
\end{aligned}$$

If the material is linear and non-dispersive, so that $\mathbf{D}$ is proportional to $\mathbf{E}$, and $\mathbf{H}$ is proportional to $\mathbf{B}$, we can write

$$\mathbf{E}\cdot(\partial_t\mathbf{D}) = (\partial_t\mathbf{E})\cdot\mathbf{D} = \frac{1}{2}\partial_t(\mathbf{E}\cdot\mathbf{D}), \tag{9.44}$$

and

$$\mathbf{H}\cdot(\partial_t\mathbf{B}) = (\partial_t\mathbf{H})\cdot)\mathbf{B} = \frac{1}{2}\partial_t(\mathbf{B}\cdot\mathbf{H}). \tag{9.45}$$

Using these results, we get

$$\int_V \mathbf{j}\cdot\mathbf{E}d\tau + \frac{1}{8\pi}\int_V \partial_t(\mathbf{E}\cdot\mathbf{D} + \mathbf{B}\cdot\mathbf{H})d\tau. = -\frac{c}{4\pi}\oint_S d\mathbf{A}\cdot(\mathbf{E}\times\mathbf{H}). \tag{9.46}$$

We interpret each term in Eq. (9.46) as follows:

- $\int_V \mathbf{j}\cdot\mathbf{E}d\tau$ is the rate at which energy is being put into the matter in the volume V.

- $\frac{1}{8\pi} \int_V \partial_t (\mathbf{E} \cdot \mathbf{D} + \mathbf{B} \cdot \mathbf{H}) d\tau$ is the rate at which energy is being put into the electromagnetic fields in the volume.

From this it follows that the energy content of the fields within a volume V is given by

$$U = \frac{1}{8\pi} \int_V (\mathbf{E} \cdot \mathbf{D} + \mathbf{B} \cdot \mathbf{H}) d\tau, \tag{9.47}$$

since its time derivative is the rate of energy change. We can use Eq. (9.47) to define an energy density of the fields given by

$$u = \frac{1}{8\pi} (\mathbf{E} \cdot \mathbf{D} + \mathbf{B} \cdot \mathbf{H}). \tag{9.48}$$

It is important to keep in mind that Eqs. (9.47) and (9.48) for electromagnetic energy hold only for linear materials.

- $\frac{c}{4\pi} \oint_S d\mathbf{A} \cdot (\mathbf{E} \times \mathbf{H})$ is the rate at which energy enters the volume through its surface.

The vector in this surface integral is called the **Poynting vector S**:

$$\mathbf{S} = \frac{c}{4\pi} \mathbf{E} \times \mathbf{H}. \tag{9.49}$$

The Poynting vector has the significance that the rate of flow of electromagnetic energy through a differential surface $d\mathbf{A}$ is given by $\mathbf{S} \cdot d\mathbf{A}$. The Poynting vector will be useful later in the text in finding the rate of emission and propagation of electromagnetic radiation.

If there is no polarizability matter present, then $\mathbf{D}$ and $\mathbf{H}$ in the above derivations are replaced by $\mathbf{E}$ and $\mathbf{B}$, respectively. Then the energy density is given more simply by

$$u = \frac{1}{8\pi} (E^2 + B^2), \tag{9.50}$$

and the Poynting vector by

$$\mathbf{S} = \frac{c}{4\pi} \mathbf{E} \times \mathbf{B}. \tag{9.51}$$

9.4.1 Potential energy in matter

When there is polarizable matter present, the energy density in terms of $\mathbf{E}$, $\mathbf{D}$, $\mathbf{B}$, and $\mathbf{H}$, in Eq. (9.48) includes the energy in the polarization ($\mathbf{P}$) and magnetization ($\mathbf{M}$)) fields as well as purely electromagnetic energy. The purely electromagnetic energy in the fields $\mathbf{E}$ and $\mathbf{B}$ Is given by Eq. (9.50). We show below how the polarization and magnetization energy densities arise.

The rate of energy input during the polarization of a molecule is given by

$$\mathbf{E} \cdot \sum_i q_i \frac{d\mathbf{r_i}}{dt} = \mathbf{E} \cdot (\partial_t \mathbf{p}), \tag{9.52}$$

where the sum is over all the bound charges in the molecule and $\mathbf{p}$ is the dipole moment of the molecule. The polarization vector $\mathbf{P}$ is the dipole moment per unit volume, so the rate of energy input per unit volume in polarizing the matter is

$$\partial_t u_{\mathbf{P}} = \mathbf{E} \cdot (\partial_t \mathbf{P}). \tag{9.53}$$

Energy is also put into the matter by the electric field acting on the bound current density, $\mathbf{j_b} = c \boldsymbol{\nabla} \times \mathbf{M}$. The rate of this energy input is

$$
\begin{aligned}
\left[\frac{dU}{dt} \right]_{\mathbf{j_b}} &= \int d\tau \mathbf{j_b} \cdot \mathbf{E} + c \int d\tau \mathbf{E} \cdot (\boldsymbol{\nabla} \times \mathbf{M}) \\
&= c \int d\tau \left[\boldsymbol{\nabla} \cdot (\mathbf{M} \times \mathbf{E}) + \mathbf{M} \cdot (\boldsymbol{\nabla} \times \mathbf{E}) \right] \\
&= c \int d\mathbf{A} \cdot (\mathbf{E} \times \mathbf{M}) - \int_V \mathbf{M} \cdot (\partial_t \mathbf{B}).
\end{aligned} \tag{9.54}
$$

The second term (because of its minus sign) is actually energy removed from molecular current loops.

Using Eqs. (9.53) and (9.54), equation (9.43) for the energy balance can be rewritten as

$$
\begin{aligned}
\int_V d\tau \mathbf{j} \cdot \mathbf{E} = &-\frac{c}{4\pi} \oint_S d\mathbf{A} \cdot [(\mathbf{E} \times \mathbf{B}) - 4\pi \cdot (\mathbf{E} \times \mathbf{M})] \\
&+ [\mathbf{B} \cdot \partial_t \mathbf{B} - 4\pi \mathbf{M} \cdot (\partial_t \mathbf{B}) - \mathbf{E} \cdot \partial_t \mathbf{E} + 4\pi \mathbf{E} \cdot (\partial_t \mathbf{P})].
\end{aligned} \tag{9.55}
$$

Equation (9.55) shows how the polarization and magnetization energies add to the purely electromagnetic energy (due to E and B) to give the full content of Eq. (9.43). Equation (9.55) holds even for nonlinear materials. If the material is linear, the field energy (including the energy due to $\mathbf{P}$ and $\mathbf{M}$) can be given by Eq. (9.46).

9.5　Magnetic Energy

We recognize the first term in Eq. (9.47) as the energy in the $\mathbf{E}$ and $\mathbf{D}$ fields that we derived in chapter 6. The second term in Eq. (9.47) is the corresponding expression for the energy in the $\mathbf{B}$ and $\mathbf{H}$ fields:

$$U_{\mathbf{BH}} = \frac{1}{8\pi} \int_V \mathbf{B} \cdot \mathbf{H} d\tau. \tag{9.56}$$

We did not derive this in chapter 8 on magnetism because we had to wait for Faraday's law to be able to relate changes in the magnetic field to electromagnetic energy.

Equation (9.56) can be broken into two parts by introducing the magnetization $\mathbf{M}$:

$$
\begin{aligned}
U_{\mathbf{BH}} &= \frac{1}{8\pi} \int_V \mathbf{B} \cdot \mathbf{H} d\tau \\
&= \frac{1}{8\pi} \int_V [\mathbf{B} \cdot \mathbf{B} - 4\pi \mathbf{B} \cdot \mathbf{M}] d\tau \\
&= \frac{1}{8\pi} \int_V B^2 d\tau - \frac{1}{2} \int_V \mathbf{B} \cdot \mathbf{M} d\tau.
\end{aligned} \tag{9.57}
$$

For a small chunk of magnetized matter, we can assume $\mathbf{B}$ does not vary over the volume of the matter, and take it out of the second integral in Eq. (9.57). This leaves

$$
U_{\mathbf{BH}} = \frac{1}{8\pi} \int_V B^2 d\tau - \frac{1}{2} \boldsymbol{\mu} \cdot \mathbf{B}, \tag{9.58}
$$

giving the energy as the sum of a purely magnetic field part and the energy of the induced dipole moment $\boldsymbol{\mu}$ of the piece of matter.

This is consistent with the result in Chapter 7 for the energy of a dipole in a magnetic field. The factor of $\frac{1}{2}$ in Eq. (9.58), which was not in Eq. (7.143), enters here because the dipole is induced. It arises in the same way as did the factor $\frac{1}{2}$ in Eq. (9.45)

Equations (9.56) and (9.58) are valid only for linear materials, and so do not apply to ferromagnetic materials. For the non-linear case, we have to go back to Eq. (9.43) because Eq. (9.45) does not apply for non-linear material. We can deduce from Eq. (9.43) that the differential work per unit volume due to changing $\mathbf{B}$ and $\mathbf{H}$ fields is

$$
dw = \frac{1}{4\pi} \mathbf{H} \cdot d\mathbf{B}. \tag{9.59}
$$

Then the work per unit volume done in the original magnetization of a ferromagnet is

$$
w = \frac{1}{4\pi} \int_0^{\mathbf{B}} \mathbf{H}(\mathbf{B}') \cdot d\mathbf{B}'. \tag{9.60}
$$

The work given by Eq. (9.60) depends on the path in the $\mathbf{H}$-$\mathbf{B}$ plane, and so cannot be used to define an energy of the magnetic field. Also, the work done around a closed $\mathbf{H}$-$\mathbf{B}$ path, such as the **hysteresis** loop in Fig. 8.1 is not zero. The work expended per unit volume in each cycle of the hysteresis loop is given by

$$
w = \frac{1}{4\pi} \oint \mathbf{H}(\mathbf{B}) \cdot d\mathbf{B}, \tag{9.61}
$$

which is just the area of the hysteresis loop.

This work goes into increasing the thermodynamic internal energy of the material, and is eventually given off as heat. This is called hysteresis loss. For this reason, soft ferromagnets with hysteresis loops having small area are used in iron core transformers to minimize this loss and its associated heat production.

We can study the magnetic energy in an electric circuit by starting with the wires of the circuit in a fixed position with no current, and then turning the current on. The power going into a circuit at any time for a given applied EMF (from a generator) $\mathcal{E}_{\mathrm{app}}$ and current I is given by

$$P = \mathcal{E}_{\mathrm{app}}I. \tag{9.62}$$

For a circuit with an inductance L and resistance R, the applied EMF is related to the current by

$$\mathcal{E}_{\mathrm{app}} = L\frac{dI}{dt} + IR. \tag{9.63}$$

This gives for the power into the circuit

$$P = LI\frac{dI}{dt} + I^2R. \tag{9.64}$$

We interpret the I^2R term as the rate of energy leaving the circuit in the form of heat. The term $LI\frac{dI}{dt}$ represents the rate of energy going into the magnetic field produced by the current. The energy of a circuit with a final current I is given by

$$U = \int P dt = \int_0^I LI' dI' = \frac{1}{2}LI^2. \tag{9.65}$$

For a collection of several circuits, the mutual inductance between them will also enter. The inductive EMF in any one of the circuits will be given by

$$\mathcal{E}_i = -L_i\frac{dI_i}{dt} - \sum_{j \neq i} M_{ij}\frac{dI_j}{dt}, \tag{9.66}$$

where the sum is over all the other circuits. This leads to a total magnetic energy of

$$U = \frac{1}{2}\sum_i L_i I_i^2 + \frac{1}{2}\sum_{i,j \neq i} M_{ij} I_i I_j. \tag{9.67}$$

9.6 Electromagnetic Momentum, Maxwell Stress Tensor.

We can determine the momentum of the electromagnetic field by considering what happens when an external field is applied to a collection of moving charges. Care must be used to include all of the forces that act when matter is present.

We consider first the electric force on charges within a volume V bounded by a surface S. This is given by

$$\mathbf{F}_{\text{elec}} = \int_V d\tau (\rho_{\text{free}} + \rho_{\text{bound}})\mathbf{E}. \tag{9.68}$$

We have added the subscript "free" to the free charge density here to distinguish it from the bound charge density. The contribution of the bound charge density is important in the case of polarizable matter. In fact, for a neutral dielectric, the only force is that on the bound charges.

From Eqs. (6.7) and (6.10), the charge densities are related to $\mathbf{E}$ by

$$\boldsymbol{\nabla}\cdot\mathbf{E} = 4\pi(\rho_{\text{free}} + \rho_{\text{bound}}). \tag{9.69}$$

Thus the electric force on the matter within the volume V can be written as

$$\mathbf{F}_{\text{elec}} = \frac{1}{4\pi} \int_V d\tau \, \mathbf{E}(\boldsymbol{\nabla}\cdot\mathbf{E}). \tag{9.70}$$

This can be put into a more useful (if a bit more complicated) form using Maxwell's equations and some vector manipulation:

$$
\begin{aligned}
\mathbf{F}_{\text{elec}} &= \frac{1}{4\pi} \int_V d\tau \, \mathbf{E}(\boldsymbol{\nabla}\cdot\mathbf{E}) \\
&= \frac{1}{4\pi} \int_V d\tau \, [\boldsymbol{\nabla}\cdot(\mathbf{E}\mathbf{E}) - (\mathbf{E}\cdot\boldsymbol{\nabla})\mathbf{E}] \\
&= \frac{1}{4\pi} \int_V d\tau \, [\boldsymbol{\nabla}\cdot(\mathbf{E}\mathbf{E}) - \frac{1}{2}\boldsymbol{\nabla}(\mathbf{E}\cdot\mathbf{E}) + \mathbf{E}\times(\boldsymbol{\nabla}\times\mathbf{E})] \\
&= \frac{1}{4\pi} \int_V d\tau \, [\boldsymbol{\nabla}\cdot(\mathbf{E}\mathbf{E}) - \frac{1}{2}\boldsymbol{\nabla}(\mathbf{E}\cdot\mathbf{E}) - \frac{1}{c}\mathbf{E}\times(\partial_t\mathbf{B})].
\end{aligned}
\tag{9.71}
$$

The magnetic force on matter within the volume V is given by the force on the bound and free currents within the volume. This is

$$\mathbf{F}_{\text{mag}} = \frac{1}{c} \int_V d\tau (\mathbf{j}_{\text{free}} + \mathbf{j}_{\text{bound}})\times\mathbf{B}. \tag{9.72}$$

Here, $\mathbf{j}_{\text{bound}}$ consists of two parts

$$\mathbf{j}_{\text{bound}} = c\boldsymbol{\nabla}\times\mathbf{M} + \partial_t\mathbf{P}. \tag{9.73}$$

The first bound current in Eq. (9.73) is given by Eq. (8.7). It is due to circulating currents that stay within each molecule. The second bound current comes from the time derivative of the electric dipole moment of a molecule

$$\frac{d\mathbf{p}}{dt} = \sum_i q_i \frac{d\mathbf{r}_i}{dt}, \tag{9.74}$$

where the sum is over all the charges in the molecule. When this is related to the polarization $\mathbf{P}$, which is the dipole moment per unit volume, it corresponds to the bound current $\partial_t \mathbf{P}$. There is a force on each of these bound currents when they are acted on by a magnetic field.

The bound and free currents are related to the magnetic field $\mathbf{B}$ by Maxwell's equation for $\mathbf{H}$, written as $\mathbf{B} - 4\pi\mathbf{M}$:

$$
\begin{aligned}
\boldsymbol{\nabla}\times\mathbf{B} &= 4\pi\boldsymbol{\nabla}\times\mathbf{M} + \frac{4\pi}{c}\mathbf{j}_{\text{free}} + \frac{1}{c}\partial_t\mathbf{D} \\
&= \frac{4\pi}{c}(\mathbf{j}_{\text{free}} + \mathbf{j}_{\text{boumd}}) - \frac{4\pi}{c}\partial_t\mathbf{P} + \frac{1}{c}\partial_t\mathbf{D} \\
&= \frac{4\pi}{c}(\mathbf{j}_{\text{free}} + \mathbf{j}_{\text{bound}}) + \frac{1}{c}\partial_t\mathbf{E}.
\end{aligned} \tag{9.75}
$$

We can use this relation to substitute for $\mathbf{j}_{\text{free}}+\mathbf{j}_{\text{bound}}$ in Eq. (9.72), so the magnetic force can be written as

$$
\begin{aligned}
\mathbf{F}_{\text{mag}} &= \frac{1}{4\pi}\int_V d\tau\left[(\boldsymbol{\nabla}\times\mathbf{B})\times\mathbf{B} - \frac{1}{c}(\partial_t\mathbf{E})\times\mathbf{B}\right] \\
&= \frac{1}{4\pi}\int_V d\tau\left[(\mathbf{B}\cdot\boldsymbol{\nabla})\mathbf{B} - \frac{1}{2}\boldsymbol{\nabla}(\mathbf{B}\cdot\mathbf{B}) - \frac{1}{c}(\partial_t\mathbf{E})\times\mathbf{B}\right] \\
&= \frac{1}{4\pi}\int_V d\tau\left[\boldsymbol{\nabla}\cdot(\mathbf{B}\mathbf{B}) - \frac{1}{2}\boldsymbol{\nabla}(\mathbf{B}\cdot\mathbf{B}) - \frac{1}{c}(\partial_t\mathbf{E})\times\mathbf{B}\right].
\end{aligned} \tag{9.76}
$$

Combining the electric force of Eq. (9.71) and the magnetic force of Eq. (9.76) gives

$$
\begin{aligned}
\mathbf{F} &= \frac{1}{4\pi}\int_V d\tau[\boldsymbol{\nabla}\cdot(\mathbf{E}\mathbf{E}) - \frac{1}{2}\boldsymbol{\nabla}(\mathbf{E}\cdot\mathbf{E}) - \frac{1}{c}\mathbf{E}\times(\partial_t\mathbf{B}) \\
&\qquad + \boldsymbol{\nabla}\cdot(\mathbf{B}\mathbf{B}) - \frac{1}{2}\boldsymbol{\nabla}(\mathbf{B}\cdot\mathbf{B}) - \frac{1}{c}(\partial_t\mathbf{E})\times\mathbf{B}] \\
&= \frac{1}{4\pi}\int_V d\tau\{\boldsymbol{\nabla}\cdot[\mathbf{E}\mathbf{E} - \frac{1}{2}\hat{\mathbf{n}}(\mathbf{E}\cdot\mathbf{E}) + \mathbf{B}\mathbf{B} - \frac{1}{2}\hat{\mathbf{n}}(\mathbf{B}\cdot\mathbf{B})] \\
&\qquad - \frac{1}{c}(\partial_t(\mathbf{E}\times\mathbf{B})\}.
\end{aligned} \tag{9.77}
$$

We define a dyadic

$$
[\mathbf{T}_{\mathbf{EB}}] = \frac{1}{4\pi}[(\mathbf{E}\mathbf{E} + \mathbf{B}\mathbf{B}) - \frac{1}{2}\hat{\mathbf{n}}(\mathbf{E}\cdot\mathbf{E} + \mathbf{B}\cdot\mathbf{B})]. \tag{9.78}
$$

Then Eq. (9.77) can be rewritten as

$$
\begin{aligned}
\mathbf{F} + \frac{1}{4\pi c}\int_V d\tau\partial_t(\mathbf{E}\times\mathbf{B}) &= \int_V d\tau\boldsymbol{\nabla}\cdot[\mathbf{T}_{\mathbf{EB}}] \\
&= \oint_S d\mathbf{S}\cdot[\mathbf{T}_{\mathbf{EB}}],
\end{aligned} \tag{9.79}
$$

where we have used the divergence theorem.

The first term ($\mathbf{F}$) on the left hand side of Eq. (9.79) is the time derivative of the mechanical momentum of the matter in the volume V. We can identify the second term as the time derivative of the momentum of the purely electromagnetic fields (**E** and **B**) in the volume. This implies that the electromagnetic field has a momentum given by

$$\mathbf{P_{EB}} = \frac{1}{4\pi c} \int_V \mathbf{E} \times \mathbf{B}\, d\tau. \tag{9.80}$$

With these identifications, we can write

$$\frac{d}{dt}(\mathbf{P_{matter}} + \mathbf{P_{EB}}) = \oint_S d\mathbf{S}\cdot[\mathbf{T_{EB}}]. \tag{9.81}$$

The dyadic $[\mathbf{T_{EB}}]$ is called **the Maxwell stress tensor.** (It is a dyadic, as was defined in Sec. 2.4.1, and we will show in Chapter 14 that it transforms as a tensor under rotation.) It has the significance that integrating the Maxwell stress tensor over any closed surface gives the time derivative of the sum of the momentum of the matter and of the purely electromagnetic fields inside the surface.

In the absence of time dependent fields, the stress tensor $[\mathbf{T_{EB}}]$ can be used in Eq. (9.79) to find the force on all matter within a closed surface of integration. This includes polarizable matter, as well as electric charges. We have used the subscript **EB** to emphasize that this form of the stress tensor is given in terms of the electric and magnetic fields **E** and **B**, even in the presence of polarizable matter.

From Eq. (9.80), we can define an electromagnetic momentum density, $\mathbf{G_{EB}}(\mathbf{r})$, by

$$\mathbf{G_{EB}} = \frac{1}{4\pi c}(\mathbf{E} \times \mathbf{B}). \tag{9.82}$$

The integral of $\mathbf{G_{EB}}$ over a volume gives the electromagnetic momentum of the **E** and **B** fields inside the volume. We can also define an electromagnetic angular momentum density by $\mathbf{r} \times \mathbf{G_{EB}}$. Then the electromagnetic angular momentum within a volume V will be given by

$$\mathbf{L_{EB}} = \int_V \mathbf{r} \times \mathbf{G_{EB}}\, d\tau = \frac{1}{4\pi c} \int_V \mathbf{r} \times (\mathbf{E} \times \mathbf{B})\, d\tau. \tag{9.83}$$

9.6.1   Momentum in the polarization and magnetization fields

In the next chapter, we will study the propagation of electromagnetic waves in neutral matter. The energy and momentum of these waves includes energy and momentum due to the polarization (**P**) and magnetization (**M**) fields, as well as that due to the pure **E** and **B** fields.

We have already discussed the potential energy contained in the **P** and **M** fields in Section (9.4.1). In this section, we study the momentum contained in the combination of these matter fields with the purely electromagnetic **E** and $\mathbf{B}$ fields. This is most easily done by redoing the previous section, at first leaving out the forces on matter that are due to the $\mathbf{E}$ and $\mathbf{B}$ fields acting on the bound densities ρ_{bound} and $\mathbf{j}_{\text{bound}}$. Instead of being included in the forces on matter, the forces on the bound charges and currents will be included as part of the change in the momentum of the electromagnetic field, now including the $\mathbf{P}$ and $\mathbf{M}$ fields.

The force on only the free charge and current densities is given by

$$\mathbf{F}_{\text{free}} = \int d\tau [\rho_{\text{free}}\mathbf{E} + \frac{1}{c}\mathbf{j}_{\text{free}} \times \mathbf{B}]. \tag{9.84}$$

This can be changed, using Maxwell's equations into

$$
\begin{aligned}
\mathbf{F}_{\text{free}} &= \frac{1}{4\pi} \int d\tau [\mathbf{E}(\boldsymbol{\nabla}\cdot\mathbf{D}) + (\boldsymbol{\nabla}\times\mathbf{H})\times\mathbf{B} - \frac{1}{c}(\partial_t\mathbf{D})\times\mathbf{B}] \\
&= \frac{1}{4\pi} \int d\tau [\boldsymbol{\nabla}\cdot(\mathbf{D}\mathbf{E}) - (\mathbf{D}\cdot\boldsymbol{\nabla})\mathbf{E} + (\mathbf{B}\cdot\boldsymbol{\nabla})\mathbf{H} - \boldsymbol{\nabla}(\mathbf{B}\cdot\mathbf{H})|_{\mathbf{B}\text{const}} - \frac{1}{c}(\partial_t\mathbf{D})\times\mathbf{B}] \\
&= \frac{1}{4\pi} \int d\tau [\boldsymbol{\nabla}\cdot(\mathbf{D}\mathbf{E}) - \boldsymbol{\nabla}(\mathbf{D}\cdot\mathbf{E})|_{\mathbf{D}\text{const}} + \mathbf{D}\times(\boldsymbol{\nabla}\times\mathbf{E}) \\
&\qquad + \boldsymbol{\nabla}\cdot(\mathbf{B}\mathbf{H}) - \boldsymbol{\nabla}(\mathbf{B}\cdot\mathbf{H})|_{\mathbf{B}\text{const}} - \frac{1}{c}(\partial_t\mathbf{D})\times\mathbf{B}] \\
&= \frac{1}{4\pi} \int d\tau [\boldsymbol{\nabla}\cdot(\mathbf{D}\mathbf{E}) - \boldsymbol{\nabla}(\mathbf{D}\cdot\mathbf{E})|_{\mathbf{D}\text{const}} - \frac{1}{c}\mathbf{D}\times(\partial_t\mathbf{B}) \\
&\qquad + \boldsymbol{\nabla}\cdot(\mathbf{B}\mathbf{H}) - \boldsymbol{\nabla}(\mathbf{B}\cdot\mathbf{H})|_{\mathbf{B}\text{const}} - \frac{1}{c}(\partial_t\mathbf{D})\times\mathbf{B}]. \tag{9.85}
\end{aligned}
$$

If the material is both linear and homogenous (ϵ and μ are not functions of position $\mathbf{r}$.), we can write

$$\boldsymbol{\nabla}(\mathbf{D}\cdot\mathbf{E})|_{\mathbf{D}\text{const}} = \boldsymbol{\nabla}(\mathbf{D}\cdot\mathbf{E})|_{\mathbf{E}\text{const}} = \frac{1}{2}\boldsymbol{\nabla}(\mathbf{D}\cdot\mathbf{E}) \tag{9.86}$$

and

$$\boldsymbol{\nabla}(\mathbf{B}\cdot\mathbf{H})|_{\mathbf{B}\text{const}} = \boldsymbol{\nabla}(\mathbf{B}\cdot\mathbf{H})|_{\mathbf{H}\text{const}} = \frac{1}{2}\boldsymbol{\nabla}(\mathbf{B}\cdot\mathbf{H}). \tag{9.87}$$

These relations also hold for anisotropic materials, but only if the permittivity and permeability tensors are symmetric. With these two relations, Eq. (9.85) becomes

$$
\begin{aligned}
\mathbf{F}_{\text{free}} &= \frac{1}{4\pi} \int d\tau [\boldsymbol{\nabla}\cdot(\mathbf{D}\mathbf{E}) - \frac{1}{2}\boldsymbol{\nabla}(\mathbf{D}\cdot\mathbf{E}) \\
&\qquad + \boldsymbol{\nabla}\cdot(\mathbf{B}\mathbf{H}) - \frac{1}{2}\boldsymbol{\nabla}(\mathbf{B}\cdot\mathbf{H}) - \frac{1}{c}\partial_t(\mathbf{D}\times\mathbf{B})]. \tag{9.88}
\end{aligned}
$$

Now, we define a stress dyadic to be

$$[\mathbf{T}_{\mathbf{DEBH}}] = \frac{1}{4\pi}[(\mathbf{D}\mathbf{E} + \mathbf{B}\mathbf{H}) - \frac{1}{2}\hat{\mathbf{n}}(\mathbf{D}\cdot\mathbf{E} + \mathbf{B}\cdot\mathbf{H})]. \tag{9.89}$$

Then Eq. (9.88) can be rewritten as

$$\mathbf{F}_{\text{free}} + \frac{1}{4\pi c} \int_V d\tau \partial_t (\mathbf{D} \times \mathbf{B}) = \int_V d\tau \boldsymbol{\nabla} \cdot [\mathbf{T_{DEBH}}]$$

$$= \oint_S d\mathbf{S} \cdot [\mathbf{T_{DEBH}}]. \qquad (9.90)$$

We can now interpret the field momentum density to be

$$\mathbf{G_{DB}} = \frac{1}{4\pi c} \mathbf{D} \times \mathbf{B}. \qquad (9.91)$$

This momentum density includes the momentum of the $\mathbf{P}$ and $\mathbf{M}$ fields in the matter as well as that of the purely electromagnetic $\mathbf{E}$ and $\mathbf{B}$ fields. This will be the appropriate momentum density to describe the momentum carried by an electromagnetic wave in matter. The corresponding angular momentum of the electromagnetic field is

$$\mathbf{L_{DB}} = \int_V \mathbf{r} \times \mathbf{G_{DB}} d\tau = \frac{1}{4\pi c} \int_V \mathbf{r} \times (\mathbf{D} \times \mathbf{B}) d\tau. \qquad (9.92)$$

There have been some objections to $[\mathbf{T_{DEBH}}]$ on the basis that it is not symmetric for a non-symmetric anisotropic material. However, we have seen in the derivation above that the use of $[\mathbf{T_{DEBH}}]$ as a stress tensor is only valid for symmetric permittivity or permeability tensors, in which case $[\mathbf{T_{DEBH}}]$ is symmetric.

Although the stress tensor $[\mathbf{T_{DEBH}}]$ can only be used for simple materials where $\mathbf{D} = \epsilon \mathbf{E}$ and $\mathbf{B} = \mu \mathbf{H}$, the identification of $\mathbf{G_{DB}}$ as the momentum density (including the momentum in $\mathbf{P}$ and $\mathbf{M}$) is still appropriate for nonlinear or inhomogeneous matter. This can be seen in the last step of Eq. (9.85), which holds for any material.

Care must be taken in applying the stress tensor to matter. To get the force on all the matter contained with a closed surface, the stress tensor $[\mathbf{T_{EB}}]$ of Eq. (9.78) in terms of $\mathbf{E}$ and $\mathbf{B}$ should be used. This will give the force on the matter including the polarization and magnetization forces. Using the stress tensor $[\mathbf{T_{DEBH}}]$ of Eq. (9.89) gives only the force on free charges and currents, and not the total force on the enclosed matter. Its use, even for that, is limited to materials that are linear and homogeneous within the bounding surface. For finding the force between charges in a material with a simple permittivity, $[\mathbf{T_{DEBH}}]$ should be used. (See problem 4.) In this case, use of $[\mathbf{T_{EB}}]$ would give the force on both the dielectric and the charges.

A third expression for the electromagnetic momentum density,

$$\mathbf{G_{EH}} = \frac{1}{4\pi c} \mathbf{E} \times \mathbf{H}, \qquad (9.93)$$

was proposed my Max Abraham in 1909, and there has been a controversity ever since over which expression, $[\mathbf{G_{DB}}]$ (first derived by Minkowski in 1908) or $[\mathbf{G_{EH}}]$, is 'correct'.

Actually, the permeability, μ, is close to 1 for most dielectrics, so Abraham's $\mathbf{G_{EH}}$ is not much different from the $\mathbf{G_{EB}}$ derived above. We have seen above that it is not a question of which is correct, but that they just apply to different cases. $\mathbf{G_{EB}}$ is the momentum density of just the electromagnetic fields in matter, while $\mathbf{G_{DB}}$ includes the momentum density of the matter fields, $\mathbf{P}$ and $\mathbf{M}$.

9.7   Application of the Stress Tensor.

It should be noted that the stress tensor does not necessarily act as a stress at a particular point on a surface, but must be integrated over the full closed surface to have the proper effect. We illustrate this by the following example of the force exerted on the planar surface of a dielectric of permittivity ϵ by a point charge, which we treated in Sec.6.4.4. The force per unit area (stress) exerted on the bound surface charge of the dielectric by a point charge q a distance $\mathbf{d}$ away is given by (The details of this example are given as a problem.)

$$\mathbf{F}/A = \frac{q^2 d(\mathbf{d} - \mathbf{x})}{2\pi(x^2 + d^2)^3}\left(\frac{\epsilon - 1}{\epsilon + 1}\right), \tag{9.94}$$

where $\mathbf{x}$ is the distance along the plane, as shown in Fig. 9.2.

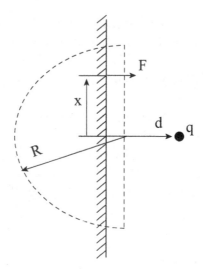

Figure 9.2: Stress and force on a dielectric plane due to a point charge.

Use of the appropriate stress tensor $[\mathbf{T_{EB}}]$ gives a different answer for the stress. The electric field just to the right of the dielectric plane is [using the image charge given in Eq. (6.57)]

$$\mathbf{E} = \frac{2q(\mathbf{x} - \epsilon \mathbf{d})}{(\epsilon + 1)(x^2 + d^2)^{\frac{3}{2}}}. \tag{9.95}$$

Evaluating $\hat{\mathbf{d}} \cdot [\mathbf{T_{EB}}]$ for this field gives

$$\hat{\mathbf{d}} \cdot [\mathbf{T_{EB}}] = \frac{q^2[\hat{\mathbf{d}}(\epsilon^2 d^2 - x^2) - 2\mathbf{x}\epsilon d]}{2\pi(\epsilon + 1)^2(x^2 + d^2)^3}, \tag{9.96}$$

which does not equal the correct stress given by Eq. (9.94).

However, the result of Eq. (9.96) does lead to the correct force on the dielectric when properly integrated over a closed surface. The surface for this integration, shown as the dashed lines in Fig. 9.2, is a plane circular surface of radius R just to the right of the dielectric, closed by a hemispherical surface of radius R, taken in the limit as $R{\to}\infty$. In this limit, the surface integral on the hemispherical surface vanishes, and the integral over the plane surface results in

$$\mathbf{F} = \left(\frac{\epsilon - 1}{\epsilon + 1}\right) \frac{q^2 \hat{\mathbf{d}}}{4d^2}. \tag{9.97}$$

This agrees with the result found in problem 3 of chapter 6 by integrating Eq. (9.94) over the plane surface, and is equal and opposite to the force on the point charge q given in Eq. (6.58).

9.8 Magnetic Monopoles.

In the absence of polarizable or magnetic matter, Maxwell's equations for the electromagnetic fields $\mathbf{E}$ and $\mathbf{B}$ are given by

$$\boldsymbol{\nabla}\cdot\mathbf{E} = 4\pi\rho \tag{9.98}$$

$$\boldsymbol{\nabla}\times\mathbf{E} = -\frac{1}{c}\partial_t\mathbf{B} \tag{9.99}$$

$$\boldsymbol{\nabla}\cdot\mathbf{B} = 0 \tag{9.100}$$

$$\boldsymbol{\nabla}\times\mathbf{B} = \frac{4\pi\mathbf{j}}{c} + \frac{1}{c}\partial_t\mathbf{E}. \tag{9.101}$$

There is an asymmetry in these equations in that there are electric charge and current densities, but no corresponding magnetic densities. This absence of magnetic charge and current densities has been noted and accepted for some time, and no such magnetic densities have ever been observed. It turns out, however, that there are some expectations from quantum field theory that magnetic charge may exist.

P. A. M. Dirac presented one intriguing suggestion (which we will examine below) for the appearance of magnetic charge in his early development of quantum electrodynamics. More recent speculative, but widely studied, quantum field theories (labeled "superstring" theories) predict the existence of magnetic charge in the form of point magnetic monopoles. There has been an extensive continuing experimental search for magnetic monopoles, but, to date, none have been observed. Here we give some outline of how they would effect classical electromagnetism.

With magnetic charge and current densities, Maxwell's equations would become

$$\nabla\cdot\mathbf{E} \;=\; 4\pi\rho_e \tag{9.102}$$

$$\nabla\times\mathbf{E} \;=\; -\frac{4\pi\mathbf{j}_m}{c} - \frac{1}{c}\partial_t\mathbf{B} \tag{9.103}$$

$$\nabla\cdot\mathbf{B} \;=\; 4\pi\rho_m \tag{9.104}$$

$$\nabla\times\mathbf{B} \;=\; \frac{4\pi\mathbf{j}_e}{c} + \frac{1}{c}\partial_t\mathbf{E}. \tag{9.105}$$

We have used the subscripts e and m to distinguish the electric and magnetic densities.

The equations for $\mathbf{E}$ and $\mathbf{B}$ are now similar. The symmetry between $\mathbf{E}$ and $\mathbf{B}$ can be utilized to introduce the **duality transformation**

$$\mathbf{E}' \;=\; \mathbf{E}\cos\theta + \mathbf{B}\sin\theta \tag{9.106}$$

$$\mathbf{B}' \;=\; -\mathbf{E}\sin\theta + \mathbf{B}\cos\theta \tag{9.107}$$

$$\rho_e' \;=\; \rho_e\cos\theta + \rho_m\sin\theta \tag{9.108}$$

$$\rho_m' \;=\; -\rho_e\sin\theta + \rho_m\cos\theta \tag{9.109}$$

$$\mathbf{j}_e' \;=\; \mathbf{j}_e\cos\theta + \mathbf{j}_m\sin\theta \tag{9.110}$$

$$\mathbf{j}_m' \;=\; -\mathbf{j}_e\sin\theta + \mathbf{j}_m\cos\theta. \tag{9.111}$$

Here, θ is an arbitrary transformation parameter and not a physical angle. Maxwell's equations with magnetic densities are invariant under this transformation of fields and source densities. If the transformation angle θ is chosen to be $90°$, the electric and magnetic fields and densities are just interchanged.

One interesting possibility is that all elementary particles have the same ratio $R_{ge} = g/e$ of magnetic charge (g) to electric charge (e). Then, an angle θ can be found for which $\rho_m' = 0$. This means that such a case of magnetic charge could be transformed away, giving the world we see today with no observed magnetic charge.

9.8.1   Dirac Charge Quantization.

Dirac[1] showed in 1931 that the existence, anywhere in the universe, of even one magnetic charge would require that electric charge be quantized in a consistent quantum mechanical theory. Since the apparent quantization of electric charge in integral multiples of the electron charge e is observed, this has been an attractive argument for the actual existence of magnetic charge. Dirac's original derivation was quantum mechanical and is not appropriate for this

---

[1]P. A. M. Dirac, Proc. Royal Soc., **A123**, 60 (1931); Phys. Rev. **74**, 817 (1948).

text. But there is a semi-classical derivation of charge quantization by M. N. Saha[2] that we detail below.

We consider the case of a single point magnetic charge g a distance $\mathbf{d}$ from single point electric charge e, as shown in Fig. 9.3.

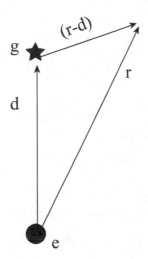

Figure 9.3: Magnetic charge and electric charge.

There is an electric field

$$\mathbf{E} = \frac{e\mathbf{r}}{r^3}, \tag{9.112}$$

and a magnetic field

$$\mathbf{B} = \frac{g(\mathbf{r} - \mathbf{d})}{|\mathbf{r} - \mathbf{d}|^3}. \tag{9.113}$$

The electromagnetic angular momentum is given by the integral

$$
\begin{aligned}
\mathbf{L}_{em} &= \frac{1}{4\pi c} \int \mathbf{r} \times (\mathbf{E} \times \mathbf{B}) d\tau \\
&= \frac{e}{4\pi c} \int \frac{\hat{\mathbf{r}} \times (\hat{\mathbf{r}} \times \mathbf{B}) d\tau}{r} \\
&= \frac{e}{4\pi c} \int \frac{[\hat{\mathbf{r}}(\hat{\mathbf{r}} \cdot \mathbf{B}) - \mathbf{B}] d\tau}{r}.
\end{aligned}
\tag{9.114}
$$

We can use the vector identity

$$
\begin{aligned}
\nabla \cdot (\mathbf{B}\hat{\mathbf{r}}) &= \hat{\mathbf{r}}(\nabla \cdot \mathbf{B}) + (\mathbf{B} \cdot \nabla)\hat{\mathbf{r}} \\
&= \hat{\mathbf{r}}(\nabla \cdot \mathbf{B}) + \frac{\mathbf{B}}{r} - \frac{(\mathbf{B} \cdot \hat{\mathbf{r}})\hat{\mathbf{r}}}{r},
\end{aligned}
\tag{9.115}
$$

---

[2]M. N. Saha, Indian J. of Physics **10**, 141 (1936); Phys. Rev. **75**, 1968 (1949).

to rewrite Eq. (9.114) as

$$\mathbf{L}_{em} = \frac{e}{4\pi c} \int [\hat{\mathbf{r}}(\mathbf{\nabla}\cdot\mathbf{B}) - \mathbf{\nabla}\cdot(\mathbf{B}\hat{\mathbf{r}})]d\tau. \qquad (9.116)$$

The second integral on the right hand side vanishes. We show this by applying the divergence theorem to get

$$\int [\mathbf{\nabla}\cdot(\mathbf{B}\hat{\mathbf{r}})]d\tau = \oint d\mathbf{S}\cdot(\mathbf{B}\hat{\mathbf{r}})$$

$$= g \int \frac{R^2 d\Omega[\hat{\mathbf{r}}\cdot(\mathbf{R} - \mathbf{d})]\hat{\mathbf{r}})}{|\mathbf{R} - \mathbf{d}|^3}. \qquad (9.117)$$

We have taken the surface integral over a large sphere of radius R. In the limit as $R\to\infty$, this integral approaches

$$\lim_{R\to\infty} \int \frac{R^2 d\Omega[\hat{\mathbf{r}}\cdot(\mathbf{R} - \mathbf{d})]\hat{\mathbf{r}})}{|\mathbf{R} - \mathbf{d}|^3} = \int d\Omega\hat{\mathbf{r}} = 0. \qquad (9.118)$$

For the first integral in Eq. (9.116), we use the fact that

$$\mathbf{\nabla}\cdot\mathbf{B} = 4\pi g\delta(\mathbf{r} - \mathbf{d}) \qquad (9.119)$$

to get

$$\begin{aligned} \mathbf{L}_{em} &= \frac{e}{4\pi c} \int \hat{\mathbf{r}}(\mathbf{\nabla}\cdot\mathbf{B})d\tau \\ &= \frac{eg}{c} \int \hat{\mathbf{r}}\delta(\mathbf{r} - \mathbf{d})d\tau \\ &= \frac{eg\hat{\mathbf{d}}}{c}. \qquad (9.120) \end{aligned}$$

We see that the combination of the magnetic and electric charges has an angular momentum that is independent of the magnitude of the distance between the charges. The magnitude of the angular momentum is fixed at eg/c no matter how far apart the two charges are.

Quantum mechanics requires the quantization of any angular momentum in units of $\hbar/2$, where $\hbar$ (pronounced aitchbar) is Planck's constant divided by 2π. This requires the relation

$$\frac{eg}{\hbar c} = n/2, \qquad (9.121)$$

where n is any positive integer. This means that both electric and magnetic charge would have to be quantized if there were at least one magnetic monopole anywhere in the universe. They each would be quantized as multiples of a smallest charge. This smallest charge would be the case for $n = 1$.

The electron charge e satisfies the condition

$$\frac{e^2}{\hbar c} = \frac{1}{137} \equiv \alpha. \qquad (9.122)$$

The dimensionless number α is called **the fine structure constant** because it corresponds to the ratio of atomic fine structure energy splittings to principle quantum number energy splittings. Its most precise determination (as of 2022) is $\alpha = 1/137.0360$. For historical reasons, and because $1/\alpha$ is so close to an integer, the numerical value of α is almost always given in terms of its inverse. In SI units, α is written as

$$\alpha = \frac{e^2}{4\pi\epsilon_0 \hbar c} = \frac{1}{137} \quad \textbf{SI}. \tag{9.123}$$

Fortunately, all systems of units have kept the same notation α for the fine structure constant. Since α is dimensionless, it has the same value $1/137$ in any system of units.

From Eq. (9.121) for $n = 1$, it follows that the smallest magnetic charge would satisfy

$$\frac{g^2}{\hbar c} = \frac{1}{4\alpha} = 137/4. \tag{9.124}$$

Because α is a small number, the basic electromagnetic interaction is relatively weak. It only has strong consequences because of its long range and the huge number of charges that can interact. Equation (9.124) shows that even one magnetic charge would have a strong interaction. This fact is used in searches for magnetic monopoles, but these have not yet found any.

9.9   Problems

1. A copper ring of radius a is at a fixed distance $\mathbf{d}$ (with $a << d$) directly above an identical copper ring. Each ring has a resistance R for circulating currents. An increasing current $I = I_0 t/\tau$ is applied in the lower ring. Neglect the self-inductance of each ring, and make appropriate approximations.

   (a) Find the dipole moment of the lower ring.

   (b) Find the magnetic flux through the upper ring.

   (c) Find the induced EMF and the current in the upper ring.

   (d) Find the induced dipole moment of the upper ring.

   (e) Show that the force between the rings is

$$F = \frac{12\pi^4 a^8 I_0^2 t}{c^4 R d^7 \tau^2}. \tag{9.125}$$

   Is this force repulsive or attractive?

2. (a) Find the mutual inductance of the two rings in problem 1 (making suitable approximations).

   (b) Use their mutual inductance to find the EMF induced in the upper ring for the current $I = I_0 t/\tau$ in the lower ring.

   (c) Find the mutual interaction energy in this system.

3. A long straight copper wire of radius a and resistance R carries a constant current I.

   (a) Find the electric and magnetic fields at the surface of the wire.

   (b) Integrate the Poynting power flux through the surface of a piece of the wire of length L to show that the power through the surface equals $I^2 R$.

   (c) Find the electromagnetic energy and momentum in this piece of the wire.

4. The upper and lower curves of the hysteresis loop of a hard ferromagnet are given by

$$\begin{aligned} B_+ &= B_0[\tanh(H/H_0 + .5) - .1], & (9.126) \\ B_- &= B_0[\tanh(H/H_0 - .5) + .1], & (9.127) \end{aligned}$$

respectively, for $-1.5 < H/H_0 < +1.5$.

   (a) Find the energy lost from the magnetic field in one cycle.

(b) If $B_0 = 3$ kilogauss, $H_0 = 1$ gauss, and the frequency is 60 Hz, find the power loss in watts.

5. Two point charges, each of charge q are a distance $2d$ apart.

   (a) Find the Maxwell stress tensor $[\mathbf{T}]$ on the plane surface midway between the charges.

   (b) Find the force on either charge by integrating $[\mathbf{T}] \cdot d\mathbf{S}$ over a plane surface, closing the surface with a large hemisphere of radius R. (Show that the integral over the hemisphere vanishes in the limit $R \to \infty$.)

   (c) Repeat parts (a) and (b) if the charges have opposite signs.

   (d) What would the force on either charge be if they were immersed in a simple dielectric of infinite extent?

6. For a point charge q a distance $\mathbf{d}$ from the planar surface of a dielectric, as shown in Fig. 9.2,

   (a) find the force per unit area on the bound surface charge density on the planar surface. (Use the surface charge density found in problem 4a of chapter 6.)

   (b) find the full electric field just outside the dielectric, due to the two charges q and q' in Fig. 6.5.

   (c) find the components $\hat{\mathbf{d}} \cdot [\mathbf{T}]$ of the Maxwell stress tensor just outside the planar surface.

   (d) integrate $d\mathbf{S} \cdot [\mathbf{T}]$ over the closed surface shown in Fig. 9.2 to find the force on the dielectric.

7. A magnetic dipole $\boldsymbol{\mu}$ is located at the center of a uniform electric charge distribution of radius R, charge e, and mass m.

   (a) Find the electromagnetic angular momentum of this configuration.

   (b) Find the value of the radius R for which the g-factor of this configuration equals 2.

8. (a) Calculate $e^2 / \hbar c$ in Gaussian units to verify Eq. (9.122).

   (b) Calculate $e^2 / (4\pi\epsilon_0 \hbar c)$ in SI units.

Chapter 10

Electromagnetic Plane Waves

10.1 Electromagnetic Waves from Maxwell's Equations.

In the absence of free charges and currents, Maxwell's equations are

$$\nabla \cdot \mathbf{D} = 0 \tag{10.1}$$

$$\nabla \times \mathbf{E} = -\frac{1}{c}\partial_t \mathbf{B} \tag{10.2}$$

$$\nabla \cdot \mathbf{B} = 0 \tag{10.3}$$

$$\nabla \times \mathbf{H} = \frac{1}{c}\partial_t \mathbf{D}. \tag{10.4}$$

For a simple medium with

$$\mathbf{D} = \epsilon \mathbf{E}, \qquad \mathbf{B} = \mu \mathbf{H}, \tag{10.5}$$

Maxwell's equations become

$$\nabla \cdot \mathbf{E} = 0 \tag{10.6}$$

$$\nabla \times \mathbf{E} = -\frac{1}{c}\partial_t \mathbf{B} \tag{10.7}$$

$$\nabla \cdot \mathbf{B} = 0 \tag{10.8}$$

$$\nabla \times \mathbf{B} = \left(\frac{\epsilon\mu}{c}\right)\partial_t \mathbf{E}. \tag{10.9}$$

Taking the curl of the curl $\mathbf{E}$ equation, we get

$$\nabla \times (\nabla \times \mathbf{E}) = \nabla(\nabla \cdot \mathbf{E}) - \nabla^2 \mathbf{E} = -\frac{1}{c}\partial_t(\nabla \times \mathbf{B}). \tag{10.10}$$

Using the $\nabla \cdot \mathbf{E}$ and $\nabla \times \mathbf{B}$ Maxwell equations, Eq. (10.10) reduces to

$$\nabla^2 \mathbf{E} = \left(\frac{\epsilon\mu}{c^2}\right)\partial_t^2 \mathbf{E}. \tag{10.11}$$

Equation (10.11) is the **wave equation** for the electric field **E**. Its solution can be written

$$\mathbf{E}(\mathbf{r},t) = \mathbf{E_0}e^{i(\mathbf{k\cdot r}-\omega t)}. \tag{10.12}$$

This is a wave travelling in the $\hat{\mathbf{k}}$ direction with **angular frequency** ω. (For any relation with complex variables, the understanding is that the value of a physical quantity, in this case $\mathbf{E}(\mathbf{r},t)$, is given by the real part of the complex expression.) The velocity of the wave is seen from the wave equation (10.11) to be

$$v = \frac{c}{\sqrt{\epsilon\mu}} = \frac{c}{n}. \tag{10.13}$$

The constant

$$n = \sqrt{\epsilon\mu} \tag{10.14}$$

is called the **index of refraction** of the material. The constants ω and k (called the **wave number**) are related by

$$k = \frac{\omega}{v} = \left(\frac{\sqrt{\epsilon\mu}}{c}\right)\omega. \tag{10.15}$$

A similar derivation shows that the magnetic field also satisfies the wave equation

$$\nabla^2\mathbf{B} = \left(\frac{\epsilon\mu}{c^2}\right)\partial_t^2\mathbf{B}. \tag{10.16}$$

with the solution

$$\mathbf{B}(\mathbf{r},t) = \mathbf{B_0}e^{i(\mathbf{k\cdot r}-\omega t)}. \tag{10.17}$$

The ∇ operator acting on the exponential factor $e^{i\mathbf{k\cdot r}}$ has the effect that

$$\nabla e^{i\mathbf{k\cdot r}} = i\mathbf{k}e^{i\mathbf{k\cdot r}}. \tag{10.18}$$

Then Maxwell's divergence equations for **E** (in the absence of charge) and **B** can be written as

$$\nabla\cdot\mathbf{E} = i\mathbf{k}\cdot\mathbf{E} = 0 \tag{10.19}$$
$$\nabla\cdot\mathbf{B} = i\mathbf{k}\cdot\mathbf{B} = 0. \tag{10.20}$$

This means that the electromagnetic wave is a **transverse wave** with **E** and **B** each perpendicular to the direction of propagation $\hat{\mathbf{k}}$.

The electric and magnetic fields are related by the $\nabla\times\mathbf{E}$ Maxwell equation:

$$\nabla\times[\mathbf{E_0}e^{i(\mathbf{k\cdot r}-\omega t)}] = -\frac{1}{c}\partial_t[\mathbf{B_0}e^{i(\mathbf{k\cdot r}-\omega t)}]$$
$$i\mathbf{k}\times\mathbf{E_0}e^{i(\mathbf{k\cdot r}-\omega t)} = i\frac{\omega}{c}\mathbf{B_0}e^{i(\mathbf{k\cdot r}-\omega t)}, \tag{10.21}$$

so

$$\mathbf{B_0} = \sqrt{\epsilon\mu}\hat{\mathbf{k}}\times\mathbf{E_0}. \tag{10.22}$$

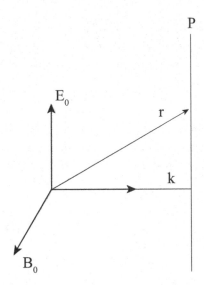

Figure 10.1: Right-handed triad of $\mathbf{E_0}$, $\mathbf{B_0}$, $\mathbf{k}$. The projection of $\mathbf{r}$ on $\mathbf{k}$ is constant when $\mathbf{r}$ ends on the plane P.

The three vectors $\mathbf{E_0}$, $\mathbf{B_0}$, $\mathbf{k}$ form a right handed triad as shown in Fig. 10.1. If ϵ and μ equal 1, then $\mathbf{E}$ and $\mathbf{B}$ are equal in magnitude if the same unit (gauss) is used for each.

The solution of the wave equation given by Eq. (10.12) is called a **plane wave**. This is because the surfaces of constant phase are plane surfaces. This is shown in Fig. 10.1. The scalar quantity $\mathbf{k \cdot r}$ is the projection of $\mathbf{r}$ on the vector $\mathbf{k}$, so $\mathbf{k \cdot r}$ is constant on any plane surface perpendicular to $\mathbf{k}$.

The complex phase of the electric field, given by $\mathbf{k \cdot r} - \omega t$, will have the same value anywhere on this plane, and the plane of constant phase will move in the $\hat{\mathbf{k}}$ direction at velocity $v = c/\sqrt{\epsilon\mu}$. In free space, where $\sqrt{\epsilon\mu} = 1$, the velocity of the wave is just equal to c, the velocity of light. This led Maxwell to the reasonable conclusion that "The velocity is so nearly that of light, that it seems we have strong reason to conclude that light itself is an electromagnetic disturbance in the form of waves propagated through the electromagnetic field according to electromagnetic laws."[1]

10.2 Energy and Momentum in an Electromagnetic Wave.

The energy density in the electromagnetic plane wave of Eq. (10.12) is given by

$$u = \frac{1}{8\pi}(\mathbf{E \cdot D} + \mathbf{B \cdot H}) = \frac{1}{8\pi}(\epsilon \mathbf{E \cdot E} + \frac{1}{\mu}\mathbf{B \cdot B}). \qquad (10.23)$$

---

[1] J. C. Maxwell, "A Dynamical Theory of the Electromagnetic Field", Royal Society Transactions, Vol. CLV (1864).

Using the relation between **B** and **E** given in Eq. (10.22), we see that the electric and magnetic parts of the energy density in Eq. (10.23) are equal . The energy density can be written purely in terms of **E** as

$$u = \frac{\epsilon}{4\pi}\mathbf{E}{\cdot}\mathbf{E}. \tag{10.24}$$

This expression for the energy density is bilinear in complex quantities. In order to get the appropriate value for the physical quantity u, we calculate using the real value for the physical electric field **E**. This gives

$$
\begin{aligned}
u &= \frac{\epsilon}{16\pi}[(\mathbf{E}+\mathbf{E}^*){\cdot}(\mathbf{E}+\mathbf{E}^*)] \\
&= \frac{\epsilon}{16\pi}(\mathbf{E}{\cdot}\mathbf{E} + \mathbf{E}^*{\cdot}\mathbf{E}^* + \mathbf{E}^*{\cdot}\mathbf{E} + \mathbf{E}{\cdot}\mathbf{E}^*).
\end{aligned}
\tag{10.25}
$$

The combination $\mathbf{E}{\cdot}\mathbf{E}$ has the time dependence $E_0^2 e^{-2i\omega t}$, and $\mathbf{E}^*{\cdot}\mathbf{E}^*$ has the time dependence $E_0^{*2}e^{2i\omega t}$. Each of these terms will oscillate with time, and be negligible if they are averaged over many periods of the function. The remaining two terms have no time dependence, so they will remain in a time average. This leaves the result

$$\bar{u} = \frac{\epsilon}{16\pi}(\mathbf{E}^*{\cdot}\mathbf{E} + \mathbf{E}{\cdot}\mathbf{E}^*) = \frac{\epsilon}{8\pi}|\mathbf{E_0}|^2 \tag{10.26}$$

for the time averaged energy density in the wave. Since the period of an electromagnetic wave is usually very short, this time average is the only physical quantity that can be measured. Note that the time averaged energy is one-half of the peak energy given in Eq. (10.24). In the above derivation, we have assumed that the constants ϵ and μ are real. The treatment with these constants complex will be considered in Chapter 11. The time averaged Poynting vector of the plane wave is given by

$$
\begin{aligned}
\overline{\mathbf{S}} &= \frac{c}{4\pi}\overline{\mathbf{E}{\times}\mathbf{H}} \\
&= \frac{c}{16\pi}(\mathbf{E}^*{\times}\mathbf{H} + \mathbf{E}{\times}\mathbf{H}^*) \\
&= \frac{c}{16\pi\mu}(\mathbf{E}^*{\times}\mathbf{B} + \mathbf{E}{\times}\mathbf{B}^*) \\
&= \frac{c}{8\pi\mu}[\mathbf{E}^*{\times}(\sqrt{\epsilon\mu}\hat{\mathbf{k}}{\times}\mathbf{E})] \\
&= \frac{c\hat{\mathbf{k}}}{8\pi}\sqrt{\frac{\epsilon}{\mu}}|\mathbf{E_0}|^2.
\end{aligned}
\tag{10.27}
$$

The time averaged Poynting vector is called the **intensity** of the electromagnetic wave. In the Gaussian system, its units are ergs/(cm²-sec). Comparing the Poynting vector with the energy density in the wave, given in Eq. (10.26), we see that

$$\overline{\mathbf{S}} = \frac{c\hat{\mathbf{k}}}{\sqrt{\epsilon\mu}}\bar{u} = v\hat{\mathbf{k}}\bar{u}, \tag{10.28}$$

where v is the wave velocity. The significance of this relation between the energy flux density $\mathbf{S}$ and the energy density u can be seen in Fig. 10.2.

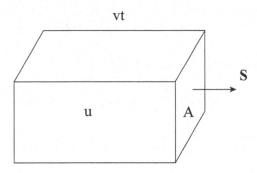

Figure 10.2: The field energy (SAt) passing through area A in time t equals the field energy ($uAvt$) in the volume Avt.

For the momentum density of an electromagnetic wave, we use the form $\mathbf{G_{DB}}$ of Eq. (9.91). This form gives the complete momentum density of the wave including the wave in the polarization and magnetization of the matter. This gives for the time averaged momentum density of a plane wave

$$
\begin{aligned}
\overline{\mathbf{G}} &= \frac{\epsilon}{16\pi c}(\mathbf{E}^*\times\mathbf{B}+\mathbf{E}\times\mathbf{B}^*) \\
&= \frac{\epsilon}{8\pi c}[\mathbf{E}^*\times(\sqrt{\epsilon\mu}\hat{\mathbf{k}}\times\mathbf{E})] \\
&= \frac{\epsilon\sqrt{\epsilon\mu}\hat{\mathbf{k}}}{8\pi c}|\mathbf{E_0}|^2.
\end{aligned}
\tag{10.29}
$$

The momentum density of the electromagnetic wave is related to its energy density and intensity by

$$
\overline{\mathbf{G}} = \sqrt{\epsilon\mu}\frac{\hat{\mathbf{k}}}{c}\overline{u} = \frac{\epsilon\mu\overline{\mathbf{S}}}{c^2}.
\tag{10.30}
$$

10.2.1   Radiation pressure

We consider the complete reflection at normal incidence of a portion of an electromagnetic wave of length vt and cross-sectional area A. The momentum of the radiation within this volume is $\overline{\mathbf{G}}Avt$. The change in the electromagnetic momentum ($\Delta\mathbf{P}_{\text{em}}$) in time t is equal and opposite to the impulse exerted on the surface, so

$$
\Delta\mathbf{P}_{\text{em}} = 2\overline{\mathbf{G}}Avt = \mathbf{F}t.
\tag{10.31}
$$

Thus, the reflected wave will exert a pressure on the surface given by

$$
p_{\text{rad}} = F/A = 2c\overline{G}/\sqrt{\epsilon\mu}.
\tag{10.32}
$$

This pressure is called the **radiation pressure**. Comparing Eqs. (10.32) to Eqs. (10.29)and (10.30) we see that the radiation pressure is related to the electric field and intensity of the incident wave by.

$$p_{\text{rad}} = \frac{\epsilon}{4\pi}|\mathbf{E_0}|^2 = \frac{2}{c}\sqrt{\epsilon\mu}\overline{S}. \tag{10.33}$$

If the radiation is completely absorbed, then the radiation pressure will be half this value.

10.3   Polarization

10.3.1   Polarized light

The **E** and **B** fields in an electromagnetic wave are in a plane perpendicular to the direction of motion $\hat{\mathbf{k}}$ of the wave. The orientation of the field vectors in this plane determines the **polarization** of the wave. The standard convention is that the polarization direction is defined to be the direction of the **E** field.

It is convenient to introduce a right handed x, y, z coordinate system with basic unit vectors given by

$$\hat{\mathbf{i}} = \hat{\boldsymbol{\epsilon}}_1, \quad \hat{\mathbf{j}} = \hat{\boldsymbol{\epsilon}}_2, \quad \hat{\mathbf{k}} = \hat{\mathbf{k}}. \tag{10.34}$$

The unit vectors $\hat{\boldsymbol{\epsilon}}_1$ and $\hat{\boldsymbol{\epsilon}}_2$ define two perpendicular polarization directions for the **E** field. A **plane polarized wave**, so called because the **E** vector remains in one plane, is described by

$$\mathbf{E} = (E_x\hat{\boldsymbol{\epsilon}}_1 + E_y\hat{\boldsymbol{\epsilon}}_2)e^{i(\mathbf{k}\cdot\mathbf{r}-\omega t)}. \tag{10.35}$$

In Eq. (10.35), the components E_x and E_y have been considered real quantities. More general types of polarization occur if the components of **E** are taken as complex. For instance, if E_x and E_y have the same magnitude, but differ in complex phase by 90°, then **E** becomes

$$\mathbf{E}_\pm = E_0(\hat{\boldsymbol{\epsilon}}_1 \pm i\hat{\boldsymbol{\epsilon}}_2)e^{i(\mathbf{k}\cdot\mathbf{r}-\omega t)}. \tag{10.36}$$

The real part of this **E** field is given by[2]

$$\mathbf{E}_\pm = E_0[\hat{\boldsymbol{\epsilon}}_1\cos(\omega t) \pm \hat{\boldsymbol{\epsilon}}_2\sin(\omega t)]. \tag{10.37}$$

This describes a **circularly polarized wave**, so called because the tip of the **E** vector describes a circle as shown on Fig. 10.3.

---

[2]For simplicity, we will take $\mathbf{r} = \mathbf{0}$ for the rest of this section.

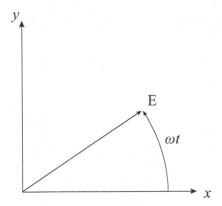

Figure 10.3: Left circularly polarized light. The **E** vector rotates as shown.

The tip of the $\mathbf{E}_+$ vector moves in a counter-clockwise circle if viewed by looking directly into the approaching wave, as in Fig. 10.3. For this reason the wave with $\hat{\boldsymbol{\epsilon}}_1 + i\hat{\boldsymbol{\epsilon}}_2$ is called **left circularly polarized**. The wave for $\mathbf{E}_-$ with $\hat{\boldsymbol{\epsilon}}_1 - i\hat{\boldsymbol{\epsilon}}_2$ is called **right circularly polarized**. This notation, which is standard in optics, is opposite to that used in high energy physics. There, photons (quantized electromagnetic waves) with the $\hat{\boldsymbol{\epsilon}}_1 + i\hat{\boldsymbol{\epsilon}}_2$ combination are called right-handed and the $\hat{\boldsymbol{\epsilon}}_1 - i\hat{\boldsymbol{\epsilon}}_2$ photon is called left handed. That is because the rotation of the **E** vector is viewed from behind, looking in the direction of motion of the wave.

If the two **E** fields are 90° out of phase, and of unequal magnitude, so that

$$E_x = a, \quad E_y = \pm ib, \tag{10.38}$$

then the **E** field will be given by

$$\mathbf{E}_\pm = a\hat{\boldsymbol{\epsilon}}_1 \cos(\omega t) \pm b\hat{\boldsymbol{\epsilon}}_2 \sin(\omega t). \tag{10.39}$$

This describes an **elliptically polarized** wave with the tip of the **E** vector tracing an ellipse in the x-y plane.

10.3.2 Circular basis for polarization

A convenient basis for the description of polarization is the **circular basis** using basic unit vectors $\hat{\boldsymbol{\epsilon}}_+$ and $\hat{\boldsymbol{\epsilon}}_-$, defined in terms of the planar basis vectors by

$$\hat{\boldsymbol{\epsilon}}_\pm = \frac{1}{\sqrt{2}}(\hat{\boldsymbol{\epsilon}}_1 \pm i\hat{\boldsymbol{\epsilon}}_2). \tag{10.40}$$

The circular basis vectors $\hat{\boldsymbol{\epsilon}}_+$ and $\hat{\boldsymbol{\epsilon}}_-$ are complex orthogonal unit vectors with the properties

$$\hat{\boldsymbol{\epsilon}}_\pm^* \cdot \hat{\boldsymbol{\epsilon}}_\pm = 1, \quad \hat{\boldsymbol{\epsilon}}_\mp^* \cdot \hat{\boldsymbol{\epsilon}}_\pm = 0. \tag{10.41}$$

A polarized wave in the circular basis is given by

$$\mathbf{E} = (E_+\hat{\boldsymbol{\epsilon}}_+ + E_-\hat{\boldsymbol{\epsilon}}_-)e^{i(\mathbf{k}\cdot\mathbf{r}-\omega t)}. \tag{10.42}$$

If E_+ and E_- are in phase with

$$E_- = rE_+, \tag{10.43}$$

the real part of $\mathbf{E}$ is given in terms of the planar basis by

$$\mathbf{E} = \frac{E_+}{\sqrt{2}}[(1+r)\hat{\boldsymbol{\epsilon}}_1\cos(\omega t) + (1-r)\hat{\boldsymbol{\epsilon}}_2\sin(\omega t)]. \tag{10.44}$$

This represents an elliptically polarized wave with the ratio $(1+r)/(1-r)$ of the major to the minor axis of the ellipse.

The most general polarized wave has an arbitrary phase between E_+ and E_-, so

$$E_- = re^{i\alpha}E_+. \tag{10.45}$$

Then the electric field is given by

$$\mathbf{E} = E_+(\hat{\boldsymbol{\epsilon}}_+ + re^{i\alpha}\hat{\boldsymbol{\epsilon}}_-)e^{-i\omega t}. \tag{10.46}$$

We modify this equation by factoring out $e^{i\alpha/2}$ so that

$$\mathbf{E} = E_+(\hat{\boldsymbol{\epsilon}}_+^{-i\alpha/2} + re^{i\alpha/2}\hat{\boldsymbol{\epsilon}}_-)e^{-i(\omega t-\alpha/2)}. \tag{10.47}$$

Introducing the planar basis vectors, and after some algebra, we can find the polarization state in terms of the two components E_x and E_y of the electric field as

$$\begin{aligned} E_x &= \frac{E_+}{\sqrt{2}}[\cos(\alpha/2)(1+r)\cos(\omega t-\alpha/2) \\ &\quad - \sin(\alpha/2)(1-r)\sin(\omega t-\alpha/2)] \tag{10.48} \\ E_y &= \frac{E_+}{\sqrt{2}}[\sin(\alpha/2)(1+r)\cos(\omega t-\alpha/2) \\ &\quad + \cos(\alpha/2)(1-r)\sin(\omega t-\alpha/2)]. \tag{10.49} \end{aligned}$$

This can be put in matrix form as

$$\begin{pmatrix} E_x \\ E_y \end{pmatrix} = (E_+/\sqrt{2})\begin{bmatrix} \cos(\alpha/2) & -\sin(\alpha/2) \\ \sin(\alpha/2) & +\cos(\alpha/2) \end{bmatrix}\begin{pmatrix} (1+r)\cos(\omega t-\alpha/2) \\ (1-r)\sin(\omega t-\alpha/2) \end{pmatrix}. \tag{10.50}$$

The square matrix in Eq. (10.13) can be recognized as the rotation matrix $[\mathbf{R}(\alpha/2)]$ corresponding to a rotation of angle $\alpha/2$ about the z-axis, as shown in Fig. 10.4. The column vector on the right hand side of Eq. (10.13) is composed

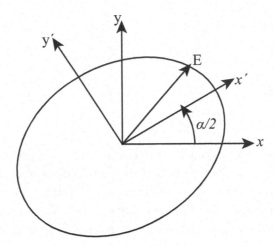

Figure 10.4: System rotated through angle $\alpha/2$ for general case of elliptically polarized light.

of the components E_x' and E_y' of $\mathbf{E}$ in the rotated system, so

$$E_x' = (E_+/\sqrt{2})(1+r)\cos(\omega t - \alpha/2) \tag{10.51}$$
$$E_y' = (E_+/\sqrt{2})(1-r)\sin(\omega t - \alpha/2). \tag{10.52}$$

This represents an elliptically polarized wave with the major axis rotated by an angle $\alpha/2$ above the original x-axis, and a ratio $(1+r)/(1-r)$ of the major to the minor axis.

We see that the polarization of a plane wave is generally elliptical. The special limiting case of $r = \pm 1$ corresponds to linear polarization, while $r = 0$ (or approaching infinity) corresponds to circular polarization.

10.3.3 Birefringence

In non-isotropic crystals, the permittivity, and hence the index of refraction, is different in different directions. This is the case when one of the eigenvalues of the permittivity tensor differs from the other two. Such a crystal is called **birefringent**. We consider a case of light propagating along the z axis, with two different indices of refraction, n_1 and n_2, in the x and y directions. Light originally polarized at 45° to the x axis will have its $\mathbf{E}$ field given by

$$\mathbf{E} = E_0(\hat{\boldsymbol{\epsilon}}_1 e^{ik_1 z} + \hat{\boldsymbol{\epsilon}}_2 e^{ik_2 z})e^{-i\omega t}. \tag{10.53}$$

There are two different wave numbers, given by

$$k_1 = \frac{n_1 \omega}{c}, \quad \text{and} \quad k_2 = \frac{n_2 \omega}{c}. \tag{10.54}$$

After the wave has travelled a distance d, the $\mathbf{E}$ field will change to

$$\mathbf{E} = E_0[\hat{\boldsymbol{\epsilon}}_1 + \hat{\boldsymbol{\epsilon}}_2 e^{i(n_2 - n_1)\omega d/c}]e^{i[k_1 d - \omega t]}. \tag{10.55}$$

If the distance d is such that

$$(n_2 - n_1)\omega d/c = \pi/2, \tag{10.56}$$

the **E** field will have become

$$\mathbf{E} = E_0(\hat{\boldsymbol{\epsilon}}_1 + i\hat{\boldsymbol{\epsilon}}_2)e^{i[k_1 d - \omega t]}. \tag{10.57}$$

This corresponds to circularly polarized light. A birefringent material of thickness d given by Eq. (10.56) is called a **quarter wave plate**, because the phase of the **E** field along one polarization direction is shifted by a quarter of a wave length with respect to the other direction. We see that a quarter wave plate can be used to change linear polarization to circular polarization. This is shown pictorially in Fig. 10.5b, where E_y has moved a quarter of a wave length ahead of E_x.

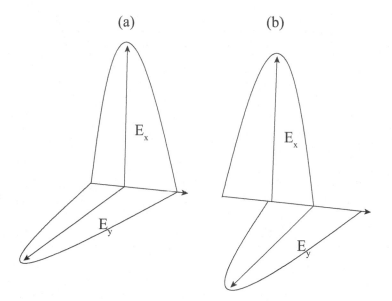

Figure 10.5: (a) Light plane polarized at 45° to the x-axis. E_x and E_y are equal in magnitude and in phase. (b) E_y has moved a quarter of a wavelength ahead of E_x, resulting in circularly polarized light.

If the thickness of the plate is such that E_y moves half a wave length ahead of E_x, the resulting wave will be plane polarized, but the plane of polarization will have rotated through 90°, to be between the x axis and the negative y axis. Such a plate is called a **half wave plate**. For a general distance, d, plane polarized light will be converted into some form of elliptically polarized light.

Birefringence is often caused by stress in the crystal. Because of this property, birefringence can be used to test material for stress. Ordinarily if a transparent object is placed between crossed polaroids, no light will pass through the system. The first polaroid polarizes the light along one plane and the second polaroid then absorbs the polarized light. Where the material is under stress and becomes birefringent, the plane polarized light will be converted into elliptically polarized light and partially pass through the second polaroid.

10.3.4 Partially polarized light

So far we have discussed only waves that are completely polarized but, more generally, radiation can be unpolarized or partially polarized.

Radiation from a single atom is in a pure state of polarization. (This will be shown in chapter 13.) The radiation electric field for unpolarized light is given by a sum of the radiation from many uncorrelated individual atoms, so no particular polarization direction can be defined.

The state of linear polarization of a radiation field can be measured with a polarizer that is composed of material that allows one direction of polarization to pass through, while absorbing radiation with the perpendicular polarization. If the polarizer is rotated in a polarized beam, only the component of **E** allowed by the polarizer will pass through. The resulting E field will now be polarized in the direction set by the polarizer, and will be decreased in magnitude by a factor $\cos\theta$ where θ is the angle between the preferred direction of the polarizer and the incident electric field.

Thus the intensity will vary like

$$S = S_0 \cos^2 \theta, \tag{10.58}$$

which in optics is called the **law of Malus**. The intensity of unpolarized radiation will not vary as the polarizer is rotated, but the transmitted radiation will now be polarized and its intensity reduced by half because the light of the other polarization has been blocked.

Partially polarized radiation is an incoherent combination of polarized and unpolarized radiation. The state of polarization is described by a polarization, Π, which is often given as a percent. The polarization is defined as the ratio of the intensity of the polarized radiation to the total radiation. The transmitted intensity for partially polarized light will vary with the angle of the polarizer as

$$S(\theta) = [\Pi \cos^2 \theta + \frac{1}{2}(1 - \Pi)]S_0. \tag{10.59}$$

The maximum intensity will be $S_{\text{max}} = \frac{1}{2}(1 + \Pi)S_0,$ \hfill (10.60)

and the minimum intensity will be $S_{\text{min}} = \frac{1}{2}(1 - \Pi)S_0,$ \hfill (10.61)

so Π can be measured by $\Pi = \dfrac{S_{\text{max}} - S_{\text{min}}}{S_{\text{max}} + S_{\text{min}}}.$

$$\tag{10.62}$$

10.4   Reflection and Refraction at a Planar Interface.

We consider an electromagnetic plane wave approaching a planar interface between two dielectrics of differing ϵ and μ, as shown on Fig. 10.6. The incident plane wave travels in the direction $\hat{\mathbf{k}}_1$. Although the wave is spread out, it is represented on the figure as a single 'ray' of radiation travelling in the $\hat{\mathbf{k}}_1$ direction. The incident ray makes an angle θ_1 with the normal direction, $\hat{\mathbf{n}}$, to the surface. In general, there will be a refracted wave, shown on the figure as a ray in direction $\hat{\mathbf{k}}_2$ propagating through medium 2 at an angle θ_2 with $\hat{\mathbf{n}}$, and a reflected wave, represented as a ray travelling back in medium 1 at an angle θ_1'.

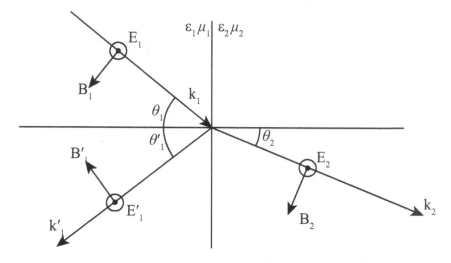

Figure 10.6: Incident, reflected, and refracted rays for light polarized perpendicular to the incident plane.

The boundary conditions on $\mathbf{E}$ (with no charges on the interface) and $\mathbf{B}$ at the interface follow from Maxwell's equations, and are given by

$$\epsilon_1 \hat{\mathbf{n}} \cdot (\mathbf{E}_1 e^{i\mathbf{k}_1 \cdot \mathbf{r}} + \mathbf{E}_1' e^{i\mathbf{k}_1' \cdot \mathbf{r}}) = \epsilon_2 \hat{\mathbf{n}} \cdot \mathbf{E}_2 e^{i\mathbf{k}_2 \cdot \mathbf{r}} \tag{10.63}$$

$$\hat{\mathbf{n}} \times (\mathbf{E}_1 e^{i\mathbf{k}_1 \cdot \mathbf{r}} + \mathbf{E}_1' e^{i\mathbf{k}_1' \cdot \mathbf{r}}) = \hat{\mathbf{n}} \times \mathbf{E}_2 e^{i\mathbf{k}_2 \cdot \mathbf{r}} \tag{10.64}$$

$$\hat{\mathbf{n}} \cdot (\mathbf{B}_1 e^{i\mathbf{k}_1 \cdot \mathbf{r}} + \mathbf{B}_1' e^{i\mathbf{k}_1' \cdot \mathbf{r}}) = \hat{\mathbf{n}} \cdot \mathbf{B}_2 e^{i\mathbf{k}_2 \cdot \mathbf{r}} \tag{10.65}$$

$$\frac{1}{\mu_1} \hat{\mathbf{n}} \times (\mathbf{B}_1 e^{i\mathbf{k}_1 \cdot \mathbf{r}} + \mathbf{B}_1' e^{i\mathbf{k}_1' \cdot \mathbf{r}}) = \frac{1}{\mu_2} \hat{\mathbf{n}} \times \mathbf{B}_2 e^{i\mathbf{k}_2 \cdot \mathbf{r}}. \tag{10.66}$$

Each of the waves in Eqs. (10.26)-(10.29) must have the same frequency so that the boundary conditions can hold for all time. For simplicity, we have omitted the common factor $e^{-i\omega t}$ from the equations.

10.4.1 Snell's law

The boundary conditions must also hold everywhere on the interface. This requires the condition that

$$\hat{\mathbf{k}}_1\cdot\mathbf{r} = \hat{\mathbf{k}}'_1\cdot\mathbf{r} = \hat{\mathbf{k}}_2\cdot\mathbf{r} \tag{10.67}$$

for $\mathbf{r}$ on the interface. We first pick $\mathbf{r}$ to be perpendicular to $\mathbf{k}_1$ so that $\hat{\mathbf{k}}_1\cdot\mathbf{r} = 0$. Then, from Eq. (10.67), $\hat{\mathbf{k}}'_1\cdot\mathbf{r}$ and $\hat{\mathbf{k}}_2\cdot\mathbf{r}$ will also equal zero. This means that the three vectors $\mathbf{k}_1$, $\mathbf{k}'_1$, $\mathbf{k}_2$ all lie in the same plane.

Next we take $\mathbf{r}$ to be in the common plane of $\mathbf{k}_1$, $\mathbf{k}'_1$, $\mathbf{k}_2$. Then

$$k_1 \sin\theta_1 = k'_1 \sin\theta'_1 = k_2 \sin\theta_2. \tag{10.68}$$

The wave numbers are related to the common frequency ω by Eq. (10.15), so

$$k = \left(\frac{\sqrt{\epsilon\mu}}{c}\right)\omega. \tag{10.69}$$

Thus Eq. (10.68) becomes

$$\sqrt{\epsilon_1\mu_1}\sin\theta_1 = \sqrt{\epsilon_1\mu_1}\sin\theta'_1 = \sqrt{\epsilon_2\mu_2}\sin\theta_2. \tag{10.70}$$

The first equality above gives the optical law that the angle of reflection (θ'_1) equals the incident angle (θ_1). Since the optical index of refraction n is related to ϵ and μ by $n = \sqrt{\epsilon\mu}$, the second equality in Eq. (10.70) is **Snell's law**

$$n_1 \sin\theta_1 = n_2 \sin\theta_2. \tag{10.71}$$

This derivation of Snell's law from Maxwell's equations is another indication that light is an electromagnetic wave.

Since the exponential factors in Eqs. (10.26)-(10.29) are all equal, we can simplify those equations by removing these exponential factors. We also replace $\mathbf{B}$ in the equations, using Eq. (10.22)

$$\mathbf{B} = \sqrt{\epsilon\mu}\,\hat{\mathbf{k}}\times\mathbf{E}. \tag{10.72}$$

Then the four boundary conditions can be written as

$$\epsilon_1\hat{\mathbf{n}}\cdot(\mathbf{E}_1 + \mathbf{E}'_1) = \epsilon_2\hat{\mathbf{n}}\cdot\mathbf{E}_2 \tag{10.73}$$

$$\hat{\mathbf{n}}\times(\mathbf{E}_1 + \mathbf{E}'_1) = \hat{\mathbf{n}}\times\mathbf{E}_2 \tag{10.74}$$

$$\sqrt{\epsilon_1\mu_1}\,\hat{\mathbf{n}}\cdot(\hat{\mathbf{k}}_1\times\mathbf{E}_1 + \hat{\mathbf{k}}'_1\times\mathbf{E}'_1) = \sqrt{\epsilon_2\mu_2}\,\hat{\mathbf{n}}\cdot(\hat{\mathbf{k}}_2\times\mathbf{E}_2) \tag{10.75}$$

$$\sqrt{\frac{\epsilon_1}{\mu_1}}\,\hat{\mathbf{n}}\times(\hat{\mathbf{k}}_1\times\mathbf{E}_1 + \hat{\mathbf{k}}'_1\times\mathbf{E}'_1) = \sqrt{\frac{\epsilon_2}{\mu_2}}\,\hat{\mathbf{n}}\times(\hat{\mathbf{k}}_2\times\mathbf{E}_2). \tag{10.76}$$

10.4.2   Perpendicular polarization.

We first consider a plane wave with polarization perpendicular to the plane of incidence, as shown in Fig. 10.6. Then, Eq. (10.73) does not enter, and the other three boundary conditions become

$$E_1 + E_1' = E_2 \tag{10.77}$$

$$\sqrt{\epsilon_1\mu_1}(E_1 + E_1')\sin\theta_1 = \sqrt{\epsilon_2\mu_2}E_2\sin\theta_2 \tag{10.78}$$

$$\sqrt{\frac{\epsilon_1}{\mu_1}}(E_1 - E_1')\cos\theta_1 = \sqrt{\frac{\epsilon_2}{\mu_2}}E_2\cos\theta_2. \tag{10.79}$$

Using Snell's law in Eq. (10.78) shows that it is equivalent to Eq. (10.77). This leaves two equations for the two unknown amplitudes E_1' and E_2, with the solutions

$$E_2 = \frac{2\sqrt{\frac{\epsilon_1}{\mu_1}}\cos\theta_1}{\sqrt{\frac{\epsilon_1}{\mu_1}}\cos\theta_1 + \sqrt{\frac{\epsilon_2}{\mu_2}}\cos\theta_2}E_1 \tag{10.80}$$

$$E_1' = \frac{\sqrt{\frac{\epsilon_1}{\mu_1}}\cos\theta_1 - \sqrt{\frac{\epsilon_2}{\mu_2}}\cos\theta_2}{\sqrt{\frac{\epsilon_1}{\mu_1}}\cos\theta_1 + \sqrt{\frac{\epsilon_2}{\mu_2}}\cos\theta_2}E_1 \tag{10.81}$$

For most dielectrics, the permeability μ is close to one, and we can replace the factor $\sqrt{\epsilon/\mu}$ by the index of refraction n. (The permeability can be put back into any of the equations in this section by making the replacement $n \to n/\mu$.) Then Eqs. (10.80) and (10.81) simplify to

$$E_2 = \left[\frac{2n_1\cos\theta_1}{n_1\cos\theta_1 + n_2\cos\theta_2}\right]E_1 \tag{10.82}$$

$$E_1' = \left[\frac{n_1\cos\theta_1 - n_2\cos\theta_2}{n_1\cos\theta_1 + n_2\cos\theta_2}\right]E_1 \tag{10.83}$$

for **E perpendicular to the plane of incidence.**

In this form, Eqs. (10.82) and (10.83) are called **Fresnel's relations** for polarization perpendicular to the plane of incidence. The term $n_2\cos\theta_2$ in the equations is given in terms of the incident angle θ_1 by Snell's law as

$$n_2\cos\theta_2 = \sqrt{n_2^2 - n_1^2\sin^2\theta_1}. \tag{10.84}$$

The transmission coefficient T is defined as the ratio of the rate at which electromagnetic energy is going into the second medium at the interface to the rate at which energy is incident on the interface. This ratio is given in terms of the time averaged intensity $\overline{\mathbf{S}}$ as

$$T = \frac{\hat{\mathbf{n}}\cdot\overline{\mathbf{S}}_2}{\hat{\mathbf{n}}\cdot\overline{\mathbf{S}}_1}. \tag{10.85}$$

The intensity of an electromagnetic wave is given by Eq. (10.27) as

$$\overline{\mathbf{S}} = \frac{c\hat{\mathbf{k}}}{8\pi}\sqrt{\frac{\epsilon}{\mu}}|\mathbf{E}|^2. \tag{10.86}$$

Thus the transmission coefficient can be written as (still assuming $\mu = 1$)

$$\begin{aligned}
T_\perp &= \frac{n_2 \cos\theta_2 |E_2|^2}{n_1 \cos\theta_1 |E_1|^2} \\
&= \frac{4n_1 n_2 \cos\theta_1 \cos\theta_2}{(n_1 \cos\theta_1 + n_2 \cos\theta_2)^2}.
\end{aligned} \tag{10.87}$$

The reflection coefficient R is the ratio of the rate of reflected energy to the incident energy. This ratio is given by

$$\begin{aligned}
R_\perp &= \frac{|E_1'|^2}{|E_1|^2} \\
&= \frac{(n_1 \cos\theta_1 - n_2 \cos\theta_2)^2}{(n_1 \cos\theta_1 + n_2 \cos\theta_2)^2}.
\end{aligned} \tag{10.88}$$

We have used the subscript $\perp$ to emphasize that these R and T coefficients are for the case of polarization perpendicular to the plane of incidence. The two coefficients satisfy the relation

$$R + T = 1, \tag{10.89}$$

which corresponds to conservation of energy at the interface.

10.4.3   Parallel polarization.

We next consider a plane wave with polarization parallel to the plane of incidence, as shown in Fig. 10.7.

Now, Eq. (10.75) does not enter, and the other three boundary conditions become

$$\epsilon_1(E_1 + E_1')\sin\theta_1 = \epsilon_2 E_2 \sin\theta_2 \tag{10.90}$$

$$(E_1 - E_1')\cos\theta_1 = E_2 \cos\theta_2 \tag{10.91}$$

$$\sqrt{\frac{\epsilon_1}{\mu_1}}(E_1 + E_1') = \sqrt{\frac{\epsilon_2}{\mu_2}}E_2. \tag{10.92}$$

Using Snell's law, equation (10.90) is equivalent to Eq. (10.92). This leaves two equations for the two unknown amplitudes E_1' and E_2, with the solutions

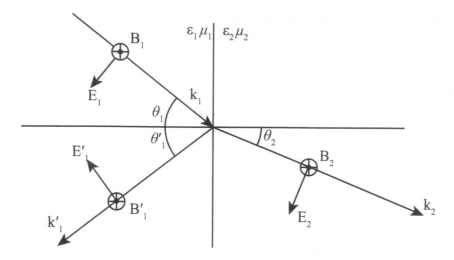

Figure 10.7: Reflection and refraction for light polarized parallel to the plane of incidence.

(letting $\mu = 1$)

$$E_2 = \left[\frac{2n_1 \cos\theta_1}{n_2 \cos\theta_1 + n_1 \cos\theta_2} \right] E_1 \tag{10.93}$$

$$E_1' = \left[\frac{n_2 \cos\theta_1 - n_1 \cos\theta_2}{n_2 \cos\theta_1 + n_1 \cos\theta_2} \right] E_1 \tag{10.94}$$

for **E parallel to the plane of incidence.** These two equations are called Fresnel's relations for polarization parallel to the plane of incidence. The equations for parallel polarization are similar in form to those for perpendicular polarization, differing only in the interchange $n_1 \leftrightarrow n_2$ in several places.

The transmission coefficient for parallel polarization is given by

$$T_{\parallel} = \frac{4n_1 n_2 \cos\theta_1 \cos\theta_2}{(n_2 \cos\theta_1 + n_1 \cos\theta_2)^2}. \tag{10.95}$$

The reflection coefficient for parallel polarization is given by

$$R_{\parallel} = \frac{(n_2 \cos\theta_1 - n_1 \cos\theta_2)^2}{(n_2 \cos\theta_1 + n_1 \cos\theta_2)^2}. \tag{10.96}$$

These relations also satisfy $R + T = 1$.

10.4.4 Normal incidence

For normal incidence, the results for electric fields perpendicular and parallel to the plane of incidence coincide and simplify to

$$E_2 = \left[\frac{2n_1}{n_1 + n_2} \right] E_1 \tag{10.97}$$

$$E_1' = \left[\frac{n_1 - n_2}{n_1 + n_2} \right] E_1. \tag{10.98}$$

We have used the sign convention for polarization perpendicular to the plane of incidence, as shown in Fig. 10.6. This has all the polarization basis vectors for the **E** fields in the same direction. When $n_1 > n_2$, the reflected field E'_1 is in phase with the incident field E_1, but for $n_1 < n_2$ there is a phase change of $180°$ upon reflection.

10.4.5 Polarization by reflection

Initially unpolarized incident light can be treated as an uncorrelated equal combination of light that is polarized perpendicular and parallel to the plane of incidence. Since the reflection coefficients for perpendicular and parallel polarization are different, the reflected light will have an unequal combination of the two polarizations. This leads to a net polarization of the reflected light.

   We define $S_\perp$ as the intensity of the reflected wave with polarization perpendicular to the plane of incidence, and $S_\parallel$ as the intensity for parallel polarization. As a polarizer is rotated, the intensity will vary like

$$S(\theta) = S_\perp \cos^2 \theta + S_\parallel \sin^2 \theta = S_\perp + (S_\parallel - S_\perp) \sin^2 \theta, \qquad (10.99)$$

where θ is the angle between normal to the plane of incidence and the preferred direction of the polarizer. Thus S_{Max} will be the larger of $S_\perp$ and $S_\parallel$, while S_{min} will be the smaller. Then, the state of polarization is given by

$$\Pi = \left| \frac{S_\perp - S_\parallel}{S_\perp + S_\parallel} \right| = \left| \frac{R_\perp - R_\parallel}{R_\perp + R_\parallel} \right|, \qquad (10.100)$$

where $R_\perp$ and $R_\parallel$ are the perpendicular and parallel reflection coefficients given by Eqs. (10.88) and (10.96), respectively.

   The polarization of the radiation reflected when visible light in air with $n = 1$ is reflected from from glass with $n = 1.5$ is plotted in Fig. 10.8 as a function of incident angle. We see that it is quite large over a range of angles and becomes 100% at about $56°$. This polarization results from $R_\parallel$ being smaller than $R_\perp$, and going to zero at one angle. The angle where $R_\parallel$ vanishes is called **Brewster's angle.**

   We can determine the Brewster angle by setting the numerator of Eq. (10.96) equal to zero, which gives the condition

$$n_2 \cos \theta_1 = n_1 \cos \theta_2. \qquad (10.101)$$

Coupled with Snell's law, this leads to the relation

$$\sin(2\theta_1) = \sin(2\theta_2). \qquad (10.102)$$

This equality can be satisfied (apart from the trivial solution) if the two angles satisfy

$$\theta_1 + \theta_2 = 90°, \qquad (10.103)$$

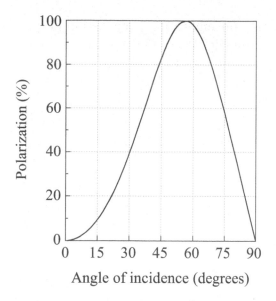

Figure 10.8: Polarization $\Pi(\theta)$ for $n_1 = 1$ and $n_2 = 1.5$.

which has the interpretation that the refracted ray is perpendicular to the reflected ray.

Snell's law can also be combined with Eq. (10.101) to give

$$(n_1/n_2)\tan\theta_1 = (n_2/n_1)\tan\theta_2. \qquad (10.104)$$

Using Eq. (10.103) in Eq. (10.104),we get the result

$$\tan\theta_B = (n_2/n_1) \qquad (10.105)$$

for the incident Brewster angle, θ_B.

This polarization of light by reflection is the basis for the use of Polaroid sunglasses. Light from the sun, reflected off the windshield of a car will become horizontally polarized since the horizontal direction is perpendicular to the plane of incidence. If the polaroid material in the sunglasses absorbs horizontally polarized light, it will preferentially absorb the glare produced by the reflected sunlight. In fact, a way to test if sunglasses are Polaroid is to look through one lense and rotate the glasses. The glare from a windshield gets worse when the lense is rotated though 90° for Polaroid glasses.

10.4.6   Total internal reflection

The reflection coefficient for unpolarized light reflected from the interface between water (n_1=1.33) and air (n_2=1) is shown in Fig. 10.9.

The reflection is seen to rise sharply for angles above 47°, and becomes 100% at about 49°. This phenomenon is called **total internal reflection**.

It is a general feature of reflection when the second index of refraction is less than the first $(n_2 < n_1)$.

Total reflection occurs at any incident angle for which application of Snell's law would give the result $\sin \theta_2 > 1$, which of course is impossible for real angles. The angle for which $\sin \theta_2 = 1$ is called the **critical angle**, θ_c. The condition for criticality is that

$$\sin \theta_2 = 1 = \frac{n_1}{n_2} \sin \theta_c, \qquad (10.106)$$

so

$$\sin \theta_c = \frac{n_2}{n_1}. \qquad (10.107)$$

For the case in Fig. 10.9, Eq. (10.107) gives $\theta_c = 49°$.

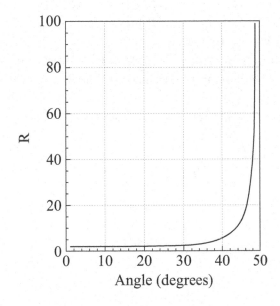

Figure 10.9: Reflection coefficient $R(\theta)$ for $n_1 = 1.33$ and $n_2 = 1$.

We may ask what happens inside the second medium for incident angles above the critical angle. This can be answered by extending the angle θ_2 in Snell's law to complex values. For incident angles above the critical angle, $\sin \theta_2$ is greater than 1, and $\cos \theta_2$ is pure imaginary, given by

$$\cos \theta_2 = i\sqrt{(n_1/n_2)^2 \sin^2 \theta_1 - 1}. \qquad (10.108)$$

The propagation of the wave in medium 2 is described by

$$
\begin{aligned}
\mathbf{E} &= \mathbf{E_2} e^{i(\mathbf{k_2} \cdot \mathbf{r} - \omega t)} \\
&= \mathbf{E_2} e^{i(k_2 r_\perp \cos \theta_2 + k_2 r_\parallel \sin \theta_2 - \omega t)} \\
&= \mathbf{E_2} e^{-k_2 r_\perp \sqrt{(n_1/n_2)^2 \sin^2 \theta_1 - 1}} e^{i(k_1 r_\parallel \sin \theta_1 - \omega t)} \\
&= \mathbf{E_2} e^{-k_1 r_\perp \sqrt{\sin^2 \theta_1 - \sin^2 \theta_c}} e^{i(k_1 r_\parallel \sin \theta_1 - \omega t)},
\end{aligned} \qquad (10.109)
$$

where $r_\perp$ and $r_\parallel$ represent distance perpendicular to and parallel to the interface. In the last step above, we have used the relation $k_2/n_2 = k_1/n_1$ to replace k_2 by the incident wave number k_1, and have used Eq. (10.107) to write the result in terms of θ_c. Equation (10.109) shows that a wave travels along the interface, but the electric field entering medium 2 falls off exponentially like

$$E \sim e^{-r_\perp/2\delta} \tag{10.110}$$

with

$$\delta = \frac{\lambda_1}{4\pi\sqrt{\sin^2\theta_1 - \sin^2\theta_c}}. \tag{10.111}$$

The constant δ is a measure of how quickly the field attenuates inside the second medium. We write it in Eq. (10.110) as 2δ so that δ will be the effective attenuation length for the intensity, which varies like E^2. We have introduced the incident wave length λ_1 (using $k = 2\pi/\lambda$) as a convenient length scale for δ. We see that the attenuation length is of the order of a wave length once the incident angle is above θ_c.

Even though the fields penetrate the interface for a short distance, no net power enters the second dielectric. We can see this by calculating the magnetic field $\mathbf{B_2}$ using Eq. (10.21)

$$\mathbf{B_2} = \frac{c}{\omega}\mathbf{k_2} \times \mathbf{E_2}. \tag{10.112}$$

Using Eq. (10.27), the time averaged power entering medium 2 is given by

$$\begin{aligned}
\overline{\hat{\mathbf{n}} \cdot \mathbf{S_2}} &= \frac{c}{8\pi\mu_2}\text{Re}\{\hat{\mathbf{n}} \cdot [\mathbf{E_2^*} \times \mathbf{B_2}]\} \\
&= \frac{c^2}{8\pi\mu_2\omega}\text{Re}\{\hat{\mathbf{n}} \cdot [\mathbf{E_2^*} \times (\mathbf{k_2} \times \mathbf{E_2})]\} \\
&= \frac{c^2}{8\pi\mu_2\omega}\text{Re}\{(\hat{\mathbf{n}} \cdot \mathbf{k_2})|\mathbf{E_2}|^2\} \\
&= 0. \tag{10.113}
\end{aligned}$$

This last result, that the time averaged power is zero, follows because $\hat{\mathbf{n}} \cdot \mathbf{k_2} = \mathbf{k_2}\cos\theta_2$ is pure imaginary. Also, looking at the Fresnel relations for the reflected $\mathbf{E_1'}$, we see that the ratio $|E_1'/E_1| = 1$, again because $\cos\theta_2$ is pure imaginary.

10.4.7 Non-reflective coating

We see from the Fresnel relations that light will always be reflected at a single interface between two substances. For the air to glass transition, the reflection coefficient is about 4% at each surface. For an optical system with several lenses, this can cause an appreciable loss of transmitted light, and can lead to 'ghost' images. For this reason, lenses are coated with a thin **non-reflective coating** to eliminate or greatly reduce the reflection.

To see how this works, we consider a system of three dielectrics with two interfaces as shown in Fig. 10.10. The middle dieletric has a thickness d. For simplicity, we take the case of light at normal incidence to the surfaces. As seen in Fig. 10.10, there are four unknown amplitudes: E_1', E_2, E_2', E_3, to be solved for in terms of the incident amplitude E_1.

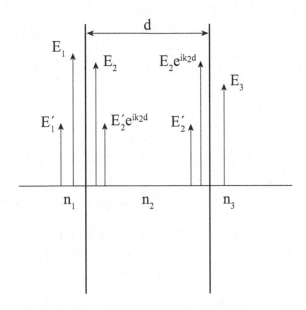

Figure 10.10: System of two interfaces a distance d apart.

We can get four equations by applying two boundary conditions at each interface. The boundary conditions are that $\mathbf{E}_\parallel$ and $\mathbf{B}_\parallel/\mu$, each be continuous across the interface. We also use Eq. (10.72)

$$\mathbf{B} = \sqrt{\epsilon\mu}\,\hat{\mathbf{k}}\times\mathbf{E}, \qquad (10.114)$$

to replace B by E in the equations. Then, the four boundary condition equations can be written as (letting $\mu=1$)

$$E_1 + E_1' = E_2 + E_2'e^{ik_2d} \qquad (10.115)$$
$$n_1(E_1 - E_1') = n_2(E_2 - E_2'e^{ik_2d}) \qquad (10.116)$$
$$E_2e^{ik_2d} + E_2' = E_3 \qquad (10.117)$$
$$n_2(E_2e^{ik_2d} - E_2') = n_3E_3. \qquad (10.118)$$

Note the term $E_2'e^{ik_2d}$ in the equations for the first interface. This represents the wave reflected at the second interface, propagating back to the first interface. Similarly, there is a term $E_2e^{ik_2d}$ in the equations for the second interface, which is the transmitted wave at the first interface, propagating to the second interface.

The four equations (10.115)-(10.118) can be solved most simply by first solving the last two for E_2 and E_2' in terms of E_3. These solutions can be substituted into the first two equations, which then become two equations for the two unknowns, E_3 and E_1'. Solving these gives (Details are given as a problem.)

$$E_3 = \left[\frac{2n_1 n_2}{n_2(n_1 + n_3)\cos(k_2 d) - i(n_1 n_3 + n_2^2)\sin(k_2 d)} \right] E_1 \quad (10.119)$$

$$E_1' = \left[\frac{n_2(n_1 - n_3)\cos(k_2 d) - i(n_1 n_3 - n_2^2)\sin(k_2 d)}{n_2(n_1 + n_3)\cos(k_2 d) - i(n_1 n_3 + n_2^2)\sin(k_2 d)} \right] E_1. \quad (10.120)$$

The transmission and reflection coefficients are then

$$T = \frac{4n_1 n_3 n_2^2}{n_2^2(n_1 + n_3)^2 \cos^2(k_2 d) + (n_1 n_3 + n_2^2)^2 \sin^2(k_2 d)} \quad (10.121)$$

$$R = \frac{n_2^2(n_1 - n_3)^2 \cos^2(k_2 d) + (n_1 n_3 - n_2^2)^2 \sin^2(k_2 d)}{n_2^2(n_1 + n_3)^2 \cos^2(k_2 d) + (n_1 n_3 + n_2^2)^2 \sin^2(k_2 d)}. \quad (10.122)$$

From Eq. (10.122), we see that R will be zero if two conditions are satisfied:

$$n_2 = \sqrt{n_1 n_3}, \quad (10.123)$$

which eliminates the second term in the numerator for R, and

$$k_2 d = \pi/2, \quad (10.124)$$

which eliminates the first. We can use this last equation to relate the thickness of the coating to the wave length of the incident light by

$$d = \frac{\pi}{2k_2} = \frac{n_1 \pi}{2n_2 k_1} = \frac{n_1 \lambda_1}{4n_2} = \frac{\lambda_1}{4}\sqrt{\frac{n_1}{n_3}}. \quad (10.125)$$

As an example of the effect of coating a glass lense with a thin layer, we have plotted in Fig. 10.11 the reflection coefficient as a function of incident wavelength for the case $n_1 = 1$, $n_3 = 1.5$. We have fixed the thickness of the coating so that $R = 0$ when the wave length is $5,000 \mathring{A}$. This keeps R well below the 4% reflection, which would occur with no coating, throughout the visible spectrum.

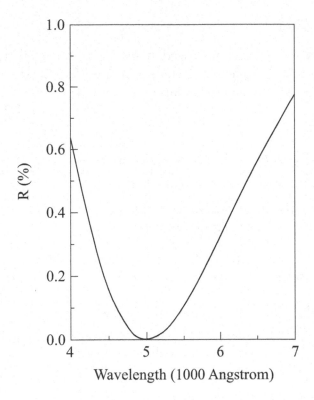

Figure 10.11: Reflection coefficient with non-reflective coating.

10.5 Problems

1. The intensity of sunlight at the Earth's surface is 12×10^5 ergs/cm^2-sec.

   (a) Find the electric field of this radiation at the Earth's surface. Express the answer in statvolts/cm and in volts/m (1 statvolt=300 volts).

   (b) Find the radiation pressure (in dynes/cm^2) if the sunlight is 100% reflected at normal incidence. Compare this to atmospheric pressure.

   (c) Find the radiation pressure if
      i. the sunlight is absorbed (no reflection).
      ii. the sunlight is diffusely reflected from white sand. (The reflected radiation has the same intensity at any angle.)

   (d) What would the intensity, electric field, and radiation pressure be at the surface of the sun?

2. A birefringent crystal has $n_1 = 1.42$, and $n_2 = 1.54$. What thickness would be required to make a quarter wave plate of this crystal for light of wave

length 5,000 Å.

3. A beam of elliptically polarized light, propagating in the z direction, passes through a polarizer in the x-y plane. The maximum transmitted intensity is $9I_0$ when the polarizer is set at 30° to the x axis. The minimum transmitted intensity is I_0 when the polarizer is set at 120° to the x axis.

   (a) Find r, α, E_+, and E_- in the circular basis.

   (b) Find E_x and E_y in the plane basis.

   (c) What would the transmitted intensities be if the polarizer were set along the x axis, and then along the y-axis?

4. Consider a beam of partially plane polarized light with the same maximum and minimum intensities as in the previous problem.

   (a) What is the percent polarization of this light?

   (b) What would the transmitted intensities be if the polarizer were set along the x axis, and then along the y-axis?

   (c) How could you tell whether an incident light beam were elliptically polarized or partially plane polarized?

5. A horizontal light ray is incident on a 60°-60°-60° glas prism ($n=1.5$) that is resting on a table. At what angle with the horizontal does the ray leave the prism. (Assume that the ray does not strike the bottom of the prism before exiting.)

6. For the prism in the preceding problem, what is the smallest angle of incidence for which a light ray will pass directly through the prism without total internal reflection?

7. You are standing in front of a rectangular fish tank filled with water (n=1.33).

   (a) Show that you can always see through the back of the tank without total internal reflection as you look through the front face.

   (b) Show that if you look through the front face toward the right side of the tank, there is an angle of view beyond which there will be total internal reflection from the right face. What is this angle?

   (c) What is the minimum index of refraction of the liquid in the tank such there that would always be total internal reflection from the right face?

   (d) Show that the fact that there is a thickness of glass (n=1.5) between the water and the air does not affect this problem.

8. Light of wavelength 5,000 Angstroms in glass with index of refraction $n = 1.5$ is incident at $45°$ on an interface with air. What is the skin depth for penetration of the light into the air?

9. (a) Solve Eqs. (10.115)-(10.118), to get the transmitted and reflected electric fields given by Eqs. (10.119) and (10.120).

   (b) Use these fields to get the transmission and reflection coefficients for the coated surface.

   (c) For an original air (n=1) to glass (n=1.5) interface, find the index of refraction and thickness of a coating that would have no reflection at an incident wave length of 5,000Å.

Chapter 11

Electromagnetic Waves in Matter

11.1 Electromagnetic Waves in a Conducting Medium

When a medium is not a perfect dielectric, but has a conductivity σ, there will be a current density $\mathbf{j}$ in the medium that is related to $\mathbf{E}$ by Ohm's law:

$$\mathbf{j} = \sigma \mathbf{E}. \tag{11.1}$$

Then Maxwell's equation for curl $\mathbf{H}$ (written in terms of $\mathbf{B}$) will be modified to

$$\nabla \times \mathbf{B} = \frac{4\pi\mu\sigma}{c}\mathbf{E} + \frac{\mu\epsilon}{c}\partial_t\mathbf{E}. \tag{11.2}$$

For $\mathbf{E}$ and $\mathbf{B}$ fields that vary with space and time like $e^{i(\mathbf{k}\cdot\mathbf{r}-\omega t)}$, Eq. (11.2) can be written as

$$
\begin{aligned}
i\mathbf{k}\times\mathbf{B} &= \frac{4\pi\mu\sigma}{c}\mathbf{E} - \frac{i\mu\epsilon\omega}{c}\mathbf{E} \\
&= -\frac{i\mu\epsilon\omega}{c}\left(1 + \frac{i4\pi\sigma}{\epsilon\omega}\right)\mathbf{E}.
\end{aligned} \tag{11.3}
$$

Also, the curl $\mathbf{E}$ Maxwell equation can be written as

$$i\mathbf{k}\times\mathbf{E} = \frac{i\omega}{c}\mathbf{B}. \tag{11.4}$$

Substituting Eq. (11.4) into Eq. (11.3) gives

$$\frac{c}{\omega}\mathbf{k}\times(\mathbf{k}\times\mathbf{E}) = -\frac{ck^2}{\omega}\mathbf{E} = -\frac{\mu\epsilon\omega}{c}\left(1 + \frac{i4\pi\sigma}{\epsilon\omega}\right)\mathbf{E}. \tag{11.5}$$

This can be solved for k, giving

$$k = k_0 \left(1 + \frac{i4\pi\sigma}{\epsilon\omega}\right)^{\frac{1}{2}} \tag{11.6}$$

where

$$k_0 = \frac{\sqrt{\mu\epsilon}\omega}{c} \tag{11.7}$$

would be the wave number for a non-conductor. It is convenient to introduce a parameter

$$\Delta = \frac{4\pi\sigma}{\epsilon\omega} \tag{11.8}$$

so that we can write

$$k = k_0(1 + i\Delta)^{\frac{1}{2}}. \tag{11.9}$$

The parameter Δ is a dimensionless measure of the effect of conductivity on the electromagnetic wave.

The wave number k is now complex. We can write it in terms of its real and imaginary parts as

$$k = \beta + i\alpha. \tag{11.10}$$

Then the propagation of the wave in the conducting medium will be

$$\mathbf{E} = \mathbf{E}_0 e^{-\alpha\hat{\mathbf{k}}\cdot\mathbf{r}} e^{i(\beta\hat{\mathbf{k}}\cdot\mathbf{r} - \omega t)}. \tag{11.11}$$

This shows an attenuation as the wave propagates, with an **attenuation length** (for the intensity) given by

$$\delta = 1/2\alpha. \tag{11.12}$$

The wavelength will depend on the real part of the wave number, so

$$\lambda = 2\pi/\beta. \tag{11.13}$$

We can solve for α and β by first squaring both sides of Eqs. (11.9) and (11.10), giving

$$k^2 = \beta^2 - \alpha^2 + 2i\alpha\beta = k_0^2(1 + i\Delta). \tag{11.14}$$

Now, equating the real and imaginary parts on each side of the equation separately gives two equations for the two constants α and β. The solutions are

$$\alpha = \frac{k_0}{\sqrt{2}}(\sqrt{1 + \Delta^2} - 1)^{\frac{1}{2}} \tag{11.15}$$

$$\beta = \frac{k_0}{\sqrt{2}}(\sqrt{1 + \Delta^2} + 1)^{\frac{1}{2}}. \tag{11.16}$$

This is a complicated result, so it is useful to look at the attenuation for the limiting cases of a poor conductor (a 'leaky' dielectric) and a good conductor. These designations are determined by the relative size of Δ compared to unity.

11.1.1 Poor conductor

For a poor conductor,

$$\Delta << 1. \tag{11.17}$$

Then, we can expand the square root in Eqs. (11.15) and (11.16) to get

$$\alpha \simeq \frac{k_0\Delta}{2} = \frac{2\pi}{c}\sqrt{\frac{\mu}{\epsilon}}\sigma, \quad \beta \simeq k_0. \tag{11.18}$$

Thus to first order in $\frac{4\pi\sigma}{\epsilon\omega}$ the wave will propagate with unchanged wavelength, but will attenuate with an attenuation length of

$$\delta = \frac{1}{2\alpha} = \sqrt{\frac{\epsilon}{\mu}}\left(\frac{c}{4\pi\sigma}\right). \tag{11.19}$$

We can see that δ is large here by writing it in terms of the wavelength (using $\lambda = 2\pi/\beta$) so that

$$\delta = \frac{\lambda}{2\pi\Delta}. \tag{11.20}$$

Since Δ is small for this case, the attenuation length will be many wavelengths long. The real part of the wave in this case of a poor conductor is given by

$$\mathbf{E} = \mathbf{E_0}e^{-\pi\Delta z/\lambda}\cos(2\pi z/\lambda - \omega t), \tag{11.21}$$

where $z = \hat{\mathbf{k}}\cdot\mathbf{r}$. This attenuated wave form is shown in Fig. 11.1 for $\Delta = 0.1$ and $t = 0$.

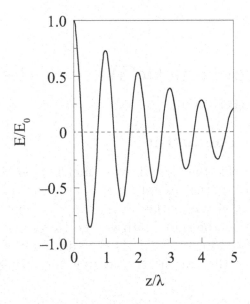

Figure 11.1: Attenuated wave form in a poor conductor.

11.1.2   Good conductor

For a good conductor,

$$\Delta \gg 1. \tag{11.22}$$

Then, keeping only the leading term the expansion for α and β results in

$$\alpha \simeq \beta \simeq k_0 \sqrt{\Delta/2} = \frac{\sqrt{2\pi\mu\sigma\omega}}{c}. \tag{11.23}$$

For this case, the attenuation is rapid with a skin depth of

$$\delta = \frac{1}{2\alpha} = \frac{c}{\sqrt{8\pi\mu\sigma\omega}} = \frac{\lambda}{4\pi}, \tag{11.24}$$

(The usual practice when δ is small is to refer to it as a **skin depth**.) The intensity will attenuate in less than a wavelength, so it will not even look like an oscillating wave.

Besides the rapid attenuation in a good conductor, the relation between **B** and **E** will be quite different than in a free wave. From Eqs. (11.4) and (11.9) (for large Δ), we see that

$$\begin{aligned}
\mathbf{B} &= \frac{ck}{\omega}\hat{\mathbf{k}} \times \mathbf{E} = \sqrt{\epsilon\mu}(i\Delta)^{\frac{1}{2}}\hat{\mathbf{k}} \times \mathbf{E} \\
&= \sqrt{\epsilon\mu\Delta}e^{i\pi/4}\hat{\mathbf{k}} \times \mathbf{E} = \sqrt{\frac{4\pi\mu\sigma}{\omega}}e^{i\pi/4}\hat{\mathbf{k}} \times \mathbf{E}.
\end{aligned} \tag{11.25}$$

Thus, in a good conductor, **B** will be much larger than **E**. Also, **B** will lag **E** in time by 45°. That is, if the time dependence of **E** is $e^{-i\omega t}$, then **B** will have the time dependence $e^{-i(\omega t-\pi/4)}$.

11.2   Electromagnetic Wave at the Interface of a Conductor

11.2.1   Perfect conductor

We now consider an electromagnetic wave incident on the surface of a conductor. We first consider the case of a perfect conductor, which we can treat by letting the conductivity σ in the previous section approach infinity. From Eq. (11.24), we see that the skin depth will go to zero, so **B** and **E** will both be zero in a perfect conductor. This is the case for the oscillating fields in electromagnetic waves that we are considering here. It is not the case for a static magnetic field.

We showed in Chapter 1 that a static **E** field must go to zero in a conductor. However, the **B** field does not have to go to zero at zero frequency. Indeed, **B** can be very large in conducting ferromagnetic metals. The reason for this can be seen in Eq. (11.23), which shows there is no attenuation in $\mathbf{B}$ at zero frequency. Equation (11.25), solved for $\mathbf{E}$, confirms that $\mathbf{E}$ would go to zero in a conductor at zero frequency.

The fields just outside the conductor can be determined from the boundary conditions at the surface of the conductor. The boundary conditions follow from Maxwell's equations, and are given by

$$\hat{\mathbf{n}}\cdot(\mathbf{D_o} - \mathbf{D}) = 4\pi\sigma_{\text{charge}} \tag{11.26}$$
$$\hat{\mathbf{n}}\times(\mathbf{E_o} - \mathbf{E}) = 0 \tag{11.27}$$
$$\hat{\mathbf{n}}\cdot(\mathbf{B_o} - \mathbf{B}) = 0 \tag{11.28}$$
$$\hat{\mathbf{n}}\times(\mathbf{H_o} - \mathbf{H}) = \frac{4\pi}{c}\mathbf{K}. \tag{11.29}$$

We have used the notation σ_{charge} to distinguish the surface charge from our use of σ for the conductivity. The subscript, o, designates the fields just outside the metal. The vector $\hat{\mathbf{n}}$ is the normal unit vector **directed out** of the surface of the conductor.

From Eqs. (11.27) and (11.28), and the fact that $\mathbf{E}$ and $\mathbf{B}$ are zero inside the conductor (for these oscillating fields), we see that $\mathbf{E}_{o\parallel} = \hat{\mathbf{n}}\times\mathbf{E} = 0$ and $B_{o\perp} = \hat{n}\cdot\mathbf{B} = 0$ at the surface of a perfect conductor. The subscripts $\parallel$ and $\perp$ denote the components parallel and perpendicular to the surface. $D_{o\perp}$ and $\mathbf{H}_{o\parallel}$ are not in general zero, but induce surface charges and currents at the surface of the metal.

11.2.2 Radiation pressure

The force exerted by $\mathbf{B}_{o\parallel}$ on the induced surface density current K leads to a radiation pressure on the surface of the conductor. (Watch out for several tricky factors of two in the following derivation.) For radiation incident normally on the conductor, the radiation pressure is given by

$$p_{\text{rad}} = -\frac{1}{2c}(\mathbf{K}\times\mathbf{B_o})\cdot\hat{\mathbf{n}}. \tag{11.30}$$

The factor of 2 in the denominator arises because only one-half the magnetic field at the surface of the conductor is produced by external currents. The other half of the field comes from the surface current itself, and does not provide a force on the current.

We can see this by the following argument: By symmetry, the $\mathbf{H}$ field due to the surface current will have the same magnitude just to the left and just to the right of the surface current, as shown in Fig. 11.2, but be in opposite directions . On the right, inside the conductor, the field due to the surface current must cancel the field due to external currents, leading to no field inside the conductor. Therefore these two fields must be of equal magnitude. Just outside the conductor, the two equal fields add, so the external $\mathbf{H}$ field is just

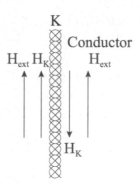

Figure 11.2: Magnetic fields inside and outside a conductor. Inside the conductor $\mathbf{H_K}$ and $\mathbf{H_{ext}}$ cancel, but they add outside the conductor.

twice as large as the field due only to external currents. The same would hold for the $\mathbf{B}$ magnetic field in a linear material.

We can replace the surface current in Eq. (11.30), using the boundary condition in Eq. (11.29), to get

$$
\begin{aligned}
p_{\text{rad}} &= -\frac{1}{2c}(\mathbf{K}\times\mathbf{B_o})\cdot\hat{\mathbf{n}} \\
&= -\frac{1}{8\pi}[(\hat{\mathbf{n}}\times\mathbf{H_o})\times\mathbf{B_o}]\cdot\hat{\mathbf{n}} \\
&= -\frac{1}{8\pi}[\mathbf{H_o}(\hat{\mathbf{n}}\cdot\mathbf{B_o}) - \hat{\mathbf{n}}(\mathbf{H_o}\cdot\mathbf{B_o})]\cdot\mathbf{n} \\
&= \frac{1}{8\pi\mu}|\mathbf{B_o}|^2.
\end{aligned} \tag{11.31}
$$

In the last step above, we have used the boundary conditon that $\hat{\mathbf{n}}\cdot\mathbf{B_o} = 0$, and replaced $\mathbf{H_o}$ by $\mathbf{B_o}/\mu$. $\mathbf{B_o}$ is the instantaneous magnetic field at the surface of the conductor. It is composed of two equal parts, corresponding to $\mathbf{B}(\text{incident})$ and $\mathbf{B}(\text{reflected})$, so

$$
\mathbf{B_o} = \mathbf{B}(\text{incident}) + \mathbf{B}(\text{reflected}) = 2\mathbf{B}(\text{incident}). \tag{11.32}
$$

$\mathbf{B}(\text{reflected})$ must equal $\mathbf{B}(\text{incident})$, because $\mathbf{E}(\text{reflected}) = -\mathbf{E}(\text{incident})$ to have $\mathbf{E}_\parallel = 0$ at the interface, and the reflected and incident waves travel in opposite directions. Therefore

$$
p_{\text{rad}} = \frac{1}{2\pi\mu}|\mathbf{B}(\text{incident})|^2. \tag{11.33}
$$

There is one more factor of 2. The result above is for the instantaneous pressure. Since, $\mathbf{B}(\text{incident})$ varies sinusoidally in the incident wave, the time averaged pressure becomes

$$
\bar{p}_{\text{rad}} = \frac{1}{4\pi\mu}[B_0(\text{incident})]^2 = \frac{\epsilon}{4\pi}[E_0(\text{incident})]^2, \tag{11.34}
$$

where we have used the notation E_0(incident) to denote the amplitude of the incident electric field. This final result is the same as we found for the time averaged radiation pressure in Eq. (10.33) using conservation of momentum

11.2.3 Interface with a good conductor

For finite conductivity, the current density and electric field are related by Ohm's law

$$\mathbf{j} = \sigma\mathbf{E}. \tag{11.35}$$

Because of this, the abstraction of a surface current $\mathbf{K}$ that appears in the boundary condition of Eq. (11.29) is no longer possible. This is because a finite surface current corresponds to an infinite current volume density $\mathbf{j}$, which would require an infinite field $\mathbf{E}$. This changes the boundary condition of Eq. (11.29) to

$$\hat{\mathbf{n}} \times (\mathbf{H_o} - \mathbf{H}) = 0. \tag{11.36}$$

$\mathbf{B}_\|$ inside the conductor will fall off with a finite attentuation length δ given by Eq. (11.24). This gives

$$\mathbf{B}_\|(z) = \mu\mathbf{H}_\|(z) = \mu\mathbf{H}_{o\|}e^{iz/2\delta}e^{-z/2\delta}, \tag{11.37}$$

where $z = \hat{\mathbf{n}} \cdot \mathbf{r}$ is the depth of penetration into the conductor. (Here, μ and ϵ are the permeability and permittive in the conductor.)

$\mathbf{E}_\|$ inside the conductor is related to $\mathbf{B}_\|$ by Eq. (11.25), so

$$\mathbf{E}_\| = \frac{e^{-i\pi/4}}{\sqrt{\epsilon\mu\Delta}}\hat{\mathbf{n}} \times \mathbf{B}_\| = \sqrt{\frac{\omega}{4\pi\mu\sigma}}e^{-i\pi/4}\hat{\mathbf{n}} \times \mathbf{B}_\|, \tag{11.38}$$

and is much smaller than $\mathbf{B}_\|$. Substituting Eq. (11.37) into Eq. (11.38) gives the z dependence of $\mathbf{E}_\|$ inside the conductor as

$$\mathbf{E}_\|(z) = \sqrt{\frac{\omega\mu}{4\pi\sigma}}\hat{\mathbf{n}} \times \mathbf{H}_{o\|}e^{-i\pi/4}e^{iz/2\delta}e^{-z/2\delta}, \tag{11.39}$$

$B_\perp$ is continuous across the interface, but is of the order of $E_\|$, and so is also small compared to $B_\|$. We can see this by looking at Snell's law, modified for the case of an interface with a good conductor. For a good conductor, using Eqs. (11.11) and (11.23) in Eq.(10.70) leads to.

$$\sin\theta_2 = \frac{e^{-i\pi/4}}{\sqrt{\epsilon\mu\Delta}}n_1\sin\theta_1. \tag{11.40}$$

The angle $\sin\theta_2$ is very small because of the factor Δ in the denominator, so we can make a small angle approximation for it. Then,

$$B_\perp = \tan\theta_2 B_\| \simeq \frac{e^{-i\pi/4}}{\sqrt{\epsilon\mu\Delta}}n_1\sin\theta_1 B_\| = n_1\sin\theta_1 E_\|. \tag{11.41}$$

In these equations, θ_1 is the angle that the incident part of the external electromagnetic wave makes with the normal to the interface. The magnitude of $n_1 \sin\theta_1$ is of the order of 1, so we see that $B_\perp$ is of the same (small) order as $E_\parallel$. This result is for an incident wave that is polarized perpendicular to the incident plane. $B_\perp$ is zero for a wave polarized parallel to the plane of incidence.

$E_{o\perp}$ can be large, but $E_\perp$ will be even smaller than $E_\parallel$ because of the relation

$$E_\perp = \tan\theta_2 E_\parallel \simeq \frac{e^{-i\pi/4}}{\sqrt{\epsilon\mu\Delta}} n_1 \sin\theta_1 E_\parallel. \tag{11.42}$$

The large difference between $E_{o\perp}$ and $E_\perp$ results in an induced surface charge on the interface with the conductor. If the incident wave were polarized perpendicular to the plane of incidence, then $E_\perp$ would be zero on both sides of the interface.

An example of typical fields inside and just outside a good conductor is shown in Fig. 11.3 for $\Delta = 100$.

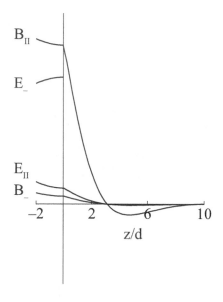

Figure 11.3: Magnetic and electric fields inside and outside a good conductor.

Energy absorption at the interface:

Because there is a non-vanishing $E_{o\parallel}$ at the interface, there will be a transmission of energy into the conductor that results in a loss of energy in the outside fields. The rate of energy loss per unit area is given by the normal component of the external Poynting vector into the surface:

$$\frac{d^2 U_{\mathrm{EM}}}{dA\,dt} = -\hat{\mathbf{n}}{\cdot}\overline{\mathbf{S}}_o = -\frac{c}{16\pi}\hat{\mathbf{n}}{\cdot}(\mathbf{E}_o^*{\times}\mathbf{H}_o + \mathbf{E}_o{\times}\mathbf{H}_o^*), \tag{11.43}$$

where we have used one form of Eq. (10.27). Using Eq. (11.39) giving $\mathbf{E}_\parallel$ inside the conductor in terms of $\mathbf{H}_{o\parallel}$, and the fact that $\mathbf{E}_\parallel$ is continuous across the interface, we get

$$\frac{d^2 U_{\text{EM}}}{dAdt} = \frac{c}{8\pi}\sqrt{\frac{\omega\mu}{8\pi\sigma}}|\mathbf{H}_{o\parallel}|^2. \tag{11.44}$$

The rate of energy loss can also be calculated from the ohmic energy loss within the conductor. This is given by

$$\frac{d^2 U_{\text{ohmic}}}{dAdt} = \frac{1}{4}\int_0^\infty (\mathbf{j}^*\cdot\mathbf{E} + \mathbf{j}\cdot\mathbf{E}^*)dz = \frac{1}{2}\int_0^\infty \sigma|\mathbf{E}|^2 dz. \tag{11.45}$$

the rate of ohmic energy loss is given by

$$\frac{d^2 U_{\text{ohmic}}}{dAdt} = \frac{\omega\mu}{8\pi}|\mathbf{H}_{o\parallel}|^2 \int_0^\infty e^{-z/\delta}dz = \frac{\omega\delta\mu}{8\pi}|\mathbf{H}_{o\parallel}|^2 \tag{11.46}$$

Using Eq. (11.24) for δ, we see that this result is the same as the rate of energy loss in Eq. (11.44), calculated using the Poynting vector.

Effective surface current:

Although the skin depth is finite for a good conductor, it is usually very small, and it is useful to define an effective surface current. This is given by

$$\begin{aligned}
\mathbf{K}_{\text{eff}} &= \int_0^\infty \mathbf{j}(z)dz = \int_0^\infty \sigma\mathbf{E}_\parallel(z)dz \\
&= \sqrt{\frac{\omega\mu}{4\pi\sigma}}\hat{\mathbf{n}}\times\mathbf{H}_{o\parallel}e^{-i\pi/4}\int_0^\infty e^{-(z/2\delta(1-i))}dz \\
&= 2\delta\sqrt{\frac{\omega\mu}{4\pi\sigma}}\hat{\mathbf{n}}\times\mathbf{H}_{o\parallel} \\
&= \frac{c}{4\pi}\hat{\mathbf{n}}\times\mathbf{H}_{o\parallel}. \tag{11.47}
\end{aligned}$$

This is the same magnitude as the surface current for a perfect conductor, but it now has a small finite depth δ The rate of energy loss can be written in terms of the surface current as

$$\frac{d^2 U_{\text{ohmic}}}{dAdt} = \frac{1}{4\sigma\delta}|\mathbf{K}_{\text{eff}}|^2. \tag{11.48}$$

This looks a little like the familiar $\frac{1}{2}I^2 R$, with the quantity $1/2\sigma\delta$ acting like an effective surface resistance.

11.3 Frequency Dependence of Permittivity

11.3.1 Molecular model for permittivity

In section 6.6.3, we found the permittivity at zero frequency in a simple molecular model. Here, we extend that model to find the permittivity as a function of the frequency of an electromagnetic wave. Newton's equation for the motion of an electron bound in a molecule with a resonant angular frequency ω_i and a damping $m\gamma_i\dot{\mathbf{r}}$ is

$$-e\mathbf{E} = m[\ddot{\mathbf{r}} + \gamma_i\dot{\mathbf{r}} + \omega_i^2\mathbf{r}]. \tag{11.49}$$

The solution of this equation for an electric field oscillating like $e^{-i\omega t}$ is

$$\mathbf{r} = \frac{-e\mathbf{E}}{m[\omega_i^2 - \omega^2 - i\omega\gamma_i]}. \tag{11.50}$$

This gives a molecular polarizability

$$\eta_i = \frac{e^2}{m[\omega_i^2 - \omega^2 - i\omega\gamma_i]}. \tag{11.51}$$

If the wave-length of the wave is longer than the distance between molecules, we can relate the microscopic electric field at the molecule to the macroscopic field $\mathbf{E}$ as we did in Sec. 6.6.3. Then, a derivation like that in Sec. 6.6.3 results in the permittivity being given by

$$\epsilon = 1 + \frac{4\pi N \sum_i z_i\eta_i}{1 - \frac{4\pi}{3} N \sum_i z_i\eta_i}. \tag{11.52}$$

This is for N molecules per unit volume, with z_i electrons in each molecule oscillating in mode i. The z_i, called **oscillator strengths**, add up to the total number (Z) of electrons in each molecule, so

$$\sum_i z_i = Z. \tag{11.53}$$

The permittivity given by Eq. (11.52) has a real and imaginary part, which we write as

$$\epsilon = \epsilon_R + i\epsilon_I. \tag{11.54}$$

For simplicity, we assume $\mu = 1$, which is usually the case in a dielectric. Then the index of refraction is given by

$$n = (\epsilon_R + i\epsilon_I)^{\frac{1}{2}}. \tag{11.55}$$

We can find the real and imaginary parts of n in the same way as we derived Eqs. (11.15) and (11.16), so

$$\text{Re } n = \frac{1}{\sqrt{2}}\left[\sqrt{\epsilon_R^2 + \epsilon_I^2} + \epsilon_R\right]^{\frac{1}{2}} \tag{11.56}$$

$$\text{Im } n = \frac{1}{\sqrt{2}}\left[\sqrt{\epsilon_R^2 + \epsilon_I^2} - \epsilon_R\right]^{\frac{1}{2}}. \tag{11.57}$$

11.3.2   Dispersion and absorption

The frequency dependence of the index of refraction n leads to **dispersion** of the different colors of light as it is bent by refraction. This frequency dependence is dominated by the resonant character of the molecular polarizability in Eq. (11.51), but it is quite complicated. It would be given by substituting Eq. (11.51) into Eq. (11.52), and this into Eqs. (11.56)and (11.57). An example with two resonances, one below and one above the visible wavelength region ($\sim 4,000 - 7,000 \, \overset{\circ}{A}$) is shown in Fig. 11.4.

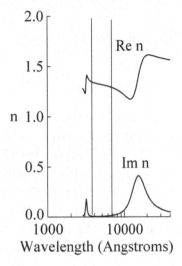

Figure 11.4: Real and imaginary parts of the index of refraction $n(\lambda)$ between two resonances. The visible region lies between the two vertical dashed lines.

The imaginary part of the index of diffraction can get quite large in the resonance peaks, correponding to strong absorption. It is small in the visible region where the medium is transparent. The real part of the index decreases slowly with wavelength in the visible region. This means that blue light will be bent more than red light in refraction, which is called **normal dispersion**.

The real part oscillates near each resonance, so there is a small region where the real part increases sharply with wave-length. This is called **anamolous dispersion** because the order of the spectrum would be reversed. However this generally occurs only where the imaginary part is resonantly large, and attenuation is rapid.

11.3.3   Conduction electrons

We saw in section 11.1 that the conductivity could be related to the imaginary part of the permittivity. That is, Eq. (11.3) can be interpreted in terms of a complex permittivity given by

$$\epsilon = \left(\epsilon_R + \frac{i4\pi\sigma}{\omega} \right) = \epsilon_R + i\epsilon_I. \tag{11.58}$$

Then the conductivity can be related to the the imaginary part of ϵ by

$$\sigma = \frac{\omega \epsilon_I}{4\pi}. \tag{11.59}$$

In conducting metals, one or more **conduction electrons** in each atom are not bound, and can move through the material. Newton's law for these electrons has no restoring force, so the equivalent polarizability is

$$\eta = \frac{-e^2}{m(\omega^2 + i\omega\gamma)}. \tag{11.60}$$

This gives a contribution to ϵ that is singular at $\omega = 0$:

$$\epsilon = \frac{i4\pi z_c N e^2}{m\omega(\gamma - i\omega)} + \epsilon_{\text{bound}}. \tag{11.61}$$

Here z_c is the number of conduction electrons per atom and ϵ_{bound} is the contribution to ϵ from the bound electrons. Note that we did not include the correction due to the action of the microscopic electric field, as in Eq. (11.52). This is because the conduction electrons move throughout the conductor, and are acted on by the space averaged macroscopic electric field. This is appropriate until the wavelength becomes comparable to the distance between atoms, at which point the concept of permittivity becomes meaningless.

The conductivity at low frequencies will be dominated by the singular part of ϵ, and is given by

$$\sigma = \frac{z_c N e^2}{m(\gamma - i\omega)}. \tag{11.62}$$

This shows that the conductivity is real at low frequencies, where the $i\omega$ in the denominator is small compared to γ. (See problem 10.6.) The classical description given above illustrates the qualitative features of conductivity at low frequencies, but the assumption that the resistive force is proportional to velocity is an over-simplification. Also, quantum mechanics must be used for a more complete calculation of conductivity.

11.4   Causal Relation Between D and E

A **dispersive medium** is one in which the permittivity depends on the frequency. As we have seen, this is generally true in simple models for the permittivity due to the polarization of bound electrons. In this case, the simple proportionality between **D** and **E** must be written at a definite frequency as

$$\mathcal{D}(\omega) = \epsilon(\omega)\mathcal{E}(\omega) = \mathcal{E}(\omega) + 4\pi\chi(\omega)\mathcal{E}(\omega). \tag{11.63}$$

We use the notation $\mathcal{D}$ and $\mathcal{E}$ to denote the fields at a definite frequency. We have separated out the non-dispersive part of the connection between $\mathcal{D}$ and

$\mathcal{E}$. The frequency dependence is in the electric susceptibility $\chi(\omega)$. We are assuming that the medum is linear. In recent years, the strong electric fields in quantum laser radiation have given rise to the new field of non-linear optics, but we will not treat that here.

For general time dependent fields, $\mathcal{D}$ and $\mathcal{E}$ are the Fourier transforms of $\mathbf{D}$ and $\mathbf{E}$. In order to relate the time dependent $\mathbf{D}(t)$ to $\mathbf{E}(t)$, we first take the Fourier transform of $\mathbf{D}(t)$ as

$$\mathcal{D}(\omega) = \frac{1}{2\pi} \int_{-\infty}^{+\infty} dt \; e^{i\omega t} \mathbf{D}(t). \tag{11.64}$$

[**Note:** The exponential $e^{i\omega t}$ used here for the Fourier transform is different than the form $e^{-i\omega t}$ used in many math books. Because of this, some of our results will differ from those in such books by a sign. We have chosen our sign for the exponent to be consistent with the time dependence $e^{-i\omega t}$ for the plane wave solutions of the wave equation.]

The Fourier transform field $D(\omega)$ is related to $E(\omega)$ by

$$\begin{aligned} \mathcal{D}(\omega) &= \mathcal{E}(\omega) + 4\pi\chi(\omega)\mathcal{E}(\omega) \\ &= \mathcal{E}(\omega) + 2\chi(\omega)\int_{-\infty}^{+\infty} dt \; e^{i\omega t}\mathbf{E}(t). \end{aligned} \tag{11.65}$$

We now can write the time dependent $\mathbf{D}(t)$ as the inverse Fourier transform of $\mathcal{D}(\omega)$:

$$\begin{aligned} \mathbf{D}(t) &= \int_{-\infty}^{+\infty} d\omega \; e^{-i\omega t}\mathcal{D}(\omega) \\ &= \mathbf{E}(t) + 2\int_{-\infty}^{+\infty} d\omega \; e^{-i\omega t}\chi(\omega) \int_{-\infty}^{+\infty} dt' \; e^{i\omega t'} \mathbf{E}(t'). \end{aligned} \tag{11.66}$$

We interchange the order of integrations, as we did in chapter 3, to get

$$\mathbf{D}(t) = \mathbf{E}(t) + \int_{-\infty}^{+\infty} dt' \left[2\int_{-\infty}^{+\infty} d\omega \; e^{-i\omega(t-t')}\chi(\omega) \right] \mathbf{E}(t'). \tag{11.67}$$

The quantity in square brackets in Eq. (11.67) is like a Green's function in time, and we can rewrite Eq. (11.67) as

$$\mathbf{D}(t) = \mathbf{E}(t) + \int_{-\infty}^{+\infty} dt' \; g_\chi(t,t')\mathbf{E}(t'), \tag{11.68}$$

with

$$g_\chi(t,t') = 2\int_{-\infty}^{+\infty} d\omega \; e^{-i\omega(t-t')}\chi(\omega). \tag{11.69}$$

At first glance, the relation between $\mathbf{D}(t)$ and $\mathbf{E}(t')$ in Eq. (11.68) is surprising. It seems as if, in order to find $\mathbf{D}(t)$, we must know $\mathbf{E}(t')$ for all times, even those times t' that are **later** than t. This would mean that we couldn't find **D** today, unless we knew **E** tomorrow. This conundrum leads to the physical principle of **causality**.

Causality says that we only need past information to calculate present quantities. In mathematical terms, this means that the function $g_\chi(t,t')$ must vanish for times t' that are later than t:

$$g_\chi(t,t') = 0 \quad \text{for } t' > t. \tag{11.70}$$

A function with this property is said to be a **causal function**.

Does the $g_\chi(t,t')$ we are using indeed have this causal property? We can answer this question by using contour integration in the complex plane to perform the integral over ω in Eq. (11.69) defining g_χ. We use the contour shown in Fig. 11.5.

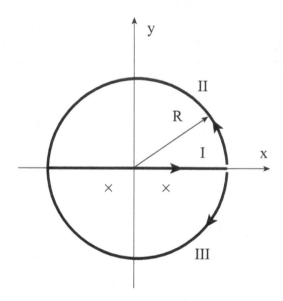

Figure 11.5: Contour for $g_\chi(t,t')$ in the complex ω plane.
The contour is closed above by II for $t' > t$ or below by III for $t' < t$.

We assume a case where the integrand has singularities in the lower half plane, but is analytic in the upper half plane. The integral we want is the part of the contour labeled I, along the real axis. To apply Cauchy's theorem, we must close the contour, which we do by a large semicircle of radius R, shown as either contour II or III. We then take the limit as R→∞, so that the integral along contour I will be from -∞ to +∞.

Jordan's lemma says that the integration on the large semicircle will go to zero in the limit R→∞ if two conditions are satisfied. First, $\chi(\omega)$ must vanish uniformly in the limit $\omega\to\infty$. The χ in our model calculation, deduced from Eq. (11.52), does have this property. Then, the contour must be closed above, by contour II, if $(t - t') < 0$ and below, by contour III, if $(t - t') > 0$. Cauchy's theorem tells us that the complete contour integral will vanish if the integrand is analytic within the closed contour in the complex plane. Thus our Green's function $g_\chi(t, t')$ will vanish for $t' > t$ (where we must close the contour by the upper semicircle) if $\chi(\omega)$ is **analytic in the upper half plane**.

The $\chi(\omega)$ from Eq. (11.52) does indeed have this property, having poles in the lower half plane, but being analytic in the upper half plane (see problem 11.3). This analyticity property of χ can be traced to the positive sign of the damping term in Eq. (11.49). Thus the fact that there is damping, retarding the motion of the bound electrons, leads to the appropriate causal behavior relating $\mathbf{D}(t)$ to $\mathbf{E}(t')$.

11.5   Wave Packets

For simplicity, we will consider a scalar wave in one space dimension in this section. A simple function representing a wave moving in the positive x direction (corresponding to a plane wave in three dimensions) is given by

$$u(x, t) = e^{i(kx - \omega t)}. \tag{11.71}$$

This function extends over the entire x-axis. We would like to describe a wave that has finite extent, but still satisfies the wave equation. We can do this by the construction

$$f(x, t) = \int_{-\infty}^{+\infty} dk \, A(k) e^{i(kx - \omega t)}. \tag{11.72}$$

The amplitude $A(k)$ is the Fourier transform of $f(x, 0)$, and is given by

$$A(k) = \frac{1}{2\pi} \int_{-\infty}^{+\infty} dx \, f(x, 0) e^{-ikx}. \tag{11.73}$$

If $f(x, t)$ extended over all space, like $u(x, t)$ above, then the Fourier integral for $A(k)$ would give a delta function, corresponding to only one wave number for the wave.

An example of a wave $f(x, t)$ that is finite in extent is shown in Fig. 11.6a. The wave form shown is called a **wave packet**. The oscillations in the wave packet have a varying amplitude, called the **envelope** of the wave packet, that is generally a smooth function of limited extent.

The particular wave packet in Fig. 11.6a has a Gaussian envelope, shown as the dashed curve in the figure. The initial form of the wave packet is

$$f(x, 0) = g(x) e^{ik_0 x} = e^{-x^2/2L^2} e^{ik_0 x}, \tag{11.74}$$

where $g(x)$ is the envelope function and $e^{ik_0 x}$ is the oscillating function within the envelope. We have plotted the real part of this wave packet. The wave packet has the nominal wave number k_0, but the Fourier transform of $f(x, 0)$

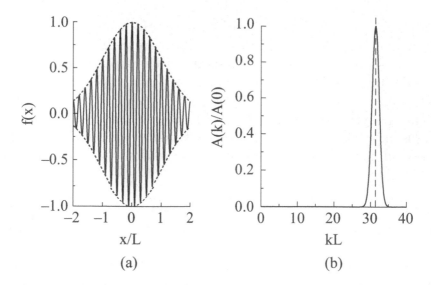

Figure 11.6: (a) Wave packet, $f(x)$, with a Gaussian envelope, shown by the dashed curves. The width of the packet is given by $L = 10\lambda$. (b) Fourier transform, $A(k)$, of the wave packet in (a). The central value k_0 is marked by the vertical dashed line.

in Fig. 11.6b shows that there is actually a spread of wave numbers in the wave packet, described by the function

$$A(k) = \frac{L}{\sqrt{2\pi}}e^{-L^2(k-k_0)^2/2}. \tag{11.75}$$

One measure of the width of the Gaussian wave packet is given as

$$\Delta x = L, \tag{11.76}$$

which is the change in x from its central value to the x value for which $|f|^2$ is $1/e$ of its peak value. Δx is usually called the **half width**, with the **full width** Γ being twice this. By the same measure, the width of the Fourier transform $A(k)$, which is also Gaussian, is given by

$$\Delta k = \frac{1}{L}. \tag{11.77}$$

Notice that the product of the two widths is

$$(\Delta x)(\Delta k) = 1, \tag{11.78}$$

and is independent of the individual width of either function. This is a general property of the Fourier transform. The narrower in x the original wave packet, the broader in k will be its Fourier transform, and vice-versa.

As a brief aside, the relation between Δx and Δk in Eq. (11.78) is the mathematical basis for the **Heisenberg uncertainty principle** in quantum mechanics. There it is given in terms of the momentum variable p, which by the deBroglie principle is given as $p = \hbar k$. Also a slightly different measure is conventionally used in quantum mechanics for the width of a wave packet, so the extension of Eq. (11.78) to quantum mechanics is

$$(\Delta x)(\Delta p) \geq \frac{\hbar}{2}.$$ (11.79)

We have used the relation $\geq$, because it turns out that the Gaussian wave packet has the minimum product of widths. For any other function cutting the wave off, the product $(\Delta x)(\Delta p)$ would be greater, but still usually of the order of $\hbar/2$.

11.5.1   Natural Line Width

The electric field emitted by the decay of an excited state of an atom with an exponential lifetime τ has the form

$$\begin{aligned} E &= E_0 e^{i(k_0 x - \omega t)} e^{(x-ct)/(2c\tau)}, \quad x < ct, \\ &= 0, \qquad\qquad\qquad\qquad\quad x > ct. \end{aligned}$$ (11.80)

The real part of this wave train is shown in Figure 11.7.
Its Fourier transform is given by

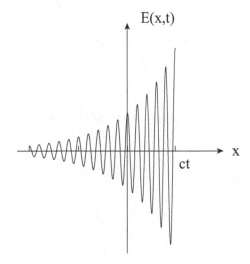

Figure 11.7: Wave train for decaying atom.

$$\mathcal{E}(k) = \frac{E_0}{2\pi[i(k_0 - k) + 1/2c\tau]}.$$ (11.81)

The absolute value squared of the Fourier transform is

$$|\mathcal{E}|^2 = \frac{E_0^2}{4\pi^2[(k - k_0)^2 + (1/2c\tau)^2]}.$$ (11.82)

A reasonable value for the width of this distribution is

$$\Delta k = \frac{1}{2c\tau},\qquad(11.83)$$

which is the difference in k from k_0 to the k value for which the distribution is one half of its peak value. This is called the **half width at half maximum** (HWHM). The full width is twice this value. The full width of the wavelength distribution is given by

$$\Gamma = 2\Delta\lambda = 2\frac{d\lambda}{dk}\Delta k = \frac{2}{2c\tau}\frac{d}{dk}\left(\frac{2\pi}{k}\right) = -\frac{1}{c\tau}\left(\frac{2\pi}{k^2}\right) = \frac{\lambda}{\omega\tau} = \frac{\lambda^2}{2\pi c\tau}.\qquad(11.84)$$

This is called the **natural line width** of the spectral line for this decay.

The discussion given above is a classical explanation for what actually is a quantum mechanical process. It holds true in this case because, in either theory, the result just depends on the mathematics of the Fourier transform.

11.6 Wave Propagation in a Dispersive Medium

11.6.1 Group velocity and phase velocity

We now consider the propagation of a wave packet in a dispersive medium. The time development of the wave packet is given by the integral over k in Eq. (11.72). The angular frequency ω in the integral is a function of k given by

$$\omega(k) = \frac{ck}{n(k)},\qquad(11.85)$$

where we have considered the index of refraction n as a function of k. Then, the integral in Eq. (11.72) is

$$f(x,t) = \int_{-\infty}^{+\infty} dk\, A(k)e^{i[kx-\omega(k)t]}.\qquad(11.86)$$

If the medium is non-dispersive, n is a constant and the integral becomes

$$f(x,t) = \int_{-\infty}^{+\infty} dk\, A(k)e^{ik[x-\frac{c}{n})t]}.\qquad(11.87)$$

Comparing this integral with that in Eq. (11.72), for $t=0$ in that integral, we see that Eq. (11.87) results in

$$f(x,t) = f\left[\left(x-\frac{c}{n}t\right),0\right].\qquad(11.88)$$

This shows that, in a non-dispersive medium, the wave packet retains its original shape, while moving with velocity $v = c/n$.

In a dispersive medium, we expand the function $\omega(k)$ about the value k_0 at which the distribution in k of the wave packet peaks:

$$\omega(k) = \omega_0 + (k - k_0)\omega_0' + ..., \tag{11.89}$$

where $\omega_0 = \omega(k_0)$ and $\omega_0' = \frac{d\omega}{dk}|_{k_0}$. We have expanded $\omega(k)$ up to first order in $(k - k_0)$. With this approximation, the wave packet integral can be written as

$$f(x,t) = e^{-i(\omega_0 - k_0\omega_0')t} \int_{-\infty}^{+\infty} dk\, A(k)e^{ik(x - \omega_0't)}. \tag{11.90}$$

Comparison of this with Eq. (11.72) now gives

$$\begin{aligned} f(x,t) &= e^{-i(\omega_0 - k_0\omega_0')t} f\left[(x - \omega_0't), 0\right] \\ &= g(x - \omega_0't)e^{i(k_0x - \omega_0t)}. \end{aligned} \tag{11.91}$$

This shows that, in a dispersive medium, the envelope of a wave packet will retain its original shape [to first order in the expansion of $\omega(k)$], and move with a **group velocity**

$$v_g = \omega_0' = \frac{d\omega}{dk}, \tag{11.92}$$

with the understanding that the derivative is taken at a central wave number k_0. The velocity at which the waves within the packet move, called the **phase velocity**, is given by

$$v_p = \frac{\omega_0}{k_0} = \frac{c}{n}. \tag{11.93}$$

The phase velocity of the oscillations within the envelope is actually the illusion of motion. If the waves go anywhere, it is at the group velocity of the envelope. I recall watching my grandmother operate a hand cranked meat grinder. As she rotated the helical blades, they seemed to be moving along, grinding the meat (or fish). Of course, they were just rotating and not moving from left to right. They were not really going anywhere. As a budding theorist, I did not perform the experiment of putting my hands in, but now I know that the 'motion' of the blades was actually the illusion of the phase velocity.

11.6.2 Spread of a wave packet

We have seen that a wave packet keeps its original shape to first order in the expansion of $\omega(k)$ about k_0. In this section, we go to the next order in this expansion and see how that results in a spread of the wave packet with time.

To second order, the expansion of $\omega(k)$ is

$$\omega(k) = \omega_0 + (k - k_0)\omega_0' + \frac{1}{2}(k - k_0)^2\omega_0'' + ..., \tag{11.94}$$

where ω_0' and ω_0'' represent the first and second derivatives of $\omega(k)$ evaluated at $k = k_0$. The second order term in this expansion has the effect of multiplying the Fourier amplitude $A(k)$ in Eq. (11.90) by a factor that changes the Fourier amplitude to

$$A_2(k) = A(k)e^{-\frac{i}{2}(k-k_0)^2\omega_0''t}, \tag{11.95}$$

where $A_2(k)$ is the Fourier amplitude, expanded to second order. This changes the Gaussian amplitude in Eq. (11.75) to

$$A_2(k) = \frac{L}{\sqrt{2\pi}}e^{-\frac{1}{2}(k-k_0)^2(L^2+i\omega_0''t)}. \tag{11.96}$$

Without going through a lot of algebra, we can see from this result that the only effect of the second order term in the expansion is to change the length parameter L^2 in the Gaussian wave packet to

$$L_2^2 = L^2 + i\omega_0''t. \tag{11.97}$$

Then, the overall result of the first and second order terms in the expansion of $\omega(k)$ is to give the Gaussian wave packet in Eq. (11.74) the time dependence

$$\begin{aligned} f_2(x,t) &= \exp\left[\frac{-(x-v_g t)^2}{2(L^2+i\omega_0''t)}\right] e^{ik_0 x - \omega_0 t)} \\ &= \exp\left[\frac{-(x-v_g t)^2}{2L^2[1+(\omega_0''t/L^2)^2]}\right] e^{ik_0(x-v_p t)}. \end{aligned} \tag{11.98}$$

We have rationalized the denominator in the first exponential, which leads to a slight variation of v_p. The main conclusion from $f_2(x,t)$ is that, to second order, the Gaussian wave packet propagates with the group velocity, while its spread Δx increases with time by

$$\Delta x \to (\Delta x)_2 = L[1+(\omega_0''t/L^2)^2]^{\frac{1}{2}}. \tag{11.99}$$

We can put the spread of the wave packet into more physical terms by introducing the wavelength by the relation

$$\omega'' = \frac{dv_g}{dk} = -\frac{\lambda}{k}\frac{dv_g}{d\lambda} = -\frac{\lambda^2}{2\pi}\frac{dv_g}{d\lambda}. \tag{11.100}$$

Then the spread in the wave packet can be written as

$$(\Delta x)_2 = L\left[1+\left(\frac{1}{2\pi}\frac{\lambda^2}{L^2}\frac{dv_g}{d\lambda}t\right)^2\right]^{\frac{1}{2}}. \tag{11.101}$$

This expression shows that the spread in the Gaussian wave packet depends on the ratio L/λ, which equals the number of oscillations in the wave train, and on the rate of change of the group velocity with wavelength. If there are many oscillations and a slow variation of v_g with λ, the spread in the wave packet will be slow. Of, course any wave packet will eventually spread in a dispersive medium because of the dependence on t.

The above analysis is precise (to second order) for a Gaussian wave packet, but an expression close to Eq. (11.101) should hold for any case where there are at least several oscillations within the envelope, and $n(\omega)$ is a smooth function near ω_0. This will be so as long as ω_0 is not near a resonance of $\epsilon(\omega)$, where $n(w)$ does have a rapid variation as seen in Fig. 11.4.

We can make two observations when ω_0 is near a resonance. First, while the variation in the real part of n is rapid, its imaginary part will be large. This causes a strong attenuation, so the wave packet would not travel far. Second, the rapid variation in the real part of n would change the shape of the wave packet, so it would no longer have a simple interpretation. In view of these observations, it is safe to say that if a wave packet travels any distance without a large change in its shape, it will move at the group velocity. In the next section, we show that, no matter how rapid the variation in n, the wave packet cannot move faster than c.

11.6.3   No electromagnetic wave travels faster than c

In most materials, n is greater than 1, so the phase velocity is less than c. In some materials, n can be less than 1 for some frequencies. Then the phase velocity will be greater than c, but, as we have emphasized, this is not a real motion of the wave packet. The velocity of the wave packet is usually the group velocity, which is given by

$$v_g = \frac{d\omega}{dk} \tag{11.102}$$

$$= \frac{c}{n}\left(1 - \frac{k}{n}\frac{dn}{dk}\right) \tag{11.103}$$

$$= \frac{c}{n}\left(1 + \frac{\lambda}{n}\frac{dn}{d\lambda}\right) \tag{11.104}$$

$$= \frac{\frac{c}{n}}{\left(1 + \frac{\omega}{n}\frac{dn}{d\omega}\right)}. \tag{11.105}$$

In the above equations, we have solved for the group velocity with n given in terms of either k, λ, or ω, so we can use whichever is convenient.

In a material with normal dispersion, where $\frac{dn}{d\lambda}$ is negative, the group velocity will be less than c. But, in the case of anomalous dispersion, $\frac{dn}{d\lambda}$ is positive. Then v_g can be greater than c. We will show however that, even in this case, a wave packet cannot move faster than c. That means that if v_g is greater than c it is **not** the velocity of the wave packet. [This is not unreasonable, since the anomalous dispersion occurs near a resonance where $n(\omega)$ varies rapidly and truncation of the expansion of $\omega(k)$ in Eq. (11.89) is not valid.]

We consider a wave packet moving in the $+x$ direction, which has not yet reached positive x at time $t=0$. For this wave packet,

$$f(x,0) = 0 \quad \text{for } x > 0, \quad \text{and } f(0,t) = 0 \quad \text{for } t < 0. \tag{11.106}$$

We describe the motion of the wave packet in terms of its Fourier frequency components by

$$f(x,t) = \int_{-\infty}^{+\infty} d\omega \; A(\omega) e^{i\omega[n(\omega)x/c - t]}. \tag{11.107}$$

This is similar to Eq. (11.72) for the wave packet, but we have used frequency rather than wave number as the integration variable.

The function $A(\omega)$ is the Fourier transform of a function, f(0,t), which is zero for $t<0$. This requires $A(\omega)$ to be analytic in the upper half ω plane. This is because the contour integral for $f(0,t)$ in Eq. (11.107) must be closed by an upper semicircle when $t<0$, and will vanish if $A(\omega)$ is analytic in the upper half plane.

We evaluate the integral in Eq. (11.107) for $f(x,t)$ by integration around the contour shown in Fig. 11.8. The integral on contour I along the real axis is the integral that gives $f(x,t)$. The contour is completed by a large semicircle of radius R, taking the limit as $R\to\infty$. The semicircle closes the contour either above (II) or below (III), as required for convergence.

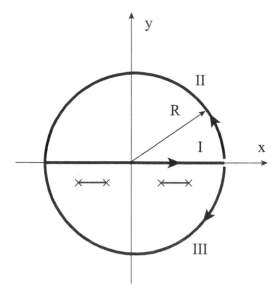

Figure 11.8: Contour for the maximum speed of a wave packet in the complex ω plane. The contour is closed above by II for $x > ct$ or below by III for $x < ct$. The integrand has branch cuts in the lower half plane, but is analytic in the upper half plane.

We will use the general properties of $n(\omega)$ to evaluate the integral. $n(\omega)$ is the square root of $\epsilon(\omega)$, which is given by Eq. (11.52). One important property is that $\epsilon \to 1$ as ω becomes infinite. Then, n will also approach 1 for large ω, and the integral along the semicircular contour II (or III) will become

$$II = \int_{II} d\omega \, A(\omega) e^{i\omega(x/c-t)}. \tag{11.108}$$

We showed in section 11.4 that ϵ, as given by Eq. (11.52), was analytic in the upper half plane, but had poles in the lower half plane. The singularities of $n(\omega)$ are somewhat more complicated, but we now show that $n(\omega)$ also is analytic in the upper half plane. We look at $\epsilon(\omega)$ near one of its resonances, where it can be approximated by

$$\epsilon(\omega) = 1 + \frac{b(\omega_i^2 - \omega^2 - i\omega\gamma_i)^{-1}}{1 - \frac{b}{3}(\omega_i^2 - \omega^2 - i\omega\gamma_i)^{-1}} \tag{11.109}$$

The constant b is given by $4\pi z_i N e^2/m$. We have neglected contributions from the other resonances.

We rewrite $\epsilon(\omega)$ as

$$\epsilon(\omega) = \frac{3(\omega_i^2 - \omega^2 - i\omega\gamma_i) + 2b}{3(\omega_i^2 - \omega^2 - i\omega\gamma_i) - b} \tag{11.110}$$

In this form, we can see that the numerator and the denominator are each quadratics in ω, so ϵ has two zeros and two poles near the resonance. The solution of each quadratic is of the form

$$\omega = -i\Gamma \pm R, \tag{11.111}$$

where Γ and R are positive real numbers. This means that each zero and each pole of ϵ is in the lower half ω plane. n, the square root of ϵ can be written as

$$n(\omega) = \left[\frac{(\omega - R_z + i\Gamma_z)(\omega + R_z + i\Gamma_z)}{(\omega - R_p + i\Gamma_p)(\omega + R_p + i\Gamma_p)} \right]^{\frac{1}{2}} \tag{11.112}$$

near each resonance. This shows there will be branch cuts in $n(\omega)$, as shown in Fig. 11.8 for one resonance, but they will all be in the lower half plane, so $n(\omega)$ will be analytic in the upper half plane.

We're still not quite through, because n appears in the exponential in Eq. (11.107). That is OK though, because the exponential of an analytic function is analytic. Since $A(\omega)$ is also analytic in the upper half plane, the entire integrand in Eq. (11.107) is analytic in the upper half plane and the integral vanishes for $x - ct > 0$. This means we finally can write

$$f(x,t) = 0 \quad \text{for } x > ct, \tag{11.113}$$

which means that **the wave packet cannot move faster than** c.

It took a bit of work, but this is an important general result.

11.7   Problems

1. A plane electromagnetic wave is incident at an angle θ from vacuum onto the flat surface of a perfect conductor.

   (a) Use conservation of momentum to find the radiation pressure on the surface of the conductor.

   (b) For a wave that is polarized perpendicular to the plane of incidence, find the radiation pressure by calculating the magnetic force on the surface current.

   (c) For a wave that is polarized parallel to the plane of incidence, find the radiation pressure by calculating the magnetic and electric forces on the surface current and surface charge.

2. Use the physical properties of copper to estimate the frequency and wave-length for which its conductivity develops an appreciable ($\sim 10\%$) imaginary part. Assume two conduction electrons per atom.

3. (a) Find the real and imaginary parts of ϵ, as given in Eq. (11.52).

   (b) Show that ϵ is analytic in the upper half of the complex ω plane. In finding any particular pole position, the contributions from other modes can be ignored since they are analytic in that region.

4. Find the Fourier transform of the function in Eq. (11.74) to verify Eq. (11.75). (Hint: Complete the square in the exponent in doing the integral.)

5. A wave packet is given by

$$f(x,0) = e^{im\pi x/L}, \quad 0 < x < L, \qquad (11.114)$$

   and $f(x,0)$ is zero elsewhere.

   (a) Plot $|f(x,0)|^2$.

   (b) Find the Fourier transform, $A(k)$, of $f(x,0)$. Plot $|A|^2$ for m=1 and m=2.

   (c) Define a reasonable Δx and Δk for this wave packet. Calculate Δx, Δk, and $\Delta x \Delta k$.

6. (a) Derive Eqs. (11.81) and (11.84).

   (b) The spectral line in the decay of an excited state of sodium has a wavelength ($\lambda = 5,893\text{Å}$), and a lifetime of 2×10^{-8} seconds. Calculate its natural line width.

7. A glass block has an index of refraction given by

$$n = 1.350 - 100\text{Å}/\lambda \qquad (11.115)$$

in the visible region.

  (a) What is the angular difference between red light (Use $\lambda = 6,500\text{Å}$) and blue light ($\lambda = 4,500\text{Å}$) in the glass if the light enters with an angle of incidence $\theta = 30°$?

  (b) What are the phase and group velocities for each color light in the glass?

8. Derive Eqs. (11.103)- (11.105) for the group velocity.

9. Solve the quadratic equations for the zeros of the numerator and the denominator of Eq. (11.110) for ϵ.

Chapter 12

Wave Guides and Cavities

12.1   Cylindrical Wave Guides

A **cylindrical wave guide** is a hollow conducting cylinder of any cross-sectional shape. We will consider electromagnetic waves of frequency ω traveling within the cylinder, along its axis in the positive z direction with wave number k. Such a wave can be written in factorized form as

$$\mathbf{E}(\mathbf{r}, t) = \mathbf{E}(\mathbf{r_T})e^{i(kz - \omega t)}. \tag{12.1}$$

The coordinate $\mathbf{r_T}$ represents the two coordinates transverse to the z direction. These would be x and y for a wave guide of rectangular cross-section, or r and θ for a circular wave guide. Equation (12.1) is for a wave with a definite frequency ω. For a more general wave, Eq. (12.1) would represent its Fourier transform.

Because of its simple z and t dependence, the electric field of Eq. (12.1) will satisfy the modified wave equation

$$(\nabla_{\mathrm{T}}^2 - k^2 + \epsilon\mu\omega^2/c^2)\mathbf{E}(\mathbf{r_T})e^{i(kz - \omega t)} = 0, \tag{12.2}$$

where ∇_{T}^2 is the Laplacian in the two transverse variables, and ϵ and μ are the permittivity and permeability of the dielectric inside the wave guide. The common factor $e^{i(kz - \omega t)}$ can be cancelled, and Eq. (12.2) rewritten as

$$\nabla_T^2\mathbf{E}(\mathbf{r_T}) = (k^2 - \epsilon\mu\omega^2/c^2)\mathbf{E}(\mathbf{r_T}). \tag{12.3}$$

This is an eigenvalue equation, with solutions and eigenvalues being determined by the boundary conditions of $\mathbf{E}(\mathbf{r_T})$ at the boundaries of the wave guide. The eigenvalue equation can be written as

$$\nabla_T^2\mathbf{E}(\mathbf{r_T}) = -\gamma^2\mathbf{E}(\mathbf{r_T}), \tag{12.4}$$

with the boundary condition

$$\hat{\mathbf{n}}\times\mathbf{E}|_{\mathbf{S}} = \mathbf{0}. \tag{12.5}$$

329

Here, $\hat{\mathbf{n}}$ is a unit vector normal to the boundary of the wave guide.

A similar derivation shows that $\mathbf{B}$ satisfies the same eigenvalue equation:

$$\nabla_T^2 \mathbf{B}(\mathbf{r_T}) = -\gamma^2 \mathbf{B}(\mathbf{r_T}), \tag{12.6}$$

with the boundary condition

$$\hat{\mathbf{n}} \cdot \mathbf{B}|_\mathbf{S} = 0. \tag{12.7}$$

Only those eigenmodes satisfying the boundary conditions can exist in the wave guide. The boundary conditions on $\mathbf{E}$ and $\mathbf{B}$ are for perfectly conducting walls. Boundary values of $\mathbf{E}$ and $\mathbf{B}$ for good, but not perfect, conductors were discussed in Section 10.6, and will also be considered later in this chapter.

12.1.1 Phase and group velocities in a wave guide

The eigenvalues γ^2 are related to k and ω by

$$\gamma^2 = \epsilon\mu\omega^2/c^2 - k^2. \tag{12.8}$$

k must be real, and therefore k^2 positive, for wave propagation in the wave guide. That means there is a cutoff frequency ω_c below which a given mode cannot propagate. This is given by

$$\omega_c = \gamma c/\sqrt{\epsilon\mu}. \tag{12.9}$$

The frequency and wave number are related by the dispersion relation

$$k^2 = \epsilon\mu(\omega^2 - \omega_c^2)/c^2. \tag{12.10}$$

The phase velocity of the wave in the wave guide is given by

$$v_p = \frac{\omega}{k} = \frac{\omega c}{\sqrt{\epsilon\mu}\sqrt{\omega^2 - \omega_c^2}} = \frac{c}{\sqrt{\epsilon\mu}\sqrt{1 - \omega_c^2/\omega^2}}. \tag{12.11}$$

The group velocity is given by

$$v_g = \frac{d\omega}{dk}. \tag{12.12}$$

From Eq. (12.10), we see that

$$\frac{d\omega}{dk} = \frac{c^2 k}{\epsilon\mu\omega}, \tag{12.13}$$

if we neglect the generally weak frequency dependence of ϵ and μ. Then, the group velocity is

$$v_g = \frac{c^2 k}{\epsilon\mu\omega} = \frac{c}{\sqrt{\epsilon\mu}}\sqrt{1 - \omega_c^2/\omega^2}. \tag{12.14}$$

These results show that the phase velocity in a wave guide is greater than $c/\sqrt{\epsilon\mu}$, and the group velocity is less than $c/\sqrt{\epsilon\mu}$. The two velocities are related by

$$v_g v_p = \frac{c^2}{\epsilon\mu}. \tag{12.15}$$

It is the group velocity, which is always less than $c/\sqrt{\epsilon\mu}$, that is the velocity of wave packets in the wave guide. The frequency dependence of the phase and group velocities for ϵ and μ equal to 1 are plotted in Fig. 12.1.

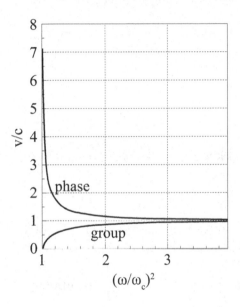

Figure 12.1: Phase and group velocities in a wave guide.

12.2   Eigenmodes in a Wave Guide

The eigenvalue equations (12.4) and (12.6) are really six separate equations, one for each vector component of $\mathbf{E}$ and $\mathbf{B}$. These various components are related by Maxwell's equations. In writing Maxwell's equations we will use the connections

$$\partial_z \rightarrow ik, \quad \partial_t \rightarrow -i\omega, \tag{12.16}$$

which follow from the simple form of the z and t dependence of the wave.

Thus, Maxwell's equations (without charges and currents) can be written as

$$(\boldsymbol{\nabla}_{\mathbf{T}} + ik\hat{\mathbf{k}})\cdot\mathbf{E}(\mathbf{r_T}) = 0 \tag{12.17}$$

$$(\boldsymbol{\nabla}_{\mathbf{T}} + ik\hat{\mathbf{k}})\times\mathbf{E}(\mathbf{r_T}) = \frac{i\mu\omega}{c}\mathbf{H}(\mathbf{r_T}) \tag{12.18}$$

$$(\boldsymbol{\nabla}_{\mathbf{T}} + ik\hat{\mathbf{k}})\cdot\mathbf{H}(\mathbf{r_T}) = 0 \tag{12.19}$$

$$(\boldsymbol{\nabla_T} + ik\hat{\mathbf{k}}) \times \mathbf{H}(\mathbf{r_T}) \;=\; \frac{-i\epsilon\omega}{c}\mathbf{E}(\mathbf{r_T}). \qquad (12.20)$$

We have followed the usual practice here of using the $\mathbf{H}$ field in the wave guide because of the mathematical symmetry in the equations for $\mathbf{E}$ and $\mathbf{H}$.

The fields can be broken up into transverse and longitudinal components as

$$\mathbf{E} = \mathbf{E_T} + \hat{\mathbf{k}}E_z, \quad \mathbf{H} = \mathbf{H_T} + \hat{\mathbf{k}}H_z. \qquad (12.21)$$

Then, Maxwell's equations can be written (in slightly different order) as

$$\boldsymbol{\nabla_T}\cdot\mathbf{E_T} \;=\; -ikE_z \qquad (12.22)$$

$$\boldsymbol{\nabla_T}\cdot\mathbf{H_T} \;=\; -ikH_z \qquad (12.23)$$

$$\boldsymbol{\nabla_T}\times\mathbf{E_T} \;=\; \frac{i\mu\omega}{c}\hat{\mathbf{k}}H_z \qquad (12.24)$$

$$\boldsymbol{\nabla_T}\times\mathbf{H_T} \;=\; \frac{-i\epsilon\omega}{c}\hat{\mathbf{k}}E_z \qquad (12.25)$$

$$ik\hat{\mathbf{k}}\times\mathbf{E_T} - \frac{i\mu\omega}{c}\mathbf{H_T} \;=\; \hat{\mathbf{k}}\times\boldsymbol{\nabla_T}E_z \qquad (12.26)$$

$$ik\hat{\mathbf{k}}\times\mathbf{H_T} + \frac{i\epsilon\omega}{c}\mathbf{E_T} \;=\; \hat{\mathbf{k}}\times\boldsymbol{\nabla_T}H_z. \qquad (12.27)$$

To avoid solving six equations in six unknowns directly, we adopt the following procedure. We first solve the eigenvalue equations (12.4) and (12.6) for the z components of the fields. Then we use the appropriate Maxwell's equations to find the transverse components from the z components.

12.2.1 TEM waves

We first look for solutions with both E_z and H_z equal to zero. The resulting waves are called **TEM waves**, which stands for **T**ransverse **E**lectric and **M**agnetic waves. In this case, $\mathbf{E_T}$ satisfies the equations

$$\boldsymbol{\nabla_T}\cdot\mathbf{E_T} = 0, \quad \boldsymbol{\nabla_T}\times\mathbf{E_T} = 0, \qquad (12.28)$$

with the boundary condition $\hat{\mathbf{n}}\times\mathbf{E_T}|_\mathbf{S} = 0$. This means that the boundary of the wave guide is an equipotential. These are just the electrostatic equations for an empty cavity in a conductor in two dimensions. We showed in Chapter 2 that these equations had the unique solution $\mathbf{E_T} = 0$. This means that, for a TEM wave to exist in the wave guide, more than one conducting surface is required.

From Eqs. (12.28) for a TEM wave, we get

$$\boldsymbol{\nabla_T}\times(\boldsymbol{\nabla_T}\times\mathbf{E_T}) = \boldsymbol{\nabla_T}(\boldsymbol{\nabla_T}\cdot\mathbf{E_T}) - \nabla_T^2\mathbf{E_T} = -\nabla_T^2\mathbf{E_T} = 0. \qquad (12.29)$$

Comparing this with Eq. (12.4) shows that the eigenvalue $\gamma^2 = 0$. Consequently, there is no cutoff frequency for TEM modes, and all frequencies are possible. Also, k and ω in a TEM wave are related by

$$k = \omega\sqrt{\epsilon\mu}/c, \tag{12.30}$$

and

$$v_p = v_g = c/\sqrt{\epsilon\mu}. \tag{12.31}$$

In a TEM wave, both E_z and H_z are zero and, from Eq. (12.26), $\mathbf{H_T}$ is related to $\mathbf{E_T}$ by

$$\mathbf{H_T} = \frac{kc}{\mu\omega}\hat{\mathbf{k}}\times\mathbf{E_T} = \sqrt{\frac{\epsilon}{\mu}}\hat{\mathbf{k}}\times\mathbf{E_T}. \tag{12.32}$$

Equations (12.28) - (12.32) for a TEM wave are the same as the corresponding equations for a plane wave. This means that a TEM wave propagates just like a plane wave, except that the TEM lines of $\mathbf{E}$ terminate on the conductors. We discuss below two common wave guides for TEM waves.

Coaxial wave guide

A **coaxial wave guide** consists of the space between two concentric circular cylinders. The solution to Eqs. (12.28) for this geometry is just that found in solving the electrostatic problem 3a in Chapter 6. The electric field is given (in polar coordinates) by

$$\mathbf{E_T} = \frac{aE_0}{r}\hat{\mathbf{r}}, \tag{12.33}$$

where E_0 is the magnitude of the field at the inner radius a of the wave guide. The $\mathbf{H}$ field inside the coaxial wave guide is

$$\mathbf{H_T} = \sqrt{\frac{\epsilon}{\mu}}\hat{\mathbf{k}}\times\mathbf{E_T} = \sqrt{\frac{\epsilon}{\mu}}\frac{aE_0}{r}\hat{\boldsymbol{\theta}}. \tag{12.34}$$

Parallel-wire wave guide

Another configuration capable of transmitting a TEM wave is that of two parallel straight wires. This configuration was used in antenna wires for TV sets. In practice, the wires can be somewhat flexible as long as the radius of curvature is much greater than the length of the wires, so the straight line approximation is still reasonable. The electric field for this configuration is the same as that for two parallel line charges of opposite charge in electrostatics. The solution for this wave is just that for the corresponding electrostatic problem 4 in Chapter 1. (If you have not solved that problem, you should do so now.)

12.2.2   TM waves

Solutions with $H_z = 0$ everywhere in the wave guide are called **TM waves**, for **T**ransverse **M**agnetic waves. For this case, we want to solve the eigenvalue equation (12.4) for the electric field component E_z. We introduce the scalar field

$$\psi(\mathbf{r_T}) = E_z(\mathbf{r_T}). \tag{12.35}$$

ψ is the solution to the eigenvalue problem

$$\nabla_T^2 \psi(\mathbf{r_T}) = -\gamma^2 \psi(\mathbf{r_T}), \tag{12.36}$$

with the boundary condition

$$\psi(\mathbf{r_T}) = 0 \quad \text{on the boundary.} \tag{12.37}$$

We will solve this problem for some specific geometries in later sections.

Once E_z is known, we can find the remaining components $\mathbf{E_T}$ and $\mathbf{H_T}$ as follows. From Eq. (12.27) (for $H_z = 0$), we get

$$\mathbf{H_T} = \frac{\epsilon \omega}{ck} \hat{\mathbf{k}} \times \mathbf{E_T}, \tag{12.38}$$

relating $\mathbf{H_T}$ to $\mathbf{E_T}$. Using this to eliminate $\mathbf{H_T}$ from Eq. (12.26) results in

$$ik\hat{\mathbf{k}} \times \mathbf{E_T} - \left(\frac{i\mu\omega}{c}\right)\left(\frac{\epsilon\omega}{ck}\right)\hat{\mathbf{k}} \times \mathbf{E_T} = \hat{\mathbf{k}} \times \boldsymbol{\nabla}_\mathbf{T} E_z. \tag{12.39}$$

We apply $\hat{\mathbf{k}}\times$ to each term, and then solve for $\mathbf{E_T}$, reducing Eq. (12.39) to

$$\mathbf{E_T} = \frac{ik}{\gamma^2} \boldsymbol{\nabla}_\mathbf{T} E_z, \tag{12.40}$$

where we have used Eq. (12.8) to introduce the eigenvalue γ^2. Equations (12.40) and (12.38) give all four remaining field components in terms of E_z for the TM case.

12.2.3   TE waves

Solutions with $E_z = 0$ everywhere in the wave guide are called **TE waves**, for **T**ransverse **E**lectric waves. For this case, we want to solve the eigenvalue equation (12.6) for the **H** field component H_z. We again introduce a scalar field

$$\psi(\mathbf{r_T}) = H_z(\mathbf{r_T}). \tag{12.41}$$

ψ is the solution to the eigenvalue problem

$$\nabla_T^2 \psi(\mathbf{r_T}) = -\gamma^2 \psi(\mathbf{r_T}). \tag{12.42}$$

The boundary condition that $\hat{\mathbf{n}} \cdot \mathbf{H} = 0$ is not directly applicable here, but we can find the appropriate boundary condition from Eq. (12.27). We first apply the operation $\hat{\mathbf{k}} \times$ to the equation, resulting in

$$\boldsymbol{\nabla}_{\mathbf{T}} H_z = ik\mathbf{H}_{\mathbf{T}} + \frac{i\epsilon\omega}{c}\hat{\mathbf{k}} \times \mathbf{E}_{\mathbf{T}}. \qquad (12.43)$$

Then, applying $\hat{\mathbf{n}} \cdot$ to this equation, we can use the fact that at the surface of a conductor $\hat{\mathbf{n}}$ dotted into $\mathbf{H}$ gives zero and $\mathbf{E}_{\mathbf{T}}$ is zero, to get

$$\hat{\mathbf{n}} \cdot \boldsymbol{\nabla}_{\mathbf{T}} H_z = 0. \qquad (12.44)$$

Interpreting this in terms of the scalar function ψ gives the boundary condition

$$\partial_n \psi(\mathbf{r}_{\mathbf{T}}) = 0 \quad \text{on the boundary}, \qquad (12.45)$$

where by $\partial_n \psi$ we mean the derivative of ψ in the direction normal to the surface of the wave guide. This is the Neumann boundary condition discussed in Chapter 3.

Once we have solved for ψ, and H_z is known, we can find the remaining components $\mathbf{H}_{\mathbf{T}}$ and $\mathbf{E}_{\mathbf{T}}$ in the same manner as we used in the TM case. This procedure results in

$$\mathbf{H}_{\mathbf{T}} = \frac{ik}{\gamma^2}\boldsymbol{\nabla}_{\mathbf{T}} H_z, \qquad (12.46)$$

and

$$\mathbf{E}_{\mathbf{T}} = \frac{-\mu\omega}{ck}\hat{\mathbf{k}} \times \mathbf{H}_{\mathbf{T}}. \qquad (12.47)$$

These two equations give all four remaining field components in terms of H_z for the TE case.

A general solution can consist of an arbitrary linear combination of all TM and TE modes having frequency ω greater than the cutoff frequency ω_c. Often, one mode can be made dominant by choosing an appropriate antenna to excite that mode in the wave guide. A single mode, that with the lowest cutoff frequency, can be guaranteed if the frequency is chosen to lie between the lowest ω_c and the next higher cutoff frequency.

12.2.4 Summary of TM and TE modes

We summarize in Table 12.1 below the essential results for TM and TE waves to emphasize their similarity and where they differ. For completeness, we have included the power P carried by the guide. We will derive the expressions given for this in Section 12.3.1.

Table 12.1: Basic equations for TM and TE modes in a wave guide.

TM	TE
$H_z = 0, \quad \psi = E_z$	$E_z = 0, \quad \psi = H_z$
$\nabla_T^2 \psi(\mathbf{r_T}) = -\gamma^2 \psi(\mathbf{r_T})$	
$\psi\|_{\text{surface}} = 0$	$\partial_n \psi\|_{\text{surface}} = 0$
$\mathbf{E_T} = \frac{ik}{\gamma^2}\boldsymbol{\nabla}_\mathbf{T}\psi$	$\mathbf{H_T} = \frac{ik}{\gamma^2}\boldsymbol{\nabla}_\mathbf{T}\psi$
$\mathbf{H_T} = \frac{\epsilon\omega}{ck}\hat{\mathbf{k}}\times\mathbf{E_T}$	$\mathbf{E_T} = \frac{-\mu\omega}{ck}\hat{\mathbf{k}}\times\mathbf{H_T}$
$P = \frac{\epsilon\omega k}{8\pi\gamma^2}\int\psi^2 dA$	$P = \frac{\mu\omega k}{8\pi\gamma^2}\int\psi^2 dA$

12.2.5   Rectangular wave guides

We consider a wave guide of rectangular cross-section with dimensions $A\times B$. The TM eigenfunction $\psi = E_z$ in this wave guide can be found using the methods of Chapter 3. Then Eqs. (12.40) and (12.38) can be used to find $\mathbf{E_T}$ and $\mathbf{H_T}$. The results are

TM modes:

$$E_z(x,y) = E_0 \sin(l\pi x/A)\sin(m\pi y/B) \tag{12.48}$$

$$E_x(x,y) = \frac{ikl\pi}{A\gamma^2}E_0 \cos(l\pi x/A)\sin(m\pi y/B) \tag{12.49}$$

$$E_y(x,y) = \frac{ikm\pi}{B\gamma^2}E_0 \sin(l\pi x/A)\cos(m\pi y/B) \tag{12.50}$$

$$H_x(x,y) = \frac{-i\epsilon\omega m\pi}{cB\gamma^2}E_0 \sin(l\pi x/A)\cos(m\pi y/B) \tag{12.51}$$

$$H_y(x,y) = \frac{i\epsilon\omega l\pi}{cA\gamma^2}E_0 \cos(l\pi x/A)\sin(m\pi y/B). \tag{12.52}$$

E_0 is the maximum magnitude of E_z in the wave guide. Each of the five vector components given above should be multiplied by $e^{i(kz-\omega t)}$ to get the full x, y, z, t dependence of the wave.

The field lines of $\mathbf{E}$ and $\mathbf{H}$ can be quite complicated, so it is more useful to characterize each mode by its **nodal surfaces**. These can be plotted on a cross-section of the wave guide as lines along the loci of zeros of the field. For rectangular wave guides, these loci are straight lines corresponding to zeros of the sines and cosines. (See problem 3.)

The mode integers l and m can each be any positive integer. The eigenvalues are given by

$$\gamma_{lm}^2 = \pi^2\left(\frac{l^2}{A^2} + \frac{m^2}{B^2}\right), \tag{12.53}$$

and the cutoff frequencies by

$$\omega_{lm} = \frac{\pi c}{\sqrt{\epsilon \mu}} \left(\frac{l^2}{A^2} + \frac{m^2}{B^2} \right)^{\frac{1}{2}}. \tag{12.54}$$

The lowest TM cutoff frequency is

$$\omega_{11} = \frac{\pi c}{\sqrt{\epsilon \mu}} \left(\frac{1}{A^2} + \frac{1}{B^2} \right)^{\frac{1}{2}}. \tag{12.55}$$

Frequencies lower than this cannot propagate in the TM mode. ω is the angular frequency of oscillation in units of sec^{-1}. Frequencies are usually stated as ν in units of **hertz** (Hz). One Hz is one full oscillation per second. This frequency is related to ω by

$$\nu = \omega/2\pi. \tag{12.56}$$

The TM cutoff frequencies in Hz are

$$\nu_{lm} = \frac{c}{2\sqrt{\epsilon \mu}} \left(\frac{l^2}{A^2} + \frac{m^2}{B^2} \right)^{\frac{1}{2}}. \tag{12.57}$$

TE modes:

The fields in the TE modes in the same rectangular wave guide are given by

$$
\begin{aligned}
H_z(x,y) &= H_0 \cos(l\pi x/A) \cos(m\pi y/B) & (12.58) \\
H_x(x,y) &= \frac{-ikl\pi}{A\gamma^2} H_0 \sin(l\pi x/A) \cos(m\pi y/B) & (12.59) \\
H_y(x,y) &= \frac{-ikm\pi}{B\gamma^2} H_0 \cos(l\pi x/A) \sin(m\pi y/B) & (12.60) \\
E_x(x,y) &= \frac{i\mu\omega m\pi}{cB\gamma^2} H_0 \cos(l\pi x/A) \sin(m\pi y/B) & (12.61) \\
E_y(x,y) &= \frac{i\mu\omega l\pi}{cA\gamma^2} H_0 \sin(l\pi x/A) \cos(m\pi y/B). & (12.62)
\end{aligned}
$$

Here H_0 is the maximum magnitude of H_z in the wave guide. The cutoff frequencies for TE modes are the same as those for corresponding TM modes, except that either l or m can now equal zero. Both cannot be zero. This would correspond to a uniform **H** field with no **E** field, and no power would be transmitted.

The lowest cutoff frequency of a rectangular wave guide will always be that of a TE mode, with l (for $A > B$) or m (for $A < B$) equalling zero.

Neither l nor m can be zero for a TM mode. For the lowest TE mode, the field components simplify to (assuming $A > B$)

$$H_z(x, y) = H_0 \cos(\pi x/A) \tag{12.63}$$

$$H_x(x, y) = \frac{-ikA}{\pi} H_0 \sin(\pi x/A) \tag{12.64}$$

$$E_y(x, y) = \frac{i\mu\omega A}{c\pi} H_0 \sin(\pi x/A), \tag{12.65}$$

with the other components being zero. The cutoff frequency for this lowest mode is

$$\nu_{10} = \frac{c}{2\sqrt{\epsilon\mu}A}. \tag{12.66}$$

12.2.6   Circular wave guides

We next consider a wave guide of circular cross-section with radius R. The eigenvalue problem for this is just the same as that for the electrostatic potential inside a cylinder of finite length that we solved in Section 4.3.3 in terms of Bessel functions. The r and θ dependence of the eigenfunctions for the circular wave guide are those given in Eq. (4.231):

$$\psi_{ml}(r, \theta) = J_m(\gamma_{ml}r) \cos(m\theta - \alpha), \tag{12.67}$$

where $J_m(\gamma_{ml}r)$ is a Bessel function of order m and γ_{ml} is the eigenvalue. The nodal surfaces (represented on a cross-section of the circular wave guide) will be radial lines at the zeros of $\cos(m\theta - \alpha)$, and circles at the zeros of the Bessel function. Changing the angle α would rotate the radial nodal lines, but have no other effect, so the simplest choice is usually $\alpha = 0$.

The eigenvalues are determined from the boundary conditions for TM and TE waves. For TM waves, the condition is $J_m(\gamma_{ml}R) = 0$. This requires that

$$\textbf{TM}: \qquad\qquad \gamma_{ml} = j_{ml}/R, \tag{12.68}$$

where j_{ml} is the lth zero of J_m, so $J_m(j_{ml}) = 0$. The first four Bessel function zeros are listed in Table 4.1 and Table 12.2 below. For TE waves, the boundary condition is that $J'_m(\gamma'_{ml}R) = 0$. For this condition, the eigenvaues are given by

$$\textbf{TE}: \qquad\qquad \gamma'_{ml} = j'_{ml}/R, \tag{12.69}$$

where j'_{ml} is the lth zero of the derivative of the Bessel function, so $J'_{ml}(j'_{ml}) = 0$. We have introduced the notation γ' to denote the eigenvalue for **TE** waves. The first four zeros of Bessel functions and their derivatives are listed in Table 12.2. Values for $l > 4$ can be estimated by observing that the spacing between succesive zeros or derivative zeros approaches π as l gets larger.

The first zero of J'_0 is for $j'_{01} = 0$. As in the rectangular TE case, this gives a constant **H** field with no **E** field and does not result in power transmission. The lowest cutoff frequency is for the TE$_{11}$ mode with $\gamma_{11} = 1.8412/R$.

Table 12.2: Bessel function zeros j_{ml} and derivative zeros j'_{ml}

l	j_{0l}	j'_{0l}	j_{1l}	j'_{1l}	j_{2l}	j'_{2l}
1	2.4048	0	3.8317	1.8412	5.1356	3.0542
2	5.5200	3.8317	7.0156	5.3314	8.4172	6.7061
3	8.6537	7.0156	10.1735	8.5363	11.6198	9.9695
4	11.792	10.1735	13.3237	11.7060	14.7960	13.1704

12.3 Power Transmission and Attenuation in Wave Guides

12.3.1 Power transmitted

The power transmitted by a wave guide is given by integrating the component of the Poynting vector pointing in the z direction over the cross-sectional area. This is

$$P = \int \mathbf{S} \cdot d\mathbf{A} = \frac{c}{8\pi} \int (\mathbf{E_T^*} \times \mathbf{H_T}) \cdot d\mathbf{A}. \tag{12.70}$$

For TM modes we can substitute for $\mathbf{E_T}$ and $\mathbf{H_T}$ from Table 12.1 to get

$$
\begin{aligned}
P &= \frac{\epsilon\omega}{8\pi ck} \int |\mathbf{E_T}|^2 dA \\
&= \frac{\epsilon\omega k}{8\pi\gamma^4} \int (\boldsymbol{\nabla_T}\psi^*) \cdot (\boldsymbol{\nabla_T}\psi) dA.
\end{aligned} \tag{12.71}
$$

The integral over the cross-sectional area can be simplified using the divergence theorem in two dimensions as follows:

$$
\begin{aligned}
\int (\boldsymbol{\nabla_T}\psi^*) \cdot (\boldsymbol{\nabla_T}\psi) dA &= \int [\boldsymbol{\nabla_T} \cdot (\psi^* \boldsymbol{\nabla_T}\psi) - \psi^* \nabla_T^2 \psi] dA \\
&= \gamma^2 \int |\psi|^2 dA.
\end{aligned} \tag{12.72}
$$

In the last step above, we used the divergence theorem and then the fact that either ψ (for TM) or $\hat{\mathbf{n}} \cdot \boldsymbol{\nabla_T}\psi$ (for TE) vanishes on the entire boundary of the wave guide. This now gives for the transmitted power

$$\mathbf{TM}: \qquad P = \frac{\epsilon\omega k}{8\pi\gamma^2} \int |E_z|^2 dA. \tag{12.73}$$

A similar derivation for TE modes involves only the replacements $E_z \to H_z$ and $\epsilon \to \mu$, so

$$\mathbf{TE}: \qquad P = \frac{\mu\omega k}{8\pi\gamma^2} \int |H_z|^2 dA. \tag{12.74}$$

For rectangular wave guides, the integral above is easily done using the fact that the average value of $\sin^2$ or $\cos^2$ is $\frac{1}{2}$. Then the transmitted power for a rectangular wave guide with dimensions $A \times B$ is

$$\textbf{TM}: \qquad P = \frac{\epsilon \omega k E_0^2 AB}{32\pi\gamma^2} \qquad\qquad (12.75)$$

$$\textbf{TE}: \qquad P = \frac{\mu \omega k H_0^2 AB}{32\pi\gamma^2} \quad (\times 2 \text{ if } \mathrm{lm} = 0), \qquad (12.76)$$

where E_0 and H_0 are the maximum magnitudes of the fields in the wave guide. If either of the mode integers l or m is zero for a TE mode, the power is twice as large because only one factor of $\frac{1}{2}$ enters.

For circular wave guides, the angular integration gives a factor of π, unless $m = 0$, when the factor is 2π. The power is then given by a radial integral of $r[J_m]^2$:

$$\textbf{TM}: \quad P = \frac{\epsilon \omega k E_0^2}{8\gamma_{lm}^2} \int_0^R [J_m(\gamma_{lm}r)]^2 r dr \quad (\times 2 \text{ if } m = 0) \qquad (12.77)$$

$$\textbf{TE}: \quad P = \frac{\mu \omega k H_0^2}{8\gamma_{lm}'^2} \int_0^R [J_m(\gamma_{lm}'r)]^2 r dr \quad (\times 2 \text{ if } m = 0). \qquad (12.78)$$

These Bessel function integrals are just the normalization integrals for Bessel functions, given in Eqs. (4.243) and (4.3.3). So for circular wave guides, the power can be written as

$$\textbf{TM}: \quad P = \frac{\epsilon \omega k R^2 E_0^2}{16\gamma_{lm}^2} [J_m'(\gamma_{lm}R)]^2 \quad (\times 2 \text{ if } m = 0) \qquad (12.79)$$

$$\textbf{TE}: \quad P = \frac{\mu \omega k R^2 H_0^2}{16\gamma_{lm}'^2} \left[1 + \left(\frac{m\pi}{\gamma_{lm}'R}\right)^2\right] [J_m(\gamma_{lm}'R)]^2 \quad (\times 2 \text{ if } m = 0). \quad (12.80)$$

12.3.2 Losses and attenuation

So far, we have assumed that the walls of the wave guide are perfect conductors for which there would be no losses. The actual case is usually that the walls are good, but not perfect, conductors, so there will be power loss as we found in Section 12.2.3. We assume that the conductivity is good enough that the large Δ approximation of Section 12.1.2 is valid, and also good enough that the homogeneous boundary conditions in Table 12.1 are still satisfied.

The power loss per surface area dA of the conducting wall is given by Eq. (12.44) to be

$$\frac{dP}{dA} = -\frac{c}{8\pi}\sqrt{\frac{\omega\mu_c}{8\pi\sigma_c}} |\hat{\mathbf{n}} \times \mathbf{H}_{\text{surface}}|^2 = -\frac{c}{8\pi}\sqrt{\frac{\omega\mu_c}{8\pi\sigma_c}} |\mathbf{H}_{\text{surface}}|^2. \qquad (12.81)$$

We have written μ_c and σ_c here to emphasize that they are the permeability and conductivity of the walls of the wave guide. The second equality in Eq.

(12.81) follows because of the boundary condition that $\hat{\mathbf{n}} \cdot \mathbf{H} = 0$ at the surface of the wave guide.

The power lost in length dz of the wave guide is given by the integral of $\frac{dP}{dA}$ around the perimeter of the wave guide, so

$$\frac{dP}{dz} = -\frac{c}{8\pi}\sqrt{\frac{\omega\mu_c}{8\pi\sigma}}\oint|\mathbf{H}_{\text{surface}}|^2 dl. \tag{12.82}$$

Both the power P and the rate of power loss $\frac{dP}{dz}$ are proportional to the field amplitude squared, and therefore are proportional to each other. This means that we can write

$$\frac{dP}{dz} = -\frac{1}{L_{\text{atten}}}P(z). \tag{12.83}$$

This differential equation has the solution

$$P(z) = P_0 e^{-z/L_{\text{atten}}}, \tag{12.84}$$

which is the characteristic equation of an attenuating wave with **attenuation length** L_{atten}. Equation (12.84) shows that the attenuation length is given by the ratio of power in the wave guide at any point to the rate of power loss per unit length at that point:

$$L_{\text{atten}} = \frac{P}{-\frac{dP}{dz}}. \tag{12.85}$$

For a rectangular wave guide, the integrals over $\sin^2$ or $\cos^2$ just give a factor of $\frac{1}{2}$, and the attenuation length can be written as (see problem 6)

$$L_{\text{atten}} = \left[\frac{\pi\mu\sigma_c}{2\epsilon\mu_c}\right]^{\frac{1}{2}}\left[\frac{\sqrt{1-\omega_c^2/\omega^2}}{\sqrt{\omega}}\right]G_{lm}, \tag{12.86}$$

$$\textbf{TM}: \quad G_{lm} = \left[\frac{AB(l^2B^2 + m^2A^2)}{l^2B^3 + m^2A^3}\right] \tag{12.87}$$

$$\textbf{TE}: \quad G_{lm} = \left[\frac{AB(l^2B^2 + m^2A^2)}{\eta AB(l^2B + m^2A) + (\omega_c/\omega)^2(l^2B^3 + m^2A^3)}\right]. \tag{12.88}$$

We see in Eq. (12.86) that the attenuation length is given by a product of three factors. The first depends on the parameters μ and ϵ of the dielectric within the wave guide, and the conductivity σ_c and permeability μ_c of the conducting walls. The second factor gives the dominant frequency dependence of the attenuation. The third factor, G, depends on the dimensions of the wave guide. This factor is given for each type of wave in Eqs. (12.87) and (12.88). For a TM wave, it has additional frequency dependence resulting from the z component of $\mathbf{H}$. The parameter η in Eq. (12.88) equals 1, unless l or m equals zero, in which case $\eta = \frac{1}{2}$.

12.4   Cylindrical Cavities

A cylindrical cavity has a constant cross-sectional shape like a cylindrical wave guide, but is cut off at each end by a planar conducting surface. The equations for the cavity are the same as those for a wave guide, except for the z dependence. Maxwell's equations for a cavity can be written as

$$\boldsymbol{\nabla_T \cdot E_T} = -\partial_z E_z \tag{12.89}$$

$$\boldsymbol{\nabla_T \cdot H_T} = -\partial_z H_z \tag{12.90}$$

$$\boldsymbol{\nabla_T \times E_T} = \frac{i\mu\omega}{c}\hat{k}H_z \tag{12.91}$$

$$\boldsymbol{\nabla_T \times H_T} = \frac{-i\epsilon\omega}{c}\hat{k}E_z \tag{12.92}$$

$$\hat{k}\times\partial_z E_T - \frac{i\mu\omega}{c}H_T = \hat{k}\times\boldsymbol{\nabla_T}E_z \tag{12.93}$$

$$\hat{k}\times\partial_z H_T + \frac{i\epsilon\omega}{c}E_T = \hat{k}\times\boldsymbol{\nabla_T}H_z. \tag{12.94}$$

12.4.1   Resonant modes of a cavity

A cavity of length L has the z dependent boundary conditions

$$\textbf{TM}: \quad \partial_z E_z = 0, \quad \text{at } z=0 \quad \text{and } z=L, \tag{12.95}$$

$$\textbf{TE}: \quad H_z = 0, \quad \text{at } z=0 \quad \text{and } z=L. \tag{12.96}$$

The boundary condition on E_z follows from Eq. (12.89) since $\mathbf{E_T}$ is zero at the conducting ends. The H_z boundary condition follows directly from Eq. (12.7).

To satisfy these boundary conditions at each end of the cavity, the z eigenfunctions must be $\cos(n\pi z/L)$ for a TM wave, and $\sin(n\pi z/L)$ for a TE wave, where n is any positive integer. $n=0$ is only allowed for the TM wave. In either case, the eigenvalues of the operator ∂_z^2 are $-(n\pi/L)^2$. With these changes, the wave equation of Eq. (12.3) becomes

$$\nabla_T^2\mathbf{E}(\mathbf{r_T}) = \left(\frac{n^2\pi^2}{L^2} - \frac{\epsilon\mu\omega^2}{c^2}\right)\mathbf{E}(\mathbf{r_T}). \tag{12.97}$$

This eigenvalue problem for the fields in a cavity has the same boundary conditions at the side walls as in Section 12.2.2 for TM waves and Section 12.2.3 for TE waves. The eigenvalue equation in the transverse variables can be written as

$$\nabla_T^2\psi(\mathbf{r_T}) = -\gamma^2\psi(\mathbf{r_T}) = \left(\frac{n^2\pi^2}{L^2} - \frac{\epsilon\mu\omega^2}{c^2}\right)\psi(\mathbf{r_T}), \tag{12.98}$$

where

$$\psi = E_z \quad \text{for TM waves}, \quad \text{and } \psi = H_z \quad \text{for TE waves}. \tag{12.99}$$

From this, we see that the resonant frequencies in a cavity are given by

$$\omega_{lmn} = \frac{c}{\sqrt{\epsilon\mu}} \left[\gamma_{lm}^2 + \left(\frac{n\pi}{L}\right)^2 \right]^{\frac{1}{2}}. \tag{12.100}$$

The fields for resonant electromagnetic waves in a cavity are given by

$$\textbf{TM}: \qquad E_z = \cos(n\pi z/L)\psi \tag{12.101}$$
$$\mathbf{E_T} = -\frac{n\pi}{L\gamma^2}\sin(n\pi z/L)\boldsymbol{\nabla_T}\psi \tag{12.102}$$

$$\mathbf{H_T} = \frac{i\epsilon\omega}{c\gamma^2}\cos(n\pi z/L)\hat{\mathbf{k}}\times\boldsymbol{\nabla_T}\psi \tag{12.103}$$

$$\textbf{TE}: \qquad H_z = \sin(n\pi z/L)\psi \tag{12.104}$$
$$\mathbf{H_T} = \frac{n\pi}{L\gamma^2}\cos(n\pi z/L)\boldsymbol{\nabla_T}\psi \tag{12.105}$$

$$\mathbf{E_T} = \frac{-i\mu\omega}{c\gamma^2}\sin(n\pi z/L)\hat{\mathbf{k}}\times\boldsymbol{\nabla_T}\psi. \tag{12.106}$$

The derivation of these equations from the cavity Maxwell equations (12.89) - (12.94) is given as a problem.

12.4.2 Rectangular cavity

We consider a rectangular cavity of dimensions $A\times B\times C$. For a TM wave, the field E_z must vanish at the side walls, and is given by

$$\textbf{TM}: \quad E_z = E_0 \sin(l\pi x/A)\sin(m\pi y/B)\cos(n\pi z/C). \tag{12.107}$$

The other TM fields are found using Eqs. (12.102) and (12.103). For a TE wave, the normal derivative of H_z must vanish at the side walls, and it is given by

$$\textbf{TE}: \quad H_z = H_0 \cos(l\pi x/A)\cos(m\pi y/B)\sin(n\pi z/C). \tag{12.108}$$

The other TE fields are found using Eqs. (12.105) and (12.106). In these equations, the time dependence $e^{-i\omega t}$ is understood.

For either TM or TE, the resonant frequencies of the cavity are

$$\nu_{lmn} = \frac{c}{2\sqrt{\epsilon\mu}} \left[\left(\frac{l}{A}\right)^2 + \left(\frac{m}{B}\right)^2 + \left(\frac{n}{C}\right)^2 \right]^{\frac{1}{2}}. \tag{12.109}$$

We see from Eq. (12.107) that only n can be zero for a TM wave. It appears from Eq. (12.108) that both l and m can be zero for a TE wave, but this is not the case. If both are zero, then all the transverse components of $\mathbf{E}$ and $\mathbf{H}$ will vanish. The result would be an oscillating H_z field, but this is not considered an electromagnetic standing wave. Thus, all cases of resonance can have at most one of the three integers l, m, n equal to zero.

The distinction between TM and TE waves in a rectangular cavity is ambiguous. A simple relabeling of the axes could change one type of mode to the other. For any particular orientation there will be degenerate TE and TM modes (two independent modes with the same frequency), except when one of the indices is zero. Degeneracies can also occur whenever the ratio of any two dimensions is a rational number. When there is a degeneracy, any linear combination of the degenerate modes can resonate in the cavity, so the fields in the cavity are not uniquely determined. The lowest resonant frequency in a rectangular cavity will have the integer corresponding to the shortest side equal to zero, with the other two integers being 1. This lowest mode will be non-degenerate, with a unique field distribution, unless two or all three sides are equal.

12.4.3   Circular cylindrical cavity

We consider a cavity of length L with circular cross-section of radius R. The eigenvalue problem in the transverse variables r and θ is the same as considered in Section 12.2.6 for a circular wave guide. The resulting solution for the fields E_z or H_z inside the cavity are

$$\textbf{TM}: \quad E_z = E_0 J_m(\gamma_{lm} r)\cos(m\theta)\cos(n\pi z/L), \quad \gamma_{lm} = j_{lm}/R \qquad (12.110)$$

$$\textbf{TE}: \quad H_z = H_0 J_m(\gamma'_{lm} r)\cos(m\theta)\sin(n\pi z/L), \quad \gamma'_{lm} = j'_{lm}/R. \qquad (12.111)$$

For simplicity, we have left out a possible $\sin(m\theta)$ term. The transverse fields for each case are given by Eqs. (12.102) - (12.103) and (12.105) - (12.106).

The resonant TM frequencies of the cavity are

$$\textbf{TM}: \quad \nu_{lmn} = \frac{c}{2\sqrt{\epsilon\mu}}[(j_{lm}/\pi R)^2 + (n/L)^2]^{\frac{1}{2}}. \qquad (12.112)$$

The TE resonant frequencies are given by the same equation with the substitution of j'_{lm} for j_{lm}, but n cannot be zero for a TE mode. The first four zeros of J_m and of J'_m are listed in Table 12.2.

The lowest frequency TM mode is

$$\nu_{100} = \frac{2.4048 c}{2\pi R\sqrt{\epsilon\mu}}. \qquad (12.113)$$

The lowest frequency TE mode is

$$\nu'_{111} = \frac{c}{2\sqrt{\epsilon\mu}}\left[\left(\frac{1.8412}{\pi R}\right)^2 + \left(\frac{1}{L}\right)^2\right]^{\frac{1}{2}}. \qquad (12.114)$$

We have used the notation ν' to denote the frequency of a TE mode. To compare the two modes, we rewrite ν'_{111} as

$$\nu'_{111} = \frac{1.8412c}{2\pi R\sqrt{\epsilon\mu}}\left[1 + 2.9914\left(\frac{R}{L}\right)^2\right]^{\frac{1}{2}}. \tag{12.115}$$

For $L < 2.06R$, the TE$_{111}$ mode will have the lowest frequency. This mode is useful because it can be fine tuned by varying the length of the cavity.

12.4.4 Electromagnetic energy in a cavity

The energy in a cavity can be found by integrating the energy density

$$u = \frac{1}{16\pi}(\mathbf{E}^*\cdot\mathbf{D} + \mathbf{B}^*\cdot\mathbf{H}) \tag{12.116}$$

over the volume of the cavity. For a TM wave, the energy is

$$\begin{aligned}\mathbf{TM}:\quad U &= \frac{1}{16\pi}\int dA \int dz[\epsilon|E_z|^2 + \epsilon|\mathbf{E_T}|^2 + \mu|\mathbf{H_T}|^2]\\ &= \frac{1}{32\pi}\int dA\left\{\epsilon|\psi|^2 + \left[\epsilon\left(\frac{n\pi}{L}\right)^2 + \mu\left(\frac{\epsilon\omega}{c\gamma^2}\right)^2\right]|\nabla\psi|^2\right\}\\ &= \frac{1}{32\pi}\int dA\left\{\epsilon|\psi|^2 + \frac{\epsilon}{\gamma^4}\left[2\left(\frac{n\pi}{L}\right)^2 + \gamma^2\right]|\nabla\psi|^2\right\}\\ &= \frac{\epsilon L}{16\pi}\left[1 + \left(\frac{n\pi}{\gamma L}\right)^2\right]\int|\psi|^2 dA \quad (\times 2 \text{ if n}=0). \end{aligned} \tag{12.117}$$

In deriving this result, we did the z integration by using the fact that $\cos^2(n\pi z/L)$ averages to $\frac{1}{2}$ unless $n=0$. We also used Eq. (12.72) to replace $|\nabla\psi|^2$ by $\gamma^2|\psi|^2$ in the integral. For the energy in a TE wave, ϵ is replaced by μ in the final result, and n cannot $= 0$, so

$$\mathbf{TE}:\quad U = \frac{\mu L}{16\pi}\left[1 + \left(\frac{n\pi}{\gamma L}\right)^2\right]\int|\psi|^2 dA \quad (n\neq 0). \tag{12.118}$$

For rectangular cavities the remaining surface integral can be done by using the average values of $\sin^2$ and $\cos^2$. This leads to a factor of $AB/4$ if neither l nor m is zero. If either l or m is zero, then the factor is $AB/2$. Thus the energy in a rectangular cavity of dimensions $A\times B\times C$ is

$$\mathbf{TM}:\quad U = \frac{\epsilon E_0^2 ABC}{64\pi}\left[\frac{\left(\frac{l}{A}\right)^2 + \left(\frac{m}{B}\right)^2 + \left(\frac{n}{C}\right)^2}{\left(\frac{l}{A}\right)^2 + \left(\frac{m}{B}\right)^2}\right] \quad (\times 2 \text{ if n}=0) \tag{12.119}$$

$$\textbf{TE}: \quad U = \frac{\mu H_0^2 ABC}{64\pi} \left[\frac{\left(\frac{l}{A}\right)^2 + \left(\frac{m}{B}\right)^2 + \left(\frac{n}{C}\right)^2}{\left(\frac{l}{A}\right)^2 + \left(\frac{m}{B}\right)^2} \right] \quad (\times 2 \text{ if } \mathrm{lm} = 0). \quad (12.120)$$

In these equations, E_0 is the amplitude of the z-component of the electric field when $H_z = 0$ (TM), while H_0 is the amplitude of the H_z when $E_z = 0$ (TE). If l, m, n are the same for each mode, the two modes will have the same frequency and may be added as degenerate modes. Then, a more general result for the energy without regard to TE or TM is

$$U = \frac{(\epsilon E_0^2 + \mu H_0^2)ABC}{64\pi} \left[\frac{\left(\frac{l}{A}\right)^2 + \left(\frac{m}{B}\right)^2 + \left(\frac{n}{C}\right)^2}{\left(\frac{l}{A}\right)^2 + \left(\frac{m}{B}\right)^2} \right] \quad (\times 2 \text{ if } \mathrm{lmn} = 0). \quad (12.121)$$

The electromagnetic energy in circular cylindrical cavities is given by (see problem 11.13)

$$\textbf{TM} \quad U = \epsilon E_0^2 \left(\frac{LR^2}{32} \right) \left[1 + \left(\frac{n\pi}{\gamma L} \right)^2 \right] [J_m'(\gamma R)]^2 \quad (12.122)$$

$$\textbf{TE} \quad U = \mu H_0^2 \left(\frac{LR^2}{32} \right) \left[1 + \left(\frac{n\pi}{\gamma' L} \right)^2 \right] \left[1 - \left(\frac{n}{\gamma' R} \right)^2 \right] [J_m(\gamma' R)]^2. \quad (12.123)$$

The energy is doubled if $m = 0$, and another factor of 2 enters if $n = 0$ for a TM mode.

12.4.5   Power loss, quality factor

The rate of energy loss in a cavity is given by integrating the power loss per unit area given by Eq. (12.81) over the surface of the cavity:

$$\frac{dU}{dt} = -\frac{c}{8\pi} \sqrt{\frac{\omega \mu_c}{8\pi \sigma_c}} \int |\mathbf{H}_{\text{surface}}|^2 dA. \quad (12.124)$$

The rate of energy loss and the energy are each proportional to the field amplitude squared, and so are proportional to each other. This proportionality can be used to define a constant factor Q, called the **Quality factor** (or just Q factor), by

$$Q = -\frac{2\pi U}{T \frac{dU}{dt}} = -\frac{\omega U}{\frac{dU}{dt}}, \quad (12.125)$$

where T is the period of oscillation of the standing wave in the cavity. The Q factor defined in this way is a dimensionless factor equal to the ratio of the energy in the cavity to 2π times the energy loss in one period of oscillation. The term "quality" is a natural one because the larger the quality factor of a cavity, the longer it will keep the electromagnetic energy contained within the cavity. The factor 2π is inserted into the definition of the quality factor to make the equations simpler in terms of the angular frequency ω.

Equation (12.125) leads to the time dependence of the energy in the cavity

$$U = U_0 e^{-\frac{\omega t}{Q}} = U_0 e^{-t/\tau}, \tag{12.126}$$

so the energy decays with a time constant

$$\tau = Q/\omega. \tag{12.127}$$

Because of this decay the electric field E (and also H) will have the time dependence

$$E(t) = E_0 e^{-i\omega_0 t} e^{-(\omega_0 t/2Q)}, \tag{12.128}$$

where ω_0 is the undamped frequency of the normal mode. In this way, the decay modifies the frequency dependence of the fields in the cavity. We can see this by taking the Fourier transform of $E(t)$:

$$\begin{aligned}
\mathcal{E}(\omega) &= \frac{E_0}{2\pi} \int_0^\infty e^{-t[i(\omega_0 - \omega) + (\omega_0/2Q)]} dt \\
&= \frac{E_0}{2\pi} \left[\frac{1}{i(\omega_0 - \omega) + (\omega_0/2Q)} \right].
\end{aligned} \tag{12.129}$$

The energy depends on $|E|^2$, so its frequency dependence will be

$$U(\omega) = \frac{U_0 (\omega_0/2Q)^2}{(\omega - \omega_0)^2 + (\omega_0/2Q)^2}. \tag{12.130}$$

The energy is plotted as a function of angular frequency in Fig. 11.2. We can see that, with decay, there is a spread of frequencies for the resonance.

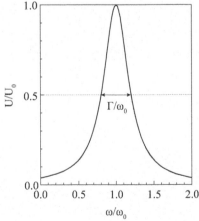

Figure 12.2: The energy in a cavity as a function of the frequency. The full width of the line shape is shown by the double arrow.

The shape of the frequency dependence in Fig. 11.2 is called a **Lorentzian**

line shape. The Lorentzian line shape is typical for any resonant state that decays exponentially. For instance, it also describes the frequency dependent shape of atomic spectral lines. Even though the atomic decay is calculated using quantum mechanics, the Fourier transform mathematics is the same, leading to the same line shape as in our classical decay.

The frequency spread of the decaying resonance is characterized by a **Full Width at Half Maximum** (**FWHM**) Γ, defined as the distance in ω between the two points on the curve where U equals half its maximum value U_0. We can see from Eq. (12.130) that $U = U_0/2$ when the two terms in the denominator are equal, so the width is given by

$$\Gamma = \omega_0/Q. \qquad (12.131)$$

Traditionally, the resonance spread is more frequently described in terms of the **Half Width at Half Maximum** (**HWHM**), $\Gamma/2$. We can see from the above results that a higher Q factor produces a resonance that is purer in frequency. However, too high a Q factor could make it difficult to tune the cavity to locate and stay on the resonance. The optimal value of Q depends on the use that is to be made of the cavity.

12.5 Problems

1. A coaxial wave guide consists of two concentric copper cylinders (the conductivity in gaussian units is related to the SI unit "siemen" by $\sigma_{\text{gaussian}} = (1/4\pi\epsilon_0)\sigma_{\text{SI}} = 0.54\times10^{18} \text{ sec}^{-1}$) with an inner cylinder of radius A=0.10 cm, and an outer cylinder of inner radius B=0.40 cm. The dielectric between the cylinders has a permittivity ϵ=2.0, and permeability μ=1.0. The outer cylinder is grounded, and a potential V_0=120 volts (convert this to esu), oscillating at a frequency ν=12 MHz, is applied at one point on the inner cylinder.

    (a) Find the fields $\mathbf{E}(r,\theta,z,t)$ and $\mathbf{H}(r,\theta,z,t)$ for a wave travelling between the cylinders in the positive z direction.

    (b) Find the power (convert it to Watts) transmitted by this wave.

    (c) Find the attenuation length for this wave.

2. A parallel wire wave guide consists of two long copper wires, each of radius $a = 0.10$ cm, a distance $2d = 0.80$ cm apart. (The algebra in this problem can get complicated. You should assume $a<<d$ wherever useful.) A potential difference V_0=120 volts (convert this to esu), oscillating at a frequency ν=12 MHz, is applied to the wires.

    (a) Find the fields $\mathbf{E}(\mathbf{r},t)$ and $\mathbf{H}(\mathbf{r},t)$ outside the wires for a wave travelling in the positive z direction.

    (b) Find the power transmitted by this wave.

    (c) Find the attenuation length for this wave.

3. A rectangular wave guide with copper walls has cross-section dimensions 4.0×6.0 mm. It is filled with air.

    (a) Find the first five TM cutoff frequencies (in Hz) for this wave guide.

    (b) Sketch nodal planes of E_z for each of these TM modes.

    (c) Find the first five TE cutoff frequencies for this wave guide.

    (d) Sketch nodal planes of H_z for each of these TE modes.

4. (a) The magnitude of E_z at the center of the wave guide in the preceding problem is 6,000 volts/meter (convert to esu). Find the power transmitted by this wave guide for a TM wave with a frequency half way between the two lowest TM cutoff frequencies.

    (b) Find the attenuation length for this wave.

5. (a) The magnitude of H_z at a corner of the wave guide in problem 12.3 is 0.20 Gauss. Find the power transmitted by this wave guide for a TE wave with a frequency half way between the two lowest TE cutoff frequencies.

(b) Find the attenuation length for this wave.

6. Derive Eqs. (12.86) - (12.88) for the attenuation length in a rectangular wave guide.

7. A circular wave guide with copper walls has a radius $R=4.0$ mm. It is filled with air.

   (a) Find the first five TM cutoff frequencies (in Hz) for this wave guide.

   (b) Sketch nodal planes of E_z for each of these TM modes.

   (c) Find the first five TE cutoff frequencies for this wave guide.

   (d) Sketch nodal planes of H_z for each of these TE modes.

8. The magnitude of E_z at the center of the wave guide in the preceding problem is 6,000 volts/meter (convert to esu). Find the power transmitted by this wave guide for a TM wave with a frequency half way between the two lowest TM cutoff frequencies.

9. Derive Eqs. (12.101) - (12.106) for the **E** and **H** fields in a circular wave guide.

10. A rectangular cavity (with ϵ and $\mu =1$) has dimensions $2.0 \times 3.0 \times 4.0$ cm.

    (a) Find the first six resonant frequencies of this cavity, and identify any degenerate modes.

    (b) Write down the **E** and **H** fields for the lowest frequency resonant mode.

11. A cubical cavity of dimensions $2 \times 2 \times 2$ cm has copper walls. Find the fundamental frequency and Q value for that frequency.

12. "And they shall make an ark of acacia-wood: two cubits and a half shall be the length thereof, and a cubit and a half the breadth thereof, and a cubit and a half the height thereof. And thou shalt overlay it with pure gold." (Exodus XXV, 10-11). Find the fundamental resonant frequency and half width for this biblical cavity. Use the length from your fingertip to your elbow as a measure of a biblical cubit.

13. A circular cavity (with ϵ and $\mu =1$) has a radius $R=2.0$ cm, and a length $L=3.0$ cm.

    (a) Find the first six resonant frequencies of this cavity, and identify the type (TM or TE) of mode.

    (b) Sketch nodal surfaces for each of these six modes.

Chapter 13

Electromagnetic Radiation and Scattering

13.1   Wave Equation with Sources.

In the presence of time dependent charge and current sources, Maxwell's equations are:

$$\boldsymbol{\nabla}\cdot\mathbf{E} \;=\; 4\pi\rho \tag{13.1}$$

$$\boldsymbol{\nabla}\times\mathbf{E} \;=\; -\frac{1}{c}\partial_t\mathbf{B} \tag{13.2}$$

$$\boldsymbol{\nabla}\cdot\mathbf{B} \;=\; 0 \tag{13.3}$$

$$\boldsymbol{\nabla}\times\mathbf{B} \;=\; \frac{4\pi\mathbf{j}}{c} + \frac{1}{c}\partial_t\mathbf{E}. \tag{13.4}$$

For simplicity, we have taken ϵ and $\mu = 1$. We can use the fact that $\mathbf{B}$ has zero divergence to introduce a vector potential $\mathbf{A}$, in terms of which $\mathbf{B}$ is given by

$$\mathbf{B} = \boldsymbol{\nabla}\times\mathbf{A}. \tag{13.5}$$

Now, the curl $\mathbf{E}$ equation can be written as

$$\boldsymbol{\nabla}\times\left(\mathbf{E} + \frac{1}{c}\partial_t\mathbf{A}\right) = 0. \tag{13.6}$$

The vanishing of this curl, permits us to relate $\mathbf{E}$ to the vector potential $\mathbf{A}$ and a scalar potential ϕ by

$$\mathbf{E} = -\boldsymbol{\nabla}\phi - \frac{1}{c}\partial_t\mathbf{A}. \tag{13.7}$$

Note that we no longer refer to $\mathbf{A}$ as the magnetic vector potential or to ϕ as the electric scalar potential, because the two potentials now relate to the combined electromagnetic field.

Putting the potentials into the two Maxwell's equations with sources, leads to source equations for the potentials

$$\nabla^2\phi + \frac{1}{c}\boldsymbol{\nabla}\cdot(\partial_t\mathbf{A}) = -4\pi\rho \qquad (13.8)$$

$$\nabla^2\mathbf{A} - \boldsymbol{\nabla}(\boldsymbol{\nabla}\cdot\mathbf{A}) - \frac{1}{c}\partial_t\boldsymbol{\nabla}\phi + \frac{1}{c^2}\partial_t^2\mathbf{A} = -\frac{4\pi}{c}\mathbf{j}. \qquad (13.9)$$

These equations are a bit complicated, with mixed space and time derivatives.

They can be simplified if the Lorenz gauge condition (first proposed by Ludwig Lorenz),

$$\boldsymbol{\nabla}\cdot\mathbf{A} + \frac{1}{c}\partial_t\phi = 0, \qquad (13.10)$$

is satisfied by the potentials. Then, Eqs. (14.163) and (14.164) simplify to

$$\nabla^2\phi - \frac{1}{c^2}\partial_t^2\phi = -4\pi\rho \qquad (13.11)$$

$$\nabla^2\mathbf{A} - \frac{1}{c^2}\partial_t^2\mathbf{A} = -\frac{4\pi}{c}\mathbf{j}. \qquad (13.12)$$

These two equations are recognized as inhomogeneous wave equations for the potentials ϕ and $\mathbf{A}$. Their solution will give rise to the electromagnetic radiation studied in this chapter.

13.2   The Lorenz Gauge.

To show that the Lorenz gauge condition can always be satisfied by the potentials, we extend the concept of gauge invariance introduced in Sect. 7.8.1. We have already seen that the gauge transformation

$$\mathbf{A} \to \mathbf{A}' = \mathbf{A} + \boldsymbol{\nabla}\psi \qquad (13.13)$$

leaves $\mathbf{B}$ unchanged. If the scalar potential ϕ also undergoes the transformation

$$\phi \to \phi' = \phi - \frac{1}{c}\partial_t\psi, \qquad (13.14)$$

then $\mathbf{E}$ will also be unchanged. The simultaneous application of these two transformations with the same gauge function ψ, leaving $\mathbf{B}$ and $\mathbf{E}$ unchanged, is called a **gauge transformation**.

If we have potentials for which the Lorenz condition is not satisfied, so

$$\boldsymbol{\nabla}\cdot\mathbf{A} + \frac{1}{c}\partial_t\phi = \chi, \qquad (13.15)$$

we can make a gauge transformation to restore the Lorenz condition. The transformed potentials will satisfy the relation

$$\boldsymbol{\nabla}\cdot\mathbf{A}' + \frac{1}{c}\partial_t\phi' = \chi + \nabla^2\psi - \frac{1}{c^2}\partial_t^2\psi. \qquad (13.16)$$

The new potentials $\mathbf{A}'$ and ϕ' will satisfy the Lorenz condition if the gauge function satisfies the equation

$$\nabla^2\psi - \frac{1}{c^2}\partial_t^2\psi = -\chi. \tag{13.17}$$

This equation is of the same type as that satisfied by the scalar potential in Eq. (13.11), which will be solved in the remainder of this chapter. Solving this equation here gives us the gauge function needed to restore the Lorenz condition.

Potentials that satisfy the Lorenz condition are still not determined uniquely. Any gauge transformation for which the gauge function satisfies the homogeous equation

$$\nabla^2\psi - \frac{1}{c^2}\partial_t^2\psi = 0, \tag{13.18}$$

will preserve the Lorenz condition. This means there can be a number of differing potentials that satisfy the Lorenz condition. Any such potential is said to be **"in the Lorenz gauge"**.

13.3   Retarded Solution of the Wave Equation.

To solve the wave equations (13.11) and (13.12), we consider sources with a periodic time dependence

$$\rho(\mathbf{r}, t) = \rho_\omega(\mathbf{r})e^{-i\omega t}, \quad \mathbf{j}(\mathbf{r}, t) = \mathbf{j}_\omega(\mathbf{r})e^{-i\omega t}. \tag{13.19}$$

These periodic sources could represent sources of definite frequency, or could be considered as Fourier components for more general time dependence. The potentials will have the same periodic time dependence as the sources.

Putting the periodic form for $\mathbf{A}(\mathbf{r}, t)$ into the wave equation, we get a spatial differential equation for $\mathbf{A}_\omega$:

$$(\nabla^2 + \omega^2/c^2)\mathbf{A}_\omega(\mathbf{r}) = -\frac{4\pi}{c}\mathbf{j}_\omega(\mathbf{r}). \tag{13.20}$$

We concentrate on finding $\mathbf{A}$, because it will lead more easily than ϕ to getting the radiation fields $\mathbf{B}$ and $\mathbf{E}$, but the same procedure can be used to solve for ϕ.

To solve Eq. (13.20), we introduce the spatial Fourier transforms $\bar{\mathbf{j}}$ and $\overline{\mathbf{A}}$ with

$$\mathbf{j}_\omega(\mathbf{r}) = \int d^3k\, e^{i\mathbf{k}\cdot\mathbf{r}}\bar{\mathbf{j}}(\mathbf{k}), \tag{13.21}$$

$$\mathbf{A}_\omega(\mathbf{r}) = \int d^3k\, e^{i\mathbf{k}\cdot\mathbf{r}}\overline{\mathbf{A}}(\mathbf{k}), \tag{13.22}$$

where d^3k represents the three dimensional differential for the three components of the vector $\mathbf{k}$.

Putting these integrals into the differential equation (13.20), we get an algebraic equation relating the Fourier transforms

$$\overline{\mathbf{A}}(k) = \frac{4\pi}{c} \frac{\overline{\mathbf{j}}(\mathbf{k})}{(k^2 - \omega^2/c^2)}. \tag{13.23}$$

Substituting this into Eq. (13.22), we get

$$\mathbf{A}_\omega(\mathbf{r}) = \frac{4\pi}{c} \int \frac{d^3 k e^{i\mathbf{k}\cdot\mathbf{r}}\overline{\mathbf{j}}(\mathbf{k})}{(k^2 - \omega^2/c^2)}, \tag{13.24}$$

with $\overline{\mathbf{j}}(\mathbf{k})$ given by the Fourier transform integral

$$\overline{\mathbf{j}}(\mathbf{k}) = \frac{1}{(2\pi)^3} \int d^3 r' e^{-i\mathbf{k}\cdot\mathbf{r}'}\mathbf{j}_\omega(\mathbf{r}'). \tag{13.25}$$

Substituting this into Eq. (13.24), the solution of Eq. (13.20) can be written as

$$\mathbf{A}_\omega(\mathbf{r}) = \frac{1}{c} \int d^3 r' G(\mathbf{r} - \mathbf{r}')\mathbf{j}_\omega(\mathbf{r}'), \tag{13.26}$$

with the Green's function, $G(\mathbf{r} - \mathbf{r}')$, being given by

$$G(\mathbf{r} - \mathbf{r}') = \frac{1}{2\pi^2} \int \frac{d^3 k e^{i\mathbf{k}\cdot(\mathbf{r}-\mathbf{r}')}}{(k^2 - \omega^2/c^2)}. \tag{13.27}$$

In general, any solution of the homogeneous wave equation could be added to this solution, but here we are only interested in the potential produced by the current distribution $\mathbf{j}_\omega(\mathbf{r})$, so we take Eq. (13.26) as our full solution.

We use spherical coordinates to calculate the Green's function:

$$\begin{aligned} G(\mathbf{r} - \mathbf{r}') &= \frac{1}{\pi} \int_0^\infty \frac{k^2 dk}{(k^2 - \omega^2/c^2)} \int_0^\pi \sin\theta d\theta e^{i(k|\mathbf{r}-\mathbf{r}'|\cos\theta)} \\ &= \frac{2}{\pi|\mathbf{r}-\mathbf{r}'|} \int_0^\infty \frac{k\sin(k|\mathbf{r}-\mathbf{r}'|)dk}{(k^2 - \omega^2/c^2)} \\ &= \frac{1}{i\pi|\mathbf{r}-\mathbf{r}'|} \int_{-\infty}^\infty \frac{k e^{i(k|\mathbf{r}-\mathbf{r}'|)}dk}{(k - \omega/c)(k + \omega/c)}. \end{aligned} \tag{13.28}$$

We have written this last integral in a suggestive form for integration around the contour shown in Fig. 13.1.

The integrand has poles on the real axis at $k = \pm\omega/c$. The integration path along the real axis is distorted as shown because this will lead to electromagnetic waves propagating outward from the oscillating charge distribution. This could also be accomplished by the 'iϵ prescription' of adding an imaginary term iϵ to ω so that $\omega \to \omega + i\epsilon$. This displaces the pole positions, and the integration path can be directly on the real axis. After the integration, the parameter ϵ can be taken to zero.

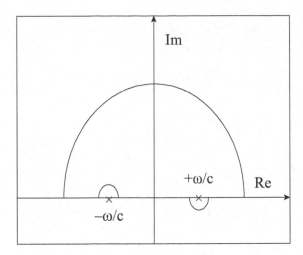

Figure 13.1: Contour for Green's function integration.

In order for the integral to converge, the contour must be closed above by a large semicircle as shown. Either method (displacing the contour or displacing the poles) leaves only the pole at $k = +\omega/c$ inside the contour, so the residue theorem gives the result

$$G(\mathbf{r} - \mathbf{r}') = \frac{e^{i\frac{\omega}{c}|\mathbf{r}-\mathbf{r}'|}}{|\mathbf{r} - \mathbf{r}'|}. \qquad (13.29)$$

Putting this Green's function into Eq. (13.26) gives the vector potential

$$\mathbf{A}_\omega(\mathbf{r}) = \frac{1}{c} \int d^3r' \frac{e^{i\frac{\omega}{c}|\mathbf{r}-\mathbf{r}'|}\mathbf{j}_\omega(\mathbf{r}')}{|\mathbf{r} - \mathbf{r}'|}. \qquad (13.30)$$

The potential $\mathbf{A}_\omega(\mathbf{r})$ in Eq. (13.30) can be considered as the Fourier frequency component of a time dependent potential $\mathbf{A}(\mathbf{r}, t)$. The current in Eq. (13.30) is the Fourier component of a time dependent current $\mathbf{j}(\mathbf{r}, t)$, and is given by

$$\mathbf{j}_\omega(\mathbf{r}) = \frac{1}{2\pi} \int dt' e^{i\omega t'} \mathbf{j}(\mathbf{r}, t'). \qquad (13.31)$$

The potential $\mathbf{A}_\omega$ is given in terms of $\mathbf{j}(\mathbf{r}, t)$ by

$$\mathbf{A}_\omega(\mathbf{r}) = \frac{1}{2\pi c} \int d^3r' \frac{e^{i\frac{\omega}{c}|\mathbf{r}-\mathbf{r}'|}}{|\mathbf{r} - \mathbf{r}'|} \int_{-\infty}^{\infty} dt' e^{i\omega t'} \mathbf{j}(\mathbf{r}', t'). \qquad (13.32)$$

The time dependent potential is then given by

$$\begin{aligned}
\mathbf{A}(\mathbf{r}, t) &= \frac{1}{2\pi c} \int_{-\infty}^{\infty} d\omega e^{-i\omega t} \int d^3r' \frac{e^{i\frac{\omega}{c}|\mathbf{r}-\mathbf{r}'|}}{|\mathbf{r} - \mathbf{r}'|} \int_{-\infty}^{\infty} dt' e^{i\omega t'} \mathbf{j}(\mathbf{r}', t') \\
&= \frac{1}{2\pi c} \int \frac{d^3r'}{|\mathbf{r} - \mathbf{r}'|} \int_{-\infty}^{\infty} dt' \mathbf{j}(\mathbf{r}', t') \int_{-\infty}^{\infty} d\omega e^{i\omega(t'-t+|\mathbf{r}-\mathbf{r}'|/c)}, \qquad (13.33)
\end{aligned}$$

where we have interchanged orders of integration. We recognize the last integral above as a delta function, which gives

$$\mathbf{A}(\mathbf{r}, t) = \frac{1}{2c} \int \frac{d^3 r'}{|\mathbf{r} - \mathbf{r}'|} \int_{-\infty}^{\infty} dt' \mathbf{j}(\mathbf{r}', t') \delta(t' - t + |\mathbf{r} - \mathbf{r}'|/c)$$

$$= \frac{1}{c} \int \frac{d^3 r' \mathbf{j}(\mathbf{r}', t - |\mathbf{r} - \mathbf{r}'|/c)}{|\mathbf{r} - \mathbf{r}'|}. \tag{13.34}$$

A similar derivation gives

$$\phi(\mathbf{r}, t) = \int \frac{d^3 r' \rho((\mathbf{r}', t - |\mathbf{r} - \mathbf{r}'|/c)}{|\mathbf{r} - \mathbf{r}'|}. \tag{13.35}$$

These solutions of the wave equation are called the **retarded potentials**. They give the potentials as integrals over the current and charge distributions evaluated at a retarded time

$$t_{\text{ret}} = t - d/c, \tag{13.36}$$

where $d = |\mathbf{r} - \mathbf{r}'|$ is the distance from the source (at the retarded time) to the observation point. This is the first indication we have seen of a finite velocity (c) of propagation for an electromagnetic effect. That is, the retarded potentials can only be affected by sources at times d/c earlier than the time the potential is evaluated.

The integrals for the retarded potentials look deceptively simple (if they can be considered to look simple). To actually perform the integrals can be difficult. It's like trying to count goldfish in a pond, while they are swimming about and you can't count them all at once. In the next section, we treat the retarded effects far from the sources where we can expand the denominator of the integrand to make the retarded integral tractable.

13.4   Radiation Solution of the Wave Equation.

We will consider radiation from charge-current distributions oscillating at a definite frequency ω. For a more general time dependence, the oscillating source distributions can be considered as Fourier transforms. To calculate electromagnetic radiation, we are interested in the potential at large distances from a finite size charge distribution. For this reason, we expand the integrand in Eq. (13.30) for $\mathbf{A}(\mathbf{r})$ (We will not bother with the subscript ω.) to first order in powers of the ratio $r'/r \ll 1$:

$$\mathbf{A}(\mathbf{r}) \simeq \frac{1}{c} \int d^3 r' \frac{e^{ik(r - \mathbf{r}' \cdot \hat{\mathbf{r}})} \mathbf{j}(\mathbf{r}')}{(r - \mathbf{r}' \cdot \hat{\mathbf{r}})}. \tag{13.37}$$

The term $\mathbf{r'}\cdot\hat{\mathbf{r}}$ is negligible in the denominator and may be dropped, but it must be kept in the numerator where it gives a finite phase factor even in the limit $r \gg r'$. Then $\mathbf{A}$ can be written as

$$\mathbf{A}(\mathbf{r}) = \frac{e^{ikr}}{cr} \int d^3r' e^{-ik\mathbf{r'}\cdot\hat{\mathbf{r}}} \mathbf{j}(\mathbf{r'})$$

$$= \frac{e^{ikr}}{ckr} \boldsymbol{\mathcal{J}}(k\hat{\mathbf{r}}), \tag{13.38}$$

where we have introduced the notation of an **effective radiation current**,

$$\boldsymbol{\mathcal{J}}(k\hat{\mathbf{r}}) = k \int d^3r' e^{-ik\mathbf{r'}\cdot\hat{\mathbf{r}}} \mathbf{j}(\mathbf{r'}). \tag{13.39}$$

The factor k is included in the definition of $\boldsymbol{\mathcal{J}}$ so that it will have the units of a current.

When we put in the exponential time dependence of Eq. (13.19), we get

$$\mathbf{A}(\mathbf{r},t) = \frac{e^{i(kr-\omega t)}}{ckr} \boldsymbol{\mathcal{J}}(k\hat{\mathbf{r}}). \tag{13.40}$$

This represents a spherical outgoing wave with surfaces of constant phase being spheres moving outward with a phase velocity $v = \omega/k = c$. We see now how choosing the contour we did in Fig. 13.1 has led to waves that are outgoing from the current distribution.

These waves are the spherical analogue of the circular waves you see if you drop a stone into calm water. Although the phase of our wave is constant on a sphere, its amplitude varies with angular direction according to $\boldsymbol{\mathcal{J}}(k\hat{\mathbf{r}})$. This is like what would happen if you dropped a long stick into the water. The waves would still be circular, but their amplitude would vary with angle.

To find the electric and magnetic fields, we have to take appropriate time and space derivatives of $\mathbf{A}(\mathbf{r}, t)$. This becomes greatly simplified in what is called the **radiation zone**, defined by

$$kr \gg 1 \quad \text{or} \quad r \gg \lambdabar, \tag{13.41}$$

where we have introduced a **reduced wavelength** $\lambdabar$, given by

$$\lambdabar = \lambda/2\pi = 1/k. \tag{13.42}$$

We see that the radiation zone is that region where the distance from the source is much larger than the reduced wavelength.

In the radiation zone, spatial derivatives that act on the $1/r$ term or the $\hat{\mathbf{r}}$ dependence in Eq. (13.40) bring down one power of r, while a space derivative on the exponential e^{ikr} introduces the factor ik. Since $k \gg 1/r$ in the radiation zone, it becomes a good approximation to replace the gradient operation on $\mathbf{A}$ in the radiation zone by

$$\nabla \rightarrow ik\hat{\mathbf{r}}. \tag{13.43}$$

Also, because of the exponential time dependence, the time derivative can be replaced by

$$\partial_t \to -i\omega. \tag{13.44}$$

This simplifies the calculation of the magnetic field to (omitting the common time factor $e^{-i\omega t}$)

$$\mathbf{B} = ik\hat{\mathbf{r}} \times \mathbf{A} = \frac{e^{ikr}}{cr} i\hat{\mathbf{r}} \times \boldsymbol{\mathcal{J}}, \tag{13.45}$$

and Maxwell's equation for the curl of $\mathbf{B}$ becomes

$$\mathbf{E} = -\hat{\mathbf{r}} \times \mathbf{B} = \frac{ie^{ikr}}{cr}[\boldsymbol{\mathcal{J}} - (\boldsymbol{\mathcal{J}} \cdot \hat{\mathbf{r}})\hat{\mathbf{r}}]. \tag{13.46}$$

The expression in square brackets is the part of the vector $\boldsymbol{\mathcal{J}}$ that is transverse to $\hat{\mathbf{r}}$, the direction of observation of the wave . We call it the **transverse radiation current**

$$\boldsymbol{\mathcal{J}}_T = \boldsymbol{\mathcal{J}} - (\boldsymbol{\mathcal{J}} \cdot \hat{\mathbf{r}})\hat{\mathbf{r}}. \tag{13.47}$$

$\boldsymbol{\mathcal{J}}_\mathbf{T}$ is the vector that determines the angular dependence of the electric and magnetic fields in the radiation zone.

To an observer in the radiation zone where $r \gg r'$, the curved surface of the spherical wave approximates a plane, and the relation between $\mathbf{E}$ and $\mathbf{B}$ is like that in a plane wave moving in the direction $\hat{\mathbf{r}}$. We can introduce a vector $\mathbf{k} = k\hat{\mathbf{r}}$, which has the same meaning as the wave vector $\mathbf{k}$ in a plane wave. The three vectors $\mathbf{k}$, $\mathbf{E}$, $\mathbf{B}$ form the same right hand triad as in a plane wave. The relation of the $\mathbf{E}$ and $\mathbf{B}$ vectors to the radiation current $\boldsymbol{\mathcal{J}}$ is shown in Fig. 13.2.

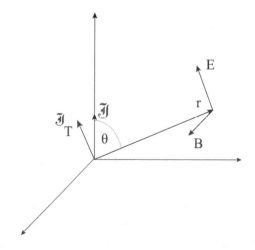

Figure 13.2: The vectors $\boldsymbol{\mathcal{J}}$, **B**, **E**, and **r**.

To an observer in the radiation zone, the spherical wave has all the characteristics of a plane wave, for instance like the light we observe from the sun. An important characteristic of the electric and magnetic fields of an oscillating current is that they fall off at large distances like $1/r$, much slower than the $1/r^2$ (or faster) behavior of static fields.

The **intensity** of the wave is given by the time averaged Poynting vector

$$\overline{\mathbf{S}} = \frac{c}{8\pi}\mathbf{E}^* \times \mathbf{B} = \frac{c}{8\pi}|\mathbf{E}|^2 = \frac{\hat{\mathbf{r}}}{8\pi c r^2}|\boldsymbol{\mathcal{J}}_T(k\hat{\mathbf{r}})|^2. \tag{13.48}$$

Note that the exponential factor $e^{i(kr-\omega t)}$ has cancelled because of the $\mathbf{E}^* \times \mathbf{B}$ combination, so there is no time dependence in $\overline{\mathbf{S}}$.

The Poynting vector represents the rate of energy flow (power) per unit area. The power through a differential surface $\mathbf{dA}$ on a sphere at radius $\mathbf{r}$ is given by

$$dP = \mathbf{S} \cdot \mathbf{dA} = \hat{\mathbf{r}} \cdot \mathbf{S} r^2 d\Omega, \tag{13.49}$$

where $d\Omega$ is the solid angle subtended by $\mathbf{dA}$. The angular distribution of the spherical wave is given by

$$\frac{dP}{d\Omega} = r^2 \hat{\mathbf{r}} \cdot \mathbf{S} = \frac{cr^2}{8\pi}|\mathbf{E}|^2 = \frac{1}{8\pi c}|\boldsymbol{\mathcal{J}}_T(k\hat{\mathbf{r}})|^2. \tag{13.50}$$

Because of the $1/r$ dependence of the fields, the intensity of electromagnetic radiation falls off as $1/r^2$, but the power into a given solid angle is independent of the distance from the source. The total power radiated by the oscillating current is then given by the integral over solid angle

$$P = \frac{1}{8\pi c}\int d\Omega |\boldsymbol{\mathcal{J}}_T(k\hat{\mathbf{r}})|^2, \tag{13.51}$$

and is also independent of the distance from the source.

This remarkable feature of electromagnetic radiation propagating over great distances with no loss of total power was the crowning achievement of Maxwell's development of electromagnetic theory. In the remainder of this chapter, we will calculate the all important transverse radiation current $\boldsymbol{\mathcal{J}}_T(k\hat{\mathbf{r}})$ for various oscillating charge and current distributions.

13.5 Center Fed Linear Antenna.

As a simple example, we consider a thin antenna of length L with current entering and leaving at the center as shown in Fig. 13.3.

To illustrate the use of Eq. (13.39) for the radiation current, we assume that the current in the wire is a standing wave of the form

$$I(z,t) = I_0 \sin(kL/2 - k|z|)e^{-i\omega t}, \quad |z| \le L/2, \tag{13.52}$$

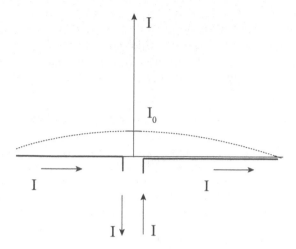

Figure 13.3: Current (dashed line) in a center-fed antenna.

with $\omega = kc$. (The actual current in an antenna can be more complicated than this, because the strong electromagnetic fields near the antenna can affect the current distribution.) The current distribution in Eq. (13.52) is indicated in Fig. 13.3 for the case $kL = \pi$, which is called a **half wave antenna**. For a thin wire with this current, the integral in Eq. (13.39) becomes

$$\boldsymbol{\mathcal{J}} = I_0 k \hat{\mathbf{z}} \int_{-L/2}^{+L/2} \sin(kL/2 - k|z|) e^{-ikz\cos\theta} dz, \tag{13.53}$$

the angle θ is measured from the axis of the antenna. We have used $\hat{\mathbf{z}}$ for the unit vector along the wire to avoid confusion with our use of the notation $\mathbf{k} = k\hat{\mathbf{r}}$.

The integral in Eq. (13.1) is elementary, giving the result

$$\boldsymbol{\mathcal{J}} = \frac{2I_0\hat{\mathbf{z}}}{\sin^2\theta} \left[\cos\left(\frac{kL}{2}\cos\theta\right) - \cos\left(\frac{kL}{2}\right)\right]. \tag{13.54}$$

The magnitude of the transverse radiation current is

$$\mathcal{J}_T = \frac{2I_0}{\sin\theta} \left[\cos\left(\frac{kL}{2}\cos\theta\right) - \cos\left(\frac{kL}{2}\right)\right]. \tag{13.55}$$

Putting this current into Eq. (13.50) gives the radiation pattern

$$\frac{dP}{d\Omega} = \frac{I_0^2}{2\pi c} \left[\frac{\cos\left(\frac{kL}{2}\cos\theta\right) - \cos\left(\frac{kL}{2}\right)}{\sin\theta}\right]^2. \tag{13.56}$$

The radiation from the wire is plane polarized with the polarization direction defined as the direction of the $\mathbf{E}$ vector in the outgoing wave. From

Eq. (13.46), we see that the plane of polarization is the plane containing the two vectors $\hat{\mathbf{r}}$ and $\boldsymbol{\mathcal{J}}$. For the linear antenna, this is the plane containing the direction of observation and the straight wire.

We examine the radiation pattern of the linear antenna for some special cases. For the half wave current distribution with $kL = \pi$, as indicated by the dashed line in Fig. 13.3, Eq. (13.56) becomes

$$\frac{dP}{d\Omega} = \frac{I_0^2}{2\pi c} \left[\frac{\cos\left(\frac{\pi}{2}\cos\theta\right)}{\sin\theta} \right]^2 , \quad kL = \pi. \qquad (13.57)$$

This angular distribution is plotted as the dotted curve in Fig. 13.4.

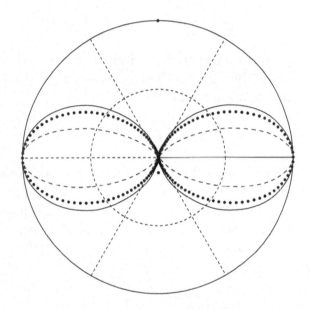

Figure 13.4: Radiation patterns for center-fed antennas: $kL = \pi$ (dotted curve), $kL = 2\pi$ (dashed curve), $kL \ll 1$, (solid curve).

The distance from the origin to the curve is proportional to the intensity of radiation in that direction.) The pattern is a toroid with no radiation along the direction of the wire (vertical in the figure), and maximum intensity in the plane perpendicular to the wire.

The intensity maximum can be sharpened by placing two half wave antennas end to end to form a **full wave antenna**. For this case, Eq. (13.56) becomes

$$\frac{dP}{d\Omega} = \frac{2I_0^2}{\pi c} \left[\frac{\cos^2\left(\frac{\pi}{2}\cos\theta\right)}{\sin\theta} \right]^2 , \quad kL = 2\pi. \qquad (13.58)$$

This pattern is shown in Fig. 13.4 as a dashed curve.. We see from the figure that the full wave antenna has a narrower radiation pattern than the half wave antenna. This is an example of how multiple antennas can be used to sharpen radiation patterns.

Another case of interest is when the antenna is much shorter than the wavelength, so $kL << 1$. Then expanding the numerator of Eq. (13.56), we get

$$\frac{dP}{d\Omega} = \frac{I_0^2(kL)^4 \sin^2\theta}{128\pi c}, \quad kL << 1. \tag{13.59}$$

This radiation pattern is plotted in Fig. 13.4 as a solid curve. It is not as sharp as the pattern for the longer antennas. The total radiated power is given by the integral of $\frac{dp}{d\Omega}$ over all solid angle. The integral for the short antenna case of Eq. (13.59) is elementary, giving

$$P = \frac{I_0^2(kL)^4}{48c}, \quad kL << 1. \tag{13.60}$$

The power dissipated by a resistive circuit is $\frac{1}{2}I^2R$, where I is the amplitude of the current entering the circuit. This connection is used to define the **radiation resistance** of an antenna by

$$P = \frac{1}{2}I^2 R_{\text{rad}}. \tag{13.61}$$

While the term 'resistance' is used here, the larger the radiation resistance the better. A larger radiation resistance means the antenna radiates more power, although the directional properties of the antenna are also important.

For the short linear antenna, the amplitude of the current entering the antenna is given by Eq. (13.52) for small kL to be $I = (kL/2)I_0$. and then the radiation resistance is

$$R_{\text{rad}} = \frac{(kL)^2}{6c}, \quad kL << 1. \tag{13.62}$$

This result is in esu units of sec/cm (or statohms). An esu resistance can be converted to the SI unit ohms by multiplying by the factor $30c$, so

$$R_{\text{rad}} = 5(kL)^2 \text{ ohms}, \quad kL << 1. \quad \textbf{SI} \tag{13.63}$$

The radiation patterns for the half wave and full wave antennas must be integrated numerically to find the total power. (See problem 13.2.)

13.6   Electric Dipole Radiation.

We have shown some examples of radiation from simple current distributions, but the current is usually more complicated, making the integration of Eq. (13.39) for the radiation current intractible. For this reason, a mutipole expansion of the potentials and fields is often used.

The radiation multipole expansion has some similarities with the static field multipole expansion we used in Chapter 2, but the physics is quite different. We have already made, in our derivation of Eq. (13.38), the expansion in r'/r that was the basis of the static multipole expansion. The radiation multipole expansion arises from a further expansion of the exponential term $e^{-ik\hat{\mathbf{r}}\cdot\mathbf{r}'}$ in the integral in over the current distribution in Eq. (13.38). This requires the new assumption that $kd \ll 1$, where d is a measure of the spatial extent of the radiating current.

The expansion of the radiation current in powers of kr' is

$$\boldsymbol{\mathcal{J}}(k\hat{\mathbf{r}}) = k \int d^3 r' \mathbf{j}(\mathbf{r}')[1 - ik\hat{\mathbf{r}}\cdot\mathbf{r}' + ...]. \tag{13.64}$$

Keeping only the first term in the expansion gives

$$\boldsymbol{\mathcal{J}}_0 = k \int d^3 r \mathbf{j}(\mathbf{r}), \tag{13.65}$$

which we see is a constant vector. This approximation, which amounts to just ignoring the exponential factor, is called the **dipole approximation** because the vector $\boldsymbol{\mathcal{J}}_0$ is closely related to the electric dipole moment of the charge-current distribution.

This follows from the continuity equation relating the charge and current distributions, as shown below using the dyadic **jr**:

$$\begin{aligned}
\int d^3 r \mathbf{j} &= \int d^3 r (\mathbf{j}\cdot\boldsymbol{\nabla})\mathbf{r} \\
&= \int d^3 r [\boldsymbol{\nabla}\cdot(\mathbf{jr}) - \mathbf{r}(\boldsymbol{\nabla}\cdot\mathbf{j})] \\
&= \int_S d\mathbf{S}\cdot\mathbf{jr} + \int d^3 r \mathbf{r}\partial_t\rho \\
&= -i\omega \int d^3 r \rho\mathbf{r}.
\end{aligned} \tag{13.66}$$

We have used the divergence theorem to convert the integral over $\boldsymbol{\nabla}\cdot(\mathbf{jr})$ to a surface integral that vanishes because the surface is outside the limited current distribution, and have used the continuity equation to replace $\boldsymbol{\nabla}\cdot\mathbf{j}$ by $-\partial_t\rho$.

We recognize the last integral in Eq. (13.66) as the electric dipole moment, so we have the result

$$\boldsymbol{\mathcal{J}}_0 = -i\omega k\mathbf{p}, \tag{13.67}$$

with the electric dipole moment **p** given by

$$\mathbf{p} = \int d^3 r \rho\mathbf{r} = \frac{i}{\omega} \int d^3 r \mathbf{j}. \tag{13.68}$$

The electric dipole moment can be found using either integral, depending on whether the charge or current distribution is known.

The dipole moment can be used in the basic radiation equations (13.45)-(13.51) to describe electric dipole radiation. The radiation magnetic field is given by

$$\mathbf{B} = \frac{k^2 e^{ikr}}{r}\hat{\mathbf{r}}\times\mathbf{p}, \tag{13.69}$$

and the electric field is

$$\mathbf{E} = -\hat{\mathbf{r}}\times\mathbf{B} = \frac{k^2 e^{ikr}}{r}[\mathbf{p} - (\mathbf{p}\cdot\hat{\mathbf{r}})\hat{\mathbf{r}}] = \frac{k^2 e^{ikr}}{r}\mathbf{p_T}. \tag{13.70}$$

This shows that electric dipole radiation is plane polarized in the plane formed by $\mathbf{p}$ and $\mathbf{r}$. The radiation pattern is

$$\frac{dP}{d\Omega} = \frac{ck^4}{8\pi}|\mathbf{p_T}|^2 = \frac{ck^4}{8\pi}|\mathbf{p}|^2\sin^2\theta, \tag{13.71}$$

where θ is the angle between $\mathbf{p}$ and $\mathbf{r}$. The total power radiated is the integral of this over all solid angle, giving

$$P = \frac{ck^4}{3}|\mathbf{p}|^2. \tag{13.72}$$

A simple example of a dipole antenna is the short center-fed linear antenna described in section 13.5. For $kL \ll 1$, the current distribution can be written as

$$I(z,t) = I_d(1 - 2|z|/L)e^{-i\omega t}, \quad |z| \le L/2, \tag{13.73}$$

where I_d is the current entering the antenna. The dipole moment is given by

$$p = \frac{i}{\omega}I_d\int_{-L/2}^{+L/2}(1 - 2|z|/L)dz = \frac{iI_d L}{2kc}. \tag{13.74}$$

The radiation pattern is

$$\frac{dP}{d\Omega} = \frac{k^2 L^2 I_d^2}{32\pi c}\sin^2\theta, \tag{13.75}$$

and the total power radiated

$$P = \frac{k^2 L^2 I_d^2}{12c}. \tag{13.76}$$

The dipole radiation pattern is the same as the $kL \ll 1$ limit of the linear antenna (taking into account that $I_d = (kL/2)I_0$.), and is shown as the same solid line on Fig. 13.4.

13.7 Radiation by Atoms.

Based on his scattering experiments, Lord Rutherford proposed a planetary model of an atom, which for hydrogen had a single electron orbiting a central nucleus (a proton) under the attractive Coulomb force. Such an atom of radius R would have an electric dipole moment rotating at the angular velocity $\boldsymbol{\omega}$ of the electron. If the electrons orbit is in the x-y plane, $\boldsymbol{\omega}$ will be in the z direction and the dipole moment would be given by

$$\mathbf{p} = -eR(\hat{\mathbf{i}} - i\hat{\mathbf{j}})e^{-i\omega t}. \tag{13.77}$$

This can be thought of as two separate oscillating dipoles at right angles to each other, 90° out of phase in time. We can use the dipole radiation formula of Eq. (13.71) to get

$$\frac{dP}{d\Omega} = \frac{e^2 R^2 \omega^4}{8\pi c^3}(\sin^2\theta_{xr} + \sin^2\theta_{yr}), \tag{13.78}$$

where θ_{xr} is the angle between $\mathbf{r}$ and the x-axis, and θ_{yr} the angle between $\mathbf{r}$ and the y-axis. Giving these angles in terms of the usual spherical coordinate angles θ and ϕ, the radiation pattern becomes

$$\begin{aligned} \frac{dP}{d\Omega} &= \frac{e^2 R^2 \omega^4}{8\pi c^3}[(1 - \sin^2\theta\cos^2\phi) + (1 - \sin^2\theta\sin^2\phi)] \\ &= \frac{e^2 R^2 \omega^4}{8\pi c^3}(1 + \cos^2\theta). \end{aligned} \tag{13.79}$$

The direction of the electric field of the radiation is given by the transverse part of the rotating dipole moment $\mathbf{p}$ given by Eq. (13.77). This is just the projection of the electron's circular orbit as it would appear if you looked at it from the observation point. That is, radiation along the axis of the atom would be circularly polarized, radiation along the plane of the atom would be plane polarized in that plane, while radiation along intermediate directions would be elliptically polarized.

The total power that would be radiated by the Rutherford atom is

$$P = \frac{2e^2 R^2 \omega^4}{3c^3}. \tag{13.80}$$

This simple result was a triumph for classical electromagnetism, but a disaster for classical physics. The radiated power would continuously take energy from the atom, leading to its inevitable collapse. Nineteenth century classical physics had no way to prevent this collapse. The conundrum was eventually resolved, first by the simple, but foundationless, Bohr model, and ultimately by quantum mechanics, the twentieth century theory that superseded classical mechanics.

How long would a classical Rutherford atom live? First, let us write Eq. (13.80) in terms of the single variable R by substituting for $\omega^2 = e^2/mR^3$ (from $F = ma$ for the atom):

$$P = \frac{2e^6}{3m^2R^4c^3}. \tag{13.81}$$

This power represents the rate of energy loss by the atom. The energy of the atom is given in terms of its radius by

$$U = \frac{1}{2}mv^2 - \frac{e^2}{R} = -\frac{e^2}{2R}. \tag{13.82}$$

Equating the radiated power to the rate of energy loss, we get

$$P = \frac{2e^6}{3m^2R^4c^3} = -\frac{dU}{dt} = \frac{-e^2}{2R^2}\frac{dR}{dt}. \tag{13.83}$$

This is a differential equation for the rate of decrease of the atom's radius. The time for R to become zero can be considered to be the lifetime of the atom. This lifetime is given by integrating $\frac{dR}{dt}$, leading to

$$T = \frac{R}{4c}\left[\frac{mc^2}{(e^2/R)}\right]^2. \tag{13.84}$$

We have written this in a simple form for computation, with the term in square brackets being a dimensionless ratio of two energies. The numerical result for hydrogen (See problem 13.4) is $T \sim 10^{-11}$ seconds. This would be the classical lifetime of a typical atom, and if this classical result were applicable, our lifetime as well. We can indeed be thankful for quantum mechanics.

Even though quantum mechanics is needed to get the correct intensity of the radiation from atoms, the angular distributions and polarization of the radiation is given correctly by the classical calculations. This is because the angular distribution and polarization depend on the properties of the spherical harmonics, and these enter in the same way for either the classical or quantum calculation.

13.8 Larmor Formula for Radiation by an Accelerating Charge.

The radiation emitted by a general time dependent current $\mathbf{j}(\mathbf{r},t)$ can be found from periodic radiation results by using Fourier analysis. The periodic current density $\mathbf{j}_\omega(\mathbf{r})$ is given by the Fourier transform

$$\mathbf{j}_\omega(\mathbf{r}) = \frac{1}{2\pi}\int e^{i\omega t}\mathbf{j}(\mathbf{r},t)dt. \tag{13.85}$$

For electric dipole radiation, the periodic vector potential $\mathbf{A}_\omega$ is given in terms of $\mathbf{j}_\omega$ by

$$\mathbf{A}_\omega(\mathbf{r}) = \frac{e^{i\omega r/c}}{cr} \int d^3r' \mathbf{j}_\omega(\mathbf{r}'). \tag{13.86}$$

Then the time dependent vector potential is the inverse Fourier transform

$$
\begin{aligned}
\mathbf{A}(\mathbf{r}, t) &= \int d\omega e^{-i\omega t} \mathbf{A}_\omega(\mathbf{r}) \\
&= \frac{1}{cr} \int d\omega e^{-i\omega(t-r/c)} \int d^3r' \mathbf{j}_\omega(\mathbf{r}') \\
&= \frac{1}{2\pi cr} \int dt' \int d\omega e^{-i\omega(t-t'-r/c)} \int d^3r' . \mathbf{j}(\mathbf{r}', t').
\end{aligned} \tag{13.87}
$$

In the last integral, we have interchanged the order of integrations over t' and ω. We can see that the ω integral gives a delta function, so we have

$$
\begin{aligned}
\mathbf{A}(\mathbf{r}, t) &= \frac{1}{cr} \int dt' \delta(t - t' - r/c) \int d^3r' . \mathbf{j}(\mathbf{r}', t') \\
&= \frac{1}{cr} \int d^3r' \mathbf{j}(\mathbf{r}', t - r/c).
\end{aligned} \tag{13.88}
$$

The vector potential is given by an integral over the current at a **retarded time** $t_r = t - r/c$. This is a causal property, because it means we need only know $\mathbf{j}(\mathbf{r}', t')$ at times earllier than the present time, t. The causality follows from our choice of contour in evaluating the integral for the Green's function in Eq. (13.28). The contour we used to get the outgoing wave solution also gives the proper causal behavior in time. Had we used a different contour, including the pole at $k = -i\omega$ inside the contour, this would have led to incoming waves and acausal behavior in Eq. (13.88).

For a point charge q following a path $\mathbf{r}(t)$, the current is given by

$$\mathbf{j}(\mathbf{r}', t) = q\mathbf{v}(t)\delta[\mathbf{r}' - \mathbf{r}(t)], \tag{13.89}$$

where $\mathbf{v} = \frac{d\mathbf{r}}{dt}$. Putting this current into Eq. (13.88), we get

$$
\begin{aligned}
\mathbf{A}(\mathbf{r}, t) &= \frac{q}{cr} \int d^3r' \mathbf{v}(t - r/c)\delta[\mathbf{r}' - \mathbf{r}(t - r/c)] \\
&= \frac{q}{cr}\mathbf{v}(t - r/c).
\end{aligned} \tag{13.90}
$$

The magnetic field is the curl of $\mathbf{A}$

$$
\begin{aligned}
\mathbf{B}(\mathbf{r}, t) &= \frac{q}{c}\boldsymbol{\nabla} \times \left[\frac{\mathbf{v}(t - r/c)}{r}\right] \\
&= -\frac{q}{c^2 r}\hat{\mathbf{r}} \times \mathbf{a}(t - r/c).
\end{aligned} \tag{13.91}
$$

We did not have the $\boldsymbol{\nabla}$ operator act on the $\frac{1}{r}$ term above because that would have led to $\frac{1}{r^2}$ and been negligible in the radiation zone. We used the chain rule for $\boldsymbol{\nabla} \times \mathbf{v}$, first taking the derivative of $\mathbf{v}$ with respect to its argument, leading to the acceleration $\mathbf{a}$, and then taking the gradient of $(t - r/c)$, while preserving the appropriate vector cross product. The electric field is given by Eq. (13.46) to be

$$\mathbf{E} = -\hat{\mathbf{r}} \times \mathbf{B} = \frac{q}{c^2 r}[\mathbf{a} - (\mathbf{a} \cdot \hat{\mathbf{r}})\hat{\mathbf{r}}], \qquad (13.92)$$

where the acceleration is evaluated at the retarded time.

Knowing the fields, the radiation pattern is given by

$$\frac{dP}{d\Omega} = \frac{cr^2}{4\pi}|\mathbf{E}|^2 = \frac{q^2}{4\pi c^3}|\mathbf{a}|^2 \sin^2 \theta, \qquad (13.93)$$

where θ is the angle between $\mathbf{a}$ and $\mathbf{r}$. This result is for the instantaneous power (at the retarded time), and so the factor 4π appears in the denominator rather than 8π, which was appropriate for the time averaged power in the oscillatory case. We see from Eq. (13.92) for $\mathbf{E}$ that the radiation is polarized in the plane formed by $\mathbf{a}$ and $\mathbf{r}$. The total radiated power is

$$P = \frac{2q^2|\mathbf{a}|^2}{3c^3}. \qquad (13.94)$$

This result, first derived by Joseph Larmor in 1897, is called the **Larmor formula** for radiation by an accelerating charge. For the Rutherford atom of the preceding section, the centripetal acceleration is $a = \omega^2 R$. Substituting this acceleration into the Larmor formula gives the same total power as in Eq. (13.80).

13.9   Magnetic Dipole Radiation

Keeping the second term in the expansion in powers of kr' in Eq. (13.64) gives the first order approximate radiation current

$$\boldsymbol{\mathcal{J}}_1(\mathbf{k}) = -ik \int (\mathbf{k} \cdot \mathbf{r}) \mathbf{j} d^3 r = -ik\mathbf{k} \cdot \int \mathbf{r} \mathbf{j} d^3 r. \qquad (13.95)$$

The dyadic $\mathbf{r}\mathbf{j}$ can be separated into symmetric and antisymmetric parts, so the first order current can be written as

$$\boldsymbol{\mathcal{J}}_1(\mathbf{k}) = \boldsymbol{\mathcal{J}}_s(\mathbf{k}) + \boldsymbol{\mathcal{J}}_a(\mathbf{k}), \qquad (13.96)$$

with

$$\boldsymbol{\mathcal{J}}_s(\mathbf{k}) = -\frac{ik}{2}\mathbf{k} \cdot \int (\mathbf{r}\mathbf{j} + \mathbf{j}\mathbf{r}) d^3 r, \qquad (13.97)$$

and

$$\boldsymbol{\mathcal{J}}_a(\mathbf{k}) = -\frac{ik}{2}\mathbf{k} \cdot \int (\mathbf{r}\mathbf{j} - \mathbf{j}\mathbf{r}) d^3 r. \qquad (13.98)$$

We will evaluate the $\mathcal{J}_a$ first. The antisymmetric integral is the same one we encountered in Eq. (7.121), which led to the static magnetic dipole moment. In the same way, we use bac-cab here to get

$$\mathcal{J}_a(\mathbf{k}) = \frac{ik}{2}\mathbf{k}\times \int (\mathbf{r}\times\mathbf{j})d^3r. \tag{13.99}$$

We define the radiation magnetic moment in the same way as the static magnetic moment so that

$$\boldsymbol{\mu} = \frac{1}{2c}\int (\mathbf{r}\times\mathbf{j})d^3r, \tag{13.100}$$

and then

$$\mathcal{J}_a(\mathbf{k}) = ick\mathbf{k}\times\boldsymbol{\mu}. \tag{13.101}$$

We can use this dipole radiation current to find the **B** and **E** fields in the radiation zone. From Eq. (13.45) for the magnetic field, we get

$$\mathbf{B}(\mathbf{r}) = \frac{e^{ikr}}{cr}i\hat{\mathbf{r}}\times \mathcal{J}_a(k\hat{\mathbf{r}}) = \frac{k^2 e^{ikr}}{r}[\boldsymbol{\mu} - (\boldsymbol{\mu}\cdot\hat{\mathbf{r}})\hat{\mathbf{r}}], \tag{13.102}$$

and the electric field is

$$\mathbf{E}(\mathbf{r}) = -\hat{\mathbf{r}}\times\mathbf{B} = -\frac{k^2 e^{ikr}}{r}\hat{\mathbf{r}}\times\boldsymbol{\mu}. \tag{13.103}$$

These equations for magnetic dipole radiation are just the same as those for electric dipole radiation with the substitutions

$$\mathbf{p} \to \boldsymbol{\mu}, \quad \mathbf{E} \to \mathbf{B}, \quad \mathbf{B} \to -\mathbf{E}. \tag{13.104}$$

The radiation pattern and the total power radiated are the same for the two types of dipole. The radiation differs only in its polarization. Electric dipole radiation is polarized in the plane of $\hat{\mathbf{r}}$ and the dipole, while magnetic dipole radiation is polarized perpendicular to that plane.

An example of a magnetic dipole is a circular loop antenna of radius R, carrying a current $I(t) = I_0 e^{-i\omega t}$. The radiation magnetic moment is $\boldsymbol{\mu} = \hat{\mathbf{z}}\pi R^2 I_0/c$. The radiation pattern will be that of Eq. (13.71) for an electric dipole polarized on the axis of the loop, so the radiation intensity will be zero on the axis of the loop and maximum in the plane of the loop. The total radiated power is

$$P = \frac{ck^4}{3}|\boldsymbol{\mu}|^2 = \frac{\pi^2 (kR)^4}{3c}I_0^2. \tag{13.105}$$

The radiation resistance, as defined in Sec. 13.5, is $20\pi^2(kR)^4$ ohms. (Recall that the esu resistance is converted to the SI unit ohms by multiplying by the factor $30c$.)

13.10   Electric Quadrupole Radiation.

We showed in Sec. 7.11 that the symmetric dyadic integral given in Eq. (13.97) for $\mathcal{J}_s$ vanished for a static charge distribution, but it does not do so for oscillating charges and currents. We repeat that derivation here, including the time variation of the charge distribution. We evaluate the integral by introducing a divergence as follows:

$$
\begin{aligned}
\int (\mathbf{rj} + \mathbf{jr}) d^3 r &= \int [\mathbf{r}(\mathbf{j}\cdot\boldsymbol{\nabla}\mathbf{r}) + (\mathbf{j}\cdot\boldsymbol{\nabla}\mathbf{r})\mathbf{r}] d^3 r \\
&= \int \boldsymbol{\nabla}\cdot(\mathbf{jrr}) d^3 r - \int (\boldsymbol{\nabla}\cdot\mathbf{j})\mathbf{rr} d^3 r. \qquad (13.106)
\end{aligned}
$$

The first integral above vanishes by using the divergence theorem to go to a surface integral outside the current distribution. In the second integral, we can use the continutity equation to get

$$
\boldsymbol{\nabla}\cdot\mathbf{j} = -\partial_t \rho = i\omega\rho. \qquad (13.107)
$$

This vanished for the static case because $\partial_t \rho$ was zero, but for an oscillating current, the symmetric integral becomes

$$
\int (\mathbf{rj} + \mathbf{jr}) d^3 r = -i\omega \int \rho\mathbf{rr} d^3 r. \qquad (13.108)
$$

This is just the integral we encountered in Sec. 2.4.2, leading to the electric quadrupole diadic. A similar result follows here.

The radiation current is

$$
\mathcal{J}_s(\mathbf{k}) = -\frac{\omega k}{2}\mathbf{k}\cdot \int \rho\mathbf{rr} d^3 r. \qquad (13.109)
$$

The radiation magnetic field [using Eq. (13.45] is

$$
\mathbf{B}(\mathbf{r}) = \frac{e^{ikr}}{cr} i\hat{\mathbf{r}}\times\mathcal{J} = -i\frac{k^3 e^{ikr}}{2r}\hat{\mathbf{r}}\times \left[\hat{\mathbf{r}}\cdot \int \rho\mathbf{r}'\mathbf{r}' d^3 r'\right]. \qquad (13.110)
$$

At this point we can introduce the electric quadrupole dyadic

$$
[\mathbf{Q}] = \frac{1}{2}\int [3\mathbf{rr} - r^2\hat{\hat{\mathbf{n}}}]\rho d^3 r. \qquad (13.111)
$$

The added term with the unit dyadic $\hat{\hat{\mathbf{n}}}$ will lead to a term in the equation for $\mathbf{B}$ of $\hat{\mathbf{r}}\times\hat{\mathbf{r}}$, which vanishes.

The quadrupole diadic is a real symmetric dyadic with zero trace (since $\hat{\mathbf{n}}\cdot\hat{\mathbf{n}} = 3$). We can now write the equation for $\mathbf{B}$ as

$$
\mathbf{B}(\mathbf{r}) = -i\frac{k^3 e^{ikr}}{3r}\hat{\mathbf{r}}\times (\hat{\mathbf{r}}\cdot[\mathbf{Q}]). \qquad (13.112)
$$

The quadrupole electric field is

$$\mathbf{E}(\mathbf{r}) = -\hat{\mathbf{r}} \times \mathbf{B} = -i\frac{k^3 e^{ikr}}{3r}\{\hat{\mathbf{r}} \cdot [\mathbf{Q}] - \hat{\mathbf{r}}(\hat{\mathbf{r}} \cdot [\mathbf{Q}] \cdot \hat{\mathbf{r}})\}, \tag{13.113}$$

and the quadrupole radiation pattern is

$$\frac{dP}{d\Omega} = \frac{ck^6}{72\pi}\{|\hat{\mathbf{r}} \cdot [\mathbf{Q}]|^2 - |\hat{\mathbf{r}} \cdot [\mathbf{Q}] \cdot \hat{\mathbf{r}}|^2\}. \tag{13.114}$$

This is a rather complicated angular distribution. It simplifies somewhat if the coordinate system is chosen so that $[\mathbf{Q}]$ is diagonal. This can always be done because $[\mathbf{Q}]$ is a real symmetric matrix. There is a considerable simplification for a quadrupole moment with axial symmetry about some axis. Then we can write (See Sec. 2.4 for details.)

$$[\mathbf{Q}] = \frac{1}{2}Q_0[3\hat{\mathbf{z}}\hat{\mathbf{z}} - \hat{\mathbf{n}}], \tag{13.115}$$

where $\hat{\mathbf{z}}$ is a unit vector along the z axis, which we have taken as the symmetry axis.

For this case, we have

$$|\hat{\mathbf{r}} \cdot [\mathbf{Q}]|^2 = \frac{1}{4}Q_0^2[3(\hat{\mathbf{r}} \cdot \hat{\mathbf{z}})\hat{\mathbf{z}} - \hat{\mathbf{r}}]^2 = \frac{1}{4}Q_0^2(1 + 3\cos^2\theta), \tag{13.116}$$

and

$$|\hat{\mathbf{r}} \cdot [\mathbf{Q}] \cdot \hat{\mathbf{r}}|^2 = \frac{1}{4}Q_0^2[3(\hat{\mathbf{r}} \cdot \hat{\mathbf{z}})^2 - 1]^2 = \frac{1}{4}Q_0^2(1 - 6\cos^2\theta + 9\cos^4\theta). \tag{13.117}$$

Putting these results into Eq. (13.114) gives

$$\frac{dP}{d\Omega} = \frac{ck^6 Q_0^2}{32\pi}\cos^2\theta\sin^2\theta. \tag{13.118}$$

This angular distribution has four lobes as shown in Fig. 13.5. The total radiated power from the symmetric quadrupole is

$$P = \frac{ck^6 Q_0^2}{60}. \tag{13.119}$$

We can get the total quadrupole power in the general case by going to the coordinate system in which $[\mathbf{Q}]$ is diagonal so that

$$[\mathbf{Q}] = Q_x\hat{\mathbf{i}}\hat{\mathbf{i}} + Q_y\hat{\mathbf{j}}\hat{\mathbf{j}} + Q_z\hat{\mathbf{k}}\hat{\mathbf{k}}. \tag{13.120}$$

We have written the diagonal elements of $[\mathbf{Q}]$ with only one subscript here to simplify the notation. In this coordinate system,

$$\langle|\hat{\mathbf{r}} \cdot [\mathbf{Q}]|^2\rangle = \frac{1}{r^2}\langle x^2 Q_x^2 + y^2 Q_y^2 + z^2 Q_z^2\rangle. \tag{13.121}$$

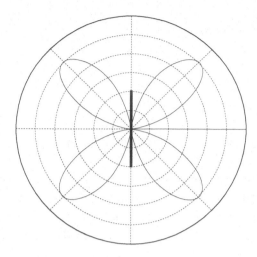

Figure 13.5: Radiation pattern for quadrupole radiation.

We have left out cross terms like $\langle xy \rangle$ because these vanish in the angular average for which we have used the notation

$$\langle f \rangle = \frac{1}{4\pi} \int f(\theta, \phi) d\Omega. \tag{13.122}$$

Each of the averages, such as $\langle (x/r)^2 \rangle$, gives a factor of $\frac{1}{3}$, so

$$\langle |\hat{\mathbf{r}} \cdot [\mathbf{Q}]|^2 \rangle = \frac{1}{3}(Q_x^2 + Q_y^2 + Q_z^2) = \frac{1}{3}\text{Tr}[\mathbf{Q}^2] \tag{13.123}$$

The other term can be written in the diagonal system as

$$\begin{aligned}
\langle |\hat{\mathbf{r}} \cdot [\mathbf{Q}] \cdot \hat{\mathbf{r}}|^2 \rangle &= \frac{1}{r^4} \langle ((x^2 Q_x^2 + y^2 Q_y^2 + z^2 Q_z^2)^2 \rangle \\
&= \frac{1}{r^4} \rangle x^4 Q_x^2 + y^4 Q_y^2 + z^4 Q_z^2 \rangle \\
&\quad + \frac{2}{r^4} \langle x^2 y^2 Q_x Q_y + y^2 z^2 Q_y Q_z + z^2 x^2 Q_z Q_x \rangle.
\end{aligned} \tag{13.124}$$

The angular averages are given by

$$\langle x^4 \rangle = \langle y^4 \rangle = \langle z^4 \rangle = \frac{1}{5} r^4, \tag{13.125}$$

and

$$\langle x^2 y^2 \rangle = \langle y^2 z^2 \rangle = \langle z^2 x^2 \rangle = \frac{1}{15} r^4, \tag{13.126}$$

so

$$\begin{aligned}
\langle |\hat{\mathbf{r}} \cdot [\mathbf{Q}] \cdot \hat{\mathbf{r}}|^2 \rangle &= \frac{1}{5}(Q_x^2 + Q_y^2 + Q_z^2) \\
&\quad + \frac{2}{15}(Q_x Q_y + Q_y Q_z + Q_z Q_x).
\end{aligned} \tag{13.127}$$

Also, since $\text{Tr}[\mathbf{Q}] = 0$,

$$
\begin{aligned}
(\text{Tr}[\mathbf{Q}])^2 &= (Q_x + Q_y + Q_z)^2 \\
&= Q_x^2 + Q_y^2 + Q_z^2 + 2(Q_x Q_y + Q_y Q_z + Q_z Q_x) = 0, \quad (13.128)
\end{aligned}
$$

so

$$
2(Q_x Q_y + Q_y Q_z + Q_z Q_x) = -(Q_x^2 + Q_y^2 + Q_z^2). \quad (13.129)
$$

Putting these results together, Eq. (13.127) can be written as

$$
\langle |\hat{\mathbf{r}} \cdot [\mathbf{Q}] \cdot \hat{\mathbf{r}}|^2 \rangle = \frac{2}{15}(Q_x^2 + Q_y^2 + Q_z^2) = \frac{2}{15}\text{Tr}([\mathbf{Q}]^2). \quad (13.130)
$$

We now have the angular averages for both terms of Eq. (13.114) for $\frac{dP}{d\Omega}$, and the total power is given by

$$
P = 4\pi \left\langle \frac{dP}{d\Omega} \right\rangle = \frac{ck^6}{90}\text{Tr}([\mathbf{Q}]^2). \quad (13.131)
$$

This general result agrees with the power given in Eq. (13.119) for a symmetric quadrupole moment with trace $\frac{3}{2}Q_0^2$. We derived the total power given in Eq. (13.131) for a diagonal quadrupole, but this result is true as well for a non-diagonal quadrupole because the trace of $[\mathbf{Q}]^2$ is invariant.

The result for a general quadrupole can be written as

$$
P = \frac{ck^6}{90}\text{Tr}([\mathbf{Q}]^2) = \frac{ck^6}{90}\sum_{i,j} Q_{ij}^2, \quad (13.132)
$$

where we have written out the double sum that gives $\text{Tr}([\mathbf{Q}]^2)$ for a symmetric matrix. This shows that the power is given in terms of the sum over the squares of all the matrix elements of the quadrupole dyadic.

13.11 Scattering of Electromagnetic Radiation.

13.11.1 Electric dipole scattering

An electromagnetic wave that is incident on a small bit of matter will induce oscillations in the matter. These oscillations will emit new electromagnetic radiation that can be interpreted as a scattering of the incident wave. We consider this scattering process in the electric dipole approximation, which is valid for objects much smaller than the wavelength of the incident radiation.

For an incident wave with electric field

$$
\mathbf{E} = \hat{\boldsymbol{\epsilon}} E_0 e^{i(kz - \omega t)}, \quad (13.133)
$$

the induced dipole moment is

$$
\mathbf{p} = \alpha \hat{\boldsymbol{\epsilon}} E_0 e^{-i\omega t}, \quad (13.134)
$$

where α is the polarizability of the matter.

From Chapter 6, the polarizability for various shapes with volume V is given by

$$\text{sphere}: \quad \alpha = \left(\frac{\epsilon - 1}{\epsilon + 2}\right)\left(\frac{3V}{4\pi}\right) \tag{13.135}$$

$$\text{disc}: \quad \alpha = \left(\frac{\epsilon - 1}{\epsilon}\right)\left(\frac{V}{4\pi}\right) \tag{13.136}$$

$$\text{needle}: \quad \alpha = (\epsilon - 1)\left(\frac{V}{4\pi}\right). \tag{13.137}$$

The disc and needle are each oriented with their axis along the incident beam. The polarizability for a general shape will be something between that for a needle and a disc.

The oscillating dipole moment radiates an electric field given by Eq. (13.70) to be

$$\mathbf{E}'(\mathbf{r}, t) = \alpha E_0 \hat{\boldsymbol{\epsilon}}_{\mathbf{T}} k^2 \frac{e^{i(kr - \omega t)}}{r}, \tag{13.138}$$

where $\hat{\boldsymbol{\epsilon}}_{\mathbf{T}}$ is the part of $\hat{\boldsymbol{\epsilon}}$ that is perpendicular to $\hat{\mathbf{r}}$. The power radiated per unit solid angle for a final polarization direction $\hat{\boldsymbol{\epsilon}}'$, is given by

$$\frac{dP}{d\Omega} = \frac{cr^2}{8\pi}|\hat{\boldsymbol{\epsilon}}' \cdot \mathbf{E}'|^2 = \frac{c\alpha^2 k^4 E_0^2}{8\pi}|\hat{\boldsymbol{\epsilon}} \cdot \hat{\boldsymbol{\epsilon}}'|^2. \tag{13.139}$$

A **differential cross section** is defined as the ratio of the radiated power per unit solid angle to the intensity of the incident beam. That is

$$\frac{d\sigma}{d\Omega} = \frac{\frac{dP}{d\Omega}}{\frac{c}{8\pi}E_0^2}. \tag{13.140}$$

For electric dipole scattering, the differential cross section is

$$\frac{d\sigma}{d\Omega} = \alpha^2 k^4 |\hat{\boldsymbol{\epsilon}} \cdot \hat{\boldsymbol{\epsilon}}'|^2. \tag{13.141}$$

Two polarization directions can be defined for the scattered wave. These are parallel and perpendicular to the plane of the scattering. From Eq. (13.141), we can see that scattered radiation will have the same type of polarization as the incident beam. That is, an incident beam polarized parallel to the scattering plane will scatter into outgoing radiation polarized parallel to the scattering plane, and similarly for perpendicular polarization.

An unpolarized incident beam can be treated by averaging the intensity for the two incident polarization directions. Thus, the differential cross sections for an unpolarized incident beam with the two different polarizations scattering through an angle θ are

$$\frac{d\sigma_\perp}{d\Omega} = \frac{1}{2}\alpha^2 k^4 \tag{13.142}$$

$$\frac{d\sigma_\parallel}{d\Omega} = \frac{1}{2}\alpha^2 k^4 \cos^2\theta, \tag{13.143}$$

If the final polarization is not measured, the differential cross section is the sum over the two possible polarizations

$$\frac{d\sigma}{d\Omega} = \frac{1}{2}\alpha^2 k^4 (1 + \cos^2\theta). \tag{13.144}$$

The **total cross section** is the integral of the differential cross section over all solid angle

$$\sigma_{\text{total}} = \int \frac{d\sigma}{d\Omega} d\Omega = \frac{8\pi}{3}\alpha^2 k^4. \tag{13.145}$$

The name 'cross section' comes from the fact that it has the units of an area. The incident beam intensity (which is the incident power per unit area) multiplied by the cross section area gives the power scattered by the beam.

The polarization Π perpendicular to the scattering plane of the scattered radiation is defined by

$$\Pi(\theta) = \frac{\frac{d\sigma_\perp}{d\Omega} - \frac{d\sigma_\parallel}{d\Omega}}{\frac{d\sigma_\perp}{d\Omega} + \frac{d\sigma_\parallel}{d\Omega}}. \tag{13.146}$$

For an unpolarized incident beam, the polarization of the scattered beam is

$$\Pi(\theta) = \frac{\sin^2\theta}{(1 + \cos^2\theta)}. \tag{13.147}$$

The electric dipole scattering cross section depends on the wave number k to the fourth power. This corresponds to a $1/\lambda^4$ dependence on the wavelength. This strong dependence on the wavelength of the scattered light is the basis for Rayleigh's explanation that the sky is blue because molecules in the atmosphere preferentially scatter light in the short wavelength, blue end of the spectrum. The blue light from the sky will also be horizontally polarized if it has scattered through an appreciable angle.

This is another advantage of polaroid sun glasses, which preferentially absorb light with horizontal polarization, thus reducing glare from the sky as well as glare from horizontal surfaces. The polarization of light from the sky can be demonstrated by rotating polaroid sun glasses while looking through them at the sky.

An important case is the scattering of an electromagnetic wave by a point charge, which could be considered as an electron without spin. A real electron with spin and a magnetic moment would have to be treated quantum mechanically. From Newton's law, the electron is displaced by the electric field, inducing an oscillating electric dipole moment with an effective polarizability

$$\alpha = \frac{e^2}{m\omega^2}. \tag{13.148}$$

Thus the differential cross section for scattering of an electromagnetic wave by a point charge is

$$\frac{d\sigma}{d\Omega} = \left(\frac{e^2}{mc^2}\right)^2 \frac{(1 + \cos^2\theta)}{2}. \tag{13.149}$$

The total cross section is (using $\langle \cos^2 \theta \rangle = 1/3$)

$$\frac{d\sigma}{d\Omega} = \frac{8\pi}{3} \left(\frac{e^2}{mc^2} \right)^2. \tag{13.150}$$

This classical result, called the **Thomson cross section**, turns out to agree with the quantum mechanical cross section in the limit of vanishing wave number k. This agreement is due essentially to dimensional considerations. In the limit $k \to 0$, the only parameters available are e, m, and c. The cross section must have the units of area, and the only way to achieve this is by the combination $(e^2/mc^2)^2$ times some dimensionless number.

The combination $e^2/m_e c^2 = 2.82$ fm is called the **classical radius**, r_e, of the electron. This nomenclature is colloquial, but not rigorous. If an electron of radius r_e scattered all the incident radiation striking it, the total cross section would be $4\pi r_e^2$, which is close to the Thomson cross section. But that is not how the scattering occurs, even classically. The "classical radius" has nothing to do with the size of the electron, which must be a point particle to actually get the Thomson cross section.

13.11.2 Scattering by a conducting sphere, magnetic dipole scattering

An electromagnetic wave incident on a conducting sphere will induce both electric and magnetic oscillating dipoles. We saw in Chaper 4 that a uniform electric field E_0 induces an electric dipole moment $\mathbf{p} = R^3 \mathbf{E_0}$ in a conducting sphere of radius R. We assume the wavelength is much larger than R so that it is reasonable to use a uniform field here. This gives an electric polarizability of $\alpha = R^3$.

The incident magnetic field of the wave, $\mathbf{B} = \hat{\mathbf{k}} \times \mathbf{E}$, will induce an oscillating magnetic moment. The problem of finding the induced magnetic moment in this case is mathematically equivalent to case of a dielectric sphere in a uniform electric field for which the induced electric dipole moment is given by Eq. (6.69). The boundary condition on the magnetic field is that the component of $\mathbf{B}$ normal to the conducting surface vanishes. This corresponds to $\epsilon = 0$ in the electrostatic case for which the induced electric dipole moment from Eq. (6.69) is $\mathbf{p} = -\frac{1}{2}R^3 \mathbf{E_0}$. Consequently, the magnetic polarizability for a conducting sphere is $\alpha_m = -\frac{1}{2}R^3$. This result can also be derived using the surface current induced on the conducting sphere. (See problem 13.11.)

Using Eq. (13.103) for magnetic dipole radiation, the two induced moments produce an electric field in the radiation zone

$$
\begin{aligned}
\mathbf{E(r},t) &= (\mathbf{p_T} - \hat{\mathbf{r}} \times \boldsymbol{\mu}) \frac{k^2 e^{i(kr - \omega t)}}{r} \\
&= \left[\hat{\boldsymbol{\epsilon}}_{\mathbf{T}} + \frac{1}{2} \hat{\mathbf{r}} \times (\hat{\mathbf{k}} \times \hat{\boldsymbol{\epsilon}}) \right] \frac{E_0 R^3 k^2 e^{i(kr - \omega t)}}{r}
\end{aligned}
$$

$$= \left[\hat{\boldsymbol{\epsilon}}_{\mathbf{T}} - \frac{1}{2}\hat{\boldsymbol{\epsilon}}(\hat{\mathbf{r}}\cdot\hat{\mathbf{k}}) + \frac{1}{2}\hat{\mathbf{k}}(\hat{\mathbf{r}}\cdot\hat{\boldsymbol{\epsilon}}) \right] \frac{E_0 R^3 k^2 e^{i(kr-\omega t)}}{r} \qquad (13.151)$$

The differential cross section for initial polarization $\hat{\boldsymbol{\epsilon}}$ and final polarization $\hat{\boldsymbol{\epsilon}}'$ is

$$\frac{d\sigma}{d\Omega} = R^3 k^4 \left| \hat{\boldsymbol{\epsilon}}\cdot\hat{\boldsymbol{\epsilon}}' \left[1 - \frac{1}{2}(\hat{\mathbf{r}}\cdot\hat{\mathbf{k}}) \right] + \frac{1}{2}(\hat{\mathbf{r}}\cdot\hat{\boldsymbol{\epsilon}})(\hat{\boldsymbol{\epsilon}}'\cdot\hat{\mathbf{k}}) \right|^2. \qquad (13.152)$$

The differential cross sections for an unpolarized incident beam and $\perp$ and $\parallel$ final polarizations are

$$\frac{d\sigma_\perp}{d\Omega} = \frac{1}{2}R^6 k^4 \left[1 - \frac{1}{2}\cos\theta \right]^2 \qquad (13.153)$$

$$\frac{d\sigma_\parallel}{d\Omega} = \frac{1}{2}R^6 k^4 \left[\cos\theta - \frac{1}{2} \right]^2. \qquad (13.154)$$

The differential cross section summed over both polarizations is

$$\frac{d\sigma}{d\Omega} = \frac{1}{8}R^6 k^4 [5(1 + \cos^2\theta) - 8\cos\theta], \qquad (13.155)$$

and the polarization of the scattered radiation is

$$\Pi = \frac{3\sin^2\theta}{5(1 + \cos^2\theta) - 8\cos\theta}. \qquad (13.156)$$

The differential cross section for the radiation scattered by the conducting sphere is plotted in Fig. 13.6. The figure also shows (as the dashed curve) the cross section for a dielectric sphere [with polarizability given by Eq. (13.135) for large ϵ], which is pure electric dipole scattering. The sharp backward peaking of the radiation scattered by the conducting sphere is due to the term $-8\cos\theta$ in Eq. (13.155), which arises from the interference of the electric and magnetic dipoles.

The polarization of the radiation scattered by each type of sphere is plotted in Fig. 13.7. Again, we can see the effect of the electric-magnetic interference producing the asymmetry (about $\cos\theta = 0$) in the polarization by the conducting sphere.

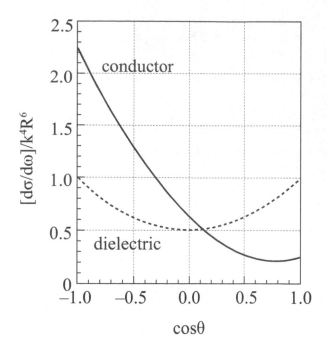

Figure 13.6: Differential scattering cross-sections for scattering by a conducting sphere and a dielectric sphere.

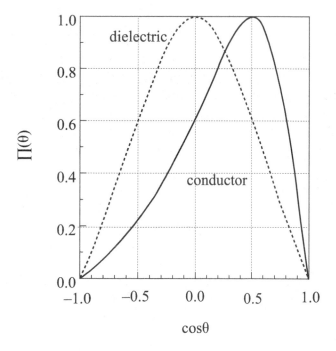

Figure 13.7: Polarization of radiation scattered by a conducting sphere and a dielectric sphere.

13.12   Problems:

1. Derive the following formulas for radiation by a center fed antenna:

   (a) Eq. (13.54).

   (b) Eq. (13.57).

   (c) Eq. (13.58).

   (d) Eq. (13.59).

2. Integrate $\frac{dP}{d\Omega}$ numerically to find the formula for the radiation resistance (in ohms) for a half wave, and for a full wave antenna. (Consider the full wave antenna as two half wave antennas in determining its current.)

3. A thin linear antenna of length L carries an oscillating current

$$I(z,t) = I_0 \sin(2\pi|z|/L)e^{-i\omega t}, \quad |z| < L/2. \tag{13.157}$$

   (a) Sketch this current as a function of z.

   (b) Find the radiation pattern for this antenna for $kL = 2\pi$, and for $kL = \pi$.

   (c) Plot the angular distribution for each of the above cases.

4. For the antenna in the previous problem:

   (a) Find the maximum $\frac{dP}{d\Omega}$, and the angle for which this occurs.

   (b) Find the total power radiated by this antenna, and the formula for its radiation resistance (in ohms).

5. Consider a model of the hydrogen atom as a heavy positive proton, surrounded by a spherical electron cloud of radius R, with a uniform charge and mass density, having the total charge and mass of an electron.

   (a) Show that the angular frequency of oscillations of this atom (with amplitude $A \leq R$) is

$$\omega = \left[\frac{(e^2/R)}{mc^2}\right]^{\frac{1}{2}} \frac{c}{R}. \tag{13.158}$$

   (b) Evaluate ω and the corresponding wavelength of radiation. (Use numerical values from the previous problem.)

   (c) Find the oscillating electric dipole moment of this atom for $A_0 = R$.

   (d) Find the total power radiated by this atom.

   (e) Find the lifetime for exponential decay of the oscillation. Evaluate the lifetime in seconds.

380 CLASSICAL ELECTROMAGNETISM

6. Evaluate the lifetime of a Rutherford hydrogen atom from Eq. (13.84). Use as data: $R = .53\text{Å}$, $mc^2 = 511$ keV, and the binding energy of hydrogen is 13.6 eV.

7. A beam of 2 keV electrons is stopped in a distance of 0.01 cm.

   (a) Calculate the total energy of the radiation emitted by each electron.

   (b) Find the ratio of the emitted energy to the initial electron energy. Use this ratio to find the radiated power for an X-ray machine with 300 Watts of power input to accelerate the electrons.

   (c) Calculate the maximum intensity of the radiation at a distance of 20 cm from the stopping target.

8. Find and sketch the absolute square of the Fourier transform of the step function in Eq. (??), and estimate its spread in ω.

9. A neutron star of mass M and radius R has a magnetic moment μ. It is rotating with angular velocity w about an axis perpendicular to μ.

   (a) Find the electomagnetic power it radiates.

   (b) Find the energy radiated in one rotation, and compare this with the rotational kinetic energy of the star.

   (c) Find the fractional change $\Delta w/w$ in one year.

10. (a) A thin linear antenna of length L carries an oscillating current

$$I(z,t) = I_0 \sin(2\pi z/L)e^{-i\omega t}, \quad |z| < L/2. \tag{13.159}$$

   Find the quadrupole moment (Q_0) of the antenna.

   (b) Find the quadrupole radiation pattern. Sketch the angular distribution.

   (c) Find the maximum intensity and the angle for which it occurs for the quadrupole.

11. A conducting sphere of radius R in an oscillating magnetic field $\mathbf{B} = \mathbf{B_0}e^{-i\omega t}$ acquires an induced magnetic moment μ.

   (a) Find the surface current $\mathbf{K}$ induced on the conducting sphere. (Use the boundary condition on $\mathbf{B}$, including the magnetic field due to the induced dipole.)

   (b) Integrate the current over the sphere to find the induced magnetic moment.

Chapter 14

Special Relativity.

14.1   The Need for Relativity.

We have seen that classical electromagnetism as fashioned by Maxwell is a powerful and compelling theory. However, we have also seen that it has a few nagging loose ends. One that we saw in the last chapter is that the classical theory predicts a short lifetime for any atom due to electromagnetic radiation, and thus is inconsistent with the stable existence of atoms.

This difficulty was circumvented by the Bohr model, leading to the birth of quantum mechanics, and extending classical mechanics and electromagnetism to the quantum realm. We will not deal with this quantum extension of electromagnetism here, except to keep it in mind and to realize that some classical results (such as radiation from atoms) may not apply in the physical world.

The other class of unresolved issues in classical electromagnetism has to do with how to treat phenomena that seem to depend on the coordinate system. We saw this in Section 7.5, in attempting to define the force between two moving particles. It also arose when we deduced the velocity c for electromagnetic waves. The question should be asked, "velocity with respect to what?". We glossed over that at the time so as not to confuse the original discussion, but an interested reader might have wondered about it.

There are two kinds of answer to this question that we can recall from our experience. For a thrown object, like a baseball, the velocity is with respect to the thrower. For sound, the velocity is independent of the speaker's motion, but is taken with respect to the medium.

Which is appropriate for light? (We use light as the typical example for electromagnetic radiation.) We will see in this chapter that, for light, neither of these previous examples holds. A third alternative, special relativity that will be developed here, has the speed of light independent of either the thrower or a medium. This may seem counter-intuitive, but only until we understand it.

Maxwell, having a complete foundation in classical mechanics, reasoned that the failure by himself and others to develop a consistent theory for the interaction of two moving particles was compelling evidence for the existence of a material ether. The final conclusion in his *Treatise on Electricity and Magnetism* was "Hence all these theories lead to the conception of a medium in which the propagation takes place."

There were numerous experimental attempts to demonstrate the existence of an ether toward the end of the nineteenth, and in the early part of the twentieth century. These tests generally depended on the velocity of motion, v, of a reference frame with respect to the ether. All of the accurate tests depended on the ratio $(v/c)^2$. The largest available velocity at the time was the 30 km/sec velocity of the Earth in its orbit around the sun. This made experimental tests of the ether hypothesis extremely difficult because the acccuracy needed was $(v/c)^2 = 10^{-8}$.

An excellent summary of the major experiments (including original references) is given in [Panofsky and Phillips]. These experiments were all either inconclusive, or were incompatible with an ether. Later, in the twentieth century, velocities of elementary particles close to the speed of light became commonplace (in high energy laboratories), and Einstein's theory of special relativity, in place of an ether hypothesis, is now confirmed thousands of times each day.

We will not dwell in detail on the early experiments because our motivation for the need for relativity as the basis for a consistent theory of electromagnetism will be based more on the awkward nature of the ether hypothesis. We briefly mention two of the early experiments here. The best known, and most accurate early test was the famous **Michelson Morley experiment** using the **Michelson interferometer** to measure the difference in the velocity of light in two perpendicular directions, presumably parallel and perpendicular to the velocity of the earthly laboratory. The lack of any difference, or of any seasonal effect as the Earth moved in different directions, was inconsistent with light traveling in a fixed ether. These experiments were first performed in 1881, and were continued for many years with increasing accuracy using various sources of light, always with a null result.

Another experiment tested the implications of Eq. (7.32) for the electromagnetic force between two moving charged particles. The magnetic force between two such charges moving with equal parallel velocities as shown in Fig. 14.1 would lead to a torque

$$
\begin{aligned}
\boldsymbol{\tau} &= \frac{qq'}{c^2 d^3}\mathbf{d}\times[\mathbf{v}\times(\mathbf{v}\times\mathbf{d})] \\
&= \frac{qq'}{c^2 d^3}(\mathbf{d}\times\mathbf{v})(\mathbf{d}\cdot\mathbf{v}) \\
&= \frac{qq'\hat{\boldsymbol{\tau}}}{2d}\left(\frac{v}{c}\right)^2\sin 2\theta.
\end{aligned}
\tag{14.1}
$$

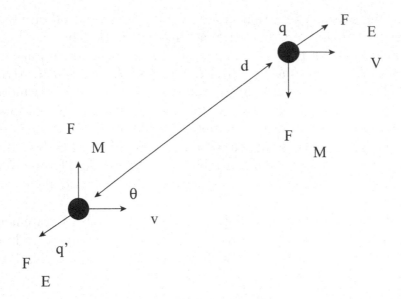

Figure 14.1: Charges q and q', each moving with velocity $\mathbf{v}$, a distance $\mathbf{d}$ apart. The electric forces, $\mathbf{F_E}$, are collinear, but the magnetic forces, $\mathbf{F_M}$, produce a torque.

Trouton and Noble attempted to measure this torque in 1903 by suspending a charged capacitor from a torsion balace, but found no indication of torque as the capacitor moved through the ether.

Einstein presumably knew of these null experiments, but based his 1905 introduction of the theory of special relativity more on the internal inconsistencies of Maxwellian classical electromagnetism. Indeed, the title of his 1905 paper was "On The Electrodynamics of Moving Bodies".

The predicted dependence of the speed of light on motion through the ether posed a serious conceptual problem, beyond its lack of experimental verification. The parameter c has a deeper significance in classical electromagnetism than just the speed at which light waves travel. It was originally introduced in 1856 as the ratio between the esu and emu units of charge. It subsequently appeared in virtually all the equations relating electric and magnetic phenomena, including Maxwell's equations.[1]

The prediction that c equaled the speed of a light wave was only one aspect of its significance in electromagnetic theory. If the parameter c on the Earth's surface could vary as the Earth moved through the ether, then the equations of electromagnetism would have a seasonal variation. While this variation could not be accurately measured for most of the nineteenth century, it made electromagnetism an unwieldy theory, conceptually. Now, with present day accuracy, almost any electromagnetic measurement could test whether c varied

---

[1]In SI units, $\frac{1}{4\pi\epsilon_0}$ is actually c^2, divided by 10^7 to compensate for the change from Gaussian to SI units.

by just noting seasonal variation in any result. Any variation of c would also make the construction of high energy accelerators impossible.

A basic premise of physics, following the great insights and achievements of Galileo and Newton, was the physical principle of Galilean relativity. This was the belief that the laws of physics were unchanged by a linear motion with constant velocity. All of classical mechanics for hundreds of years confirmed this physical principle. Until the nineteenth century, all known interactions between groups of particles depended only on relative velocities, as a manifestation of the physics of Galilean relativity. We saw in Chapter 7 that the classical theory of the magnetic force between charged particles violated this principle.

Actually, there is very little difference between Galileo's statement of his principle of relativity and that proposed by Einstein, many years later. Einstein's first postulate for the theory of relativity introduced in his 1905 paper was

> **"The same laws of electrodynamics and optics will be valid for all frames of reference for which the equations of mechanics hold good."**[2]

That is not much different than what Galileo said almost 300 years earlier: In what is called his 'parable of the ship', he described a number of mechanical experiments on a moving ship, concluding with

> **"Have the ship proceed with any speed you like, so long as the motion is uniform and not fluctuating this way and that. You will discover not the least change in all the effects named, nor could you tell from any of them whether the ship was moving or standing still".**

The only difference in these two statements is that Galileo restricted his statement to mechanical experiments because, in 1632, he did not know that "effects" would be extended to include electromagnetic effects many years later. I think that Galileo would have been horrified to see that the fundamental laws of electromagnetism could depend on a parameter, c, that varied with the movement of the ship.

Einstein's first postulate, would have put Galileo's mind at ease by adding electromagnetism to Galileo's principle of relativity. In this sense, Einstein just brought Galilean relativity up to date, so Einstein's theory of special relativity is not really that revolutionary.

A number of postulates for special relativity have been given in papers and books since 1905, all agreeing somewhat, but different in wording. We

---

[2]It would probably have been better for Einstein to have said, "...all frames of reference moving with constant relative velocities", since his theory of relativity would change most of "the equations of mechanics".

present here what we consider a concise, yet sufficient, postulate combining the postulates of Galileo and Einstein:

> **The laws of physics are the same in all coordinate systems moving with constant relative velocity with respect to each other.**

One important aspect of this postulate is that it constrains what are "the laws of physics". That is, in order to be a law of physics, the law must hold in all coordinate systems moving with constant relative velocity with respect to each other.

The constancy of the speed of light was added as a separate postulate by Einstein but that is unnecessary, since, as we discussed above, variation of the parameter, c, would make the basic equations of electromagnetism frame dependent, and we have shown that Maxwell's equations lead to c being the speed of light.

With this background, we show in the next section that extending the physical principle of Galilean invariance to include electromagnetism requires replacing the mathematical Galilean transformation of spatial coordinates with the Lorentz transformation of space-time coordinates, and greatly modifying Newton's laws of mechanics. Galileo and Newton had the right physical principles, but should be excused for not producing the mathematical transformations required for their application to electromagnetism several hundred years later.

14.2 Mathematical Basis of Special Relativity, the Lorentz Transformation.

We consider two Cartesian coordinate systems aligned with axes parallel to each other as in Fig. 14.2. System $S'(x', y', z')$ is moving with velocity $\mathbf{v}$ in the x direction of system $S(x, y, z)$. We will speak of system S as the fixed system and system S' as the moving system. We should keep in mind, though, that the physical principle of relativity requires that these designations be interchangeable. That is, none of the physics can change if S is chosen as the moving system with S' the fixed system.

The transformation equations proposed by Galileo to connect the coordinates in the two systems were

$$x' = x - vt \tag{14.2}$$
$$y' = y \tag{14.3}$$
$$z' = z, \tag{14.4}$$

where t is a universal time variable.

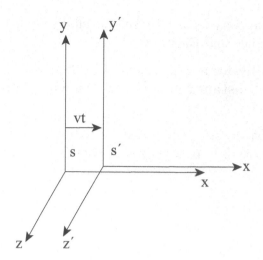

Figure 14.2: Coordinate system S', moving with velocity v along the x-axis of coordinate system S. The x-axis has been displaced slightly for clarity.

These equations lead to the standard classical rule for the addition of velocities that the velocity $\mathbf{u}'$ in system S' of a particle moving with velocity $\mathbf{u}$ in system S is given by

$$\mathbf{u}' = \mathbf{u} - \mathbf{v}. \qquad (14.5)$$

There was no reason to expect that this result should not apply to the velocity of a light wave, which led to the problems with electromagnetism discussed in the previous section.

If we agree with Einstein, and all who followed his lead, the Galilean coordinate transformations and the velocity addition that follows from them must be altered. That is, the physical principle of Galilean relativity is inconsistent with the mathematical Galilean transformation equations when applied to electromagnetism. To preserve the physics, the mathematics has to be altered.

The culprit is the assumption, made implicitly by Galileo and Newton, of a universal time. They could not be faulted for this because the assumption served well for a long time, and even today some people shake their heads at the implications of a time variable that changes from system to system. Dropping the universality of time, and including its transformation as just one other coordinate in the transformation equations, is essential to the implementation of special relativity.

We build up the transformation equations between the systems S and S' step by step:

1. We keep the two transverse coordinates unchanged, as in Eqs. (14.3) and (14.4). One reason for this is a desire to change the Galilean equations as little as possible. A more compelling reason is that we require that the

inverse transformation from S' to S be the same as the direct transformation from S to S', which could not be accomplished if the transverse coordinates changed.

2. We put into mathematics the statement that "system S' moves with velocity **v** in the x direction with respect to system S." This means that the origin of S' moves with velocity **v** in system S, so the locus $x - vt = 0$ must correspond to $x' = 0$. We accomplish this by making x' proportional to $x - vt$, writing the transformation equation as

$$x' = \gamma(x - vt), \qquad (14.6)$$

where γ is a constant to be determined. We have chosen a linear relation between x and x' so that a uniform velocity in S will transform into a uniform velocity in S'. (This will be shown shortly.) That is, the motion of system S' with constant velocity should not induce acceleration in S' if there was none in system S.

3. The key element in extending Galilean relativity to electromagnetism is to drop the notion of a universal time, and write a transformation equation for t. We take

$$t' = \overline{\gamma}(t - \beta x/c), \qquad (14.7)$$

We have put c into the equation in order to make the constant β dimensionless. The t transformation equation is linear, again to avoid introducing acceleration in S' if there was none in S.

4. Because of the postulate that there is no physical difference between the systems S and S', the transformation equations from S' to S must be the same as the original transformation from S to S'. The only difference is that $v \rightarrow -v$, because system S is moving with velocity $-v$ with respect to S'. This leads the identity

$$x = \gamma(x' + vt') = x(\gamma^2 - \beta\gamma\overline{\gamma}v/c) + vt(\gamma\overline{\gamma} - \gamma^2), \qquad (14.8)$$

where we have used Eqs. (14.6) and (14.7) to substitute for x' and t'. For Eq. (14.8) to be an identity requires

$$\overline{\gamma} = \gamma \quad \text{and} \quad \gamma^2(1 - \beta v/c) = 1. \qquad (14.9)$$

5. We can derive a transformation law for velocities by taking derivatives of Eqs. (14.6) and (14.7). If $u = \frac{dx}{dt}$ is a velocity in the x direction in S, the corresponding velocity u' in S' will be given by

$$u' = \frac{dx'}{dt'} = \frac{\gamma(dx - vdt)}{\overline{\gamma}(dt - \beta dx/c)}$$

$$= \frac{(u - v)}{(1 - \beta u/c)}. \qquad (14.10)$$

6. We require that, if the velocity in S is the speed of light c, then the velocity in S' also be c. If it were otherwise, then, as we have argued above, the physics would be different in the two systems. This requirement gives us

$$c = \frac{(c - v)}{(1 - \beta)}. \tag{14.11}$$

We divide by c to get

$$1 = \frac{(1 - v/c)}{(1 - \beta)}. \tag{14.12}$$

This equation must hold for any velocity v, so we have $\beta = v/c$.

Thus the x and t transformation equations become

$$x' = \gamma(x - vt) \tag{14.13}$$
$$t' = \gamma(t - vx/c^2), \tag{14.14}$$

with γ given by

$$\gamma = \frac{1}{\sqrt{1 - v^2/c^2}}. \tag{14.15}$$

The equations we have just derived are called the **Lorentz transformation** equations. We have derived them here for a transformation velocity along the x axis, but any axis could be chosen with similar results. For a transformation velocity in a general direction, the Lorentz transformation can be written in vector form as

$$t' = \gamma(t - \mathbf{r}\cdot\mathbf{v}/c^2) \tag{14.16}$$
$$\mathbf{r}'_{\parallel} = \gamma(\mathbf{r}_{\parallel} - \mathbf{v}t) \tag{14.17}$$
$$\mathbf{r}'_{\perp} = \mathbf{r}_{\perp}. \tag{14.18}$$

The Lorentz transformation was originally derived by H. A. Lorentz as part of a failed attempt to give a classical description of the electromagnetic and mechanical properties of the electron. His interpretation was that the equations represented real physical distortions of matter and time, whereas the interpretation here is that the equations do not refer to matter, but to the coordinates used to described the orientation of matter. A physical manifestation of this difference is that by Lorentz's original interpretation, stresses could be set up by the distortion of matter, but in the relativistic interpretation no stress is imposed on the matter, since there can be no physical consequence of the constant velocity motion of S'.

The physical basis of special relativity is that the laws of physics are unchanged by any Lorentz transformation from one coordinate system to another one moving with constant relative velocity. We will refer to any set of coordinate systems that are connected by Lorentz transformations as **Lorentz systems**.

14.3 Spatial and Temporal Consequences of the Lorentz Transformation.

The Lorentz transformation equations are

$$t' = \gamma(t - xv/c^2) \tag{14.19}$$
$$x' = \gamma(x - vt) \tag{14.20}$$
$$y' = y \tag{14.21}$$
$$z' = z, \tag{14.22}$$

with

$$\gamma = \frac{1}{\sqrt{1 - v^2/c^2}}. \tag{14.23}$$

In this section, we derive some of the remarkable consequences of these equations.

14.3.1 Relativistic addition of velocities

We have already seen in Eq. (14.10) that the relativistic law for addition of parallel velocities is modified from the simple Galilean addition. We write it here in its more usual form of an addition by finding the velocity $\mathbf{u}$ in system S of a particle moving with velocity $\mathbf{u}'$ in a system S' that is moving with velocity $\mathbf{v}$ with respect to S. Equation (14.10) shows that the component of velocity parallel to the transformation velocity transforms as

$$\mathbf{u}_\parallel = \frac{\mathbf{u}'_\parallel + \mathbf{v}}{1 + \mathbf{u}' \cdot \mathbf{v}/c^2}. \tag{14.24}$$

The transverse coordinates don't change in the Lorentz transformation, but the time does transform. This causes the transverse velocities to transform as

$$\mathbf{u}_\perp = \frac{\mathbf{u}'_\perp}{\gamma_v(1 + \mathbf{u}' \cdot \mathbf{v}/c^2)}, \tag{14.25}$$

where the denominator results from the time transformation of Eq. (14.19). We have added the subscript v to γ to emphasize that it is the γ for the transformation velocity.

For small velocities, the relativistic equations reduce to the Galilean velocity addition, but for larger velocities there are new features. No addition of velocities can lead to a velocity larger than c, and adding a velocity c to any velocity results in the velocity c. (See problem 14.2.) Also two velocities that are parallel in one system, may not be parallel in a moving system. (See problem 14.9.) We see that some things we were taught as 'obvious' in a general physics class, do not hold if things move too fast.

The fact that no addition of velocities can exceed c means that there can be no way of accelerating a particle to a velocity faster than that of light in vacuum. There is no known violation of this relativistic speed limit. (No speeder has been caught.) Some conjectures in quantum field theory permit particles, called **tachyons**, to be produced at superluminal velocities. But there is no experimental evidence or strong theoretical expectation that tachyons exist, and we will not consider them here.

We can find transformation equations for acceleration by taking an additional time derivative of the velocities in Eq. (14.24) and (14.25). The result is (See problem 14.3.)

$$\mathbf{a}_\parallel = \frac{\mathbf{a}'_\parallel}{\gamma_v^3(1 + \mathbf{u}'\cdot\mathbf{v}/c^2)^3} \tag{14.26}$$

$$\mathbf{a}_\perp = \frac{\mathbf{a}'_\perp + (\mathbf{u}'\times\mathbf{a}')\times\mathbf{v}/c^2}{\gamma_v^2(1 + \mathbf{u}'\cdot\mathbf{v}/c^2)^3}. \tag{14.27}$$

Note that, although we cannot treat accelerating coordinate systems in special relativity, there is no problem in treating accelerations within coordinate systems. That is, the velocities $\mathbf{u}$ and $\mathbf{u}'$ can be functions of time and have acceleration, but the transformation velocity $\mathbf{v}$ must be time independent.

14.3.2   Lorentz contraction

We next consider the measurement of the length of a moving object. For simplicity we will consider a stick of length L'. We first have to define precisely how to measure the length of a fixed or a moving object. To do so, we can place the object on the x axis of the coordinate system, and record x_1 and x_2, the x values of each of its ends. For a fixed stick, we can do this at our leisure, first recording x_1 and then walking to the other end of the stick to record x_2. But to measure the length of a moving stick, this procedure would lead to obvious error. The stick would have moved before we could get to record the second x value. To measure the length of a moving stick, we have to measure x_1 and x_2 **simultaneously**.

This turns out to be the key step in measuring the length of a moving object. To put this into mathematics, we place the stick at rest along the x' axis in system S', so $L' = x'_1 - x'_2$. Then, we measure its length in system S by recording x_1 and x_2, being careful to keep $t_1 = t_2$. (Note that t'_1 will not equal t'_2.)

The Lorentz transformation of the coordinates is

$$x'_1 = \gamma(x_1 - vt_1) \tag{14.28}$$

$$x'_2 = \gamma(x_2 - vt_2). \tag{14.29}$$

The two length measurements are related by

$$L' = x'_1 - x'_2 = \gamma(x_1 - x_2) = \gamma L. \tag{14.30}$$

We got this result because the times t_1 and t_2 were equal and cancelled in taking the difference.

The relation between L and L' is usually written as

$$L = \frac{L'}{\gamma}. \tag{14.31}$$

In words, this means that a measurement of the length of a stick of length L' will result in $L = L'/\gamma$ if the stick is moving. Since γ is always greater than 1, a length measurement of a moving stick will give a shorter length than its length at rest. This phenomenon is generally called **Lorentz contraction**, although, as we have emphasized, it is not a change in the moving object as Lorentz envisioned, but a result of coordinate transformation. (This point is discussed further in Section 14.4.2.)

The apparent contraction of the moving object is also called Fizgerald-Lorentz contraction, because George Fitzgerald suggested a physical contraction of the moving arms of Michelson's interferometer as the cause of his null result. But again, it is not a physical contraction of the moving object. Cartoons you may have seen of a space ship collapsing like an accordian because it goes too fast, are just cartoons.

One suggested paradox by doubters of relativity arises if we consider a meter stick at rest on a moving relativistic train at it passes a station where an identical meter stick is at rest. People in the train and those in the station will each measure the others' stick as shorter. This occurs because each group will claim that the other group measured the two ends of the moving (for them) stick at different times. Due to the Lorentz tranformation of time, although t_1 and t_2 in Eqs. (14.28) and (14.29) are equal, t_1' and t_2' are not. This lack of absolute simultaneity for events that are separated in space is an important consequence of special relativity.

14.3.3 Time dilation

We consider a clock at rest in a system S' that is moving with respect to a fixed system S. What we mean by "at rest" is that in system S', the x' coordinate of the clock does not change with time, so $x_1' = x_2'$ even if $t_1' \neq t_2'$.

A time interval Δt is defined as the difference between two measured times t_1 and t_2. The times in each system are related by the Lorentz transformation equations

$$
\begin{align}
t_1 &= \gamma(t_1' + x_1'v/c^2) \tag{14.32} \\
t_2 &= \gamma(t_2' + x_2'v/c^2). \tag{14.33}
\end{align}
$$

We have used the inverse transformation so as to be able to implement the condition that $x_1' = x_2'$. Time intervals in each system will be related by

$$\Delta t = t_1 - t_2 = \gamma(t_1' - t_2'), \tag{14.34}$$

so a time interval for a moving clock is given by

$$\Delta t = \gamma \Delta t'. \tag{14.35}$$

Since $\gamma > 1$, this means that a time interval in system S where the clock is moving will be longer than the corresponding time interval in system S' where the clock is at rest. This phenomenon is called **time dilation**. In words, it is sometimes referred to as "Moving clocks slow down." But it is due to the transformation of the time variable between moving systems, and not to any change in the physical mechanism of the clock. (This point is discussed further in Section 14.5.2.)

14.4 Mathematics of the Lorentz Transformation.

The Lorentz transformation equations of Eqs. (14.19) - (14.22) can be put in a more symmetric form by introducing the coordinates

$$x_0 = ct \tag{14.36}$$
$$x_1 = x \tag{14.37}$$
$$x_2 = y \tag{14.38}$$
$$x_3 = z. \tag{14.39}$$

Then the Lorentz transformation can be written as

$$x_0' = \gamma(x_0 - \beta x_1) \tag{14.40}$$
$$x_1' = \gamma(x_1 - \beta x_0) \tag{14.41}$$
$$x_2' = x_2 \tag{14.42}$$
$$x_3' = x_3, \tag{14.43}$$

with

$$\beta = v/c, \qquad \gamma = 1/\sqrt{1 - \beta^2}. \tag{14.44}$$

The four dimensional space of the variables x_0, x_1, x_2, x_3 is called **space-time**. The time variable, written now as x_0, is just one of the four variables of space-time, and transforms along with the other coordinates. The Lorentz transformation of the four space-time coordinates can be interpreted as a generalized rotation of coordinates in four dimensions. To see this, we first review the mathematical properties of rotations in three dimensions.

14.4.1 Three dimensional rotations

The transformation equations for a rotation through an angle θ about the z axis (See Fig. 14.3) are given by

$$x' = x\cos\theta + y\sin\theta \tag{14.45}$$

$$y' = -x \sin\theta + y \cos\theta \tag{14.46}$$
$$z' = z \tag{14.47}$$

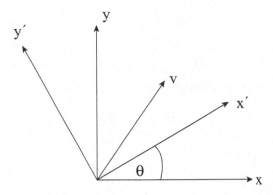

Figure 14.3: Rotation of coordinates by an angle θ.

Letting $x_1 = x$, $x_2 = y$, $x_3 = z$, these equations can be written as

$$x'_i = \sum_{j=1}^{3} R_{ij} x_j, \quad i = 1, 2, 3. \tag{14.48}$$

In order to simplify summation equations, we introduce the **Einstein summation convention** by which terms with repeated indices are to be summed over the repeated indices. With this convention we can write

$$x'_i = R_{ij} x_j. \tag{14.49}$$

This notation can be deceptively simple, so we have to keep in mind that this is really three equations, each with three terms. Equation (14.49) can be written (for a rotation about the z axis) as the matrix equation

$$\mathbf{x}' = [\mathbf{R}^{(z)}(\theta)]\mathbf{x}, \tag{14.50}$$

with the transformation coefficients R_{ij} being the ij elements of a rotation matrix

$$[\mathbf{R}^{(z)}(\theta)] = \begin{pmatrix} \cos\theta & \sin\theta & 0 \\ -\sin\theta & \cos\theta & 0 \\ 0 & 0 & 1 \end{pmatrix}. \tag{14.51}$$

A rotation transforms an orthogonal coordinate system into another orthogonal coordinate system. (By 'orthogonal', we will generally mean orthonormal.) The basic unit vectors of an orthogonal coordinate system are defined by

$$n_i^{(r)} = \delta_{ri}, \tag{14.52}$$

where $n_i^{(r)}$ is the ith component of unit vector r, and δ_{ri} is the Kronecker delta. The orthogonality of each system leads to

$$\delta_{rs} = n_i'^{(r)} n_i'^{(s)} = R_{ij} R_{ik} n_j^{(r)} n_k^{(s)} = R_{ij} R_{ik} \delta_{rj} \delta_{sk} = R_{ir} R_{is}. \tag{14.53}$$

Thus a rotation matrix must satisfy the condition

$$R_{ir} R_{is} = \delta_{rs}. \tag{14.54}$$

As a matrix equation, this condition is

$$[\tilde{\mathbf{R}}][\mathbf{R}] = [\mathbf{I}], \tag{14.55}$$

where $[\tilde{\mathbf{R}}]$ is the transpose of $[\mathbf{R}]$, and $[\mathbf{I}]$ is the unit marix. This means that $[\tilde{\mathbf{R}}]$ is the inverse of $[\mathbf{R}]$, which is the definition of a **Real Orthogonal Transformation (R.O.T.)**.

A **proper rotation** transforms a right handed coordinate system into a right handed coordinate system. A right handed coordinate system is defined by

$$\hat{\mathbf{n}}^{(\mathbf{r})} \cdot (\hat{\mathbf{n}}^{(\mathbf{s})} \times \hat{\mathbf{n}}^{(\mathbf{t})}) = \epsilon_{rst}. \tag{14.56}$$

For the transformed system to also be right handed, we require

$$+1 = \hat{\mathbf{n}}'^{(\mathbf{1})} \cdot (\hat{\mathbf{n}}'^{(\mathbf{2})} \times \hat{\mathbf{n}}'^{(\mathbf{3})}) = R_{1j} R_{2k} R_{3l} \hat{\mathbf{n}}^{(\mathbf{j})} \cdot (\hat{\mathbf{n}}^{(\mathbf{k})} \times \hat{\mathbf{n}}^{(\mathbf{l})}) = \epsilon_{jkl} R_{1j} R_{2k} R_{3l}. \tag{14.57}$$

The right hand result in Eq. (14.57) is an algebraic expression for the determinant of the matrix $[\mathbf{R}]$, so the condition for a real orthogonal transformation (R.O.T.) to be a proper rotation (ROT) is that its determinant is $+1$. That is, if an R.O.T. has determinant $+1$, we can remove the periods and make it a ROT.

An improper rotation changes the handedness of the system, so its determinant is -1. Any improper rotation can be written as the product of a proper rotation and the parity transformation

$$[\mathbf{P}] = \begin{pmatrix} -1 & 0 & 0 \\ 0 & -1 & 0 \\ 0 & 0 & -1 \end{pmatrix}, \tag{14.58}$$

which takes each coordinate into its negative.

The components of the position vector $\mathbf{r}$ are just the coordinates x_i, so its components transform like the coordinates. We can take this to be true of any vector, so any vector will transform as

$$V_i' = R_{ij} V_j. \tag{14.59}$$

With this transformation, a vector, such as vector $\mathbf{V}$ in Fig. 14.3 will remain fixed in space. That is, the components of the vector transform in such a way so as to keep the vector unchanged by the rotation. Note that the magnitude of the vector also does not change under the rotation of coordinates, nor will the dot product of two vectors change. For any two vectors $\mathbf{U}$ and $\mathbf{V}$, we get

$$U_i'V_i' = R_{ij}R_{ik}U_jV_k = \delta_{jk}U_jV_k = U_jV_j. \tag{14.60}$$

If the vectors $\mathbf{U}$ and $\mathbf{V}$ are the same vector, then this equation shows its length is invariant under the rotation.

We can make the transformation equation (14.59) the defining property of a **vector**. That is, **a vector is composed of three components that transform like the coordinates.** We can also define a **scalar to be quantity that is invariant under a transformation**. An example of a scalar is the dot product (also called the scalar product) of two vectors as shown in Eq. (14.60). From this equation, we can see that the dot product will be invariant under improper as well as proper rotations.

One more transformation of interest is that of the cross product of two vectors. This is given by

$$A_i' = (\mathbf{U}'\times\mathbf{V}')_i = \epsilon_{ijk}U_j'V_k' = \epsilon_{ijk}R_{jl}R_{km}U_lV_m. \tag{14.61}$$

This does not look like the transformation of a vector. We can make it look more like a vector by using the property of the determinant of $[\mathbf{R}]$ that

$$\epsilon_{njk}R_{ns}R_{jl}R_{km} = \epsilon_{slm}\det[\mathbf{R}]. \tag{14.62}$$

We can introduce this identity into Eq. (14.61) by the following use of the orthogonality of $[\mathbf{R}]$

$$\begin{aligned}
A_i' &= \epsilon_{ijk}R_{jl}R_{km}U_lV_m \\
&= \epsilon_{njk}\delta_{ni}R_{jl}R_{km}U_lV_m \\
&= \epsilon_{njk}R_{ns}R_{is}R_{jl}R_{km}U_lV_m \\
&= \det[\mathbf{R}]R_{is}\epsilon_{slm}U_lV_m \\
&= \det[\mathbf{R}]R_{is}A_s. \tag{14.63}
\end{aligned}$$

The appearance of the determinant of $[\mathbf{R}]$ in the transformation means that the cross product will transform as a vector under proper rotations, but the transformation will have a minus sign for improper rotations. This defines the transformation property of a **pseudovector**. The terminology **axial vector** is also used, while a true vector is also called a **polar vector**. If the vector $\mathbf{V}$ shown in Fig. 14.3 were an axial vector, then its coordinates would not change under the parity transformation, but that would result in the transformed pseudovector pointing in the opposite direction in space.

More complicated transformations are possible. A dyadic, being a sum of products of vectors would transform as

$$D_{ij}' = R_{ik}R_{jl}D_{kl}. \tag{14.64}$$

This is called the transformation law of a second rank tensor, because the transformation matrix appears twice. This terminology is generalized so that a third rank tensor would have three indices transformed by three $[\mathbf{R}]$ matrices. Looking at Eq. (14.62) shows that ϵ_{ijk} transforms like a third rank pseudotensor. Even though it transforms that way, its components still have the same value in any system. This property makes ϵ_{ijk} an **idempotent** pseudotensor. Similarly the Kronecker δ_{ij} can be considered an idempotent second rank tensor. In tensor terminology, a vector is a first rank tensor and a scalar is a zeroth rank tensor.

14.4.2   Four Dimensional Rotations in Space-Time

We can extend the transformation of three spatial coordinates by rotations to the Lorentz transformation of the four space-time coordinates

$$x'_\mu = L_{\mu\nu} x_\nu. \tag{14.65}$$

Here we have followed the standard notation that a Greek subscript stands for the four space-time indices 0,1,2,3, while Roman subscripts are restricted to the three spatial coordinates. The transformation of coordinates in Eq. (14.65) are also the transformation equations of the components of a four-vector. As we did with three-vectors, we can make the transformation equation (14.65) the defining property of a **four-vector**. That is, **a four-vector is composed of four components that transform like the coordinates.**

The matrix equation for a Lorentz transformation is

$$\mathbf{x}' = [\mathbf{L}]\mathbf{x}. \tag{14.66}$$

The matrix for a Lorentz transformation with velocity $v = \beta c$ in the x direction is

$$[\mathbf{L}^{(x)}(\beta)] = \begin{pmatrix} \gamma & -\beta\gamma & 0 & 0 \\ -\beta\gamma & \gamma & 0 & 0 \\ 0 & 0 & 1 & 0 \\ 0 & 0 & 0 & 1 \end{pmatrix}. \tag{14.67}$$

There is a major complication in defining the length of a four-vector. For a three-vector, the sum $x_i x_i$ is invariant under rotations, but for a four vector, $x_\mu x_\mu$ is not invariant under Lorentz transformations. The combination that is invariant for four-vectors is $x_0^2 - x_1^2 - x_2^2 - x_3^2$. That is, under the Lorentz transformation of Eq. (14.67)

$$\begin{aligned} x_0'^2 - x_1'^2 - x_2'^2 - x_3'^2 &= \gamma^2[(x_0 - \beta x_1)^2 - (x_1 - \beta x_0)^2] - x_2^2 - x_3^2 \\ &= x_0^2 - x_1^2 - x_2^2 - x_3^2. \end{aligned} \tag{14.68}$$

There are several ways of incorporating these minus signs in the mathematics. The standard method is the introduction of a metric tensor or matrix.

The **metric matrix** for special relativity is

$$[G] = \begin{pmatrix} 1 & 0 & 0 & 0 \\ 0 & -1 & 0 & 0 \\ 0 & 0 & -1 & 0 \\ 0 & 0 & 0 & -1 \end{pmatrix}. \tag{14.69}$$

This is said to be an **indefinite metric** because its diagonal elements do not all have the same sign. With a metric, the 'length' of a four-vector is given by the matrix equation

$$\mathrm{x} \cdot \mathrm{x} = \mathrm{x}^2 = \tilde{\mathrm{x}}[G]\mathrm{x}. \tag{14.70}$$

($\tilde{\mathrm{x}}$ is a row matrix that is the transpose of the column matrix x.) By $\mathrm{x} \cdot \mathrm{x}$ and x^2, we mean the scalar product of the vector with itself, which is the length squared of the four-vector. To distinguish between four-vectors and three-vectors, the usual practice is to write a three-vector in boldface, and a four-vector in normal type.

The character of a space-time is determined by its metric. In general relativity, the metric can become quite complicated, and in most cases still has not been found. Space-times having the constant diagonal metric given above are called **Minkowski spaces**.

In tensor notation a **metric tensor** $g_{\mu\nu}$ is used. The components of $g_{\mu\nu}$ are the $\mu\nu$ elements of the metric matrix. Two sets of components are introduced for each four-vector, with the connection

$$x_\mu = g_{\mu\nu}x^\nu. \tag{14.71}$$

This procedure is called **lowering**. In Minkowski spaces, the lowering operation reverses the sign of the spatial components, leaving the time components unchanged . Specifically, if $x^\mu = (ct, \mathbf{x})$, then $x_\mu = (ct, -\mathbf{x})$, and the length squared of the four-vector is

$$\mathrm{x}^2 = x_\mu x^\mu = c^2 t^2 - \mathbf{x}^2. \tag{14.72}$$

The set of components with upper indices are called the **contravariant** components, and those with lower indices are called the **covariant** components of the four vector. With the use of the two sets of components, the summation convention applies to repeated indices only when one is upper and the other lower. That is, $x_\mu x_\mu$ should not be summed. There is also a corresponding **raising** operation

$$x^\mu = g^{\mu\nu}x_\nu, \tag{14.73}$$

which changes covariant components into contravariant components. The metric tensor $g^{\mu\nu}$ has the same matrix elements as $g_{\mu\nu}$, so raising also changes the sign of the spacelike components of a four-vector.

The scalar product of two different four-vectors is given by

$$\mathbf{x} \cdot \mathbf{y} = x_\mu y^\mu = x^\mu y_\mu, \tag{14.74}$$

and, in matrix notation by

$$\mathbf{x} \cdot \mathbf{y} = \tilde{\mathbf{x}}[G]\mathbf{y}. \tag{14.75}$$

The invariance of the scalar product can be shown by Lorentz transforming it

$$\mathbf{x}' \cdot \mathbf{y}' = (\widetilde{[L]\mathbf{x}})[G][L]\mathbf{y} = \tilde{\mathbf{x}}(([\tilde{L}][G][L])\mathbf{y}. \tag{14.76}$$

The invariance follows because the Lorentz matrix is a generalized real orthogonal matrix with the property (See problem 14.8.)

$$[\tilde{L}][G][L] = [G]. \tag{14.77}$$

The operation of setting two indices (one upper and one lower) equal and summing on them, as in Eq. (14.74) is called **contraction**. The contraction theorem states that **the contraction of an mth rank tensor with an nth rank tensor is a new tensor of rank $|m-n|$.** We have proven this above for the contraction of two vectors. The proof for a higher order tensor expression is given as a problem.

The generalized orthogonality property in Eq. (14.77) can be taken as the definition of a Lorentz matrix. That is, any matrix that satisfies Eq. (14.77) is a Lorentz matrix. The set of Lorentz matrixes, so defined, form a mathematical group, called **the Lorentz group**. They satisfy the four basic properties of a group:

1. There is an identity: the unit matrix $[I]$.

2. Every Lorentz matrix has an inverse: $[L(\beta)]^{-1} = [L(-\beta)]$.

3. The matrix multiplication is associative: $([L_1][L_2])[L_3] = [L_1]([L_2][L_3])$.

4. The product of two Lorentz matrices is a Lorentz matrix:
   $[\tilde{L}_2][\tilde{L}_1][G][L_1][L_2] = [G]$.

Spatial rotation matrices in four dimensional space-time are given by

$$[R] = \begin{pmatrix} 1 & 0 & 0 & 0 \\ \hline 0 & & & \\ 0 & & [\mathbf{R}] & \\ 0 & & & \end{pmatrix}, \tag{14.78}$$

where $[\mathbf{R}]$ is a three dimensional rotation matrix. The rotation matrices satisfy Eq. (14.77), and the four group properties, so they are a sub-group of the Lorentz group. A pure Lorentz transformation, having no spatial rotation, is called a **boost**. A general Lorentz transformation could be a product of a boost and a rotation. This can become quite complicated, and if possible it is best to treat the boost and rotation in two steps.

Thus far, we have used only a pure Lorentz transformaton along the x axis. This is easily generalized to other spatial axes by simply interchanging rows and columns appropriately. A boost in a general direction can be described by first rotating the spatial coordinates to make one coordinate the axis of the boost, performing the boost, and then rotating back with the inverse of the original rotation. This gives

$$[\mathrm{L}(\boldsymbol{\beta})] = [\mathrm{R}]^{-1}[\mathrm{L}^{(x_i)}(\beta)][\mathrm{R}]. \tag{14.79}$$

The resulting matrix, $[\mathrm{L}(\boldsymbol{\beta})]$, is quite complicated, so it is a good idea to try to orient the coordinates to keep the boost along a coordinate axis. In fact the best way to use a Lorentz transformation is to avoid it by using invariance relations to relate the physics in two different Lorentz frames. We will see examples of this in the following sections.

From the generalized orthogonality relation in Eq. (14.77), it follows that the determinant of the matrix [L] must be +1 or -1. If the determinant is +1, the transformation said to be a **proper Lorentz transformation**, and if -1, the transformation is an **improper Lorentz transformation**. There are two kinds of improper transformation. One kind involves the parity operator P on the spatial components, and the second kind involves the **time reversal operator** T, which does nothing but change the time variable to its negative. There is also a type of proper Lorentz transformation that reverses all time and space coordinates. Transformations that do not reverse time are called **orthochronous**, and we will usually consider only proper orthochronous Lorentz transformations.

The matrix equations can also be written in tensor notation, but are usually a bit more complicated. In tensor notation, the Lorentz transformation is

$$x'^{\mu} = L^{\mu}_{\nu} x^{\nu}. \tag{14.80}$$

Notice that care must be taken with the upper and lower indices. The general rule is that the same index on opposite sides of an equation must be in the same position, while repeated indices on the same side of an equation must be in opposite positions. This requires the Lorentz matrix element L^{μ}_{ν} to have one upper and one lower index. The transformation of covariant components is

$$x'_{u} = L^{\nu}_{\mu} x_{\nu}. \tag{14.81}$$

We see that following the rules for indices requires that the covariant components transform with the transpose of the Lorentz matrix. However, the Lorentz matrix is symmetric, so the covariant and contravariant components have the same transformation. For more complicated or non-linear transformations, as occur in general relativity, the two types of components have different transformations.

For a general change in coordinates from x^μ to functions $x'^\nu(x^\mu)$, the transformation of a contravariant vector is defined to be the same as that of the calculus rule for the coordinate differentials

$$dx'^\mu = (\partial_\nu x'^\mu)dx^\nu. \tag{14.82}$$

The transformation of a covariant vector is defined to be the same as that of the calculus rule for the partial derivatives

$$\partial'_\mu = (\partial'_\mu x^\nu)\partial_\nu. \tag{14.83}$$

For the Lorentz transformation, these two transformatons reduced to the linear forms, and the transformation is the same for each type of component.

You can usually guess the transformation properties of components by how they appear, but this does have to be checked. For instance, $g_{\mu\nu}$ are the components of an idempotent covariant second rank tensor. This can be seen by writing Eq. (14.77) in tensor form as

$$L^\rho_\mu L^\sigma_\nu g_{\rho\sigma} = g_{\mu\nu}, \tag{14.84}$$

which is the transformation law of a second rank covariant tensor.

Two important Lorentz scalars are the four dimensional volume element and delta function. The volume element d^4x transforms by the calculus rule

$$d^4x' = |\det[L]|^{-1}d^4x, \tag{14.85}$$

where $|\det[L]|^{-1}$ is the Jacobian of the transformation from the variables x^μ to x'^ν. Since the magnitude of the determinant of a Lorentz matrix is 1, it follows that d^4x transforms as a Lorentz scalar. That the four dimensional delta function is a scalar follows from its integral representation

$$\delta^{(4)}(x) = \frac{1}{(2\pi)^4}\int e^{-ik_\mu x^\mu}d^4x. \tag{14.86}$$

We have seen that the transformation between coordinate systems in relative motion is a generalized rotation in space-time. This gives new insight into the 'Lorentz contraction' phenomenon we discussed earlier. The physical length of a stick can be measured by placing it along the x-axis and measuring the difference, $x_2 - x_1$, of the x-coordinates of the ends of the stick. If there is a three dimensional rotation, as in Figure 14.3, the transformed difference $x'_2 - x'_1$ changes, but this difference is only for one coordinate of a vector. The length of the stick does not change, and it can only be measured by the difference of its x coordinates when it lies along the x-axis.

The same thing happens in the four dimensional rotation in space-time that is the Lorentz transformation. The length of a stick can be measured in its rest frame by placing it along the x-axis, and measuring the difference $x_2 - x_1$ between the coordinates of each end of the stick. The transformed difference $x'_2 - x'_1$ changes in a Lorentz transformation, but the difference is only for one coordinate of a four-vector.

Just as the length of a stick doesn't change in the three dimensional rotation of Figure 14.3, its length does not change in a four dimensional rotation. If the stick is moving, making the Lorentz tranformation to its rest system is the only way to get a reliable measurement of its length. This is generally true of intrinsic properties of physical objects that are not Lorentz invariants.

14.5 Relativistic Space-time

A point in space-time is called an **event**, and both the spatial location and time must be designated. The square of the distance between two events, x and y, in space time is an invariant given by

$$s^2 = (x - y)^2 = (x - y)_0^2 - (\mathbf{x} - \mathbf{y})^2. \tag{14.87}$$

If s^2 is negative, the distance between the two events is said to be a **spacelike** interval. Since s^2 is invariant, a spacelike interval will remain spacelike under Lorentz transformation, although the separate time and space components may change. In particular, the time difference $(x-y)_0$ may change from positive to negative or to zero, so designations like 'earlier', 'later', or 'simultaneous' can be different in different Lorentz frames. Testimony in a relativistic court of law as to "Who shot first?" would be valid only if the witness and the two gunmen were all at rest in the same Lorentz frame.

If s^2 is positive, the interval between the two events is said to be **timelike**. A timelike interval will remain timelike in any Lorentz frame. Also, since $(x_0 - y_0)$ can never go through zero, the relative time order cannot change for an orthochornous transformation, so 'earlier' and 'later' have invariant meanings for a timelike interval.

14.5.1 The light cone

An important case is $s^2=0$, which is called a **null interval**. Because of the indefinite metric, this does not mean the four-vector is identically zero, but just that

$$(\mathbf{x} - \mathbf{y})^2 = (x_0 - y_0)^2. \tag{14.88}$$

This equality defines a four dimensional hyper-cone, called the **light cone**. A sectional view of your light cone is shown in Fig. 14.4 with its apex at your present position in space-time.

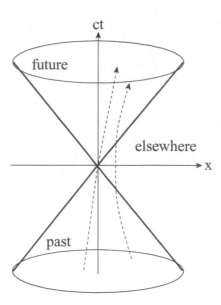

Figure 14.4: Your light cone with your world line and that of a friend (shown as dashed lines).

Equation (14.88) represents the propagation of a pulse of light emitted at space point $\mathbf{y}$ at time $t_0 = y_0/c$. The light front will be on an expanding sphere whose radius is given by Eq. (14.88). On the space-time diagram in Fig. 14.4, the light will move along the light cone in the positive direction. Any object moving from $(y_0, \mathbf{y})$ with speed less than c will be restrained to move within the future light cone, and can never get to points outside.

The light cone divides space-time into three separate regions,as indicated on Fig. 14.4. The inside of the light cone, in the positive time direction is **the future**. If you start at the apex of the light cone, you can only reach or affect points inside your future light cone. The only people who can see you (using light) must be on the surface of your future light cone.

The inside of the light cone in the negative time direction is **the past**. Points inside your past light cone are inaccesible to you, but are the only points that can affect you here and now. There is no orthochronous Lorentz transformation that connects points in the past and future light cones, so time travel (to your past) is not possible in special relativity.

The rest of space-time, outside the light cone is always a spacelike separation from the apex of the cone, and can be called **elsewhere**. There is no way that you can influence or be influenced by points outside your light cone. But those points are still very important. They are the only space-time points where other people can be at the same time as you. At this very moment, everybody you know is outside your light cone. Only solipsists have nobody outside their light cones.

Although there are people outside your light cone right now, you can only see them when they are on your past light cone. (Think a bit about that.) For intervals outside the light cone, there is no meaning to absolute future or past, because even an orthochronous Lorentz transformation can interchange past and future. This is not time travel however, because these points are inaccessible to you and can't affect you.

Each person, and every material object, follows a path in space-time, called a **world line**. The path must always be upward (in the positive time direction), with a slope (on Fig. 14.4) that is always greater than 1. Two world lines are shown on Fig. 14.4. Your world line moves up the inside of a light cone whose apex is at your present position in space-time, coming from your past and moving into your future. Your friend's world line cannot pass through the apex of your light cone (unless she sits on your lap), but will come from your past, move elsewhere and then enter your future. You will only see her as she leaves your past light cone, and she will only see you as she enters your future light cone. You and your friend can only exist at the same time if she is elsewhere. Relativity is very clear about that.

14.5.2   Proper Time

The squared differential displacement of a moving object, $ds^2 = dx_0^2 - \mathbf{dr}^2$, is a Lorentz scalar. If we factor out dx_0^2, we can write ds^2 as

$$ds^2 = dx_0^2(1 - \beta^2). \tag{14.89}$$

In the rest system of the object ($\beta = 0$), $ds^2 = dx_0^2$ so ds/c is the rest system time differential, which we denote as $d\tau$. Then Eq. (14.89) can be written as

$$d\tau = dt\sqrt{1 - \beta^2} = dt/\gamma, \tag{14.90}$$

which relates a time interval in the rest system with a time interval when the object is moving with velocity $v = c\beta$. This is in agreement with our earlier derivation of time dilation in Eq. (14.35). The time interval $d\tau$ is not just the rest frame time interval, but, in addition, is a relativistic scalar (called the **proper time**).

Earlier, we referred to the Lorentz transformation of a time interval as "time dilation" or by the statement "Moving clocks slow down." Let us reexamine these interpretations in terms of the proper time interval for a moving clock.

We present a simple example in the two dimensional space of a road map. Consider two twins who each buy a new automobile in Philadelphia. One twin drives the car directly to New York where its odometer shows 150 km. The other twin detours to the beach at Atlantic City, and then proceeds to New York, putting 250 km on the odometer. This situation is shown on Fig. 14.5,

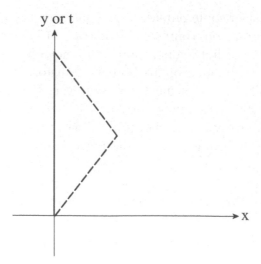

Figure 14.5: Two paths in x-y space or in x-t space-time.

where the first twin drives directly up the y axis, while the other twin takes the longer dashed path. (Of course, both paths are oversimplified.) The two cars are in the same parking lot in New York, but have different odometer readings. There is no problem in understanding this, since one twin covered more distance on the map.

Now consider two twins, one staying at rest on Earth, the other traveling out into space and back. We now take the vertical axis in Fig. 14.5 to be the time axis in the Earth system. The vertical line shows the world line of the twin at rest. The dashed path is now the world line of the twin travelling in space-time. It looks like this twin has taken a longer path, but the indefinite metric of special relativity changes this interpretation. The proper time interval on the travelling twin's clock is given by $\int d\tau$, where $d\tau^2 = dt^2 - \mathbf{dr}^2$. Because of this minus sign, the travelling twin's clock shows less elapsed time than that of the stationary twin. This is not because of any difference in the clocks, but because the travelling twin covered less distance in space-time (as defined by $\int ds$ or $\int d\tau$). The moving clock did not run slower. It just traveled for a shorter proper time.

14.6   Relativistic Kinematics.

So far we have only found one four-vector, $s^\mu = (x_0, \mathbf{x})$ (choosing y to be at the space-time origin), and one scalar s^2 (or the proper time interval $d\tau$). To discuss relativistic physics, we will need more four-vectors and scalars corresponding to physical variables.

In finding physical four-vectors and scalars, we can be guided by the long success of non-relativistic physics in describing the world for small velocities. That is, many of the familiar three-vectors can be extended to four dimensions with the requirement that the result be a relativistic four-vector that reduces at low velocity to the three-vector. This usually gives a unique four-vector, but the ultimate test is agreement with relativistic experiments.

14.6.1   Four-velocity

We look first for a relativistic extension of velocity. It is clear that the velocity components $\frac{d\mathbf{r}}{dt}$ cannot be part of a four-vector, because their transformation properties in Eqs. (14.24) and (14.25) are certainly not those of a four-vector. The problem is that $\frac{d\mathbf{r}}{dt}$ divides three components of a vector by a fourth component. We can construct a four-vector for velocity by dividing a four-vector by a scalar.

If we divide the space-time four-vector dx^μ by the scalar $d\tau$ we will have a four-vector for velocity, given by

$$U^\mu = \frac{dx^\mu}{d\tau}. \tag{14.91}$$

The four-velocity is a relativistic four-vector, which makes it useful, but it is not the actual velocity of an object. In any Lorentz frame, an object's velocity is still given by $\mathbf{u} = \frac{d\mathbf{r}}{dt}$. The four-velocity, written in terms of the actual velocity, is

$$U^\mu = (\gamma c, \gamma \mathbf{u}). \tag{14.92}$$

In the non-relativistic limit $\beta \ll 1$, the spacelike components of U^μ approach the velocity $\mathbf{u}$.

We can verify that U^μ is the appropriate relativistic four-vector velocity by seeing what how it Lorentz transforms. In a Lorentz transformation of $U^\mu = \gamma_u(c, \mathbf{u})$, we get

$$U'^0/c = \gamma_{u'} \quad = \quad \gamma_v \gamma_u (1 - \mathbf{u} \cdot \mathbf{v}/c^2) \tag{14.93}$$

$$\mathbf{U}'_\parallel = \gamma_{u'} \mathbf{u}'_\parallel \quad = \quad \gamma_v \gamma_u (\mathbf{u}_\parallel - \mathbf{v}) \tag{14.94}$$

$$\mathbf{U}'_\perp = \gamma_{u'} \mathbf{u}'_\perp \quad = \quad \gamma_u \mathbf{u}_\perp. \tag{14.95}$$

Dividing Eq. (14.94) by Eq. (14.93) gives

$$\mathbf{u}'_\parallel = \frac{\mathbf{u}_\parallel - \mathbf{v}}{1 - \mathbf{u} \cdot \mathbf{v}/c^2}, \tag{14.96}$$

and dividing Eq. (14.95) by Eq. (14.93) gives

$$\mathbf{u}'_\perp = \frac{\mathbf{u}_\perp}{\gamma_v(1 - \mathbf{u} \cdot \mathbf{v}/c^2)}. \tag{14.97}$$

These are the same velocity transformation equations as Eqs. (14.24) and (14.25), confirming that U^μ is the appropriate four-vector of velocity.

14.6.2 Energy-momentum four-vector

The non-relativistic momentum is given by $\mathbf{p}_{nr} = m\mathbf{v}$. We can extend this to a relativistic four-vector by choosing

$$p^\mu = mU^\mu = (\gamma mc, \gamma m\mathbf{v}). \tag{14.98}$$

This has the appropriate non-relativistic limit, $\mathbf{p}_{nr} = m\mathbf{v}$. That p^μ is the correct relativistic momentum requires confirmation in relativistic experiments. First, we investigate the physical significance of the time-like component $p^0 = \gamma mc$. This has the non-relativistic expansion

$$p^0 = mc(1 - v^2/c^2)^{-\frac{1}{2}} = mc + \frac{1}{2}m\mathbf{v}^2/c + ... \tag{14.99}$$

We recognize the second term in the expansion as the non-relativistic kinetic energy divided by c, so we can identify the combination $(cp^0 - mc^2)$ as the relativistic extension of the kinetic energy (T). The term mc^2 does not vary, so we can define a relativistic energy for an object as

$$E = mc^2 + T = cp^0 = \gamma mc^2 = \frac{mc^2}{\sqrt{1 - v^2/c^2}}. \tag{14.100}$$

The relativistic momentum four-vector can now be written as

$$p^\mu = (E/c, \mathbf{p}). \tag{14.101}$$

The invariant 'length' of the four-vector cp^μ is

$$c^2 p^2 = E^2 - c^2\mathbf{p}^2 = m^2\gamma^2(1 - \mathbf{v}^2/c^2) = (mc^2)^2. \tag{14.102}$$

This shows that the energy is related to the three-vector momentum $\mathbf{p}$ by

$$E = c\sqrt{\mathbf{p}^2 + m^2 c^2}, \tag{14.103}$$

This has the non-relativistic expansion

$$E = mc^2 + \frac{\mathbf{p}^2}{2m} + ..., \tag{14.104}$$

which again shows that the non-relativistic limit of cp^0 is the kinetic energy plus the constant mc^2.

The fact that $E^2 - c^2\mathbf{p}^2$ is an invariant can be used to determine the mass of a decaying particle from its decay products. For example, the Λ hyperon has no charge and leaves no track until it decays into a proton and a π^- meson. These two tracks looked like a Greek Λ in some early pictures, and this gave the particle its name. The square root of the combination $(E_p + E_\pi)^2/c^4 - (\mathbf{p}_p + \mathbf{p}_\pi)^2$ for each event is called the 'apparent mass' of the combination. If the number of events is plotted against the apparent mass, a peak in the distribution shows that a particle of that mass has decayed into the proton and pion. This can be used to detect the neutral particle and to measure its mass. This procedure works for any number of decay products by summing over all of them.

14.6.3   E=mc^2

We have not yet discussed the significance of the constant mc^2, although some of you may have seen it before. From Eq. (14.100) or (14.104), we see that an object at rest has energy given by

$$E_{\text{rest}} = mc^2. \tag{14.105}$$

This relation is sometimes considered as an "equivalence" of mass and energy, but it is not a true equivalence. The energy and mass of an object are closely related but not equivalent. To within factors of c, the energy is the time-like component of the momentum four-vector, while the mass is the invariant length of the four-vector. The energy in the rest system equals mc^2 but, while in motion, the energy varies with velocity as in Eq. (14.100).

In early expositions of special relativity, and still in some popularizations, the mass is considered to vary as $m = m_0/\sqrt{1 - v^2/c^2}$, with m_0 called the "rest mass". Then $E = mc^2$ even in motion. However this interpretation goes against the principle that the mass, as an intrinsic property of an object, should be the same in all Lorentz frames. Also, the relativistic mass of an object would increase as we walked past it, and observed it from a moving frame. In the modern interpretation, m is the **invariant mass** of an object, and the equation $E = mc^2$ holds only in the rest system.

The equation $E = mc^2$ has also been widely held (to Einstein's distaste) as responsible for the atomic bomb via a conversion of mass into energy. But this, too, is not really appropriate. There is only a slight connection between $E = mc^2$ and the atomic, or even the hydrogen bomb.

In the fission of a nucleus, the total number of protons, neutrons and electrons does not change, and the energy release is due to the difference in binding energies of the original nucleus and its fission products.

The energy release in fission is not different in character from the energy release in chemical combustion, which also is the difference in binding energies. The fuss about $E = mc^2$ arose because the energy difference in fission is large enough to be measureable in the masses of the nuclei, while for chemical combustion that difference is too small to be noticed in the masses.

In nuclear fusion, two light nuclei fuse to form a heavier nucleus. A typical fusion reaction is a deuteron fusing with a tritium nucleus to form helium and a neutron: d+t→He+n. In this case, the energy release is the difference in nuclear energies of the nuclei involved. The binding energy difference is greater (per nucleon) than in fission, but still no matter is converted into energy.

Matter can be converted into energy and *vice-versa*, but not in fission or fusion. An example of matter being converted to energy is in the decay of elementary particles. For instance in the decay of a Λ hyperon into a proton and a pi meson ($\Lambda \rightarrow p + \pi^-$), the difference in initial and final masses goes into the kinetic energy of the decay particles. An example of energy being converted to matter is in a production process like $p + p \rightarrow p + p + \pi^0$.

The theory that describes the conversion of energy and matter is **Quantum Field Theory (QFT)**, which was developed in the 1930's and since. In QFT, relativity can only be combined with quantum mechanics in a consistent manner if quantum mechanical operators that create and destroy particles are introduced. So relativity did eventually lead to matter-energy conversion, but only as a part of quantum mechanics, and not in fission or fusion.

14.7   Doppler Shift and Stellar Aberration

The phase of an electromagnetic plane wave, given by $e^{i(\mathbf{k}\cdot\mathbf{r}-\omega t)}$, must be an invariant for a light wave to look the same in all Lorentz frames. Since the space-time components $x^\mu = (ct, \mathbf{r})$ form a four-vector, the components $k^\mu = (\omega/c, \mathbf{k})$ must also form a four-vector. This is an example of the **division theorem**, which states, **If the contraction of the four components of an arbitrary four-vector with another set of four components results in a scalar, then those four components are also the components of a four-vector.** The division theorem is the converse of the contraction theorem, and also holds for higher rank contractions. The proof of this theorem is left as an exercise.

The fact that the components k^μ form a four-vector makes it easy to find the effect of a Lorentz transformation on the frequency and direction of a light ray. We consider a star moving with velocity $\mathbf{v}$, relative to an earthly observer, while emitting light of angular frequency ω in the star's rest system. The Lorentz transformed wave-vector in the observer's rest system is given by

$$\omega' = \gamma(\omega + \mathbf{v}\cdot\mathbf{k}) = \gamma\omega(1 + v\cos\theta/c) \qquad (14.106)$$

$$\mathbf{k}'_\parallel = \gamma(\mathbf{k}_\parallel + \mathbf{v}\omega/c^2) \qquad (14.107)$$

$$\mathbf{k}'_\perp = \mathbf{k}_\perp, \qquad (14.108)$$

where $\mathbf{k}_\parallel$ and $\mathbf{k}_\perp$ are the components of $\mathbf{k}$ parallel and perpendicular to the star's velocity $\mathbf{v}$, and θ is the angle between $\mathbf{k}$ and $\mathbf{v}$, as shown in Fig. 14.6. The **Doppler shift** for the frequency is given by Eq. (14.106). The Doppler shifted frequency increases if a star is coming toward us, and decreases if the star is receding. The effect on the wave length is the opposite ($\lambda \sim 1/\omega$), and the fact that the light from distant galaxies has a **red shift** increasing with distance is taken as evidence that the universe is expanding.

The frequency shift differs from the formula for sound from a source moving towards an observer by the factor γ. Another difference from the Doppler effect for sound is that the relativistic Doppler effect depends only on the relative velocity between the source and the observer. The Doppler effect for sound is different when the observer moves than when the source moves. This is because sound travels in a material substance, while, for light waves, relativity dispenses with a material medium.

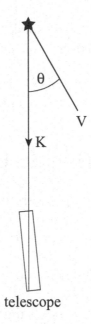

Figure 14.6: Light with wave vector **k** emitted by a star moving with velocity **v**. The angle the telescope makes with **v** is slightly smaller than the angle θ between **k** and **v**.

Because of the γ factor, there is also a **transverse Doppler effect** for light. That is, even if $\theta = 90°$, so the star is moving transverse to your line of sight, there is a Doppler shift of the frequency.

The angle at which light from a star enters a telescope is shifted from the angle of observation of the star. This effect, called **stellar aberration**, is given from Eqs. (14.107) and (14.108), by

$$\tan \theta' = \frac{k'_\perp}{k'_\parallel} = \frac{\sin \theta}{\gamma(\cos \theta + v/c)}. \tag{14.109}$$

This shows that the angle θ' at which a telescope should be pointed differs slightly from the angle of observation θ of the star. In Fig. 14.6, it appears that light would not go straight through the telescope. But, in the Lorentz system in which the star was at rest, the telescope would be moving from right to left. Then, because of the finite velocity of light, light entering the center of the objective lense of the telescope, would pass through the center of the eyepiece if the telescope were angled as shown (exaggerated) in the figure.

A simple nonrelativistic aberration calculation gives the formula $\phi = \frac{v}{c} \sin \theta$ for the angle at which the telecope must turned from the direction to the star. This agrees with the nonrelativistic reduction of Eq. (14.109). Since the aberration depends on the first power of v/c, it is readily observable, and was first discovered in 1729 by the English astronomer James Bradley. For stars, the effect is mainly due to the Earth's orbital velocity of 30 km/sec, which results in a maximum aberration angle of $\phi = 20''$.

Although the relativistic part is too small to be observed in stellar aberration, Einstein's derivation of aberration was important in the early acceptance of special relativity. A competing theory, suggesting that the Earth "dragged" the ether, thus explaining null attempts to detect earthly motion though the ether, was inconsistent with stellar aberration.

14.8   Natural Relativistic Units, No More c

The constant c originated in classical electromagnetism as the ratio of the esu unit of charge in electrostatic formulas to the emu unit of charge in magnetic formulas, and c was an important parameter in many of the basic equations that were developed in the 19th century.

With the introduction of special relativity, the number c took on a different significance. In three dimensions, rotational invariance implies that all three coordinates x, y, z should have the same units. In four dimensional space-time, the Lorentz transformation forms linear combinations of time-like and space-like components.

The c that appears in the relativistic equations is now there as a conversion factor, needed when different units are used for the two sets of components. With the advent of special relativity, c is no longer a physical constant to be measured, but just one more conversion factor between two sets of units. This fact is now recognized by the adoption of a fixed value for c used to define the unit of length in terms of the unit of time. That is, the accepted value of the meter is that 299,792,458 meters is the distance traveled by light in 1 second.

If space and time components had the same units, then the conversion constant c would not appear. In fact, this is a common practice in astronomy where both time and distance are measured in years (although the distance unit is called "light-years"), and no conversion constant is needed.

The implementation is very simple. The constant c need not appear in any equation. The common unit used for space and time can be whatever is convenient. Then, if a numerical answer is in units we don't want, the conversion factor c can be used as necessary. For instance, if a distance comes out in seconds, multiplying by c cm/sec will produce the result in cm.

We will henceforth leave c out of equations, and only use its numerical value if necessary to put final answers into appropriate units. Writing equations without the appearance of c is referred to as the use of **natural units**. It is sometimes described as "setting c equal to one." However, it is more fundamental than that. c need not be introduced in the first place if consistent time and space units are used.

The simplification of leaving c out also applies to the energy-momentum four-vector p^μ because a Lorentz transformation mixes the energy and momentum components. The relativistic connection between energy and momentum becomes

$$E = m\gamma = \sqrt{\mathbf{p}^2 + m^2}. \tag{14.110}$$

The three-vector momentum is given by

$$\mathbf{p} = m\mathbf{v}\gamma, \tag{14.111}$$

and the velocity is related to momentum and energy by

$$\mathbf{v} = \frac{\mathbf{p}}{E}. \tag{14.112}$$

Note that velocity is dimensionless, and must be less than 1 for an object with mass. Since particle energies are often much larger than their mass, it is sometimes useful to take the large energy limit of Eq. (14.112), getting

$$v = \frac{p}{E} = \frac{\sqrt{E^2 - m^2}}{E} \simeq 1 - \frac{m^2}{2E^2}. \tag{14.113}$$

The common unit for energy and momentum can be chosen for convenience. For the physics of elementary particle interactions, the energy unit MeV (or GeV) can be used for both mass and momentum as well as for energy, and the conversion c need never appear. (Usage is still made of the unit MeV/c for momentum. This is more to distinguish momentum from energy, and not to use the numerical value of c.)

14.9 Relativistic 'Center of Mass'

In non-relativistic mechanics, a transformation to the center of mass system is useful in studying two body interactions. This is true also in relativistic mechanics, but a center of mass is not well defined and the appropriate transformation is to the system where the total momentum is zero. The strict terminology for this system is the **barycentric system**, but we will follow the common usage and still call it the **center of mass system**. The relativistic transformation is based on the Lorentz transformation, but it is usually easier to use invariance principles rather than the explicit Lorentz transfomation. We illustrate this below with several examples.

In a **fixed target** experiment, a moving particle interacts with a target that is at rest in the laboratory. For a particle of mass m, incident with energy E_L on a target of mass M, the total energy and momentum in the laboratory system are given by

$$E = E_L + M, \quad \mathbf{p} = \mathbf{p_L}. \tag{14.114}$$

The quantity $E^2 - \mathbf{p}^2$ is an invariant and, in the center of mass system where $\mathbf{p} = \mathbf{0}$, it is equal to the square of the total center of mass energy W. So we can write

$$W^2 = (E_L + M)^2 - \mathbf{p_L}^2 = M^2 + m^2 + 2ME_L, \tag{14.115}$$

where we have used the relation $E_L^2 - \mathbf{p_L}^2 = m^2$.

Equation (14.115) gives the total center of mass energy in terms of the incident energy. The beam energy is usually specified in terms of the kinetic energy, $T_L = E_L - m$, of the incident particle, in which case the center of mass energy is given by

$$W^2 = (M + m)^2 + 2MT_L. \qquad (14.116)$$

This equation shows that, at high energies, the center of mass energy increases as the square root of the beam energy for a fixed target. For this reason, recent accelerators designed for large center of mass energy have been colliding beam accelerators, where particles of equal mass are made to collide with equal, but opposite velocities so that they are already in the center of mass system. For instance, in the LHC (Large Hadron Collider) at CERN, 7 TeV protons collide with 7 Tev protons, so the center of mass energy is 14 TeV. A fixed target machine would require a beam energy of

$$T_L = \frac{W^2 - (M + m)^2}{2M} \simeq W^2/2M = \frac{(14)^2}{2 \times 10^{-3}} = 10^5 \text{ TeV.} \qquad (14.117)$$

(The proton mass is 938 Mev, which we have approximated by 1 Gev for this calculation.)

The total center of mass energy is important in determining the threshold for particle production. If a new particle is produced in a collision of two other particles, the lowest possible center of mass energy is when all three final particles are at rest in the center of mass system. Thus the minimum center of mass energy for production is $W = M + m + \mu$ to produce a particle of mass μ. Using Eq. (14.117), the threshold beam energy for production is given by

$$T_L = \frac{(M + m + \mu)^2 - (M + m)^2}{2M} = \mu + \frac{\mu^2}{2M} + \frac{\mu m}{M}. \qquad (14.118)$$

The velocity ($\mathbf{V}$) to transform to the center of mass system can be found by considering the two particles, incident and target, as one composite object. The momentum of this object is $\mathbf{p_L}$, and its energy is $E_L + M$, so its velocity is

$$\mathbf{V} = \frac{\mathbf{p_L}}{E_L + M}. \qquad (14.119)$$

It is also useful to know the energy of each particle and their common momentum value in the center of mass. We can find these by writing the Lorentz parameter γ two different ways as

$$\gamma = \frac{E_L + M}{W} = \frac{E_M}{M}. \qquad (14.120)$$

The first expression for γ is the energy over the mass for the composite object. The second expression is the same ratio for the target particle, whose velocity in the center of mass system is equal to the velocity to transform to the center of mass.

The energy E_M is the energy of the target particle in the center of mass. We can solve for it by using Eq. (14.115) for E_L, to find

$$E_M = \frac{W^2 + M^2 - m^2}{2W}. \tag{14.121}$$

The center of mass energy of the incident particle is given by

$$E_m = W - E_M = \frac{W^2 + m^2 - M^2}{2W}. \tag{14.122}$$

Note that these two expressions are similar, because their is no distinction (other than their masses) between the two particles in the center of mass. The momentum of either particle in the center of mass is given (after some algebra) by

$$p_{cm} = \sqrt{E_M^2 - M^2} = \sqrt{E_m^2 - m^2} \tag{14.123}$$

$$= \frac{\sqrt{[W^2 - (M+m)^2][W^2 - (M-m)^2]}}{2W}. \tag{14.124}$$

Applications of the formulas derived in this section are given in several problems.

14.10   Covariant Electromagnetism.

In this section, we address a question that was intractable before the introduction of special relativity. That is, how are the equations of the first 13 chapters affected by going to a moving coordinate system? As a pleasant surprise, we will see that almost all of the equations we have been using still hold unchanged in any Lorentz coordinate system.

14.10.1   Charge-current four-vector j^μ

First we develop the basic four-vectors and invariants of electromagnetism. We start with the continuity equation

$$\partial_t \rho + \boldsymbol{\nabla}{\cdot}\mathbf{j} = 0, \tag{14.125}$$

This equation is equivalent to the conservation of electric charge, and so it should hold in all Lorentz frames. As we saw in Eq. (14.83), the partial derivatives ∂_μ form the four-vector

$$\partial_\mu = (\partial_t, \boldsymbol{\nabla}). \tag{14.126}$$

If we form the charge and current densities into the combination

$$j^\mu = (\rho, \mathbf{j}), \tag{14.127}$$

the continuity equation can be written in covariant form as

$$\partial_\mu j^\mu = 0. \tag{14.128}$$

We will show below that the components j^μ are the contravariant components of a four-vector. Then, the continuity equation corresponds to the vanishing of the four-divergence of the current four-vector j^μ, and will hold in any Lorentz system.

Equation (14.128) is said to be **manifestly covariant**, because its Lorentz invariance is manifest in its form. (Each side of the equations transforms in the same way.) Equation (14.125) also holds in all coordinate system, but to show this directly would require Lorentz transforming each of its elements. Note that the individual components ρ and $\mathbf{j}$ change in a Lorentz transformation, but the continuity equation retains its form, and charge is conserved in every Lorentz system. Reducing equations to a manifestly covariant form, as we did above, is generally an easier way to show invariance than actually Lorentz transforming each of the elements of the equation.

To show that j^μ actually is a four-vector, we will write it in covariant form. We first consider a point charge q following a path $\mathbf{s}(t)$, for which the j^μ components are given by

$$j^\mu(\mathbf{r}, t) = [q, q\mathbf{v}(t)]\delta[\mathbf{r} - \mathbf{s}(t)]. \tag{14.129}$$

This can be written in covariant form by introducing the charged particle's four-velocity U^μ in an integral over the particle's proper time τ:

$$j^\mu(x) = q \int d\tau U^\mu(\tau)\delta^{(4)}[x - s(\tau)], \tag{14.130}$$

where x and s represent the coordinates $(t, \mathbf{r})$ and $(t_s, \mathbf{s})$. The differential $d\tau$ and $\delta^{(4)}$ are scalars, and U^μ is a four-vector, so the resulting j^μ is a four-vector. Since a continuous charge-current distribution is actually composed of large numbers of point charges, the continuous density j^μ will also be a four-vector.

14.10.2 Lorentz invariance of charge

We showed in Chapter 6 that the continuity equation, $\partial_\mu j^\mu = 0$, was equivalent to the conservation of electric charge. Now we apply Gauss's law in four dimensional space-time to show that the vanishing of the four-divergence of j^μ leads to the total charge being Lorentz invariant. That is, the total charge given by $\int \rho\, d^3r$ is the same in all Lorentz systems.

The quadrilateral ABCD in Fig. 14.7 represents a four dimensional hypervolume in system S. Think of the y and z axes as both coming out of the paper

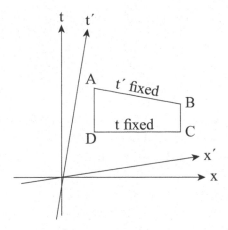

Figure 14.7: The quadrilateral ABCD represents a hypervolume for application of Gauss's law in space-time.

in four dimensions.

In a Lorentz transformation, the x, t axes in Fig. 14.7 transform to the x', t' axes. Note that for a Lorentz transformation, the x and t axes rotate by the same angle, but in opposite directions. We apply Gauss's law to the hypervolume inside the quadrilateral, resulting in integrals over the hypersurfaces represented by the sides of ABCD. The integral over sides AD and BC are integrals in $dtdydz$ with x fixed. (There are additional integrals, with y fixed and z fixed, that are not pictured.) If the sides AD and BC are removed to spatial infinity, and the current density vanishes at infinity, the integrals over those sides (and the integrals with y and z fixed) will vanish. Then the full surface integral will be given by the two integrals over sides AB and CD.

The side AB represents a three dimensional spatial volume at constant t', while the side CD represents a spatial volume at constant t. The integral over side AB is the volume integral $\int \rho' \, d^3r'$, which equals the total charge at fixed time t' in system S. Similarly, the integral over CD equals the total charge at a fixed time t in system S. Due to the vanishing of the four-divergence of j^μ, the integral over all surfaces of the hypervolume is zero, which means that the integrals over sides AB and CD are equal. Thus, we have shown that the total charge must be the same in each Lorentz system. This is true even though the integral $\int \rho \, d^3r$ does not look like a Lorentz scalar.

The surface represented by side AB has two interpretations. It is one surface of the hypervolume in system S (with t varying along the surface), but is also a surface at fixed t' in system S'. If the hypersurface AB had been chosen to be at fixed t, instead of fixed t', we would have proven that the total charge is a constant in time in a single system.

14.10.3   The four-potential A^μ

The wave equations for the potentials ϕ and $\mathbf{A}$ are

$$(\partial_t^2 - \nabla^2)\phi = 4\pi\rho, \quad (\partial_t^2 - \nabla^2)\mathbf{A} = 4\pi\mathbf{j}. \tag{14.131}$$

The differential operator $(\partial_t^2 - \nabla^2)$ is called the **D'Alembertian**, and we will write it as $\Box^2$. It is a four dimensional generalization of the Laplacian ∇^2. The D'Alembertian is a Lorentz scalar, since it is the invariant length $\partial_\mu\partial^\mu$ of the four vector ∂_μ.[3]

The wave equation must be the same in all Lorentz frames, because this is the basis for the speed of light c being the same. To write the wave equation in manifestly covariant form, we combine the scalar and vector potentials in the combination

$$A^\mu = (\phi, \mathbf{A}). \tag{14.132}$$

Then the wave equations for ϕ and $\mathbf{A}$ can be written as

$$\partial_\nu\partial^\nu A^\mu = \Box^2 A^\mu = 4\pi j^\mu, \tag{14.133}$$

which is given here in manifestly covariant form. Since j^μ is a four-vector and $\Box^2$ is a Lorentz scalar, Eq. (14.133) shows that the components A^μ form a four-vector.

In deriving the wave equation from Maxwell's equations in Chapter 10, we had to impose the Lorenz gauge condition

$$\nabla\cdot\mathbf{A} + \partial_t\phi = 0. \tag{14.134}$$

This can be put into manifestly covariant form as

$$\partial_\mu A^\mu = 0. \tag{14.135}$$

14.10.4   The electromagnetic field tensor $F^{\mu\nu}$

The electromagnetic fields $\mathbf{E}$ and $\mathbf{B}$ are given in terms of the potentials as

$$\mathbf{E} = -\nabla\phi - \partial_t\mathbf{A}, \quad \mathbf{B} = \nabla\times\mathbf{A}. \tag{14.136}$$

These do not look anything like Lorentz invariant equations. A hint of how to make these equations Lorentz invariant is given by writing out the $\mathbf{B} = \nabla\times\mathbf{A}$ equation in component form:

$$B_1 = \partial_2 A_3 - \partial_3 A_2, \quad \text{and cyclic.} \tag{14.137}$$

---

[3]Some books use the opposite sign for $\Box^2$. We have chosen our sign so that it will be the positive length squared of ∂_μ, which leads to fewer negative signs.

This looks like the components of **B** are matrix elements of a second rank tensor whose spacelike elements are $\partial_i A_j - \partial_j A_i$. The extension to a four-dimensional tensor is

$$F^{\mu\nu} = \partial^\mu A^\nu - \partial^\nu A^\mu. \tag{14.138}$$

It is left as an exercise (Problem 14.18) to show that the matrix representation of $F^{\mu\nu}$ is

$$[\mathrm{F}] = \begin{pmatrix} 0 & -E_x & -E_y & -E_z \\ E_x & 0 & -B_z & B_y \\ E_y & B_z & 0 & -B_x \\ E_z & -B_y & B_x & 0 \end{pmatrix}. \tag{14.139}$$

That $F^{\mu\nu}$ is a second rank tensor follows from the fact that it is composed of direct products of two four-vectors. This shows that the electromagnetic fields **E** and **B** are not separate relativistic vectors, but are combined in an antisymmetric second rank tensor, $F^{\mu\nu}$. Thus relativity requires even a stricter unification of the electric and magnetic fields than was given in Maxwell's classical equations. Although this was demonstrated over a century ago, the full acceptance of this fact has not reached many lower level texts.

Under a Lorentz transformation, the tensor $F^{\mu\nu}$ transforms in matrix notation as

$$[\mathrm{F}'] = [\mathrm{L}][\mathrm{F}][\tilde{\mathrm{L}}], \tag{14.140}$$

or in tensor notation as

$$F'^{\mu\nu} = L^\mu_\rho L^\nu_\sigma F^{\rho\sigma}. \tag{14.141}$$

This results in the following transformations on the **E** and **B** fields for a boost in the x direction with velocity β:

$$\begin{aligned} E'_x &= E_x & B'_x &= B_x \\ E'_y &= \gamma(E_y - \beta B_z) & B'_y &= \gamma(B_y + \beta E_z) \\ E'_z &= \gamma(E_z + \beta B_y) & B'_z &= \gamma(B_z - \beta E_y). \end{aligned} \tag{14.142}$$

These equations can be written in vector form as

$$\begin{aligned} \mathbf{E}'_\parallel &= \mathbf{E}_\parallel & \mathbf{B}'_\parallel &= \mathbf{B}_\parallel \\ \mathbf{E}'_\perp &= \gamma(\mathbf{E}_\perp + \boldsymbol{\beta}\times\mathbf{B}) & \mathbf{B}'_\perp &= \gamma(\mathbf{B}_\perp - \boldsymbol{\beta}\times\mathbf{E}). \end{aligned} \tag{14.143}$$

The effect of these transformations on the electromagnetic interactions of charged objects in motion will be treated in the next chapter.

How do Maxwell's equations behave under a Lorentz transformation? We could answer that by separately transforming the **E** and **B** fields, and the partial derivatives, but it is better to see if the equations can be put into manifestly covariant form.

Consider the equation

$$\boldsymbol{\nabla}\cdot\mathbf{E} = \partial_i E^i = \partial_i F^{i0} = 4\pi\rho. \tag{14.144}$$

This is three parts of an invariant four-divergence, which can be written

$$\partial_\mu F^{\mu 0} = 4\pi\rho. \tag{14.145}$$

The added term $\partial_0 F^{00}$ vanishes because $F^{\mu\nu}$ is antisymmetric. The charge density ρ is the zeroth component of the four vector j^μ. This suggests that we can extend the four-divergence to the other components of $F^{\mu\nu}$. Doing so results in

$$\partial_\mu F^{\mu\nu} = 4\pi j^\nu. \tag{14.146}$$

The space-like components of this reduce to the $\boldsymbol{\nabla}\times\mathbf{B}$ Maxwell equation (See Prob. 14.18.),

$$\boldsymbol{\nabla}\times\mathbf{B} = 4\pi\mathbf{j} - \partial_t\mathbf{E}. \tag{14.147}$$

Thus, we have written the two inhomogeneous Maxwell equations in manifestly covariant form with each side a four-vector, confirming that the equations hold in any Lorentz frame. The fact that the left hand side of Eq. (14.146) is a four-vector follows from the contraction theorem.

The two homogeneous Maxwell equations can be put into manifestly covariant form by first forming the associated tensor

$$\mathcal{F}^{\mu\nu} = \epsilon^{\mu\nu\rho\sigma} F_{\rho\sigma}, \tag{14.148}$$

where $F_{\rho\sigma}$ is the covariant form of $F^{\mu\nu}$, formed by the double lowering

$$F_{\rho\sigma} = g_{\rho\mu} g_{\sigma\nu} F^{\mu\nu}. \tag{14.149}$$

The tensor $\epsilon^{\mu\nu\rho\sigma}$ is the four dimensional generalization of ϵ_{ijk}, and is also completely antisymmetric. The elements of $\epsilon^{\mu\nu\rho\sigma}$ are defined by

$$\epsilon^{0123} = 1, \tag{14.150}$$

with the other components determined by the antisymmetry to be ± 1 or 0.

With this definition, $\epsilon^{\mu\nu\rho\sigma}$ has the same value in all Lorentz frames. That $\epsilon^{\mu\nu\rho\sigma}$ is a actually a tensor is confirmed by the identity for the determinant of a matrix that

$$\epsilon'^{\mu'\nu'\rho'\sigma'}\text{Det}[\mathrm{L}] = \frac{1}{2} L_\mu^{\mu'} L_\nu^{\nu'} L_\rho^{\rho'} L_\sigma^{\sigma'} \epsilon^{\mu\nu\rho\sigma}. \tag{14.151}$$

This shows that $\epsilon^{\mu\nu\rho\sigma}$ is an idempotent pseudotensor of rank four. It is called the **Levi-Civita tensor density**, after Tullio Levi-Civita, who developed much of the mathematical formalism for general relativity. The term **density** refers to any tensor whose transformation involves a power of the determinant of the transformation matrix.

The tensor $\mathcal{F}^{\mu\nu}$ is called the **dual tensor** of $F^{\mu\nu}$. Its matrix form is

$$[\mathcal{F}] = \begin{pmatrix} 0 & -B_x & -B_y & -B_z \\ B_x & 0 & E_z & -E_y \\ B_y & -E_z & 0 & E_x \\ B_z & E_y & -E_x & 0 \end{pmatrix}. \tag{14.152}$$

The components of $\mathcal{F}^{\mu\nu}$ are seen to follow from $F^{\mu\nu}$ by the transformation

$$\mathbf{E} \to \mathbf{B}, \qquad \mathbf{B} \to -\mathbf{E}. \tag{14.153}$$

This is a special case (with $\theta = 90°$) of the duality transformation of Section 9.8. The four-divergence of $\mathcal{F}^{\mu\nu}$ vanishes due to symmetry considerations. This can be seen from

$$\partial_\mu \mathcal{F}^{\mu\nu} = \partial_\mu \epsilon^{\mu\nu\rho\sigma} (\partial_\rho A_\sigma - \partial_\sigma A_\rho) = 0. \tag{14.154}$$

The four-divergence vanishes because it reduces to a sum of products of symmetric terms $\partial_\mu \partial_\rho$ and the antisymmetric terms in $\epsilon^{\mu\nu\rho\sigma}$. Written in terms of the fields $\mathbf{E}$ and $\mathbf{B}$, the vanishing of the four-divergence $\partial_\mu \mathcal{F}^{\mu\nu}$ gives the two homogeneous Maxwell equations

$$\mathbf{\nabla} \cdot \mathbf{B} = 0, \qquad \mathbf{\nabla} \times \mathbf{E} + \partial_t \mathbf{B} = 0. \tag{14.155}$$

We have shown that all four of Maxwell's equations, which are the basis of classical electromagnetism, can be written in manifestly covariant form. This means that Maxwell's equations, even in three-vector form, are the same in all Lorentz systems. The situation is simpler than we might have imagined, and simpler than Maxwell contemplated in his final conclusions. Almost all of the equations in classical electromagnetism can be written in the same way in any Lorentz frame, without bothering to transform each of the elements. The 19th century electromagnetists had all the correct relativistic equations, but didn't know it.

It is a bit ironic that special relativity, which was introduced because earlier electromagnetism had internal inconsistencies, left most of the electromagnetic relations unchanged. On the other hand, classical mechanics had been an internally consistent theory, (Its disagreements with experiment could not be observed in the 19th century.) but virtually all of its equations involving motion were changed by relativity.

14.11 What About D and H?

Can the $\mathbf{D}$ and $\mathbf{H}$ fields be put into a relativistic tensor,

$$[G] = \begin{pmatrix} 0 & -D_x & -D_y & -D_z \\ D_x & 0 & -H_z & H_y \\ D_y & H_z & 0 & -H_x \\ D_z & -H_y & H_x & 0 \end{pmatrix}, \tag{14.156}$$

as was done for $\mathbf{E}$ and $\mathbf{B}$ in the tensor $F^{\mu\nu}$? In 1908, Hermann Minkowski assumed (without proof) that $\mathbf{D}$ and $\mathbf{H}$ could be put into a tensor $G^{\mu\nu}$ just like the tensor $F^{\mu\nu}$ containing $\mathbf{E}$ and $\mathbf{B}$, and he developed a formalism based on that assumption. Since then, a number of papers and textbooks have discussed consequences of Minkowski's assumption. In the following, we will address the question of whether $\mathbf{D}$ and $\mathbf{H}$ can, in fact, be components of a relativistic tensor.

Minkowski considered the case of a simple linear polarizable material with $\mathbf{D} = \epsilon\mathbf{E}$ and $\mathbf{B} = \mu\mathbf{H}$, for which Maxwell's equations are

$$\boldsymbol{\nabla}\cdot(\epsilon\mathbf{E}) = 4\pi\rho \tag{14.157}$$

$$\boldsymbol{\nabla}\times\mathbf{E} = -\partial_t\mathbf{B} \tag{14.158}$$

$$\boldsymbol{\nabla}\cdot\mathbf{B} = 0 \tag{14.159}$$

$$\boldsymbol{\nabla}\times(\mathbf{B}/\mu) = 4\pi\mathbf{j} + \partial_t(\epsilon\mathbf{E}). \tag{14.160}$$

Scalar and vector potentials can still be derived from the homogeneous equations, and wave equations can be derived for the potentials in the Lorenz gauge, but they are somewhat more complicated.

We follow the derivation in section 14.10 leading to the definition of the tensor $F^{\mu\nu}$, but now with permittivity ϵ and permeability μ. We can still use the fact that $\mathbf{B}$ has zero divergence to introduce a vector potential $\mathbf{A}$, in terms of which $\mathbf{B}$ is given by

$$\mathbf{B} = \boldsymbol{\nabla}\times\mathbf{A}, \tag{14.161}$$

and the curl $\mathbf{E}$ equation still leads to

$$\mathbf{E} = -\boldsymbol{\nabla}\phi - \partial_t\mathbf{A}. \tag{14.162}$$

We note that, if the assumption that $F^{\mu\nu}$ in Eq. (14.139) and $G^{\mu\nu}$ in Eq. (14.156) are relativistic tensors is correct, then Eqs. (14.157)-(14.162) would be the same in every Lorentz system, even though they are not manifestly covariant.

Putting Eqs. (14.161) and (14.162) into the two Maxwell's equations with sources, now leads to new equations for the potentials,

$$\boldsymbol{\nabla}\cdot(\epsilon\mathbf{E}) = 4\pi\rho = -\epsilon\boldsymbol{\nabla}\cdot(\boldsymbol{\nabla}\phi + \partial_t\mathbf{A}),$$
$$4\pi\rho = -\epsilon[\nabla^2\phi + \boldsymbol{\nabla}\cdot(\partial_t\mathbf{A})]. \tag{14.163}$$
$$\boldsymbol{\nabla}\times(\mathbf{B}/\mu) = 4\pi\mathbf{j} + \partial_t(\epsilon\mathbf{E}),$$
$$4\pi\mathbf{j} = \frac{1}{\mu}\boldsymbol{\nabla}\times(\boldsymbol{\nabla}\times\mathbf{A}) + \epsilon\partial_t[\boldsymbol{\nabla}\phi + \partial_t\mathbf{A}],$$
$$4\pi\mathbf{j} = \frac{1}{\mu}[\boldsymbol{\nabla}(\boldsymbol{\nabla}\cdot\mathbf{A}) - \nabla^2\mathbf{A}] + \epsilon[\partial_t\boldsymbol{\nabla}\phi + \partial_t^2\mathbf{A}]. \tag{14.164}$$

The Lorenz gauge condition of Eq. (14.134) can still be applied, and reduces Eqs. (14.163) and (14.164) to

$$\epsilon[\nabla^2\phi - \partial_t^2\phi] = -4\pi\rho \tag{14.165}$$

$$\frac{1}{\mu}[\nabla^2\mathbf{A} - \epsilon\mu\partial_t^2\mathbf{A} - (1-\epsilon\mu)\boldsymbol{\nabla}(\boldsymbol{\nabla}\cdot A)] = -4\pi\mathbf{j}. \tag{14.166}$$

These two equations are wave equations for the scalar and vector potentials, modified to include the permittivity and permeability constants.

For ϵ and μ equal 1, we used Eq. (14.133) to show that the potentials ϕ and $\mathbf{A}$ formed a four vector because the d'Alembertian on the left-hand side was a Lorentz scalar. However, while the combination $(\rho, \mathbf{j})$ on the right-hand side of Eqs. (14.165) and (14.166) forms a four-vector, the operators acting on ϕ and $\mathbf{A}$ on the left-hand sides do not constitute a four-scalar. Consequently, Eqs. (14.165) and (14.166) show that the combination $(\phi, \mathbf{A})$ cannot be a four-vector.

If A^ν is not a relativistic four-vector, the product, $\partial^\mu A^\nu$, of A^ν with the relativistic four-vector, ∂^μ, cannot be part of a second rank tensor. Consequently, $F^{\mu\nu} = \partial^\mu A^\nu - \partial^\nu A^\mu$ is not a tensor for a linearly polarizable material. This means that even $\mathbf{E}$ and $\mathbf{B}$ cannot be Lorentz transformed to a moving frame. Although the electromagnetic field matrix $F^{\mu\nu}$ may appear to be Lorentz transformed to a matrix $F'^{\mu\nu}$, the vectors $\mathbf{E}'$ and $\mathbf{B}'$ in $F'^{\mu\nu}$ would not be the electric and magnetic fields in the moving frame. The matrix formed by $\mathbf{D}$ and $\mathbf{H}$ also cannot be a tensor, because they, too, are based on the product, $\partial^\mu A^\nu$, just with each term modified by either ϵ or $1/\mu$.

We see that $\mathbf{E}$ and $\mathbf{B}$, as well as $\mathbf{D}$ and $\mathbf{H}$, cannot be Lorentz transformed to a moving frame. There is no way in special relativity to calculate the electromagnetic fields in any case where there is a moving linearly polarizable medium. Even calculating the $\mathbf{E}$ and $\mathbf{B}$ fields in the rest system of the polarizable medium, and then trying to Lorentz transforming the fields to a moving system would not work because the $\mathbf{E}$ and $\mathbf{B}$ fields themselves cannot be combined in a relativistic tensor in the presence of a polarizable medium.

For a polarizable medium in motion, all electromagnetic calculations have to be done in the rest frame of the medium. Any mechanical results could then be Lorentz transformed to a moving system.

14.12   Problems

1. (a) Use the relativistic velocity addition equations to prove that c added in any direction to any velocity results in velocity c.

   (b) Prove that adding any two velocities cannot give a velocity greater than c.

2. A space ship moving with velocity $0.6\,c$ shoots a projectile with muzzle velocity $0.8\,c$. Find the projectile's velocity with respect to a fixed target if it is fired

   (a) straight ahead.

   (b) straight backwards.

   (c) at right angles to the ship's velocity.

   (d) Redo parts a,b,c if the muzzle velocity were c.

3. Derive Eqs. (14.26) and (14.27) for the transformation of acceleration.

4. A space ship 100 meters long flies through an open hangar 80 meters long.

   (a) How fast must the space ship fly in order to permit both doors of the hangar to be briefly closed at the same time while the space ship is inside?

   (b) How would the pilot of the space ship interpret what happens as the doors close and then reopen?

5. An object falls with a constant acceleration (in its rest system) of g.

   (a) Find its velocity after a time t. (Ans: $v = gt/\sqrt{1 + g^2 t^2/c^2}$

   (b) What is its velocity after one year?

   (c) Find p/m of the object at time t. sketch graphs of $v(t)$ and $p(t)/m$.

6. A jet plane circles the Earth at the equator with a speed of Mach 4. By how much time is the pilot's watch slow when he lands?

7. A space ship leaves Earth for a distant star. It has an acceleration g in its rest system for five Earth years, and then decelerates at g, coming to rest. It then reverses the process to return to Earth.

   (a) How far from Earth does it come to rest?

   (b) The trip takes 20 Earth years. How much older are the astronauts when they return?

   (Do this problem using light years (LY). Calculate c and g in those units.)

8. (a) Show that matrix equation $[\tilde{L}][G][L] = [G]$ holds for the Lorentz matrix $[\mathbf{L}^{(\times)}(\beta)]$.

   (b) Find the Lorentz matrix for a Lorentz transformation of velocity $\mathbf{v} = \beta c(\hat{\mathbf{i}} + \hat{\mathbf{j}})/\sqrt{2}$.

9. Two space ships are traveling along parallel paths at different speeds, u_1 and u_2.

   (a) Show that the paths will no longer be parallel in a Lorentz system moving with velocity $\mathbf{v}$ (not parallel to $\mathbf{u_1}$ and $\mathbf{u_2}$) with respect to the original system. (Calculate $\mathbf{u}_1' \times \mathbf{u}_2'$.)

   (b) Now that their paths are not parallel, can the two space ships ever collide? Explain.

10. (a) Prove the contraction theorem for the contraction of a second rank tensor with a vector.

   (b) Prove the division theorem for the contraction of an arbitrary four-vector V^μ with four components W_μ, leading to a scalar.

   (c) Prove the division theorem for the contraction of an arbitrary tensor $T^{\mu\nu}$ with four components W_ν, leading to a vector.

11. (a) Derive the nonrelativistic result for stellar aberration by considering the motion (with $v \ll c$) of a telescope transverse to incoming light.

   (b) Show that, for $v \ll c$, the relativistic formula for aberration agrees with the nonrelativistic result.

   (c) Calculate the maximum aberration angle ϕ due to the Earth's orbital velocity of 30 km/sec.

12. Find the velocity of 1 TeV protons, and of 20 GeV electrons. (Express your answer as $1-v$.)

13. A proton of mass 938 MeV has a kinetic energy 469 MeV.

   (a) Find its total energy E.

   (b) Find its momentum p.

   (c) Find its velocity v.

   (d) Find its Lorentz parameter γ.

14. Antiprotons were first produced by colliding a beam of protons with protons at rest (fixed target) in the reaction p+p→p+p+p+p̄.

   (a) What minimum beam energy is required to produce antiprotons in this process?

(b) In the actual experiment, the protons were bound inside heavy nuclei, which resulted in some of them having a kinetic energy of 40 MeV. What incident energy would be required to produce antiprotons off these protons if their velocity was directed toward the beam?

15. (a) A $\pi^+(140)$ meson decays at rest into a $\mu^+(106)$ lepton and a $\nu(0)$, neutrino. (The mass in MeV is given in parentheses for each particle.) Find the energy, momentum and velocity of the produced muons. Muons produced in this way are identified by the fact that they all have the same energy.

    (b) A neutron at rest decays in the beta decay process n(939.6)$\rightarrow$p(938.3)+e(0.51)+ν(0). Find the electron's energy if there were no neutrino emitted in this decay. The fact that the decay electrons did not all have this energy was the first clue to the existence of the neutrino.

16. A $\Lambda(1116)$ hyperon with a kinetic energy of 2400 MeV decays in flight into a proton(938) and a pion(140). The pion has its maximum laboratory energy when it is emitted in the same (forward) direction as the Λ. What is this maximum pion energy?

17. The muon has a lifetime of $\tau = 2.2 \times 10^{-6}$ sec. Muons are produced by cosmic rays hitting the upper atmosphere at 80 km above the Earth. What energy must a muon have to be able to reach the Earth within its lifetime? The fact that large numbers of such muons do reach the Earth is a verification of the time dilation prediction of special relativity.

18. (a) Derive the matrix elements of $F^{\mu\nu}$ in Eq. (14.139).

    (b) Show that the covariant equation (14.146) is equivalent to the $\nabla \cdot \mathbf{E}$ and $\nabla \times \mathbf{B}$ Maxwell equations.

    (c) Show that the covariant equation $\partial_\mu \mathcal{F}^{\mu\nu} = 0$ corresponds to the two homogeneous Maxwell equations.

19. Derive Eqs. (14.142) for the transformation of the electric and magnetic fields.

20. (a) Show, by contracting bilinear combinations of $F^{\mu\nu}$ and $\mathcal{F}^{\mu\nu}$ that $\mathbf{E} \cdot \mathbf{B}$ and $\mathbf{E}^2 - \mathbf{B}^2$ are Lorentz invariants.

    (b) Show, using the vector transformations of Eqs. (14.17) and (14.18), that $\mathbf{E} \cdot \mathbf{B}$ and $\mathbf{E}^2 - \mathbf{B}^2$ are Lorentz invariants.

    (c) If $\mathbf{E}$ and $\mathbf{B}$ are perpendicular, and $|\mathbf{E}| > |\mathbf{B}|$, find the velocity β for a Lorentz transformation to a system in which $\mathbf{B} = \mathbf{0}$.

Chapter 15

The Electrodynamics of Moving Bodies.

15.1 Relativistic Electrodynamics

15.1.1 Covariant extension of $\mathbf{F} = m\mathbf{a}$

The Newtonian equation $\mathbf{F} = m\mathbf{a}$ is not relativistically invariant. To find a relativistic generalization, we replace $m\mathbf{a}$ by the relativisitic four-vector $\frac{dp^\mu}{d\tau}$. For the electromagnetic force on a charged particle we want to write the force as a relativistic four-vector that combines the electromagnetic fields with the state of motion of the particle. The contraction $U_\mu F^{\mu\nu}$ accomplishes this, and then the covariant dynamical equation for a charged particle is

$$\frac{dp^\nu}{d\tau} = qU_\mu F^{\mu\nu} = \frac{q}{m}p_\mu F^{\mu\nu}. \tag{15.1}$$

This reduces, in vector notation, to

$$\frac{dW}{dt} = q\mathbf{v}\cdot\mathbf{E} \tag{15.2}$$

$$\frac{d\mathbf{p}}{dt} = q\left[\mathbf{E} + \mathbf{v}\times\mathbf{B}\right]. \tag{15.3}$$

These two equations have had numerous experimental verifications, some of which will be discussed later in this chapter. We have used W here for the energy to avoid confusion with the electric field $\mathbf{E}$, and are using units with $c=1$.

The vector forms of the dynamical equations are just what was used before relativity. The fact that they can be written in manifestly covariant form in Eq. (15.1) means that even the vector forms are the same in any Lorentz system. In order to have this invariance, we had to replace $m\mathbf{a}$ by $\frac{d\mathbf{p}}{dt}$.

If we try to use acceleration in a dynamical equation, we get

$$\frac{d\mathbf{p}}{dt} = m\frac{d}{dt}\left[\frac{\mathbf{v}}{\sqrt{1-\mathbf{v}^2}}\right]$$
$$= m[\gamma\mathbf{a} + \gamma^3\mathbf{v}(\mathbf{v}\cdot\mathbf{a})] = m\gamma^3[\mathbf{a} + \mathbf{v}\times(\mathbf{v}\times\mathbf{a})]. \quad (15.4)$$

Apart from the factors of γ, this equation shows that the acceleration may not even be in the direction of an applied force (if force is defined by $\frac{d\mathbf{p}}{dt}$) for a moving particle.

This complicated result is an example of a general rule that relativistic equations are simpler when written in terms of energy and momentum, rather than velocity and acceleration. Also, since high energy particles travel close to the speed of light with acceleration close to zero, momentum and its time derivative are more meaningful variables than velocity and acceleration.

As a consequence, the important variables for physics of the 21st century and beyond are momentum and energy, with velocity and acceleration being of secondary importance. This is also true for the transition from classical to quantum mechanics, which essentially requires the use of momentum rather than velocity.

There are three possible definitions of force in Special Relativity:

1.   The covariant **Minkowski force**, defined as

$$\mathcal{F}^\mu = \frac{dp^\mu}{d\tau}, \quad (15.5)$$

2.   A three-vector force, defined by

$$\mathbf{F} = \frac{d\mathbf{p}}{dt}. \quad (15.6)$$

The Minkowski force can be written in terms of the three-vector force as

$$\mathcal{F}^\mu = \left[\gamma\frac{dW}{dt}, \gamma\mathbf{F}\right]. \quad (15.7)$$

3.   A force that follows from a simplification of Eq. (15.4) for acceleration in the direction of velocity. Then

$$\mathbf{F_N} = \frac{d\mathbf{p}}{dt} = m\mathbf{a}\gamma^3 = m\mathbf{a}', \quad (15.8)$$

where $\mathbf{a}'$ is the particle's acceleration in its instantaneous rest frame. This could be considered a relativistic form of Newton's force equation, $\mathbf{F} = m\mathbf{a}$.

Using any force definition, the dynamical equations of motion could be written in two steps, as was traditional in non-relativistic physics. However, nothing would be gained by this, except writing two equations to replace one. The concept of force in modern physics became increasingly irrelevant in the 20th century, and will not have much use in the relativistic (special and general) and quantum physics of the 21st century.

In retrospect, we can argue that its use could even have been dispensed with in Newtonian dynamics (although it is convenient for statics). The concept of force introduced by Newton at the time was useful in relating motion to everyday experience, like the force of an apple striking a head. But even $\mathbf{F} = m\mathbf{a}$ is nothing more than a definition of force rather than a physical law. In electromagnetism the dynamical physical law is given by Eq. (15.1), given in vector form in Eqs. (15.2) and (15.3) with no mention of force. For any other theory, the form of the four-vector on the right hand side would be determined by the theory.

15.1.2 Motion in a magnetic field

In chapter 7 we showed that a non-relativistic charged particle moving perpendicular to a constant magnetic field followed a circular path. The relativistic equation of motion of such a particle is given by Eq. (15.3) as

$$\frac{d\mathbf{p}}{dt} = \frac{q\mathbf{p}}{\gamma m} \times \mathbf{B}. \tag{15.9}$$

With no electric field, the energy, and thus the magnitude $|\mathbf{p}|$, will be constant. For a constant $\mathbf{B}$ field the vector $\mathbf{p}$ rotates at a constant angular frequency

$$\omega = \frac{qB}{\gamma m}. \tag{15.10}$$

The velocity $\mathbf{v}$ of the particle will rotate at the same angular frequency, leading to circular motion with radius R determined by

$$R = \frac{v}{\omega} = \frac{p}{qB}. \tag{15.11}$$

This shows that the radius of curvature of a charged particle in a magnetic field is proportional to its momentum, which makes this a good way to measure a particle's momentum.

As an example of how to use natural units, we find the radius for an electron with p=1,200 MeV, in a perpendicular magnetic field of 2,000 gauss. The calculation is

$$R = \frac{p}{eB} = \frac{1,200\,\text{MeV}}{e \times 2,000\,\text{gauss}} = 6 \times 10^5\,\frac{\text{volts}}{\text{gauss}} \times \frac{1\,\text{statvolt}}{300\,\text{volts}} = 2,000\,\text{cm}. \tag{15.12}$$

We used the definition of an MeV that 1 MeV/e=10^6 volts, and then the fact that 1 gauss=1 statvolt/cm. We didn't have to use a numerical value for e, and c never entered because of our use of the unit MeV for momentum.

The angular frequency of the circular motion varies with γ, as shown in Eq. (15.10). This limits the use of cyclotrons to only accelerating particles to velocities that are small compared with 1. This led to the development of synchrocyclotrons, in which the frequency of oscillation of the accelerating voltage was varied according to Eq. (15.10). Even with this synchronization, the cyclotron configuration with two large D's, as in Fig. 7.7, is impractical for very high energy particles, for which the radius can become quite large.

Modern high energy circular accelerators, called synchrotons (or phasotrons), have particles kept at a fixed radius by an increasing strength of magnetic field, while they are accelerated by electric fields.

15.1.3   Linear accelerator

In a linear accelerator, a charged particle's energy is increased from rest by a uniform electric field via Eq. (15.2):

$$\frac{dW}{dt} = qvE. \tag{15.13}$$

We can convert the left hand side to

$$\frac{dW}{dt} = \frac{dW}{dx}\frac{dx}{dt} = v\frac{dW}{dx}. \tag{15.14}$$

Then, we have

$$\frac{dW}{dx} = qE, \tag{15.15}$$

and the particle's kinetic energy at any distance L along the accelerator is given by

$$T = W - m = \int_0^L qE\,dx. \tag{15.16}$$

. This is just the same as the non-relativistic expression for the increase in kinetic energy as the work done by the electric field. Since $W >> m$ in a high energy accelerator (This is what is meant by 'high energy'.), the momentum approaches the energy and the velocity approaches 1 during the acceleration.

The final energy in the unit eV of a particle with a charge of magnitude e, the electron charge, is numerically equal to the voltage difference across the length of the accelerator. That is a '20 GeV' linear accelerator has a net voltage difference of 20×10^9 volts. (The energy is high enough that we do not need to distinguish between the total or kinetic energy, or the momentum of the particle.)

Because the energy unit eV is given in terms of SI units, the SI system is simpler in this case. In Gaussian units, a 20 GeV accelerator would have a voltage difference of $(20/300)\times10^9$ statvolts. In any case, the required voltage does not depend on the mass of the accelerated particle, but would change if the particle had a different charge.

15.2 Lagrange's and Hamilton's Equations for Electrodynamics

15.2.1 Non-relativistic Lagrangian

In this section, we briefly review non-relativistic Lagrangian dynamics, and then extend its application to relativistic electrodynamics. A more throrough discussion of variational calculus is given in most books on Mathematical Physics or a graduate Mechanics text, such as [Goldstein].

Lagrange's equations for a dynamical theory follow from a variational principle applied to a function (the **Lagrangian**) of generalized coordinates $q_i(t)$, their time derivatives $\dot{q}_i(t)$, and the time t. The dynamical equations of motion follow from the **principle of least action**. This states that the action, defined by

$$\mathcal{A} = \int_{t_1}^{t_2} L[q_i(t), \dot{q}_i(t), t] dt, \tag{15.17}$$

is a minimum (or, more generally, an extremum) with respect to infinitesimal variations $\delta q_i(t)$ of the functions $q_i(t)$. The variables $q_i(t)$ are specified at the times t_1 and t_2, so the variations $\delta q_i(t)$ must vanish at the endpoints of the action integral.

The variation of the action is given by

$$
\begin{aligned}
\delta A &= \int_{t_1}^{t_2} \left[\frac{\partial L}{\partial q_i} \delta q_i(t) + \frac{\partial L}{\partial \dot{q}_i} \delta \dot{q}_i(t) \right] dt \\
&= \int_{t_1}^{t_2} \left[\frac{\partial L}{\partial q_i} - \frac{d}{dt} \left(\frac{\partial L}{\partial \dot{q}_i} \right) \right] \delta q_i(t) dt.
\end{aligned} \tag{15.18}
$$

In deriving this result, we used the fact that

$$\delta \dot{q}_i(t) = \frac{d[\delta q_i(t)]}{dt}, \tag{15.19}$$

and then integrated the second term in the integral by parts. Since the variation is arbitrary, the action integral will vanish only if the integrand is identically zero, which gives

$$\frac{\partial L}{\partial q_i} - \frac{d}{dt} \left(\frac{\partial L}{\partial \dot{q}_i} \right) = 0. \tag{15.20}$$

These differential equations are called the Euler-Lagrange equations, although physicists usually refer to them simply as Lagrange's equations. They comprise the equations of motion for any dynamical theory. For a single particle with the position variable $\mathbf{r}$ as the generalized coordinate and $\mathbf{v}$ as its time derivative, Lagrange's equation can be written in vector notation as

$$\boldsymbol{\nabla} L(\mathbf{r}, \mathbf{v}, t) = \frac{d}{dt} \boldsymbol{\nabla}_{\mathbf{v}}[L(\mathbf{r}, \mathbf{v}, t)]. \tag{15.21}$$

We use the notation $\nabla_{\mathbf{v}}$ to denote the gradient with respect to the variable $\mathbf{v}$.

The procedure for finding the appropriate Lagrangian for any theory is not always straightforward. The ultimate test of a Lagrangian is whether, when put into Langrage's equations, it gives equations of motion that agree with experiment.

For non-relativistic electrodynamics, the equation of motion is $\mathbf{F} = m\mathbf{a}$, with $\mathbf{F}$ being the Lorentz force. To find the proper Lagrangian, we write the Lorentz force in terms of the scalar and vector potentials, so Newton's equation becomes

$$
\begin{aligned}
m\frac{d\mathbf{v}}{dt} &= q[-\nabla\phi - \partial_t\mathbf{A} + \mathbf{v}\times(\nabla\times\mathbf{A})] \\
&= q[-\nabla\phi - \partial_t\mathbf{A} - (\mathbf{v}\cdot\nabla)\mathbf{A} + \nabla(\mathbf{v}\cdot\mathbf{A})] \\
&= \nabla[-q\phi + q\mathbf{v}\cdot\mathbf{A}] - q\frac{d\mathbf{A}}{dt}.
\end{aligned}
\tag{15.22}
$$

In the last step above, we used the calculus rule for the time derivative of a function of $\mathbf{r}$ and t that

$$
\frac{d\mathbf{A}}{dt} = (\mathbf{v}\cdot\nabla)\mathbf{A} + \partial_t\mathbf{A}.
\tag{15.23}
$$

We see from Eqs. (15.21) and (15.22) that a suitable Lagrangian for non-relativistic electrodyamics is

$$
L = \frac{1}{2}m\mathbf{v}^2 - q\phi + q\mathbf{v}\cdot\mathbf{A}.
\tag{15.24}
$$

If this Lagrangian is put into Eq. (15.21), the result is equation of motion Eq. (15.22).

15.2.2 Relativistic Lagrangian

The relativistic action should be a Lorentz scalar so that the resulting equations of motion will be the same in all Lorentz systems. To do this, we rewrite Eq. (15.17) for the action in terms of the proper time τ

$$
\mathcal{A} = \int_{\tau_1}^{\tau_2} L\gamma d\tau = \int_{\tau_1}^{\tau_2} \mathcal{L} d\tau.
\tag{15.25}
$$

If the function $\mathcal{L}$ is a Lorentz scalar, then the action will be a Lorentz scalar, and be the same in all Lorentz systems. The scalar function $\mathcal{L}$ is not itself the Lagrangian. The actual Lagrangian to be put into Lagrange's equation (15.21) is given by

$$
L = \mathcal{L}/\gamma.
\tag{15.26}
$$

We will refer to $\mathcal{L}$ as the **covariant Lagrangian**, and to L as the **relativistic Lagrangian**.

Now we want to find the appropriate functional form of $\mathcal{L}$ for electrodynamics. There should be one term that depends only on the velocity. This term was the kinetic energy in the non-relativistic case, and, by itself, would be the full Lagrangian for a free particle. We have only one four-vector associated with velocity, the four-velocity U^μ. The only scalar we can get from that is its length $U_\mu U^\mu$ which just equals 1. So the only possible covariant Lagrangian is a constant. The Lagrangian should have the units of energy, and the only scalar associated with a particle is its mass. (Recall that, in units with $c=1$, mass and energy have the same units.) So, we have

$$\mathcal{L} = -m, \quad L = -m/\gamma, \tag{15.27}$$

for the Lagrangian of a free particle. We introduced the minus sign to agree with the sign convention we will use for the interaction Lagrangian.

In order to have linear equations of motion, the Lagrangian for the electromagnetic interaction should be linear in the potentials, the particle's charge, and in its four-velocity (which is the only four-vector associated with its motion). Thus, the only scalar combination we can form for the covariant interaction Lagrangian is

$$\mathcal{L}_{\text{int}} = -q U_\mu A^\mu. \tag{15.28}$$

Then, the relativistic Lagrangian for the particle's electromagnetic interaction is

$$
\begin{aligned}
L &= -m/\gamma - (q/\gamma) U_\mu A^\mu \\
&= -m/\gamma - q\phi + q\mathbf{v}\cdot\mathbf{A}.
\end{aligned}
\tag{15.29}
$$

The interaction part of this relativistic Lagrangian is the same as that in Eq. (15.24) for the non-relativistic interaction Lagrangian, but the free particle part is no longer the kinetic energy. When the relativistic Lagrangian is put into Lagrange's equation (15.21), the relativistic equation of motion are produced in agreement with Eq. (15.4).

It is interesting to note that in the relativistic approach, we have been led to the correct form of the relativistic Lagrangian by constructing the only Lorentz scalars available to us. Any other form for the Lagrangian would require new quantities to be introduced into the theory. We see that the relativistic development of electrodynamics is simpler and more compelling than the earlier construction of the non-relativistic Lagrangian (which we just picked to give the right answer).

Maxwell condensed much of Electromagnetism into the four Maxwell's equations and the Lorentz force equation. With Einstein's Special Relativity, these five equations become two,

$$\Box^2 A^\mu = 4\pi j^\mu \tag{15.30}$$

$$\mathcal{L} = -m - q U_\mu A^\mu. \tag{15.31}$$

Notice that in the Lagrangian formulation, there is no need to introduce the electromagnetic fields, either in the vector forms **E** and **B**, or the field tensor $F^{\mu\nu}$. The dynamics can be described just in terms of the potential A^{μ}. This is an alternate presentation, compared to the earlier development where the fields were the physical quantities and the potential a mathematical convenience. In the Lagrangian formulation, the potential is the physical quantity and the electromagnetic fields are not even mathematical conveniences– they are just not present.

The principle of gauge invariance is still important, but it has a different motivation. In the earlier treatment, a gauge transformation on the potential did not change the fields. Now, a gauge transformation does change the Lagrangian, but does not change the dynamical equations of motion. To see this, we make the gauge transformation

$$A^{\mu} \rightarrow A'^{\mu} + \partial^{\mu}\Lambda. \tag{15.32}$$

This changes the Lagrangian to

$$L \rightarrow L'^{\mu} = L - q\partial_t\phi + q(\mathbf{v}\cdot\boldsymbol{\nabla})\mathbf{A}. \tag{15.33}$$

When this new Lagrangian is put into Lagrange's equation, the added terms do not change the dynamical equation of motion. (See problem 5c.)

15.2.3   Hamiltonian for Electrodynamics

Passage from the Lagrangian to the Hamiltonan involves a change of independent variables from the **r** and **v** in the Lagrangian (for one particle) to **r** and a **canonical momentum** $\mathcal{P}$ defined by

$$\mathcal{P} = \boldsymbol{\nabla}_{\mathbf{v}}L. \tag{15.34}$$

Lagrange's equation can now be written in two steps: the first is Eq. (15.34) defining $\mathcal{P}$, with the second being

$$\frac{d\mathcal{P}}{dt} = \boldsymbol{\nabla}L. \tag{15.35}$$

For the relativistic Lagrangian of Eq. (15.29), the canonical momentum is given by

$$\begin{aligned}
\mathcal{P} &= \boldsymbol{\nabla}_{\mathbf{v}}[-m/\gamma - q\phi + q\mathbf{v}\cdot\mathbf{A}] \\
&= m\mathbf{v}\gamma + q\mathbf{A} \\
&= \mathbf{p} + q\mathbf{A}, \tag{15.36}
\end{aligned}$$

where **p** is the relativistic mechanical momentum.

The Hamiltonian is defined in terms of the Lagrangian by the **Legendre Transformation**[1]

$$H(\mathbf{r}, \boldsymbol{\mathcal{P}}, t) = \mathbf{v} \cdot \boldsymbol{\mathcal{P}} - L(\mathbf{r}, \mathbf{v}, t). \tag{15.37}$$

That H is a function of the variables $\mathbf{r}$ and $\boldsymbol{\mathcal{P}}$ (and t if L depends on t) is seen by finding the differential dH using Eq. (15.37),

$$
\begin{aligned}
dH &= \mathbf{v} \cdot d\boldsymbol{\mathcal{P}} + \boldsymbol{\mathcal{P}} \cdot d\mathbf{v} - d\mathbf{v} \cdot \boldsymbol{\nabla}_v L - d\mathbf{r} \cdot \boldsymbol{\nabla} L - (\partial_t L) dt \\
&= \mathbf{v} \cdot d\boldsymbol{\mathcal{P}} - d\mathbf{r} \cdot \boldsymbol{\nabla} L - (\partial_t L) dt,
\end{aligned}
\tag{15.38}
$$

where we have used the definition of $\boldsymbol{\mathcal{P}}$ to cancel the two $d\mathbf{v}$ terms. Since H is a function of $\mathbf{r}$ and $\boldsymbol{\mathcal{P}}$, its differential can also be written as

$$dH = d\boldsymbol{\mathcal{P}} \cdot \boldsymbol{\nabla}_{\boldsymbol{\mathcal{P}}} H + d\mathbf{r} \cdot \boldsymbol{\nabla} H + (\partial_t H) dt. \tag{15.39}$$

Comparing the two equations for dH, we see that

$$\boldsymbol{\nabla}_{\boldsymbol{\mathcal{P}}} H = \mathbf{v} \tag{15.40}$$

$$\boldsymbol{\nabla} H = -\frac{d\boldsymbol{\mathcal{P}}}{dt}, \tag{15.41}$$

These two equations are called **Hamilton's canonical equations**, or more briefly just Hamilton's equations. They lead to the same equations of motion as Lagrange's equation. If L depends on t, we have the additional equation

$$\partial_t H = -\partial_t L. \tag{15.42}$$

The Hamiltonian for electrodynamics is given by

$$
\begin{aligned}
H &= \mathbf{v} \cdot \boldsymbol{\mathcal{P}} + m/\gamma + q\phi - q\mathbf{v} \cdot \mathbf{A} \\
&= \mathbf{v} \cdot (\boldsymbol{\mathcal{P}} - q\mathbf{A}) + m/\gamma + q\phi \\
&= \frac{(\boldsymbol{\mathcal{P}} - q\mathbf{A})^2 + m^2}{m\gamma} + q\phi \\
&= \sqrt{(\boldsymbol{\mathcal{P}} - q\mathbf{A})^2 + m^2} + q\phi.
\end{aligned}
\tag{15.43}
$$

In the last step above, we used the relation $m\gamma = \sqrt{(\boldsymbol{\mathcal{P}} - q\mathbf{A})^2 + m^2}$, which holds since each is the particle's relativistic energy. We see that the electrodynamic Hamiltonian is the particle's total energy, including its potential energy $q\phi$. Putting this Hamiltonian into Hamilton's equations gives Eq. (15.3), the relativistic equation of motion for a charged particle.

---

[1]See, for instance, Section 8.1 of [Goldstein].

15.3   Fields of a Charge Moving with Constant Velocity

To find the electromagnetic fields of a charged particle moving with constant velocity, we will first find the fields in the rest system of the particle, and then Lorentz transform them to a system where the particle is moving. We have to be careful to transform both the field components and the variables on which the fields depend.

We start with Coulomb's law for the electric field of a point charge at rest in in a system S′,

$$E' = \frac{qr'}{r'^3}, \tag{15.44}$$

and transform this field to a system S where the point charge has velocity **v**, using the transformation equations [from Eqs. (14.142) in the text with **B′ = 0**]

$$E_\parallel = E'_\parallel = \frac{qr'_\parallel}{r'^3} \tag{15.45}$$

$$E_\perp = \gamma E'_\perp = \frac{q\gamma r'_\perp}{r'^3} \tag{15.46}$$

$$B = v \times E. \tag{15.47}$$

The subscript $\parallel$ in these equations signifies the component of a vector parallel to **v**, and the subscript $\perp$ signifies the components perpendicular to **v**.

These fields are still in terms of the primed variables of system S′. The primed variables are given in terms of the unprimed variables by the Lorentz transformation:

$$r'_\parallel = \gamma r_\parallel \tag{15.48}$$
$$r'_\perp = r_\perp \tag{15.49}$$
$$r' = \gamma r_\parallel + r_\perp \tag{15.50}$$
$$r'^2 = \gamma^2 r_\parallel{}^2 + r_\perp{}^2. \tag{15.51}$$

The fields in Eqs. (15.45-15.47) then become

$$E = \frac{q\gamma r}{(r_\perp{}^2 + \gamma^2 r_\parallel{}^2)^{\frac{3}{2}}} = \frac{qr}{\gamma^2[r^2 - (v \times r)^2]^{\frac{3}{2}}} \tag{15.52}$$

$$B = v \times E = \frac{qv \times r}{\gamma^2[r^2 - (v \times r)^2]^{\frac{3}{2}}}. \tag{15.53}$$

These two equations are the relativistic expressions for the fields due to a charge moving with constant velocity. They resolve (at least, for constant velocity) the classic nineteenth century conundrum of how to describe the electromagnetic fields of moving charges, which we discussed in Chapter 7.

The electric and magnetic fields are drastically changed for large velocities. The electric field of a fast moving charge is still in the radial direction, but is not spherically symmetric. The field is stronger in the direction perpendicular to the charge's velocity. This is shown in Fig. 15.1, which depicts the lines of force of the electric field, which is seen to become stronger in the transverse direction.

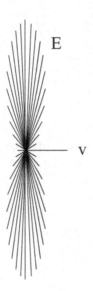

Figure 15.1: Electric field lines of a charge moving with constant velocity v=0.9. The length of each field line is proportional to the magnitude of **E**.

15.3.1   Energy loss of a moving charge

For a particle of charge q moving along the x axis with velocity v, the electric field at a point on the y axis, as given by Eq. (15.52), is

$$E_x = \frac{-q\gamma vt}{[y^2 + (\gamma^2 v^2 t^2)]^{3/2}} \tag{15.54}$$

$$E_y = \frac{q\gamma y}{[y^2 + (\gamma^2 v^2 t^2)]^{3/2}}. \tag{15.55}$$

These parallel (x) and perpendicular (y) components of the electric field are plotted as a function of time in figure 15.2. The field E_y, perpendicular to the velocity, peaks sharply in time as the motion becomes relativistic (v approaching 1). The field E_x, in the direction of **v**, shows a rapid single oscillation, with a smaller magnitude. The figure shows the transverse pulse to be quite short for a large velocity (v=0.9), but more gradual for a low velocity (v=0.1).

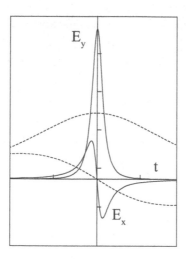

Figure 15.2: Electric field components E_x (longitudinal) and E_y (transverse) of a moving unit charge as a function of time. The solid curves are for v=0.9, and the dashed curves for v=0.1

When a charged particle travels though matter, the sharp pulse of its transverse electric field can ionize electrons in the matter. The energy given to such electrons will cause the moving charged particle to lose energy. The electric force on an electron (of charge -e) in the matter, will produce a momentum change of

$$
\begin{aligned}
\Delta p &= -\int_{-\infty}^{\infty} e E_y dt = -\int_{-\infty}^{\infty} \frac{eq\gamma y}{(y^2 + \gamma^2 v^2 t^2)^{\frac{3}{2}}} dt \\
&= \frac{-2eq}{vy}.
\end{aligned}
\tag{15.56}
$$

This assumes that **v** of the moving particle remains constant during the interaction. This will be true because the moving particle is interacting with many evenly distributed electrons at the same time.

If the electron in the matter is initially at rest, the energy given to it is

$$
\Delta W(y) = \frac{\Delta p^2}{2m_e} = \frac{2e^2 q^2}{m_e v^2 y^2}.
\tag{15.57}
$$

We have used the non-relativistic form for the kinetic energy, which is a good approximation for most of the electrons.

By conservation of energy, the energy transferred to the electron will equal the energy loss of the moving particle. The energy lost by such a particle in traversing a slab of matter of thickness dx, due to all the electrons in the slab is

$$
dW = 2\pi N Z dx \int_0^\infty \Delta W(r) r dr.
\tag{15.58}
$$

where, N is the number of atoms per unit volume and Z is the average atomic number, and r is the distance of each from the particle's path.

The integral diverges logarithmically at both the lower and upper limits. This is because approximations we have made break down for electrons that are too close or too far away from the particle's path. For electrons that are too close to the path, the energy loss ΔW calculated above could exceed the maximum possible energy loss by the particle. Also, the force on the electron can become strong enough that the electron moves during the interaction, or its velocity becomes high enough that relativistic corrections modify Eq. (15.57) for the energy given to the electron.

For electrons far from the particle's path, the impulse to the electron becomes so weak that the orbital motion of the electron in the atom, and the quantization of its energy levels (including a minimum ionization energy) modify the energy given to the electron. The combination of all these effects are usually taken into account by putting phenomenological limits r_{Max} and r_{min} on the integral, leading to a rate of energy loss

$$\begin{aligned} \frac{dW}{dx} &= 2\pi N Z \int_{r_{\text{min}}}^{r_{\text{Max}}} \frac{2e^2 q^2 r dr}{m_e v^2 r^2} \\ &= \frac{4\pi N Z q^2 e^2}{m_e v^2} \ln(r_{\text{Max}}/r_{\text{min}}). \end{aligned} \qquad (15.59)$$

This classical formula for the rate of energy loss by a charged particle was first derived by Neils Bohr. The ratio $(r_{\text{Max}}/r_{\text{min}})$ leads to a logarithmic increase of $\frac{dW}{dx}$ with energy for relativistic particles with $v \sim 1$. For non-relativistic particles, the $1/v^2$ dependence dominates the energy loss. The ionization of electrons causes tracks to form in particle detectors like emulsions, cloud chambers or bubble chambers. The density of the tracks is proportional to the amount of ionization. The ionization, and the track density, decreases slightly with distance along the track for high energy particles (with $W \gg M$), since they have velocity $v \simeq 1$. Near the end of a track, the ionization increases, because v starts to decrease. This shows up as a thickening of the track near its end.

15.3.2 Interaction between moving charges

The relativistic field equations (15.52) and (15.53) permit us to write the relativistic equation for the force (as measured by $\frac{d\mathbf{p}}{dt}$) between two charges moving with different velocities. We just have to let the relativistic electric and magnetic fields given by Eqs. (15.52) and (15.53) act on a particle of charge q' a distance $\mathbf{r}$ away, getting

$$\begin{aligned} \frac{d\mathbf{p}'}{dt} &= q' \left[\mathbf{E} + \mathbf{v}' \times \mathbf{B}\right] \\ &= \frac{q q' [\mathbf{r} + \mathbf{v}' \times (\mathbf{v} \times \mathbf{r})]}{\gamma_v^2 [\mathbf{r}^2 - (\mathbf{v} \times \mathbf{r})^2]^{\frac{3}{2}}}. \end{aligned} \qquad (15.60)$$

438 CLASSICAL ELECTROMAGNETISM

This equation almost solves the nineteenth century problem of how to treat forces between moving particles. It also resolves the question of what system the velocities are with respect to, since the equation holds in any Lorentz system. Remarkably, this is true even though the force equation does not involve the relative velocity between the particles.

You must have noticed the word "almost" in the above discussion. The reason for this is twofold. First, as we will see later in this chapter, there is an additional contribution to the fields due to acceleration of the charge producing the field. This acceleration term depends on the retarded time, and hence on the past trajectory of the particle, which itself depends on the force equation.

If you think this is a circular and fairly intractable procedure, you are right. In deriving the force equation (15.60), we needed to use the fact that the charge's trajectory was a straight line at constant velocity. Otherwise, we could not have gotten a closed formula. This means that Eq. (15.60) strictly applies only if the particle producing the force is massive enough that it has negligible acceleration during the interaction.

The second reason for "almost" is the fact that Quantum Mechanics must be used for most interactions between relativistic particles. This makes moot much of the classical theoretical effort around the end of the nineteenth century on how to treat interactions between moving particles. Only in some special cases, generally if there is not too much effect on the particles' trajectories, can Eq. (15.60) be usefully implemented.

An interesting feature of the electromagnetic force between two moving particles is that the force does not satisfy Newton's third law, or even conservation of mechanical momentum. The force on the particle with charge q and velocity $\mathbf{v}$, which produced the fields acting on charge q', is given by the replacements

$$q \rightleftharpoons q', \quad q', \quad \mathbf{v} \rightleftharpoons \mathbf{v}', \quad \mathbf{r} \to -\mathbf{r} \tag{15.61}$$

in Eq. (15.60). Then, we get

$$\frac{d\mathbf{p}}{dt} = \frac{-qq'[\mathbf{r} + \mathbf{v}\times(\mathbf{v}'\times\mathbf{r})]}{\gamma_{v'}^2[\mathbf{r}^2 - (\mathbf{v}'\times\mathbf{r})^2]^{\frac{3}{2}}}, \tag{15.62}$$

where $\mathbf{r}$ is still the vector from charge q to charge q'. The total mechanical momentum $\mathbf{p} + \mathbf{p}'$ of the two particles is not conserved. This is because the electromagnetic field momentum has to be included, and it is the sum of this and the mechanical momentum that is conserved. This was shown in Chapter 9 as a general theorem, but is difficult to show explicitly in this case because of the complicated fields in Eqs. (15.52) and (15.53).

Although the forces on the two particles do not conserve mechanical momentum, it is conserved in any scattering where the particles start and end up far apart from each other. That is because the field momentum vanishes (neglecting radiation for now) as the particle separation becomes infinite, so that it is the same (zero) before and after the scattering. A simple example of this is given in Problem 7.

There is another apparent problem with the force between two moving charges. Remember the Trouton-Noble experiment (in Chapter 14), where no torque was observed on a moving charged capacitor? The direction of the force between two charges ($+q$ and $-q$, as on capacitor plates) moving with the same velocity on parallel paths is given by the numerator of Eq. (15.60) to be

$$\frac{d\mathbf{p}}{dt} \propto \mathbf{r} + \mathbf{v} \times (\mathbf{v} \times \mathbf{r}) = \mathbf{r}(1 - v^2) + \mathbf{v}(\mathbf{v} \cdot \mathbf{r}), \qquad (15.63)$$

which is not in the $\mathbf{r}$ direction. Won't this still predict a rotation of the moving capacitor, coming from $\mathbf{r} \times \mathbf{F} \neq \mathbf{0}$? The surprising answer is NO.

If we equate Eq. (15.4), relating $\frac{d\mathbf{p}}{dt}$ to the acceleration, to Eq. (15.60) (with $\mathbf{v}'=\mathbf{v}$), we get

$$\frac{d\mathbf{p}}{dt} = m\gamma^3[\mathbf{a} + \mathbf{v} \times (\mathbf{v} \times \mathbf{a})] = \frac{-q^2[\mathbf{r} + \mathbf{v} \times (\mathbf{v} \times \mathbf{r})]}{\gamma^2[\mathbf{r}^2 - (\mathbf{v} \times \mathbf{r})^2]^{\frac{3}{2}}}. \qquad (15.64)$$

This is a complicated equation for $\mathbf{a}$, but it has the relatively simple solution

$$\mathbf{a} = \frac{-q^2\mathbf{r}}{m\gamma^5[\mathbf{r}^2 - (\mathbf{v} \times \mathbf{r})^2]^{\frac{3}{2}}}. \qquad (15.65)$$

So we see that the relativistic acceleration is in the direction of the vector distance between the charges, and there will be no tendency to rotate. Special relativity has passed the Trouton-Noble test. Charged capacitors will not tend to rotate, either at rest or in motion. To see this, we have had to modify another bit of non-relativistic intuition. If we want 'Torque' to mean 'tendency to rotate' then it cannot be $\mathbf{r} \times \mathbf{F}$, but must be proportional to $\mathbf{r} \times \mathbf{a}$. We have also seen that 'acceleration' and 'rate of change of momentum' have quite a complicated connection.

Another lesson from the above example is that it is usually easier to study a system of two (or more) particles in their barycentric (the fancy word for center of mass) system. For a charged capacitor at rest, there is obviously no tendency to rotate. In the system in which the capacitor is moving, we eventually reached the same conclusion, but it took some tricky work.

In general, if the results of a calculation in the center of mass system can be described in terms of Lorentz invariants, then the results will hold in any Lorentz system without further calculation. It pays to believe in Special Relativity. Note that this Lorentz invariance still holds in Relativistic Quantum Mechanics, so quantum results calculated in one system will also hold in other Lorentz systems.

15.4 Electromagnetic Fields of a Moving Charge

15.4.1 The Liénard-Wiechert Potentials and Fields

The Liénard-Wiechert potentials and fields of a point charge moving with arbitrary velocity and acceleration were originally derived by Alfred-Marie Liénard in 1898 and Emil Wiechert in 1900. Although derived before special relativity, the derivation and its results were fully relativistic (as was much of classical electromagnetism).

We start our derivation of the Liénard-Wiechert potentials with Eq. (??) from the derivation of the retarded vector potential in Chapter 13:

$$\mathbf{A}(\mathbf{r}, t) = \frac{1}{2\pi} \int \frac{d^3 r'}{|\mathbf{r} - \mathbf{r}'|} \int_{-\infty}^{\infty} dt' \mathbf{j}(\mathbf{r}', t') \int_{-\infty}^{\infty} d\omega e^{i\omega(t' - t + |\mathbf{r} - \mathbf{r}'|)},$$

$$= \int \frac{d^3 r'}{|\mathbf{r} - \mathbf{r}'|} \int_{-\infty}^{\infty} dt' \mathbf{j}(\mathbf{r}', t') \delta(t' - t + |\mathbf{r} - \mathbf{r}'|). \tag{15.66}$$

The current distribution for a point charge q moving on a trajectory $\mathbf{s}(t)$ is

$$\mathbf{j}(\mathbf{r}, t) = q\mathbf{v}(t)]\delta^3[\mathbf{r} - \mathbf{s}(t)], \tag{15.67}$$

where

$$\mathbf{v}(t) = \frac{d\mathbf{s}(t)}{dt}. \tag{15.68}$$

With this current distribution, we get

$$\mathbf{A}(\mathbf{r}, t) = \int_{-\infty}^{\infty} dt' \int \frac{d^3 r' q\mathbf{v}(t')\delta^3[\mathbf{r}' - \mathbf{s}(t')]\delta(t' - t + |\mathbf{r} - \mathbf{r}'|)}{|\mathbf{r} - \mathbf{r}'|}$$

$$= \int_{-\infty}^{\infty} \frac{q\mathbf{v}(t')\delta(t' - t + |\mathbf{r} - \mathbf{s}(t')|)dt'}{|\mathbf{r} - \mathbf{s}(t')|}$$

$$= \int_{-\infty}^{\infty} \frac{q\mathbf{v}(t')\delta(t' - t_r)dt'}{|\mathbf{r} - \mathbf{s}(t')| \left[\frac{d[t' - t + |\mathbf{r} - \mathbf{s}(t')|]}{dt'} \right]_{t' = t_r}}$$

$$= \frac{q\mathbf{v}(t_r)}{|\mathbf{r} - \mathbf{s}(t_r)| \left[\frac{d[t' - t + |\mathbf{r} - \mathbf{s}(t')|]}{dt'} \right]_{t' = t_r}}$$

$$= \frac{q\mathbf{v}(t_r)}{|\mathbf{r} - \mathbf{s}(t_r)| \left[1 - \frac{[\mathbf{r} - \mathbf{s}(t_r)] \cdot \mathbf{v}(t_r)}{|\mathbf{r} - \mathbf{s}(t_r)|} \right]}$$

$$\mathbf{A}(\mathbf{r}, t) = \frac{q\mathbf{v_r}}{r_r - \mathbf{r_r} \cdot \mathbf{v_r}}, \tag{15.69}$$

where $t_r = t - r_r$ is the retarded time, $\tag{15.70}$

and $\mathbf{r_r} = \mathbf{r} - \mathbf{s}(t_r)$ $\tag{15.71}$

is the distance from the retarded position of the point charge to the point of

observation, and $\mathbf{v_r} = \dfrac{d\mathbf{s}(t_r)}{dt_r}. \tag{15.72}$

A similar derivation with charge density

$$\rho(\mathbf{r}, t) \;=\; q\delta^3[\mathbf{r} - \mathbf{s}(t)] \tag{15.73}$$

gives the Liénard-Wiechert scalar potential as

$$\phi(\mathbf{r}, t) \;=\; \frac{q}{r_r - \mathbf{r_r}\!\cdot\!\mathbf{v_r}}. \tag{15.74}$$

These equations for the potentials are deceptively simple in appearance, but their dependence on the retarded time makes their application quite complicated. In order to find the **E** and **B** fields, we have to take the derivatives

$$\mathbf{E}(\mathbf{r}, t) \;=\; \left[\frac{q}{(r_r - \mathbf{r_r}\!\cdot\!\mathbf{v_r})^2}\right]\boldsymbol{\nabla}(r_r - \mathbf{r_r}\!\cdot\!\mathbf{v_r}) - \partial_t\left[\frac{q\mathbf{v}}{(r_r - \mathbf{r}\!\cdot\!\mathbf{v_r})}\right] \tag{15.75}$$

$$\mathbf{B}(\mathbf{r}, t) \;=\; \left[\frac{q}{(r_r - \mathbf{r_r}\!\cdot\!\mathbf{v_r})^2}\right]\mathbf{v_r}\times\boldsymbol{\nabla}(r_r - \hat{\mathbf{r}}_\mathbf{r}\!\cdot\!\mathbf{v_r}) + \left[\frac{q\boldsymbol{\nabla}\times\mathbf{v_r}}{(r_r - \mathbf{r_r}\!\cdot\!\mathbf{v_r})}\right] \tag{15.76}$$

where the vector functions $\mathbf{r_r}$ and $\mathbf{v_r}$ are evaluated at the retarded time.

The $\boldsymbol{\nabla}$ operator here is the gradient with respect to the variable $\mathbf{r}$, but it also acts on $\mathbf{r_r}$ and $\mathbf{v_r}$ through their dependence on t_r. Taking the indicated derivatives requires great care in treating the retarded time, which is only given implicitly by the two equations

$$t_r \;=\; t - r(t_r), \tag{15.77}$$

$$\mathbf{r_r} \;=\; \mathbf{r} - \mathbf{s}(t_r). \tag{15.78}$$

An important first step is to evaluate the partial derivative

$$\partial_t(t_r) = \frac{1}{1 - \hat{\mathbf{r}}_\mathbf{r}\!\cdot\!\mathbf{v_r}}. \tag{15.79}$$

The successive steps in the derivation of this result, which are typical of the steps required in working with the retarded time, are listed below:

$$
\begin{aligned}
\partial_t(t_r) &= \partial_t(t - r_r) = 1 - \partial_t(r_r),\\
\partial_t(r_r) &= \partial_t[(\mathbf{r_r}\!\cdot\!\mathbf{r_r})^{\frac{1}{2}}] = \hat{\mathbf{r}}_\mathbf{r}\!\cdot\![\partial_t(\mathbf{r_r})],\\
\partial_t(\mathbf{r_r}) &= \partial_t(\mathbf{r} - \mathbf{s}) = -\partial_t[\mathbf{s}(t_r)] = -\partial_t(t_r)\frac{\mathbf{ds}}{dt_r} = -\partial_t(t_r)\mathbf{v_r},\\
\partial_t(r_r) &= \hat{\mathbf{r}}_\mathbf{r}\!\cdot\![\partial_t(\mathbf{r_r})] = -\hat{\mathbf{r}}_\mathbf{r}\!\cdot\!\mathbf{v_r}\partial_t(t_r),\\
\partial_t(t_{r_r}) &= 1 - \partial_t(r_r) = 1 + \hat{\mathbf{r}}_\mathbf{r}\!\cdot\!\mathbf{v}\partial_t(t_{r_r}) = \frac{1}{1 - \hat{\mathbf{r}}_\mathbf{r}\!\cdot\!\mathbf{v_r}}. \tag{15.80}
\end{aligned}
$$

The steps for deriving the other derivatives needed to find the **E** and **B** fields from the retarded potentials are listed in Appendix B. The results we need are:

$$\partial_t(\mathbf{v_r}) \;=\; \mathbf{a_r}/d \tag{15.81}$$

$$\partial_t(r) \;=\; -\hat{\mathbf{r}}\cdot\mathbf{v}/d \tag{15.82}$$

$$\partial_t(\mathbf{r_r}\cdot\mathbf{v_r}) \;=\; (\mathbf{r_r}\cdot\mathbf{a_r} - \mathbf{v_r}^2)/d \tag{15.83}$$

$$\boldsymbol{\nabla}(r_r) \;=\; (\hat{\mathbf{r}}_\mathbf{r} + \mathbf{v_r})/d \tag{15.84}$$

$$\boldsymbol{\nabla}(\mathbf{r_r}\cdot\mathbf{v_r}) \;=\; [\hat{\mathbf{r}}_\mathbf{r}\mathbf{v_r}^2 - \mathbf{v_r}(\hat{\mathbf{r}}_\mathbf{r}\cdot\mathbf{v_r}) + \mathbf{v_r}(\hat{\mathbf{r}}_\mathbf{r}\cdot\mathbf{a_r})]/d. \tag{15.85}$$

In these equations, $\mathbf{v_r}$ and $\mathbf{a_r}$ are the charged particle's velocity and acceleration at the retarded time. The denominator function is given by $d = 1 - \hat{\mathbf{r}}\cdot\mathbf{v_r}$.

Using Eqs. (15.81)-(15.85) in Eq. (15.75), the electric field of a moving charge is given (after more algebra) by

$$\mathbf{E}(\mathbf{r},t) \;=\; \frac{q(\hat{\mathbf{r}}_\mathbf{r} - \mathbf{v_r})}{r_r^2\gamma_r^2(1 - \hat{\mathbf{r}}_\mathbf{r}\cdot\mathbf{v_r})^3} + \frac{q(\hat{\mathbf{r}}_\mathbf{r} - \mathbf{v_r})\times(\hat{\mathbf{r}}_\mathbf{r}\times\mathbf{a_r})}{r_r(1 - \hat{\mathbf{r}}_\mathbf{r}\cdot\mathbf{v_r})^3}. \tag{15.86}$$

A similar calculation for the magnetic field results in the connection

$$\mathbf{B}(\mathbf{r},t) \;=\; \hat{\mathbf{r}}_\mathbf{r}\times\mathbf{E}(\mathbf{r},t) \tag{15.87}$$

between the two fields.

We see from Eq. (15.86) that the fields are given in two separate parts as 'velocity fields', independent of $\mathbf{a_r}$, and 'acceleration fields' that depend linearly on $\mathbf{a_r}$. The velocity fields fall off with the power $1/r^2$, while the acceleration fields have a slower $1/r$ dependence. Because of this, it is only the acceleration fields that will lead to outgoing radiation

15.4.2   Constant velocity fields

The first term in Eq. (15.86) gives the electric field due to a charge moving with constant velocity. This equation using retarded time should give the same $\mathbf{E}$ field as that given in terms of the actual time by Eq. (15.52). We can see that it does by studying Fig. 15.3, which is a graphic example of the use of retarded time. A charged particle traveling from point O with velocity $\mathbf{v}$ along the x axis for a time t will be at point B, with the distance $\overline{OB}=vt$. The retarded position of the particle is at point A, a distance vr to the left of B, and $r = \overline{AP}$ is the distance from the retarded position A to the observation point P. The distance $\overline{OA} = \mathbf{s}$ is the distance the particle has traveled in the time $t_\mathrm{r} = t - r$, so $\mathbf{s}$ is the position of the particle at the retarded time.

To evaluate the retarded time expression for $\mathbf{E}$, we write it in the form

$$\mathbf{E}(\mathbf{x},t) = \left\{ \frac{q(\mathbf{r} - \mathbf{v}r)}{\gamma^2(r - \mathbf{r}\cdot\mathbf{v})^3} \right\}_\mathrm{ret}. \tag{15.88}$$

We want to express the denominator term $(r - \mathbf{r}\cdot\mathbf{v})$ in terms of the actual time t. From the figure, we can see that $-\mathbf{r}\cdot\mathbf{v}$ is the projection of the length $\overline{AB} = vr$

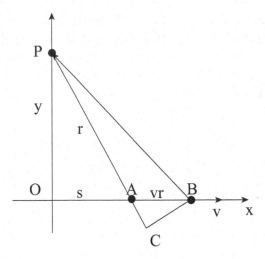

Figure 15.3: Graphic connection between t and t_r.

onto the line AC. Thus the length $\overline{AC} = -\mathbf{r}\cdot\mathbf{v}$, and the length $\overline{PC} = (r - \mathbf{r}\cdot\mathbf{v})$. Line PC is one leg of the right triangle PBC, whose hypotenuse is given by $\overline{PB}^2 = v^2 t^2 + y^2$. The other leg of the triangle has length $\overline{BC} = vy$. This follows from the similar triangles POA and BCA, leading to the equality of ratios $\overline{BC}/vr = y/r$. Now, applying Pythagorus' theorem, we get

$$
\begin{aligned}
\overline{PC}^2 &= \overline{PB}^2 - \overline{BC}^2 \\
\left\{(r - \mathbf{r}\cdot\mathbf{v})^2\right\}_{\text{ret}} &= (v^2 t^2 + y^2) - v^2 y^2,
\end{aligned}
\tag{15.89}
$$

which equals the term $(r^2 - (\mathbf{r}\times\mathbf{v})^2)$ in the denominator of Eq. (15.52). The vector $\mathbf{r} - \mathbf{v}r$ in the numerator of Eq. (15.88) is the vector BP, and BP is the vector that was called $\mathbf{r}$ in Eq. (15.52). Thus, the constant velocity electric field as given by the retarded field, and as given by Lorentz transformation from rest, are the same, even though the expressions for them are quite different.

15.5 Observed Electromagnetic Radiation

Once $\mathbf{E}$ and $\mathbf{B}$ are known, the rate at which electromagnetic energy passes through a solid angle $d\Omega_r$ a distance $\mathbf{r}_r$ from a radiating charge is given by

$$
\frac{d^2 W}{dt\, d\Omega_r} = [\mathbf{r}_r{}^2 (\hat{\mathbf{r}}_r\cdot\mathbf{S})],
\tag{15.90}
$$

where $\mathbf{S}$ is the Poynting vector, given by

$$
\mathbf{S} = \frac{1}{4\pi}(\mathbf{E}\times\mathbf{B}).
\tag{15.91}
$$

The observed radiation is given by Eq. (15.90) as the derivative with respect to the present time, t, at which time the radiation is observed, while all the variables in the Liénard-Wiechert $\mathbf{E}$ and $\mathbf{B}$ fields depend on the point charge's position at the retarded time, t_r, when the radiation was emitted.

In evaluating the Poynting vector, we will use only the acceleration fields from Eqs. (15.86) and (15.87). The velocity fields fall off like $1/r^2$, so their contribution to $r^2[\mathbf{E}\times\mathbf{B}]$ becomes negligible at large distance. Putting the acceleration fields into Eq. (15.90) gives

$$\frac{dP}{d\Omega_r} = \frac{q^2}{4\pi} \frac{\{\hat{\mathbf{r}}_\mathbf{r}\times[(\hat{\mathbf{r}}_\mathbf{r} - \mathbf{v_r})\times\mathbf{a_r}]\}^2}{(1 - \hat{\mathbf{r}}_\mathbf{r}\cdot\mathbf{v_r})^6} \tag{15.92}$$

for the angular distribution of the observed radiation from an accelerating point charge.

This relativistic expression differs from the non-relativistic radiation distribution in Eq. (??) by the explicit appearance of the velocity $\mathbf{v_r}$. The denominator $(1 - \hat{\mathbf{r}}_\mathbf{r}\cdot\mathbf{v_r})^6$ especially causes a strong forward pitch in the relativistic angular distribution.

Equation (15.92) gives the angular distribution observed at a distance r_r from the radiating charge. The angular distribution will be different at different distances because it depends on the retarded velocity and acceleration of the radiating particle. Also, the distance has be large enough so that the contribution from the velocity terms in the Liénard-Weichert fields are negligible.

15.5.1   Radiation with acceleration parallel to velocity

For $\mathbf{a_r}$ parallel to $\mathbf{v_r}$, the radiation pattern in Eq. (15.92) reduces to

$$\frac{dP}{d\Omega_r} = \frac{q^2\mathbf{a_r}^2}{4\pi} \left[\frac{\sin^2\theta}{(1 - v_r\cos\theta)^6}\right], \tag{15.93}$$

where θ is the angle measured from the common direction of $\mathbf{v_r}$ and $\mathbf{a_r}$. This angular distribution is plotted in Fig. 15.4 for a high velocity and for a low velocity particle, showing how the radiation pattern shifts forward for high velocity. There is also a large increase in the intensity of the radiation as v_r increases.

The forward pitch of the radiation pattern becomes very sharp as $v_r \rightarrow 1$ and γ_r grows large. In this case, the small angle approximations $\sin\theta \simeq \theta$ and $\cos\theta \simeq 1 - \theta^2/2$ can be implemented. Then Eq. (15.93) reduces to

$$\frac{dP}{d\Omega_r} \simeq \frac{8q^2a_r^2\gamma_r^8}{\pi} \left\{\frac{(\gamma_r\theta)^2}{[1 + (\gamma_r\theta)^2]^6}\right\}. \tag{15.94}$$

This describes a universal angular distribution in terms of a **scaling variable** $x_s=\gamma_r\theta$. That is, for $\gamma_r>>1$, the angular distribution is given by a single function, $f = x_s^2/(1 + x_s^2)^6$, for any particle energy. This universal curve for $\mathbf{a_r}\|\mathbf{v_r}$ is plotted as the solid curve in Fig. 15.5.

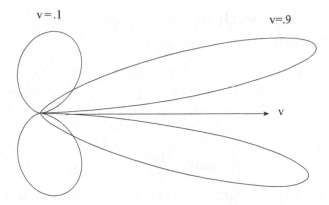

Figure 15.4: Radiation pattern for **a** parallel to **v**, for v=0.1 and 0.9 with the same acceleration. The v=0.1 distribution has been multiplied by a factor of 100.

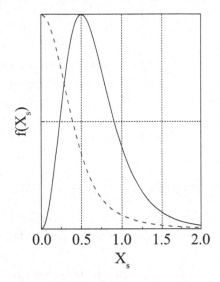

Figure 15.5: Universal curves for angular distribution of radiation. The solid curve is for **a** parallel to **v**, and the dashed curve is the distribution for **a** perpendicular to **v**, averaged over ϕ.

The intensity is zero at x_s=0, but rises sharply to a maximum at $x_s = \frac{1}{2}$, followed by a rapid falloff with increasing x_s. For large γ_r the radiation is essentially confined to a narrow cone of angle $\theta = \frac{1}{2\gamma_r}$. The angle of the cone becomes so small that the direction of the radiation is practically indistinguishable from the forward direction.

15.5.2   Radiation with acceleration perpendicular to velocity

When a charged particle is in a circular orbit, or when a beam is deflected by an electric or magnetic field, the acceleration is perpendicular to the particle's velocity. In that case, using bac-cab to expand the numerator of Eq. (15.92), we get

$$\frac{dP}{d\Omega} = \frac{q^2\mathbf{a}^2}{4\pi(1 - v\cos\theta)^4} \left[1 - \frac{\sin^2\theta\cos^2\phi}{\gamma^2(1 - v\cos\theta)^2}\right], \tag{15.95}$$

where θ and ϕ are the usual spherical coordinate angles shown in Figt. 15.6.

This angular distribution has a ϕ dependence, but still shows a sharp forward peaking due to the $(1 - v\cos\theta)$ factors in the denominator. For relativistic particles, with v close to 1, we can make a small angle approximation for θ (but not for ϕ) to get the approximate form

$$\frac{dP}{d\Omega} = \frac{2q^2\mathbf{a}^2\gamma^6}{\pi} \left[\frac{1 + x_s^4 - 2x_s^2\cos(2\phi)}{(1 + x_s^2)^6}\right], \tag{15.96}$$

where $x_s = \gamma\theta$.

The universal function in square brackets, averaged over ϕ, is plotted as the dashed curve in Fig. 15.5. The two curves in the figure have been normalized to have the same maximum value. We see that the distribution for $\mathbf{v}\perp\mathbf{a}$ is sharper than that for $\mathbf{v}\|\mathbf{a}$. The factor of x^4 in the numerator of Eq. (15.96) does not become appreciable until the function has already become very small.

15.5.3   Radiation from a circular orbit.

Radiation for accleration perpendicular to the velocity is important for synchrotons, where particles are kept in circular orbit with relatively small acceleration along their path.

Equation (15.95) gives the radiation pattern for an infinitesimal length of time in a circular orbit, but this is not what would be seen for radiation over a longer period. For the energy radiated over a finite time, the variation of $\mathbf{v}$, $\mathbf{a}$, and the particle's changing position have to be taken into account. We do that here for the case of a charged particle in a circular orbit of radius R. The angular variables do not have a clear meaning if the particle moves too much in the time interval considered, so we need the condition that $r \gg R$, where r is the observation distance from the circle.

We consider circular motion in the x-z plane in Fig. 15.6 with angular velocity $\boldsymbol{\omega} = \omega\hat{\mathbf{j}}$. The velocity and acceleration vectors shown in the figure will then rotate with angular velocity ω about the y axis. We want to observe the radiation at an angle ψ above the plane of the circular motion, so we choose the position vector $\mathbf{r}$ to be in the y-z plane making the angle ψ with the z

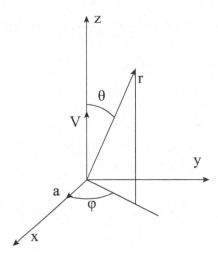

Figure 15.6: Angular orientation for $\mathbf{a} \perp \mathbf{v}$.

axis. In terms of the fixed Cartesian unit vectors $\hat{\mathbf{i}}$, $\hat{\mathbf{j}}$, $\hat{\mathbf{k}}$, the relevant vectors are given by

$$\hat{\mathbf{r}} = \hat{\mathbf{j}} \sin\psi + \hat{\mathbf{k}} \cos\psi \qquad (15.97)$$

$$\hat{\mathbf{v}} = \hat{\mathbf{i}} \sin\omega t_r + \hat{\mathbf{k}} \cos\omega t_r \qquad (15.98)$$

$$\hat{\mathbf{a}} = \hat{\mathbf{i}} \cos\omega t_r - \hat{\mathbf{k}} \sin\omega t_r. \qquad (15.99)$$

We use the retarded time t_r, which is the appropriate time variable for the motion of the particle. Using these unit vectors in Eq. (15.86), we get

$$
\begin{aligned}
\mathbf{E} = \ & \frac{qa_r}{r_r}\{\hat{\mathbf{i}}[v_r\cos\psi - \cos\omega t_r] - \hat{\mathbf{j}}\sin\psi\cos\psi\sin\omega t_r \\
& +\hat{\mathbf{k}}[\sin^2\psi\sin\omega t_r - (1-v_r)\cos\psi\sin\psi\cos\omega t_r]\} \\
& /[1 - v_r\cos\psi\cos\omega t_r]^3
\end{aligned}
\qquad (15.100)
$$

for the time dependence of $\mathbf{E}$. In this equation, the $\hat{\mathbf{i}}$ and $\hat{\mathbf{k}}$ terms correspond to polarization in the plane of the circle, and the $\hat{\mathbf{j}}$ term to polarization perpendicular to the plane. We will keep the two polarizations separate in the remainder of the calculation.

We are interested in radiation from high energy particles for which $v \simeq 1$, and the radiation is strongly peaked in the direction of $\mathbf{v}$. From Eq. (15.96), we see that $\frac{dP}{d\Omega}$ will only be appreciable for $\gamma_r\psi$ and $\gamma_r\omega t_r$ of order unity. This means that we can make small angle expansions in ψ and ωt_r, and also use the approximations $v_r \simeq 1$ and $1 - v_r \simeq 1/2\gamma_r^2$. With these approximations, $\mathbf{E}$ becomes

$$\mathbf{E} \simeq \frac{4qa_r\gamma_r^4}{r_r}\left[\frac{\hat{\mathbf{i}}(\gamma_r^2\omega^2 t_r^2 - \gamma_r^2\psi^2 - 1) - 2\hat{\mathbf{j}}\gamma_r^2\psi\omega t_r}{(1 + \gamma_r^2\psi^2 + \gamma_r^2\omega^2 t_r^2)^3}\right], \qquad (15.101)$$

where we have kept only those terms in the square brackets that are finite in the limit $\gamma_r \to \infty$.

The angular distribution of the radiation observed in one circular orbit is given by

$$\frac{dW}{d\Omega_r} = \frac{1}{4\pi} \int_{-\pi/\omega_r}^{+\pi/\omega_r} |r_r \mathbf{E}|^2 dt. \tag{15.102}$$

The integration is over the present time, t, but the charged particle's motion and the electric field are given in terms of the retarded time, t_r. Consequently, we make the replacement

$$dt = d(t_r + r_r) + dt_r = dt_r + \hat{\mathbf{r}}_\mathbf{r} \cdot \mathbf{dr_r} \tag{15.103}$$

$$= dt_r(1 - \hat{\mathbf{r}}_\mathbf{r} \cdot \mathbf{v_r}), \tag{15.104}$$

with the minus sign coming because $\mathbf{r_r} = \mathbf{r} - \mathbf{s}(t_r)$, as in Eq. (15.78).

Then,

$$\frac{dW}{d\Omega_r} = \frac{1}{4\pi} \int_{-\pi/\omega}^{+\pi/\omega} |r_r \mathbf{E}|^2 (1 - \hat{\mathbf{r}}_\mathbf{r} \cdot \mathbf{v_r}) dt_r$$

$$= \frac{2q^2 a_r^2 \gamma_r^5}{\pi \omega} \int_{-\infty}^{+\infty} \frac{[(u^2 - \gamma_r^2 \psi^2 - 1)^2 + 4\gamma_r^2 \psi^2 u^2] du}{(u^2 + 1 + \gamma_r^2 \psi^2)^5}. \tag{15.105}$$

We have introduced t_r as the integration variable, and then use the variable $u = \gamma_r \omega t_r$ for the actual integration. The integrations over u are straightforward, but numerous (too many even for a problem), and lead to the result

$$\frac{dW}{d\Omega_r} = \frac{q^2 \gamma_r^5}{16R(1 + \gamma_r^2 \psi^2)^{\frac{7}{2}}} [7(1 + \gamma_r^2 \psi^2) + 5\gamma_r^2 \psi^2]. \tag{15.106}$$

The first term in the numerator corresponds to polarization parallel to the circular orbit, and the second term to the perpendicular polarization. We have used the fact that $v_r \simeq 1$ to replace the factor a_r^2/ω by the approximation

$$\frac{a_r^2}{\omega} = \frac{v_r^4/(4R^2)}{v_r/R} \simeq \frac{1}{4R}. \tag{15.107}$$

Equation (15.106) can be written in terms of the scaling variable $x_s = \gamma_r \psi$ as

$$\frac{dW}{d\Omega_r} = \frac{q^2 \gamma_r^5}{16R} \left[\frac{12x^2 + 7}{(x^2 + 1)^{\frac{7}{2}}} \right] = \frac{q^2 \gamma_r^5}{16R} f(x). \tag{15.108}$$

The scaling function, $f(x)$, is plotted as the solid line in Fig. 15.7. For comparison, the dashed curve gives the instantaneous power distribution of Eq. (15.96). We use $\phi = 90°$ for that curve so that the angle θ used in Eq. (15.95) will be the angle from the plane of the orbit. We see from the figure that the radiated energy, which is what would ordinarily be observed, has a somewhat sharper angular distribution with respect to the plane of the orbit than does the instantaneous power.

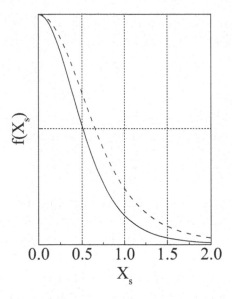

Figure 15.7: Universal curves for angular distribution of radiation from a circular orbit. The solid curve is for the energy radiated in one orbit. The dashed curve is for the instantaneous power. The curves are normalized to agree at $x_s=0$.

15.6   Power Radiated by an Accelerating Point Charge

The power, $\frac{dW}{dt}$, represented in Eq. (15.92) is the rate at which electromagnetic energy emitted at the retarded time, t_r, is observed at a later time of observation, t. The rate at which the power is emitted by the point charge at the retarded time is the derivative of the electromagnetic energy with respect to the retarded time,

$$
\begin{aligned}
\frac{dP_{\text{rad}}}{d\Omega_r} &= \frac{d^2W}{d\Omega_r dt_r} = \left(\frac{dt}{dt_r}\right)\frac{d^2W}{d\Omega_r dt} \\
&= (1 - \hat{\mathbf{r}}_\mathbf{r}{\cdot}\mathbf{v_r})\frac{d^2W}{d\Omega_r dt} \\
&= \frac{q^2}{4\pi}\frac{\{\hat{\mathbf{r}}_\mathbf{r}{\times}[(\hat{\mathbf{r}}_\mathbf{r} - \mathbf{v_r}){\times}\mathbf{a_r}]\}^2}{(1 - \hat{\mathbf{r}}_\mathbf{r}{\cdot}\mathbf{v_r})^5}.
\end{aligned}
\tag{15.109}
$$

Comparison of equations (15.92) and (15.110) shows the angular distribution of the radiation observed at a distance, r_r, from an accelerating charged particle is different than the angular distribution of the radiation it emits at the retarded time.

The total radiated power passing through a sphere of radius, R, centered at the retarded position of the accelerating point charge is given by

$$P_{\text{rad}} = \frac{q^2}{4\pi} \int \frac{\{\hat{\mathbf{r}}_{\mathbf{r}} \times [(\hat{\mathbf{r}}_{\mathbf{r}} - \mathbf{v_r}) \times \mathbf{a_r}]\}^2 d\Omega_r}{(1 - \hat{\mathbf{r}}_{\mathbf{r}} \cdot \mathbf{v_r})^5}. \tag{15.110}$$

However, this result is ambiguous because it depends on the radius R. The variables in the integral are evaluated at the retarded time, which is given by

$$t_r = t - r_r = t - R. \tag{15.111}$$

That means that the acceleration and velocity appearing in Eq. (15.110) could be any acceleration and velocity from the past motion of the accelerating particle, depending on what was chosen as the radius of observation, R_{rad}, of the radiation.

In order to find the power radiated at the present time and position of the charged particle, we take the limit $R \to 0$ instead of large R. This makes the retarded time $t_r \to t$, and the variables in Eq. (15.110) will be evaluated at the present time, so

$$P_{\text{rad}} = \frac{q^2}{4\pi} \int \frac{\{\hat{\mathbf{r}} \times [(\hat{\mathbf{r}} - \mathbf{v}) \times \mathbf{a}]\}^2 d\Omega}{(1 - \hat{\mathbf{r}} \cdot \mathbf{v})^5}. \tag{15.112}$$

With the variables evaluated at the present time, the velocity term in Eq. (15.86) corresponds to the Lorentz transformation of the Coulomb electric field from the charged particle's rest frame to the frame in which it is moving with velocity $\mathbf{v}$. Consequently, it does not contribute to the radiated power, and it has been left out of the integral.

15.6.1　Radiation with acceleration parallel to velocity

For $\mathbf{a}$ parallel to $\mathbf{v}$, the integral in Eq. (15.112) becomes

$$\begin{aligned} P_{\parallel} &= \frac{q^2 a^2}{4\pi} \int \frac{\sin^2 \theta d\Omega}{(1 - v \cos \theta)^5} \\ &= \frac{2}{3} q^2 a^2 \gamma^6. \end{aligned} \tag{15.113}$$

This reduces to the non-relativistic Larmor formula of Eq. (??) for $\gamma = 1$, but seems to have a large γ^6 enhancement for relativistic particles.

However, we can relate the acceleration to $\frac{d\mathbf{p}}{dt}$ by using Eq. (15.4):

$$\begin{aligned} \frac{d\mathbf{p}}{dt} &= m\frac{d}{dt}\left[\frac{\mathbf{v}}{\sqrt{1 - \mathbf{v}^2}}\right] \\ &= m[\gamma \mathbf{a} + \gamma^3 \mathbf{v}(\mathbf{v} \cdot \mathbf{a})] \tag{15.114} \\ &= m\mathbf{a}\gamma^3, \tag{15.115} \end{aligned}$$

for $\mathbf{a} \| \mathbf{v}$. Then, the radiated power is

$$P_\| = \frac{2}{3} \frac{q^2}{m^2} \left(\frac{d\mathbf{p}}{dt} \right)^2 . \tag{15.116}$$

This formula, written in terms of the momentum, is the same for the relativistic or non-relativistic case.

Equation (15.116) gives the power radiated by relativistic electrons in **brems-strahlung** (German for 'braking radiation'). We can rewrite it in terms of the electron energy W by using the kinematical relation

$$\frac{dp}{dt} = \frac{dW}{dx}. \tag{15.117}$$

Then

$$P = \frac{2}{3} \frac{e^2}{m^2} \left(\frac{dW}{dx} \right)^2 . \tag{15.118}$$

We saw in Eq. (15.59) that the rate of energy loss with distance is nearly constant for high energy particles (with v close to 1) for most of their stopping distance. This means we can write the radiated power as

$$P = \frac{2}{3} \frac{e^2 W_0^2}{m^2 d^2}, \tag{15.119}$$

where W_0 is the initial electron energy and d is its stopping distance. The energy radiated by the electron in stopping is given by

$$W_{\text{rad}} = \int P dt = \int_0^d \frac{P}{v} dx \simeq \frac{2}{3} \frac{e^2 W_0^2}{m^2 d}, \tag{15.120}$$

where we have let $v{=}1$, which is a good approximation for most of the stopping distance for high energy electrons. We can write the ratio of radiated energy to the electron's initial energy as

$$\frac{W_{\text{rad}}}{W_0} = \frac{2}{3} \frac{e^2}{md} \frac{W_0}{m} = \frac{2}{3} \frac{r_c}{d} \gamma_0, \tag{15.121}$$

with r_c (the classical radius of the electron)=2.82 fm. Since $d \gg r_c$, for any reasonable d, the radiated energy is a tiny fraction of the electron's initial energy, even for the highest energy electrons.

15.6.2   Radiation with acceleration perpendicular to velocity

When the acceleration is perpendicular to the particle's velocity, Eq. (15.92), Eq. (15.110) becomes

$$P_\perp = \frac{q^2 a^2}{4\pi} \int \left[1 - \frac{\sin^2 \theta \cos^2 \phi}{\gamma^2 (1 - v \cos \theta)^2} \right] \frac{d\Omega}{(1 - v \cos \theta)^3} \tag{15.122}$$

$$= \frac{2}{3} q^2 a^2 \gamma^4 . \tag{15.123}$$

Using $\frac{d\mathbf{p}}{dt} = m\mathbf{a}\gamma$ from Eq. (15.115), this becomes

$$P_\perp = \frac{2}{3}\frac{q^2\gamma^2}{m^2}\left(\frac{d\mathbf{p}}{dt}\right)^2. \tag{15.124}$$

This power is a factor γ^2 larger than the power radiated for $\mathbf{a}\|\mathbf{v}$, or for the non-relativistic Larmor formula for the same applied force (as measured by $\frac{d\mathbf{p}}{dt}$). Using the acceleration instead of $\frac{d\mathbf{p}}{dt}$ in the power formulas had given a misleading indication of the relative importance of the radiation.

The time for one orbit in a synchrotron of radius R is $t = 2\pi R/v$, and the centripetal acceleration is $\mathbf{v}^2/R$. Thus, the energy radiated in one orbit is

$$\Delta W = \frac{4\pi}{3}\frac{q^2}{R}\mathbf{v}^4\gamma^4 \simeq \frac{4\pi}{3}\frac{q^2}{R}\left(\frac{W}{m}\right)^4, \tag{15.125}$$

where the last equality holds for relativistic particles with $v \simeq 1$. This energy loss can limit the maximum energy of synchrotons, especially for electrons. The numerical value of the energy loss per revolution for electrons is

$$\Delta W(\text{keV}) = 88.5\left\{\frac{[W(\text{GeV})]^4}{R(\text{meters})}\right\}. \tag{15.126}$$

The energy given to the particles in each revolution has to be greater than this radiation loss for acceleration to continue. For particles kept circulating in storage rings, the input energy must equal the radiated energy. For a proton synchroton, the energy loss is down from that in Eq. (15.126) by a factor of about 10^{13}, so they can go to higher energy.

15.6.3 Relativistic Larmor formula

The total power radiated for a general direction of $\mathbf{a}$ and $\mathbf{v}$ can be found by integrating Eq. (15.112) over all angles. But, in fact, we have already done that integral in previous subsections. Looking at Eq. (15.112), we see that the result of the angular integration can only depend on three possible spatial scalars, $\mathbf{a}\cdot\mathbf{a}$, $\mathbf{v}\cdot\mathbf{a}$, and $\mathbf{v}\cdot\mathbf{v}$, and must be bilinear in $\mathbf{a}$. This means the resulting integral can be written as a function of v^2 times a linear combination of $\mathbf{a}_\|^2$ and $\mathbf{a}_\perp^2$. ($\mathbf{a}_\|$ and $\mathbf{a}_\perp$ are the components of $\mathbf{a}$ parallel and perpendicular to $\mathbf{v}$.)

We have already done the angular integrals for $\mathbf{a}_\|$ and $\mathbf{a}_\perp$ separately above. The result of angular integration in the general case is simply the sum of Eq. (15.113) (for $\mathbf{a}_\|$) and Eq. (15.123) (for $\mathbf{a}_\perp$), giving

$$P = \frac{2}{3}q^2\gamma^4[\gamma^2\mathbf{a}_\|^2 + \mathbf{a}_\perp^2]. \tag{15.127}$$

This can also be written as

$$P = \frac{2}{3}q^2\gamma^6[\mathbf{a}^2 - v^2\mathbf{a}_\perp^2] = \frac{2}{3}q^2\gamma^6[\mathbf{a}^2 - (\mathbf{v}\times\mathbf{a})^2]. \tag{15.128}$$

This relativistic extension of the Larmor formula was first derived by Alfred-Marie Liénard in 1898.

The radiated power, as given by the relativistic Liénard formula of Eq. (15.128) is actually a Lorentz scalar, which can be written as

$$P = -\frac{2}{3}\frac{q^2}{m^2}\frac{dp^\mu}{d\tau}\frac{dp_\mu}{d\tau}. \tag{15.129}$$

To show this, we have to reduce the four-vector scalar product to vector notation. We start with

$$\frac{dp^\mu}{d\tau}\frac{dp_\mu}{d\tau} = \left(\frac{dW}{d\tau}\right)^2 - \left(\frac{d\mathbf{p}}{d\tau}\right)^2 = \gamma^2\left[\left(\frac{dW}{dt}\right)^2 - \left(\frac{d\mathbf{p}}{dt}\right)^2\right]. \tag{15.130}$$

The time derivatives are.

$$\frac{dW}{dt} = \frac{d}{dt}(m\gamma) = m\frac{d}{dt}\left(\frac{1}{\sqrt{1-\mathbf{v}^2}}\right) = m\gamma^3(\mathbf{v}\cdot\mathbf{a}) = m\gamma^3 v a_\parallel \tag{15.131}$$

$$\frac{d\mathbf{p}}{dt} = \frac{d}{dt}(m\mathbf{v}\gamma) = m[\gamma\mathbf{a} + \gamma^3\mathbf{v}(\mathbf{v}\cdot\mathbf{a})] = m\gamma^3(\mathbf{a}_\parallel + \mathbf{a}_\perp/\gamma^2). \tag{15.132}$$

When these derivatives are put into Eq. (15.130), we get

$$\frac{dp^\mu}{d\tau}\frac{dp_\mu}{d\tau} = m^2\gamma^8[(\mathbf{v}^2 - 1)\mathbf{a}_\parallel{}^2 - \mathbf{a}_\perp{}^2/\gamma^4] = -m^2\gamma^4(\gamma^2\mathbf{a}_\parallel{}^2 + \mathbf{a}_\perp{}^2). \tag{15.133}$$

With this expression for the scalar product, the manifestly invariant form for the power agrees with the vector form in Eq. (15.127), demonstrating that the radiated power is a Lorentz scalar.

15.7 Radiation Reaction

In this section we derive the radiation reaction on an accelerating point charge resulting from the radiation it emits during its acceleration. For the power radiated by an accelerating point charge, we use Larmor's formula as extended to relativity by Liénard in Eq. (15.128). For acceleration parallel to the velocity, this reduces to

$$\frac{dW_{\text{rad}}}{dt} = \frac{2}{3}q^2a^2\gamma^6 = \frac{2}{3}q^2a'^2, \tag{15.134}$$

where $\mathbf{a}'$ is the charged particle's acceleration in its instantaneous rest frame.

We want to relate $\frac{dW_{\text{rad}}}{dt}$ to its effect on the motion of an accelerating point charge. For an accelerating particle of mass m, the rate of change of its kinetic energy is given (for $\mathbf{a}$ parallel to $\mathbf{v}$) by

$$\frac{dW_{\text{mat}}}{dt} = \frac{d(m\gamma - m)}{dt} = m\gamma^3(\mathbf{v}\cdot\mathbf{a}) = mva'. \tag{15.135}$$

If an external force acts on the charged particle, the electromagnetic power produced by the external force will increase the sum of the particle's kinetic energy and the radiated electromagnetic energy at the rate

$$\frac{dW_{\text{ext}}}{dt} = \frac{dW_{\text{mat}}}{dt} + \frac{dW_{\text{rad}}}{dt}. \tag{15.136}$$

This would reduce the particle's velocity and acceleration by

$$mva' = m\bar{v}\bar{a}' - \frac{2}{3}q^2a'^2. \tag{15.137}$$

where $\bar{v}$ and $\bar{a}'$ are the velocity and rest frame acceleration an uncharged partical would have.

Equation (15.137) is a quadratic equation for a', with the solution

$$a' = \frac{2m\bar{v}\bar{a}'}{\left[mv + \sqrt{m^2v^2 + (8/3)q^2m\bar{v}\bar{a}'}\right]}, \tag{15.138}$$

$$va' = \frac{2\bar{v}\bar{a}'}{\left[1 + \sqrt{1 + \left(\frac{q^2}{2m}\right)\left(\frac{16\bar{v}\bar{a}'}{3v^2}\right)}\right]}. \tag{15.139}$$

Equations (15.137), (15.138), and (15.139) each show the decrease in the charged particle's acceleration due to the diversion of some of the applied energy into radiated electromagnetic energy.

We note that the increase in electromagnetic energy does not produce an added force on the charged particle, but is just electromagnetic energy produced in space by the external force. The energy going into the electromagnetic field reduces the increase of the mechanical energy of the particle, which reduces its acceleration.

The external force acts on two separate entities, the accelerating particle and the electromagnetic field, so the energy put into the electromagnetic field should not be considered as a separate force on the accelerating particle. A particle, even if accelerating, cannot exert force on itself.

To get an idea of the relative size of the radiation reaction, we consider the case of a constant external force acting on a particle starting from rest. An uncharged particle would have acceleration $\bar{a}'$ that is constant in the particle's instantaneous rest frame.

The distance traveled by a particle starting from rest with uniform acceleration was derived in the solution to Problem 14.7 [Eq. (14.16)] as

$$x = \frac{(\bar{\gamma} - 1)}{\bar{a}'} = \frac{\bar{\gamma}^2\bar{v}^2}{(\bar{\gamma} + 1)\bar{a}'}. \tag{15.140}$$

Then, Eq. (15.139) becomes

$$va' = \frac{2\bar{v}\bar{a}'}{\left[1 + \sqrt{1 + \left(\frac{r_c}{x}\right)\left[\frac{16\bar{\gamma}^2\bar{v}^3}{3v^2(\bar{\gamma}+1)}\right]}\right]}. \tag{15.141}$$

We have taken the accelerating particle to be an electron, and have introduced the 'classical radius' of the electron, $r_c = \frac{q^2}{2m} = 2.82$ fm.

It is interesting to look at the non-relativistic and the extreme relativistic limits of Eq.(15.141) . The non-relativistic limit is

$$\bar{v} << 1, \quad va' \simeq \frac{2\bar{v}\bar{a}'}{\left[1 + \sqrt{1 + \left(\frac{r_c}{x}\right)\left[\frac{8\bar{v}^3}{3v^2}\right]}\right]},$$

$$a' \simeq \frac{\bar{a}'}{\left[1 + \left(\frac{r_c}{x}\right)\left(\frac{2\bar{v}}{3}\right)\right]}. \tag{15.142}$$

The relativistic limit is

$$\bar{\gamma} >> 1, \quad a' = \frac{\bar{a}'}{\left[1 + \left(\frac{4\bar{\gamma}r_c}{3x}\right)\right]}. \tag{15.143}$$

We can see from each of Eqs. (15.141), (15.142), and (15.143) that radiation reaction on a charged particle, accelerating parallel to its velocity, has a negligible effect on the acceleration of the particle. The distance, $r_c = 2.82$ fm, is much less than any reasonable distance traveled by the particle. Even for a highly relativistic particle, the ratio $\bar{\gamma}r_c/x$ is too small to give an observable effect. For the 50 GeV electrons at the SLAC linear collider (SLC), $\bar{\gamma}r_c$ would be about 30 nm.

15.8   Problems

1. (a) Reduce the covariant form of dynamical equation (15.1) to its vector forms.

   (b) Derive Eq. (15.4) for the acceleration of a charged particle.

2. Find the radius of curvature in a perpendicular magnetic field of 2,000 gauss for muons with momentum $p = 30$ MeV, and for electrons with energy $W = 1.50$ MeV.

3. If the relativistic law of gravity were $\frac{d\mathbf{p}}{dt} = -mg\hat{\mathbf{j}}$, find the acceleration components a_x and a_y for a particle moving in the x-y plane with velocity $\mathbf{v} = v_x\hat{\mathbf{i}}$.

4. (a) The LHC accelerates protons to an energy of 7 TeV in an orbit 27 km in circumference. What magnetic field is required to keep the 1 TeV protons in orbit?

   (b) Electrons are accelerated to 50 GeV in the 2 miles long SLAC linear accelerator. What is the average electric field (in gauss and in volts/meter)?

5. (a) Verify that the Lagrangian in Eq. (15.24) produces the nonrelativistic equation of motion for a charged particle in electric and magnetic fields.

   (b) Show the the relativistic Lagrangian in Eq. (15.29) leads to the relativistic equation of motion.

   (c) Show that a gauge transformation on the potential A^μ changes the Lagrangian, but does not change the dynamical equation of motion.

   (d) Show that Hamilton's equations with the Hamiltonian of Eq. (15.43) leads to the relativistic equations of motion.

6. Show that the track length in a bubble chamber of a particle of mass M and initial energy E is proportional to $(E^2 - M^2)/E$. [Write both sides of Eq. (15.59) in terms of γ, and integrate.]

7. (a) A particle of charge q' is held at rest at a point $y = a$ on the y axis. Find the time dependent force on this particle due to a particle of charge q moving at constant velocity v along the x axis.

   (b) Find the force on the moving particle of charge q.

   (c) Integrate each force over time to show that the total impulse on both particles vanishes. (Assume that the particle on the y axis does not move.)

8. A $\pi^+(140)$ meson is produced in a bubble chamber. It travels a short distance, before coming to rest. It then decays into a $\mu^+(106)$ and a neutrino(0). If the μ^+ track has the same length as the π^+ track, what was the initial kinetic energy of the π^+?

9. Two electrons are initially at rest a distance d apart.

   (a) Show that the energy of each electron as they escape to large distance is $E = m(1 + 3e^2/md)^{\frac{1}{3}}$. (Use $\gamma = E/m$ and $\mathbf{v} = \mathbf{p}/E$ to find dE/dx, and then integrate.)

   (b) Calculate the kinetic energy (in MeV) of each electron for initial distance d=1Å, and for d=1 fm. Compare each result to the non-relativistic answer.

10. (a) For the case of $\mathbf{a}$ parallel to $\mathbf{v}$, show that the angle at which the radiated intensity is a maximum is given by

$$\cos\theta_{\text{Max}} = \frac{\sqrt{1 + 15v^2} - 1}{3v}. \qquad (15.144)$$

   (b) Show that for $\gamma \gg 1$, $\theta_{\text{Max}} \to 1/2\gamma$, and the maximum intensity increases like γ^8.

   (c) Calculate θ_{Max} for v=0.1, v=0.9, and for a 20 GeV electron.

11. (a) Derive the numerical result for energy loss in Eq. (15.126) in the text.

   (b) What energy input per revolution must be put into 50 GeV electrons in a storage ring of radius R=200 meters?

   (c) What energy input is required for 1 TeV protons in a storage ring of radius R= 40 meters?

12. Calculate the Larmor power (in Watts) radiated by 2 MeV electrons in

   (a) a linear accelerator with an electric field of 16×10^6 volts per meter.

   (b) a circular orbit of radius 16 meters.

Chapter 16

Classical EM in a Quantum World.

Classical Electromagnetism was for the most part a 19th century theory. In this chapter, we will try to bring it up to date.

16.1 Looking Back

Classical Electromagnetism started in 1820 when Hans Christian Oersted observed the deflection of a magnetic needle by an electric current (the **Oersted effect**). This was the first indication that the two ancient phenomena of Electricity (E) and Magnetism (M) were, in fact, different aspects of the same physical entity: **Electromagnetism**, the title (well, half) of this text.

We have seen in the preceding chapters the connection between E & M deepen, to where the & became superfluous, and the theory became just **EM**. In 1831, Michael Faraday observed that a changing magnetic field induced currents in a closed electric circuit (**Faraday's law**). Now it was understood that electricity could produce magnetism (Oersted), and magnetism could produce electricity (Faraday). But, although it was becoming clear that E could affect M and M could affect E, they each had their own set of unrelated units, esu and emu.

The next step toward correlating electricity and magnetism was the determination of the ratio of the two charge units in 1856 by Wilhelm Weber and Friedrich Kohlrausch. They found that

$$\frac{1 \text{ abcoulomb}}{1 \text{ statcoulomb}} = 3.1 \times 10^{10} \text{cm/sec} \equiv c. \qquad (16.1)$$

This permitted the introduction of the Gaussian system of units that provided the first theoretical unification of (at least the units of) E and M. The constant c took on the status of a dimensional constant connecting E and M.

The close approximation of the ratio abcoulomb/statcoulomb with the speed of light led to conjectures that light was a result of combined E and

M effects. Faraday was the first to record this idea in 1857, after he demonstrated the rotation of the plane of polarization of light by a magnetic field. Then Maxwell showed that, just as $\partial_t \mathbf{B}$ produced an $\mathbf{E}$ field, so would $\partial_t \mathbf{E}$ produce a $\mathbf{B}$ field. This further unification of E and M led to Maxwell's derivation in 1864 of electromagnetic waves that propagated at the speed of light. Heinrich Hertz's experimental verification in 1888 of Maxwell's theory completed the 19th century unification of E & M into EM.

Classical EM at the close of the 19th century was an impressive, but not quite consistent theory. It could not handle the question of how to describe the electrodynamics of moving bodies. All proposed equations for the fields due to moving charges, or to forces between moving charges could not handle the fact of their movement. What was the meaning of velocity? Velocity with respect to what? How could a force, such as the magnetic force, change so drastically or disappear, simply by moving the coordinate system?

There were two competing approaches to the 19th century problem of moving charged bodies. One approach, which we have not discussed at all, was 'action at a distance', championed mainly by Weber. It tried to described the force between two moving charges in terms of only their relative velocity (and acceleration), as was considered necessary in the earlier, highly successful, development of classical mechanics. Inconsistencies in Weber's formulation led Maxwell to "the conception of a medium (the ether) in which propagation takes place." as the other alternative. With Maxwell's death in 1879, further theoretical efforts made no real progress toward a resolution as the century ended. Experiments could not confirm salient points of any theory of moving charges. The playing field was open for a new century.

In Chapters 14 and 15, we discussed how Einstein's Special Relativity solved the problem of how to treat velocity in Electromagnetism. It also virtually completed the unification of the electric and magnetic fields. Maxwell's equations had connected $\mathbf{E}$ to $\mathbf{B}$ and *vice-versa*, but they were still separate spatial vectors. Only by combining $\mathbf{E}$ and $\mathbf{B}$ into a single four dimensional tensor could the electrodynamics of moving bodies be understood. Any movement of a source or coordinate system mixes $\mathbf{E}$ and $\mathbf{B}$, just as rotating your head changes your view of the world.

This unification also occurs if the EM potentials are considered. The separate scalar electric potential ϕ and vector magnetic potential $\mathbf{A}$ must be unified in a single four-vector electromagnetic potential A^μ in order to treat moving charges. Einstein's Special Relativity paper, "On the Electrodynamics of Moving Bodies", was written in 1905, but can be considered as the final chapter of 19th century physics. With the inclusion of Special Relativity in what we call Classical Electromagnetism, it was now a consistent theory. BUT...

The now consistent theory of EM predicted the end of the world, or at least of atoms as they were understood, in one tenth of a nanosecond. We saw that in Chapter 13. EM was a strong enough theory that this fortunately unrealized prediction could not be compromised within the theory. Other phenomena like Black Body Radiation, the Photoelectric Effect, the Balmer Series of Spectral Lines also could not be explained within EM theory. The upshot was that the other pillar of theoretical physics, Classical Mechanics, had to be superseded by the 20th century revolution that was Quantum Mechanics (QM). We will see that this revolution (as revolutions do) also eventually replaced (or should we say extended?) Classical EM with Quantum Electrodynamics (QED).

Twentieth century physics also saw the discovery of new forces, in addition to Gravity and Electromagnetism: A Weak Interaction, seen first in nuclear beta decay, and a Strong Interaction that held the nucleus together. The names of these interactions came from their strength relative to Electromagnetism. A theoretical trend, continuing into the 21st century, has been the unification of these new interactions with the electromagnetic interaction. We will study this unification in suceeding sections to see what place Classical Electromagnetism will have in the new century.

16.2   Electromagnetism as a Gauge Theory

Interestingly, a unifying principle in Quantum Mechanics, Local Gauge Invariance, provides a foundation for Classical EM. One thing that EM theory lacks, as we have so far developed it, is some reason why. Why is there a force between charged particles like the Lorentz force, or any other force? At first it seems that this is too strong a requirement to ask of any theory, but QM does answer it for EM.

In Classical EM, Coulomb's law, and the $q\mathbf{v}\times\mathbf{B}$ law for magnetic force were introduced *ad hoc* to agree with experiment, and were originally unconnected. Relativity combined them in the Lorentz force, but gave no deeper understanding of their origin. Some insight into the force's origin, referred to as **minimal electromagnetic interaction**, arose in the early development of the Hamiltonian formulation of QM. (grossly simplified below)

The basic QM wave equations can be written down by taking classical equations for the energy, and making the transformations

$$\mathbf{p} \to -i\hbar\boldsymbol{\nabla}, \qquad E \to i\hbar\partial_t, \tag{16.2}$$

with the derivatives acting on a wave function $\Psi(\mathbf{r}, t)$. For instance, starting with the nonrelativistic energy of a free particle, we get

$$E = \frac{\mathbf{p}^2}{2m} \quad \to \quad i\hbar\partial_t\Psi = \frac{-\hbar^2\boldsymbol{\nabla}^2}{2m}\Psi, \tag{16.3}$$

the nonrelativistic Schrodinger wave equation for a free particle.

The EM interaction for a particle of charge q can be added to the Schrödinger equation by the simple replacements

$$\mathbf{p} \to \mathbf{p} - q\mathbf{A}, \qquad E \to E - q\phi \tag{16.4}$$

before going to the quantum derivatives. This method of incorporating Classical EM into nonrelativistic QM in a hybrid theory (half quantum-half classical) was very successful in the early quantum theory of atoms. Even in the later transition to relativistic one particle QM (the Dirac equation), the same prescription incorporated the correct EM interaction, given in relativistic notation as

$$p^\mu \to p^\mu - qA^\mu. \tag{16.5}$$

This success led to the suggestion that, in a complete theory, the full electromagnetic interaction would be included by just this minimal electromagnetic substitution. It was striking that whenever an electromagnetic interaction was observed that did not seem minimal, a new development of the QM incorporated it minimally. For instance, the magnetic moment of the electron did not seem like a minimal EM interaction. But, when the relativistic Dirac wave equation was introduced, the minimal substitution rule of Eq. (16.5) produced a magnetic moment with the correct g value of 2. (See Section 7.11.6.)

Then the proton and neutron were found to have large anomalous magnetic moments that were not minimal. But the quark model showed that Dirac quarks with minimal EM interactions could explain the nucleon magnetic moments, as well as those of other baryons. It may still be an open question whether all EM interactions are truly minimal, but most theoretical opinion favors the minimal hypothesis.

While minimal electromagnetic interaction is a neat way of combining EM with QM, it still does not answer the question of why there should be any EM interaction (apart from the fact that we need it). The answer comes from a somewhat deeper look at QM.

A basic principle of QM is that the state of a system is fully described by a complex wave function $\Psi(\mathbf{r}, t)$. All physical observables, however, are complex bilinear in Ψ, depending on the combination $\Psi^*\Psi$ (or, more generally, $\Psi^*\mathbf{O}\Psi$, where $\mathbf{O}$ is an operator on Ψ). This leads immediately to the trivial observation that changing Ψ by a constant phase in the transformation

$$\Psi \to \Psi' = e^{i\phi}\Psi \tag{16.6}$$

can have no physical consequence.

This property could be called 'phase invariance', but has been given the fancier name of **Global Gauge Invariance**. The 'global' refers to the fact (seemingly obvious) that the same phase angle ϕ is used globally thoughput space. The "gauge invariance" arose because of the connection we will see with gauge invariance in EM.

What if there were **Local Gauge Invariance (LGI)**? That is, what if we demanded that there be no physical consequence of changing the phase of Ψ by a different angle at each point of space and time? What are the consequences of the transformation

$$\Psi \to \Psi' = e^{i\chi(\mathbf{r},t)}\Psi? \tag{16.7}$$

But, that won't work. Without going too deeply into QM, all we need to know is that the basic Schrodinger Equation (either nonrelativistic or relativistic extensions) involves the time and space derivatives ∂_t and ∇ acting on Ψ. In a local gauge transformation, we would get (We use the relativistic formulation, which is simpler.)

$$\partial_\mu \Psi(x) \to \partial_\mu[e^{i\chi(x)}\Psi(x)] = e^{i\chi}[\partial_\mu\Psi + i(\partial_\mu\chi)\Psi]. \qquad (16.8)$$

For a local gauge transformation, the added term $i(\partial_\mu\chi)\Psi$ would change the wave equation for Ψ markedly.

How can we restore LGI to QM? The form of the problem is suggestive of the solution. We need a mechanism to cancel the unwanted $i(\partial_\mu\chi)\Psi$. We can do this by two steps:

1. Modify the derivative operators by the substitution

$$\partial_\mu \to \partial_\mu - iA_\mu(x), \qquad (16.9)$$

    where $A_\mu(x)$ is a four-vector function of x.

2. When the wave function is modified by a local gauge transformation $\Psi \to e^{i\chi}\Psi$, make the corresponding gauge transformation

$$A_\mu \to A_\mu + \partial_\mu\chi. \qquad (16.10)$$

These two steps solve the problem. Now a complete local gauge transformation results in

$$
\begin{aligned}
(\partial_\mu - iA_\mu)\Psi \;&\to\; (\partial_\mu - iA_\mu - i\partial\chi)(e^{i\chi}\Psi) \\
&=\; e^{i\chi}(\partial_\mu - iA_\mu)\Psi + (i\partial_\mu\chi - i\partial_\mu\chi)e^{i\chi}\Psi.
\end{aligned} \qquad (16.11)
$$

Since the last two terms cancel, LGI is preserved.

The above procedure can be generalized by introducing an arbitrary constant q without affecting the mechanism. Then a full local gauge transformation, as usually written, first introduces the substitution

$$p_\mu \to p_\mu - qA_\mu. \qquad (16.12)$$

Here p_μ can be considered either as the classical four-momentum before the passage to QM, or the actual QM operator $-i\partial_\mu$. Now, a local gauge transformation consists of the simultaneous transformations

$$\Psi \to e^{iq\chi}\Psi, \qquad A_\mu \to A_\mu + \partial_\mu\chi. \qquad (16.13)$$

The QM wave equation will be invariant under these combined operations for any value of the constant q.

Now, look at what we have. Equation (16.12) is just minimal electromagnetic substitution, and the second step in Eq. (16.13) is the same gauge transformation we have been using in Classical Electromagnetism. We have deduced the need for Classical Electromagnetism from the principle of LGI of Quantum Mechanics. In order to achieve LGI, a particle must have a charge q and interact by minimal interaction with a vector field A_μ.

This anwers the question of why there is an electromagnetic field. Of course, that raises the new question of why is LGI so important in QM. But that is the nature of physics, answers always bring new questions, but perhaps more fundamental ones. In any event, it will turn out that LGI of QM brings us more than just Electromagnetism, as we will see in Section 16.4.

One aspect of the preceding discussion is that the electric and magnetic vector fields, $\mathbf{E}$ and $\mathbf{B}$, never entered. The 'field' we have been referring to is the four-potential field, A^μ, and its scalar and three-vector potential fields, ϕ and $\mathbf{A}$.

In the historical development of the first 15 chapters, the electromagnetic vector fields, $\mathbf{E}$ and $\mathbf{B}$, were the primary fields for treating clasical electromagnetism. The potential fields, ϕ and $\mathbf{A}$, were introduced for simplification, and in order to develop wave equations with sources. The gauge transformation was introduced as a convenient way to modify the potentials without affecting the final result.

Here we have seen that the gauge transformation is a vital step in introducing the electromagnetic potentials and all of classical electromagnetism in order to preserve local gauge invariance in quantum mechanics.

16.3 Classical Electromagnetism and Quantum Electrodynamics

How is Classical Electromagnetism (CEM) related to Quantum Electrodynamics (QED)? We have seen in a previous section how CEM could be deduced from Local Gauge Invariance in QM, but here we will study its closer connection as the classical limit of the fundamental theory of Electromagnetism–Quantum Electrodynamics (QED).

Actually, CEM is more comparable to Quantum Mechanics than it is to Classical Mechanics. The fundamental equation of both CEM and QM is a wave equation, $\Box^2 A_\mu = 4\pi j_\mu$ for CEM, and the Schrodinger equation for QM. For each theory, the 'wave function', A^μ for CEM and Ψ for QM, contains all the physical information. Although the wave function $\Psi(\mathbf{r}, t)$ in QM is often faulted for not giving a particle's position precisely, that is not a failure of QM, but because it is just the wave solution of a wave equation.

The same 'fault' (if that is what it is) holds for the wave solution A^μ in CEM. In each case, to ask for the 'position' of a wave is just to ask the wrong question. In this context, the combination in atomic physics of Quantum Mechanics with Classical Electromagnetism is not a hybrid, but really a consistent combination of two wave theories. There does seem to be a difference in the character of CEM and QM in that Ψ is the wave function of a particle. Does A^μ describe the motion of a particle? We will answer that question below.

Each of the theories, CEM and QM, were found deficient in the 1930's, in that they could not cope with new phenomena. Quantum Mechanics could not describe the creation and decay of particles, while the combination of CEM and QM could not explain nonlinear effects such as the Lamb shift of degenerate hydrogen levels, and an anomalous magnetic moment of the electron. The resolution of these deficiencies led to a more fundamental form of QM called Quantum Field Theory (QFT).

We will try to explain the foundations of QFT without the details. Ordinary Quantum Mechanics takes the classical canonical variables $\mathbf{r}$ and $\mathbf{p}$, and considers them non-commuting operators. This leads, after a bit of complication, to the quantum mechanical Schrodinger wave equation.

Quantum Field Theory extends this process by considering the wave function Ψ as a canonical variable, and deduces $\partial_t \Psi$ as its canonically conjugate momentum variable. Their lack of commutativity leads to the result (leaving out the details) that the old wave function Ψ is now a quantum mechanical operator that can create and destroy particles.

The old original QM is just the case where only one particle has been created. With this interpretation of Ψ as a QM operator, all the early arguments (unfortunately, some of them continuing) about the meaning of the wave function, without including its operator character, are beside the point.

The above procedure in Quantum Mechanics is called 'Second Quantization', because it quantizes what was already considered a quantum wave equation. Classical Mechanics is now seen as a two step reduction from the fundamental QFT. First, only the one particle sector of the theory is considered, resulting in the original version of Quantum Mechanics. This leads to quantized energy levels and angular momenta. In the limit of high quantum level (usually characterized by large quantum numbers), the quantum results look like the old classical mechanics. For instance, the moon is in a Bohr orbit, but the principal quantum number is so high that no quantum effects are noticeable.

Now we do this with Electromagnetism. The wave equation for A^μ in CEM is equivalent to the Schrodinger equation in QM. To go to QED, which is the QFT for EM, we consider the vector potential $\mathbf{A}$ as a QM variable. (It is easier to quantize EM using three-vector variables, with time as a separate variable.)

In the quantization, $\mathbf{A}$ becomes a QM operator that can create and destroy particles. But what particles? It turns out that these particles have zero mass, and are called **photons**. So the particle associated with the EM wave equation is the photon. An electromagnetic wave is a state composed of one or more photons. You can think of the wave as a continuous wave whose amplitude can only take on discrete values, corresponding to the number of coherent photons in the state. The Classical Electromagnetism we have been discussing in this book is the limit of QED for two different cases

1. The state has so many photons that the amplitudes seem continuous, and we have Classical Electromagnetic waves.

2. The state has no photons. Then we have static Electromagnetism.

We had to do a bit too much QM (with too little detail), but we have illustrated that **Classical Electromagnetism is a limiting form of Quantum Electrodynamics**.

16.4　Local Gauge Invariance as the Grand Unifier of Interactions

The charge q in the local gauge transformation of A^μ is arbitrary. But, if we extend the principle of LGI by demanding that the wave functions of all particles are invariant under the same local gauge transformation, then we have another argument (in addition to Dirac's magnetic monopole argument of Chapter 9) for all charged particles to have the same charge. We take the value e for the charge so that it will be the magnitude of the electron charge. (You have to be careful in reading some books, because the gauge transformation is sometimes applied to the electron, which has charge $-e$. Then all the signs are reversed.)

But, all particles do not have charge e. First of all, we have to explain that LGI applies only to really elementary particles. For instance, the neutron or neutral pi meson are not elementary, but are composed of elementary particles called quarks. (Unfortunately the designation 'elementary particle' was applied too early to many particles that are not truly elementary, in the same way that the term atom, Greek for indivisible, was applied too early to atoms.)

In the Standard Model of elementary particles (accepted as a working hypothesis by most theoretical physicists), there are 24 truly elementary point particles of spin $\frac{1}{2}$ obeying Fermi statistics (so called Fermions). We will consider first the 12 particles called **leptons**. Six of these (the electron, muon, and tau leptons, and their respective antiparticles) have charge $+e$ or $-e$. The occurrence of two signs for the charge is natural for the gauge factor e^{iqx}, because it is expected that antiparticles should obey the complex conjugate of the wave equation for particles.

However, neutrinos, the other six leptons, have no charge. How can their wave equation satisfy LGI? Only if they have an interaction other than electromagnetism, with a neutrino charge for that interaction.

Look back at the previous section. We were not completely general in introducing a single constant e into the gauge transformation e^{ieX}. We can introduce a linear combination of several constants. This is most easily done by using a 2×2 matrix in the transformation

$$\Psi \rightarrow \Psi' = e^{i[\mathbf{M}]}\Psi. \qquad (16.14)$$

If the matrix $[\mathbf{M}]$ is Hermitian and traceless, its exponential will be a 2×2 unitary matrix with determinant $+1$. Such matrices form a group called SU(2) (for Special Unitary 2×2 matrices, where the 'Special' means they have determinant $+1$). What do the matrices act on? They act on 2 component column matrices that link each neutrino with its corresponding charged lepton (the electron neutrino with the electron, etc.).

Thus, the LGI principle can be extended to the neutral neutrinos only by linking them in an interaction with the charged leptons. Because this interaction changes particle type (whereas in EM, an electron always stays an electron), there is no classical counterpart of neutrino interactions, and it is an intrinsically quantum theory. Because a traceless Hermitian matrix like the $[\mathbf{M}]$ in Eq. (16.14) has three independent matrix elements, there are three vector fields in Weak Interactions, W_μ^+, W_μ^-, Z_μ^0, whereas Electromagnetism had only one, A_μ.

The theory we have described (with great oversimplification) is the Glashow, Salam, Weinberg theory of Electro-Weak interactions. The two words, electro and weak, are combined here because, in its implementation, the theory requires that the electromagnetic and the weak fields be combined (The physical fields Z_μ^0 and A_μ are actually linear combinations of two separate neutral fields.), just as the electric and magnetic fields were combined by Maxwell and Einstein. Thus LGI, extended now to neutral particles, has continued the unification of forces from Electric and Magnetic to Electromagnetic to Electro-Weak.

Are we through? Not quite. Why not consider 3×3 Special Unitary matrices in a group called SU(3)? Doing so leads to 8 vector fields (from the 8 independent matrix elements of a 3×3 traceless Hermitian matrix). The 3×3 matrices act on 3 component column matrices composed of three different states (called red, blue, and green) of each quark. This theory of the Strong Interaction, called Quantum Chromodynamics (QCD) (although it has nothing to do with actual color), is somewhat like Quantum Electrodynamics (QED), the quantum extension of Classical Electromagnetism, but has several interesting new properties.

There are 8 fields in QCD, which can interact with each other, rather than just the one field of Electromagnetism. Because of this, the QCD equivalent of the electric charge varies significantly with distance. This has the effect that the QCD potential energy between two quarks is roughly Coulombic at 'moderate' distances ($\sim$1 fm), but does not diverge for small distances, and grows linearly at large distances. This latter property is a mechanism for quark confinement, perhaps explaining why free quarks have never been observed. The $\pm\frac{1}{3}$ and $\pm\frac{2}{3}$ electric charges of quarks, which seemed problematic, arise from the electric charge operator in QCD.

We see that the principle of LGI in QM is quite powerful, leading to the rationale and the mechanism for three kinds of force, Electromagnetic, Weak, and Strong. Leaving aside Gravity for the moment, the LGI principle has accounted for three of the four known physical forces, and has unified two of them in the Electro-Weak theory.

What about QCD? It seems able to stand on its own, whereas the Weak Interaction had to be unified with Electromagnetism to work as a gauge theory. An outstanding question at the end of the 20th century was whether QCD could be united with the Electro-Weak theory in what is referred to as a **Grand Unified Theory (GUT)**. The group SU(5) of 5×5 Special Unitary matrices is available to put into the Local Gauge Transformation for this unification. The group would combine all 24 particles, 12 leptons and 12 quarks (including antiparticles), in one unified interaction. A major problem with this speculation is that it would predict a decay of the proton that has not been observed.

It is interesting that the 20th century began with the outstanding problem (among others) of the Electromagnetism prediction of rapid decay of atoms, whose resolution led to the revolution of Quantum Mechanics. Now, the 21st century begins with the problem (among others) of a predicted decay of the proton. This prediction is not quite as dramatic, since the predicted lifetime is only slightly shorter that the current experimental lower limit of 10^{31} years for the proton lifetime. Whether this discrepancy will lead to a new revolution or some simpler resolution (either theoretical or experimental) is for young 21st century physicists to discover.

What about gravity, the fourth of the known interactions in physics? Is it also a gauge theory that can be unified by a local gauge transformation with the other three forces? This does not seem likely, because the gravitational field in the General Relativity version of gravity is a second rank tensor, rather than a vector.

As you may have noticed, all of the gauge theories had vector fields that combined with the momentum four-vector to produce the interaction. Whether gravity can be unified with the other forces or will remain separate as just a space-time metric in which to embed the other fields is another outstanding problem for 21st century physicists to solve.

The fields we discussed in section 16.4 are actually all quantum fields with their associated particles, called **gauge Bosons**. The gauge Bosons all have spin one, resulting from their vector character. They form symmetric combinations ('Bose-Einstein' statistics), which allows for large numbers of Bosons in the same state.

For the Electro-Weak theory, the gauge Bosons are the W^+, W^-, Z^0, and the photon (or γ). A triumph of the Electro-Weak theory was the discovery in the 1980's of the W and Z particles with masses of 80 and 91 GeV, respectively. The eight gauge Bosons of QCD are called gluons. They each have zero mass, and are very much like the photon, except that they have strong interactions with each other.

This makes a total of 12 gauge Bosons. Along with the elementary 24 Fermions, that makes a total of 36 fundamental elementary particles in the Standard Model of the 20th century. This seems like an awful lot, but in a unified theory the number shrinks. That is, in the Electro-Weak theory the 12 leptons can be considered as 12 different states of a single particle, and the 4 gauge Bosons as 4 states of a single Boson. The 12 quarks and 8 gluons of QCD can be considered as 1 quark and 1 Boson in several states. In a GUT like SU(5), all the Fermions would be different states of a single Fermion, and similarly for a single gauge Boson.

16.5   Natural Units

(Note: the term 'natural units' has been used for various combinations of units. The units discussed here are not, for instance, the natural units related to Newton's gravitational constant G that were introduced by Max Planck.)

In this section we discuss the future of units in the 21st century. (Although we advertise the units discussed below as units for the 21st century, they have actually been in common use in elementary particle physics since the 1950's.) We have seen the steady progression in physics toward unification of interactions and even unification between space and time. How have the units used kept up with this unification?

Electromagnetism originally had two sets of units, one for electric phenomena and one for magnetic phenomena. These two sets of units were combined into one system, the Gaussian system, but were still a bit separate, with the parameter c still distinguishing them. The advent of relativity showed that space and time were just two different directions in a unified space-time so that the parameter c could be dispensed with in our equations. It became just a conversion constant between cm and sec. For instance, if we calculate a lifetime, and it comes out as 6 cm, we could convert it to 6 cm/[3×10^{10}(cm/sec)] $= 2\times10^{-10}$ sec.

Quantum Mechanics has brought about a similar connection between space and momentum (or time and energy) that can be used to simplify units, even for classical theories. In Quantum Mechanics, the position and momentum obey the commutation relation

$$[x, p] = xp - px = i\hbar. \tag{16.15}$$

The constant $\hbar$ has the value 6.58×10^{-22} MeV-sec in one set of units.

Since this numerical value depends on the units chosen, we can always choose units in which $\hbar$ would equal one. For instance, a simple choice would be giving momentum in units of cm^{-1}, which are actually the usual units of wave number k for a wave.

Although x and p are not as physically connected as are x and t in Special Relativity, we can still use any convenient set of units for them. If we choose units for which $\hbar$ equals one, we need never put $\hbar$ into our equations in the first place. Just as it was with c, we only need $\hbar$ in equations if we use incompatible units.

This is the content of **natural units**:

1. The dimensioned constants c and $\hbar$ need never appear in equations.

2. The physical quantities

$$m \sim E \sim p \sim \omega \sim \frac{1}{x} \sim \frac{1}{t} \tag{16.16}$$

   are all equivalent dimensionally, although, for convenience, any desired unit can be used for any physical quantity.

3. If the final numerical answer is not in desired units, the conversion constants

$$1 = c = 2.99792458 \times 10^{10} cm/sec = \hbar = 6.58 \times 10^{-22} MeV - sec \tag{16.17}$$

   can be used to put the answer into any desired units. (Other numerical values could be used if other units, say meters, were desired.)

Another useful conversion constant is the combination

$$1 = \hbar c = 197.33\,MeV - fm = 1.9733\,keV - Å. \tag{16.18}$$

Natural units could be called the '1,2,3' system of units, because its as easy as 1,2,3. All you have to know is

- $\hbar{=}c{=}1$ in all equations,

- $\hbar c{=}2$ keV-Å,

- $c{=}3{\times}10^{10}cm/sec$.

In many applications, the approximate numbers 2 keV-Å and 3×10^{10} cm/sec have suffficient accuracy

One dimensionless number is important in the use of natural units, and in all of physics. The square of the electron charge, e^2, has the units of energy×distance, as does the combination $\hbar c$. Consequently, the ratio $e^2/\hbar c$, or just e^2 in natural units, is dimensionless. Calculating in the Gaussian system, we get

$$e^2 = \frac{(4.8 \times 10^{-10})^2}{1.05 \times 10^{-27} \times 3 \times 10^{10}} = 1/137.036 \equiv \alpha. \qquad (16.19)$$

(We have used more accurate numbers to calculate e^2 than the approximate values shown.) The same dimensionless number, $1/137.036$, would occur in any other system of units, although factors of 4π and ϵ_0 have to be inserted in some systems. For instance, in SI units, the dimensionless constant α is defined by

$$\alpha = \frac{e^2}{4\pi\epsilon_0\hbar c} = 1/137.036. \quad \textbf{SI} \qquad (16.20)$$

The constant α is called the **fine structure constant**, because it originally arose as the ratio of the fine structure splitting in spectral lines to the distance between lines. Fortunately, it has the same numerical value in all systems of units (so far). So, although e^2 may vary with the system used, α is still the same.

Its numerical value is almost always given in the inverse form $1/137$ (which is a good approximate value). This is probably for historical reasons. Arthur Eddington made a theoretical suggestion (never clearly documented) that α should equal *exactly* $1/137$. Over the years, exerimental measures of $1/\alpha$ slowly and dramatically climbed down toward the integral value. However, it now seems to have settled at the accurate value of $\alpha=1/137.03599$, just a little above an integer, while no good basis for Eddington's conjecture has emerged.

The practical use of natural units is best seen in examples:

- The classical radius of the electron is

$$r_c = \frac{e^2}{m} = \left(\frac{1}{137 \times .511\,\text{MeV}}\right) \times 200\,\text{MeV} - \text{fm} = 2.82\,\text{fm}. \qquad (16.21)$$

- The radius of the first Bohr orbit (from QM) in hydrogen is

$$a_0 = \frac{1}{me^2} = \left(\frac{137}{511\,\text{keV}}\right) \times 2\,\text{keV} - \text{Å} = 0.529\,\text{Å}. \qquad (16.22)$$

- The potential energy of an electron in the first Bohr orbit of hydrogen is

$$\frac{-e^2}{a_0} = \left(\frac{-1}{137 \times .529\,\text{Å}}\right) \times 2\,\text{keV} - \text{Å} = -27.2\,\text{eV}. \qquad (16.23)$$

In each case, we have shown the round number 2 (or 200) but have used the more accurate number for the final answer. Note that cumbersome powers of 10 are absent in these calculations.

16.6 α

The dimensionless constant α is the most important number in physics. It determines the structure of all matter. (We will see shortly that it applies to more than just Electromagnetism.) Because it is dimensionless, it will be the same in all cultures and in all galaxies that obey the same physics that we do.

On the other hand, no dimensioned constant can be fundamental. A constant with dimensions is merely a conversion between different arbitrary units, its value determined only by the international body that devised the particular units. Such a constant is manmade, and not natural or organic.

Why α does equal $1/137.036$ is one of the major challenges of the 21st century. On the contrary, there is no need to explain dimensioned constants like c, $\hbar$, ϵ_0, μ_0. They just have the values some committee gave them in their choice of units.

Now we consider the unifying hypothesis that the three forces, Electromagnetic, Weak, and Strong all actually have the same coupling constant. It does turn out in the Electro-Weak theory that α is the coupling constant for both the electromagnetic and the weak sectors of the theory.

The reason the weak sector has been called weak is that the weak interaction was first observed in nuclear beta decay. In that process, the effective interaction strength was $\alpha(M_{\mathrm{proton}}/M_W)^2$, which is about $\alpha/7,000$. For a process like high energy scattering of neutrinos by electrons, the 'weak' and 'electromagnetic' forces are of comparable strength.

The effective coupling constant in QCD, often called α_s (for α strong), varies considerably with the momentum transfer in a scattering process. This can be related, via the Fourier transform, to a variation with r for an effective potential energy $V(r)$. We discussed that variation earlier.

Because there are 8 gluons, corresponding to the 8 vector fields in QCD, α_s decreases with increasing momentum transfer. (The mechanism for this is quite complicated.) The electromagnetic α also varies with momentum transfer, although not as markedly as does α_s. Because there is only a single photon in QED, the variation in α is slower, and in the other direction from α_s. That is, the QED α slowly increases with increasing momentum transfer.

In terms of distance, α gets stronger as you approach a point charge. This is interpreted in QED as a polarization of the vacuum. For a point charge electron, the strong field close to the charge excites electron-positron pairs from the vacuum. The electrons escape to infinity,while the positrons cluster around the electron, partially shielding its charge. The charge we see at large distances, which we measure as e, is smaller than the actual charge at the origin. As we get closer to the point charge, we get partially inside the positron cloud and the effective α increases.

This phenomenon is called **vacuum polarization**, and the fact that the magnitude of the charge we see is not the actual point charge is called **charge**

renormalization. Although these are quantum effects, they would affect Classical Electromagnetism at small distances. They could be interpreted, either as a change in effective charge with distance or as a nonlinear permittivity of free space due to the vacuum polarization. (That is why we objected to the common designation of the ϵ_0 of SI units as the "permittivity of free space".)

The reason we did not include vacuum polarization in any of the previous chapters is that the distances at which it would become important are so small that Quantum Mechanics would have to be used from the beginning. Classical Electromagnetism cannot be applied at the close distances where vacuum polarization could be important. The value of α that is appropriate for CEM is that evaluated at zero momentum, or infinite distance, which is the precise value 1/137.036 we have been using. At momenta or distances where α started to increase significantly from that value, we would have to abandon classical physics and use QM.

The value of α_s suitable for calculating quark bound states is about 0.65, much larger than α. That is why it is called α strong. The strong QCD fields near a quark also excites quark-antiquark pairs. But gluons are also produced, and the gluon effect is antishielding. With 8 gluons, their cumulative effect overshadows the quark-antiquark effect, which is why α_s decreases as you approach a quark, or as momentum transfer increases in scattering.

The decrease in the QCD $\alpha_s(Q)$, where Q is momentum transfer in a scattering process, is large. α is down to 0.12 for Q=91 GeV, the mass of the Z^0 boson. The QED α varies more slowly, increasing to 1/128=.0078 at the mass of the Z^0. That is up from 1/137=0.0073 at zero momentum. Although still far apart in magnitude, it is significant that the two α's are approaching each other.

In GUT theories, such as the SU(5) theory we discussed earlier, the two α's approach equality at some very large value of Q, we could call the 'unification mass'. In that case, all the coupling constants of physics (except, possibly, that of gravity) would be given by the electron charge of Electromagnetism, suitably evolved to high Q. In fact, such unification of charge, all three charges (weak, EM, and strong) being equal in some symmetry limit, is required if LGI is to be considered universal.

Because the electromagnetic coupling α is dimensionless, it makes sense to talk about its size. The numbers, 4.8×10^{-10} esu or 1.6×10^{-19} coulomb, give no indication of whether they are large or small, even though they look very small. The value of α=1/137 is clearly much less than 1. This tells us that Electromagnetism is, to a good approximation, a linear theory, and the superposition we used in Chapter 1 and all through the text was permissible.

In QED, the nonlinearity can be treated in perturbation theory because of the small size of α. On the other hand, α_s is of order 1 for low energy quark interactions, so QCD is a nonlinear theory at low energies. At very high energies, α_s decreases, and perturbation theory can be used.

What is the place of SI units in the 21st century? Unfortunately, although adopted by majority vote at an international congress, SI units seem opposed to every advance in Electromagnetism, Classical or Quantum. The units impede the implementation of the unification of electricity and magnetism. SI emphasizes the differences between $\mathbf{E}$ and $\mathbf{B}$, where we have seen there is no difference. SI units treat $\mathbf{E}$ and $\mathbf{D}$ very differently, so it becomes difficult to see that they are each electric fields, just coming from different charge distributions. Similarly for $\mathbf{B}$ and $\mathbf{H}$.

SI units introduce a fictitious permittivity and permeability of free space that have nothing to do with the properties of free space. One of the bases for the SI system was the mistaken idea that electric charge was a new dimension for physics (the Coulomb), whereas electric charge is intrinsincally dimensionless. SI units make the teaching of elementary EM confusing to students, and complicates more advanced aspects of EM. By the end of the 21st century, I hope that the system of units in common usage in electromagnetism is appropriate to the physics.

APPENDIX A

Conversion of Units

In this appendix we list conversions between Gaussian and SI units. The most common names are given here for the Gaussian units. In some usages, the letters "stat" are appended to the front of the SI unit for the corresponding Gaussian unit. Sometimes the symbol "esu" is simply used for any Gaussian unit, with the context telling what it is. (This is the simplest usage for Gaussian units, and may be preferable. There is no need to memorize names for units.) In the table, the constant c has been shown as 3×10^{10} cm/sec. If more accuracy is desired, the exact value of $c=2.99792458 \times 10^{10}$ cm/sec should be used.

Table 16.1: Conversions between Gaussian and SI units

Quantity	Symbol	Gaussian		SI
Force	$\mathbf{F}$	10^5 dynes	=	1 newton
Energy	W or U	10^7 ergs	=	1 joule
Power	P	10^7 ergs/sec	=	1 watt
Charge	q	3×10^9 statcoulombs	=	1 coulomb
Current	I	3×10^9 statamperes	=	1 ampere
Potential	ϕ or V	1 statvolt	=	300 volts
Electric field	$\mathbf{E}$	1 gauss	=	3×10^4 volts/meter
Magnetic field	$\mathbf{B}$	10^4 gauss	=	1 tesla
Magnetic flux	Φ	10^8 gauss-cm^2	=	1 weber
Resistance	R	1 sec/cm	=	9×10^{11} ohms
Capacitance	C	9×10^{11} cm	=	1 farad
Inductance	L	9×10^{11} sec^2/cm	=	1 henry

APPENDIX B

Derivatives of Retarded Variables

In this appendix, we calculate the derivatives needed to find the $\mathbf{E}$ and $\mathbf{B}$ fields from the potentials ϕ and $\mathbf{A}$, given in terms of the retarded time. The retarded time t_r is related to the other variables by the two implicit relations

$$t_r = t - r(t_r) \tag{16.24}$$

$$\mathbf{r_r} = \mathbf{r} - \mathbf{s}(t_r). \tag{16.25}$$

For completeness, we start with the partial derivative $\partial_t(t_r)$, which we derived in Chapter 15. This derivative is important for finding all the other derivatives. We just list the successive steps. The reader can fill in the explanation for each step. Many steps are listed, so we will only number the equations for final results.

$$\partial_t(t_r) = \partial_t(t - r_r) = 1 - \partial_t(r_r),$$

$$\partial_t(r_r) = \partial_t[(\mathbf{r_r \cdot r_r})^{\frac{1}{2}}] = \hat{\mathbf{r}}_\mathbf{r} \cdot [\partial_t(\mathbf{r_r})],$$

$$\partial_t(\mathbf{r_r}) = \partial_t(\mathbf{r} - \mathbf{s}) = -\partial_t[\mathbf{s}(t_r)] = -\partial_t(t_r)\frac{\mathbf{ds}}{dt_r} = -\partial_t(t_r)\mathbf{v_r},$$

$$\partial_t(r_r) = \hat{\mathbf{r}}_\mathbf{r} \cdot [\partial_t(\mathbf{r_r})] = -\hat{\mathbf{r}}_\mathbf{r} \cdot \mathbf{v_r}\partial_t(t_r),$$

$$\partial_t(t_r) = 1 - \partial_t(r_r) = 1 + \hat{\mathbf{r}}_\mathbf{r} \cdot \mathbf{v_r}\partial_t(t_r) = \frac{1}{1 - \hat{\mathbf{r}}_\mathbf{r} \cdot \mathbf{v_r}}. \tag{16.26}$$

We list below useful results from the above steps[1], writing d for the denominator $(1 - \hat{\mathbf{r}} \cdot \mathbf{v})$:

$$\partial_t(t_r) = 1/d \tag{16.27}$$

$$\partial_t(\mathbf{r}) = -\mathbf{v}/d \tag{16.28}$$

$$\partial_t(r) = -\hat{\mathbf{r}} \cdot \mathbf{v}/d. \tag{16.29}$$

Note that we can use the result for $\partial_t(t_r)$ to give the general formula

$$\partial_t[f(t_r)] = \frac{1}{d}\frac{df}{dt_r}.$$

This gives, immediately,

$$\partial_t(\mathbf{v}) = \mathbf{a}/d \tag{16.30}$$

$$\partial_t(\mathbf{r} \cdot \mathbf{v}) = (\hat{\mathbf{r}} \cdot \mathbf{a} - \mathbf{v}^2)/d. \tag{16.31}$$

---

[1] For simplicity, we drop the subscript $\mathbf{r}$ from $\mathbf{r_r}$ and $\mathbf{v_r}$ in the following steps.

477

The derivatives with $\boldsymbol{\nabla}$ (the gradient with respect to $\mathbf{x}$) are a bit more complicated, because they also involve vector manipulation. We start with

$$
\begin{aligned}
(\boldsymbol{\nabla} r) &= \boldsymbol{\nabla}[(\mathbf{r}\cdot\mathbf{r})^{\frac{1}{2}}] = \frac{1}{2r}\boldsymbol{\nabla}(\mathbf{r}\cdot\mathbf{r}) = \hat{\mathbf{r}}\times(\boldsymbol{\nabla}\times\mathbf{r}) + (\hat{\mathbf{r}}\cdot\boldsymbol{\nabla})\mathbf{r} \\
\boldsymbol{\nabla}\times\mathbf{r} &= \boldsymbol{\nabla}\times[\mathbf{x} - \mathbf{s}(t_{\mathrm{r}})] = -\boldsymbol{\nabla}\times\mathbf{s}(t_{\mathrm{r}}) = -(\boldsymbol{\nabla} t_{\mathrm{r}})\times\mathbf{v} = (\boldsymbol{\nabla} r)\times\mathbf{v} \\
\hat{\mathbf{r}}\times(\boldsymbol{\nabla}\times\mathbf{r}) &= -\hat{\mathbf{r}}\times[\mathbf{v}\times(\boldsymbol{\nabla} r)] = (\hat{\mathbf{r}}\cdot\mathbf{v})(\boldsymbol{\nabla} r) - \mathbf{v}[\hat{\mathbf{r}}\cdot(\boldsymbol{\nabla} r)] \\
(\hat{\mathbf{r}}\cdot\boldsymbol{\nabla})\mathbf{r} &= (\hat{\mathbf{r}}\cdot\boldsymbol{\nabla})[\mathbf{x} - \mathbf{s}(t_{\mathrm{r}})] = \hat{\mathbf{r}} - (\hat{\mathbf{r}}\cdot\boldsymbol{\nabla})[\mathbf{s}(t_{\mathrm{r}})] \\
&= \hat{\mathbf{r}} - [\hat{\mathbf{r}}\cdot(\boldsymbol{\nabla} t_{\mathrm{r}})]\mathbf{v} = \hat{\mathbf{r}} + \mathbf{v}[\hat{\mathbf{r}}\cdot(\boldsymbol{\nabla} r)] \\
(\boldsymbol{\nabla} r) &= \hat{\mathbf{r}}\times(\boldsymbol{\nabla}\times\mathbf{r}) + (\hat{\mathbf{r}}\cdot\boldsymbol{\nabla})\mathbf{r} \\
&= (\hat{\mathbf{r}}\cdot\mathbf{v})(\boldsymbol{\nabla} r) - \mathbf{v}[\hat{\mathbf{r}}\cdot(\boldsymbol{\nabla} r)] + \hat{\mathbf{r}} + \mathbf{v}[\hat{\mathbf{r}}\cdot(\boldsymbol{\nabla} r)] \\
&= (\hat{\mathbf{r}}\cdot\mathbf{v})(\boldsymbol{\nabla} r) + \hat{\mathbf{r}}.
\end{aligned}
$$

Finally

$$
(\boldsymbol{\nabla} r) = -(\boldsymbol{\nabla} t_{\mathrm{r}}) = \hat{\mathbf{r}}/d. \tag{16.32}
$$

We use some of the intermediate steps above to get our last and most difficult gradient $\boldsymbol{\nabla}(\mathbf{r}\cdot\mathbf{v})$:

$$
\begin{aligned}
\boldsymbol{\nabla}(\mathbf{r}\cdot\mathbf{v}) &= \mathbf{v}\times(\boldsymbol{\nabla}\times\mathbf{r}) + (\mathbf{v}\cdot\boldsymbol{\nabla})\mathbf{r} + \mathbf{r}\times(\boldsymbol{\nabla}\times\mathbf{v}) + (\mathbf{r}\cdot\boldsymbol{\nabla})\mathbf{v} \\
\mathbf{v}\times(\boldsymbol{\nabla}\times\mathbf{r}) &= \mathbf{v}\times[(\boldsymbol{\nabla} r)\times\mathbf{v}] = \mathbf{v}\times(\hat{r}\times\mathbf{v}) = [\mathbf{v}^2\hat{\mathbf{r}} - (\hat{\mathbf{r}}\cdot\mathbf{v})\mathbf{v}]/d \\
(\mathbf{v}\cdot\boldsymbol{\nabla})\mathbf{r} &= \mathbf{v} + \mathbf{v}[\mathbf{v}\cdot(\boldsymbol{\nabla} r)] = \mathbf{v} + \hat{\mathbf{r}}\cdot\mathbf{v}/d = \mathbf{v}/d \\
\mathbf{r}\times(\boldsymbol{\nabla}\times\mathbf{v}) &= \mathbf{r}\times[\boldsymbol{\nabla}(t_{\mathrm{r}})\times\mathbf{a}] = -[\mathbf{r}\times(\hat{\mathbf{r}}\times\mathbf{a})]/d = [r\mathbf{a} - \mathbf{r}(\hat{\mathbf{r}}\cdot\mathbf{a})]/d \\
(\mathbf{r}\cdot\boldsymbol{\nabla})\mathbf{v} &= [\mathbf{r}\cdot(\boldsymbol{\nabla} t_{\mathrm{r}})]\mathbf{a} = -r\mathbf{a}/d.
\end{aligned}
$$

Adding these four terms up, we get

$$
\boldsymbol{\nabla}(\mathbf{r}\cdot\mathbf{v}) = [\hat{\mathbf{r}}\mathbf{v}^2 - \mathbf{v}(\hat{\mathbf{r}}\cdot\mathbf{v}) + \mathbf{v} - \mathbf{r}(\hat{\mathbf{r}}\cdot\mathbf{a})]/d. \tag{16.33}
$$

Now, back to the text.

BIBLIOGRAPHY

This is not a comprehensive bibliography, but rather a short collection of books I have found helpful. This text is meant to be self contained, but some of these books may help to clarify a particular point for you. Of course, your professor can recommend other books that are suited to your needs.

ARFKEN, George *Mathematical Methods for Physics*, Any Edition, Academic Press- A good math resource for physics. Later editions include additional authors.

BUTKOV, Eugene, *Mathematical Physics*, Addison-Wesley- A good math resource for physics.

GRIFFITHS, David J., *Introduction to Electrodynamics*, 3rd Edition, Prentice Hall- An advanced undergraduate text.

HALZEN, Francis, and MARTIN, Alan -A good book to read after completing this text and a Quantum Mechanics course. H & M gives a good, high level introduction to gauge theories.

JACKSON, John David, *Classical Electrodynamics*, Any Edition, John Wiley and Sons- A comprehensive graduate text with a number of topics not covered in this text. The 3rd Edition mixes SI and Gaussian unts.

LANDAU, L. D. and LIFSCHITZ, E. M., *Classical Theory of Fields*, 4th Edition, Pergammon Press- A good book for advanced topics, but difficult to learn from.

MAXWELL, James Clerk, *A Treatise on Electricity and Magnetism* (3rd Edition, 1891. Two volumes, edited by J. J. Thomson), Dover Publications- A beautiful book with everything but relativity.

PANOFSKY, Wolfgang K. H., & PHILLIPS, Melba, *Classical Electricity and Magnetism*, 2nd Edition, Addison-Wesley- A graduate text with some topics not covered in this text, all in SI units. A good section on the experimental search for the ether.

SCHWARTZ, Melvin, *Principles of Electrodynamics*, Dover Publications, 1972- A very nice graduate text, with some new slants.

SIMPSON, Thomas K., *Maxwell on the Electromagnetic Field*, Rutgers University Press, 1977- A valuable book giving the physical import of much of Maxwell.

ANSWERS TO
ODD-NUMBERED PROBLEMS

Chapter 1

Problem 1.1:

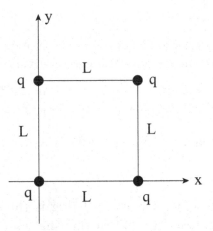

(a) The force on the upper right hand charge due to the other three charges at the corners of a square with sides of length L is

$$\mathbf{F} = \frac{q^2}{L^2}\hat{\mathbf{i}} + \frac{q^2}{L^2}\hat{\mathbf{j}} + \frac{q^2}{2L^2}\left(\frac{\hat{\mathbf{i}}+\hat{\mathbf{j}}}{\sqrt{2}}\right)$$

$$= q^2\left[\frac{1+2\sqrt{2}}{2\sqrt{2}L^2}\right](\hat{\mathbf{i}}+\hat{\mathbf{j}}). \tag{1.1}$$

The magnitude of this force on any of the four charges is

$$F = q^2(1+2\sqrt{2})/2L^2. \tag{1.2}$$

(b) If the four charges are released from rest, we can write for the acceleration of any of the charges

$$\frac{F}{m} = a = \frac{dv}{dt} = \frac{dv}{dL'}\frac{dL'}{dt}, \tag{1.3}$$

where L' is the side length of the square at any time during the motion. The center of the square remains fixed, and the distance, r, of a charge from the center is related to L' by $L' = \sqrt{2}r$, so

$$\frac{dL'}{dt} = \sqrt{2}\frac{dr}{dt} = \sqrt{2}v. \quad \text{Then, } a = \sqrt{2}v\frac{dv}{dL'}, \quad (1.4)$$

and the squared velocity after a long time is given by

$$V^2 = \int_L^\infty \sqrt{2}a\, dL' = \left[\frac{q^2(1+2\sqrt{2})}{\sqrt{2}m}\right]\int_L^\infty \frac{dL'}{L'^2} = \frac{q^2(1+2\sqrt{2})}{\sqrt{2}mL}. (1.5)$$

The velocity is the square root of this:

$$V = q\left[\frac{(1+2\sqrt{2})}{\sqrt{2}mL}\right]^{1/2}. \quad (1.6)$$

Problem 1.3:

By symmetry, the electric field of a long straight wire is perpendicular to the wire. The field a distance **r** from the wire is given by

$$\mathbf{E(r)} = \int\frac{(\mathbf{r} - \mathbf{r'})dq'}{|\mathbf{r} - \mathbf{r'}|^3} = \lambda\mathbf{r}\int_{-\infty}^\infty \frac{dz}{(z^2 + r^2)^{3/2}} = \frac{\lambda\hat{\mathbf{r}}}{r}\int_{-\pi/2}^{\pi/2}\cos\theta d\theta = \frac{2\hat{\mathbf{r}}\lambda}{r}. \quad (1.7)$$

We made the substitution $z = r\tan\theta$ in doing the integral.

Problem 1.5:

(a) Every point on the uniformly charged ring is the same distance from a point a distance z along the axis of the ring, and a line from any point on the ring makes the same angle θ with the z-axis. Thus the electric field at z is

$$E_z = \frac{Q\cos\theta}{r^2} = \frac{Qz}{(z^2 + R^2)^{\frac{3}{2}}}. \quad (1.8)$$

(b) The disk has a surface charge density $\sigma = Q/\pi R^2$. It can be considered as a collection of rings, each of radius r' with a charge

$$dq = 2\pi r'\sigma dr' = \frac{2Qr'dr'}{R^2}. \quad (1.9)$$

The electric field a distance z along the axis of the disk is given as (using part a)

$$E_z = \int_0^R \frac{2Qr'}{R^2}\frac{zdr'}{(z^2 + r'^2)^{\frac{3}{2}}}$$

$$= \frac{2Q}{R^2}\left[1 - \frac{z}{\sqrt{z^2 + R^2}}\right]. \quad (1.10)$$

(c)   (i) For $z = 0_+$ (just above the disk),

$$E_z = \frac{2Q}{R^2} = 2\pi\sigma. \tag{1.11}$$

(ii) For $z >> R$, we write E as

$$E_z = \frac{2Q}{R^2}\left[1 - \left(1 + \frac{R^2}{z^2}\right)^{-\frac{1}{2}}\right]. \tag{1.12}$$

Using the binomial theorem, we get

$$E_z = \frac{2Q}{R^2}\left[1 - \left(1 - \frac{R^2}{2z^2} + \cdots\right)\right].$$

$$\simeq \frac{Q}{z^2}. \tag{1.13}$$

This limit, equal to the field of a point charge Q, can be used as a check on the original result.

Problem 1.7:

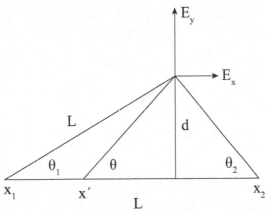

The parallel component of the electric field a distance d from a uniformly charged straight wire of length L is given by the integral

$$E_x = \frac{Q}{L}\int_{x_1}^{x_2}\frac{dx'\,\cos\theta}{r^2}, \tag{1.14}$$

where x_1 and x_2 are the two endpoints of the wire. Let

$$x' = -d\cot\theta$$
$$dx' = d\csc^2\theta\,d\theta$$
$$r = d\csc\theta. \tag{1.15}$$

Then

$$E_x = \frac{Q}{Ld}\int_{\theta_1}^{\pi-\theta_2}\cos\theta\,d\theta = \frac{Q}{Ld}(\sin\theta_2 - \sin\theta_1), \tag{1.16}$$

where θ_1 and θ_2 are the angles shown.

For the perpendicular component of $\mathbf{E}$, the same substitution for z, leads to

$$E_y = \frac{Q}{Ld}\int_{\theta_1}^{\pi-\theta_2}\sin\theta\,d\theta = \frac{Q}{Ld}(\cos\theta_2 + \cos\theta_1). \tag{1.17}$$

Problem 1.9:

(a) Each charge q, at the corner of a square with sides of length L is a distance $\sqrt{z^2 + L^2/2}$ from a point z along the perpendicular axis of the square. (See the figure in Prob. 1.2.)

Thus the potential at the point z is $\qquad \phi = \dfrac{4q}{\sqrt{z^2 + L^2/2}}.$ $\qquad\qquad$ (1.18)

(b) The electric field is

$$E_z = -\partial_z \phi = \frac{4qz}{(z^2 + L^2/2)^{\frac{3}{2}}}. \qquad\qquad (1.19)$$

Problem 1.11:

(a) Every point on the uniformly charged ring is the same distance $\sqrt{z^2 + R^2}$ from a point a distance z along the axis of the ring, so the potential at z is

$$\phi = \frac{Q}{\sqrt{z^2 + R^2}}. \qquad\qquad (1.20)$$

(b) The uniformly charged disk of radius R has a surface charge density $\sigma = Q/\pi R^2$. It can be considered as a collection of rings, each of radius r' with a charge

$$dq = 2\pi r' \sigma dr' = \frac{2Qr'dr'}{R^2}. \qquad\qquad (1.21)$$

The potential a distance z along the axis of the disk is given as (using part a)

$$\phi \; = \; \frac{2Q}{R^2} \int_0^R \frac{r'dr'}{\sqrt{z^2 + r'^2}} = \frac{2Q}{R^2} \left[\sqrt{z^2 + R^2} - z \right]. \qquad (1.22)$$

(c) The electric field of the ring is

$$E_z = -\partial_z \left[\frac{Q}{\sqrt{z^2 + R^2}} \right] = \frac{Qz}{(z^2 + R^2)^{\frac{3}{2}}}. \qquad\qquad (1.23)$$

The electric field of the disk is

$$E_z = -\frac{2Q}{R^2} \partial_z \left[\sqrt{z^2 + R^2} - z \right] = \frac{2Q}{R^2} \left[1 - \frac{z}{\sqrt{z^2 + R^2}} \right]. \qquad (1.24)$$

Problem 1.13:

(a) The charge density inside the uniformly charged sphere is

$$\rho = \frac{3Q}{4\pi R^3}. \tag{1.25}$$

By symmetry, the electric field is in the radial direction. We consider a Gaussian sphere of radius r, concentric with the uniformly charged sphere. Then, for $r \leq R$, Gauss's law becomes

$$\oint \mathbf{E} \cdot \mathbf{dA} = 4\pi Q$$
$$4\pi r^2 E = \left[\frac{3Q}{4\pi R^3}\right]\left[\frac{4\pi r^3}{3}\right]$$
$$E = \frac{Qr}{R^3}. \tag{1.26}$$

For $r \geq R$, Gauss's law is

$$\oint \mathbf{E} \cdot \mathbf{dA} = 4\pi Q$$
$$4\pi r^2 E = 4\pi Q$$
$$E = \frac{Q}{r^2}. \tag{1.27}$$

(b) The electric field for $r \geq R$ is the same as that for a point charge, so the potential is the same as the potential of a point charge:

$$\phi(r) = \frac{Q}{r}, \quad r \geq R. \tag{1.28}$$

For $r \leq R$,

$$\phi(r) - \phi(R) = -\int_R^r \mathbf{E} \cdot \mathbf{dr}$$
$$\phi(r) = \frac{Q}{R} - \frac{Q}{R^3}\int_R^r r\,dr = \frac{Q}{R}\left[1 - \frac{1}{2R^2}(r^2 - R^2)\right]$$
$$= \frac{Q}{2R^3}(3R^2 - r^2). \tag{1.29}$$

Problem 1.15:

(a) The long straight wire has radius R and uniform charge density ρ. By symmetry, the electric field is perpendicular to the axis of the wire. We consider a Gaussian cylinder of length L and radius r, concentric with the wire. Then, for $r \leq R$, Gauss's law becomes

$$\oint \mathbf{E} \cdot \mathbf{dA} = 4\pi Q$$
$$2\pi r L E = 4\pi^2 r^2 \rho L$$
$$E = 2\pi\rho r, \quad r \leq R. \tag{1.30}$$

For $r \geq R$, Gauss's law is

$$
\begin{aligned}
\oint \mathbf{E} \cdot \mathbf{dA} &= 4\pi Q \\
2\pi r L E &= 4\pi^2 R^2 \rho L \\
E &= \frac{2\pi \rho R^2}{r}, \quad r \geq R.
\end{aligned}
\tag{1.31}
$$

(b) We have to pick some radius for which $\phi = 0$. We choose $\phi(R) = 0$, so that the surface of the wire is at zero potential. Then, for $r \leq R$,

$$
\phi(r) = -\int_R^r \mathbf{E} \cdot \mathbf{dr} = -\int_R^r 2\pi \rho r \, dr = \pi \rho (R^2 - r^2).
\tag{1.32}
$$

For $r \geq R$,

$$
\phi(r) = -\int_R^r \mathbf{E} \cdot \mathbf{dr} = -\int_R^r \frac{2\pi R^2 \rho \, dr}{r} = 2\pi \rho R^2 \ln(R/r).
\tag{1.33}
$$

Problem 1.17:

(a) For $\phi = q e^{-\mu r} / r$, the electric field is

$$
\begin{aligned}
\mathbf{E} &= -q \boldsymbol{\nabla}(e^{-\mu r}/r) = -q e^{-\mu r} \boldsymbol{\nabla}\left(\frac{1}{r}\right) - \frac{q}{r}\boldsymbol{\nabla}(e^{-\mu r}) \\
&= -q e^{-\mu r}\left(-\frac{\hat{\mathbf{r}}}{r^2} - \frac{\mu \hat{\mathbf{r}}}{r}\right) = \frac{q \hat{\mathbf{r}} e^{-\mu r}}{r^2}(1 + \mu r).
\end{aligned}
\tag{1.34}
$$

(b) The charge density is given by

$$
\rho = \frac{1}{4\pi} \boldsymbol{\nabla} \cdot \mathbf{E} = \frac{q}{4\pi} \boldsymbol{\nabla} \cdot \left[e^{-\mu r}\left(\frac{\mathbf{r}}{r^3} + \frac{\mu \mathbf{r}}{r^2}\right) \right],
\tag{1.35}
$$

where we have written $\mathbf{E}$ in a more convenient form for taking its divergence. We differentiate each term in turn, leading to

$$
\begin{aligned}
\rho &= \frac{q}{4\pi}\left\{ e^{-\mu r}\left[\boldsymbol{\nabla} \cdot \left(\frac{\mathbf{r}}{r^3}\right) + \boldsymbol{\nabla} \cdot \left(\frac{\mu \mathbf{r}}{r^2}\right) \right] + \left[\frac{\mathbf{r}}{r^3} + \frac{\mu \mathbf{r}}{r^2}\right] \cdot \boldsymbol{\nabla}(e^{-\mu r}) \right\} \\
&= \frac{q e^{-\mu r}}{4\pi}\left[4\pi \delta(\mathbf{r}) + \frac{3\mu}{r^2} - \frac{2\mu}{r^2} - \frac{\mu}{r^2} - \frac{\mu^2}{r} \right] \\
&= q \delta(\mathbf{r}) - \frac{\mu^2 q e^{-\mu r}}{4\pi r}.
\end{aligned}
\tag{1.36}
$$

Note that above we isolated the term $\boldsymbol{\nabla} \cdot (\mathbf{r}/r^3)$, because we knew that it equals $4\pi \delta(\mathbf{r})$. This correctly accounted for the singular behavior at the origin. The term $q\delta(\mathbf{r})$ corresponds to a point charge q at the origin. The other part of Eq. (1.36) represents a negative charge distribution surrounding the point charge.

(c) Gauss's law is

$$\oint \mathbf{E}\cdot\mathbf{dA} = 4\pi Q. \tag{1.37}$$

We choose a sphere of radius R as our Gaussian surface. The integral of $\mathbf{E}\cdot\mathbf{dS}$ over the surface of the sphere is

$$\oint \mathbf{E}\cdot\mathbf{dA} = 4\pi R^2 E_r(R) = 4\pi q e^{-\mu R}\left(1 + \mu R\right), \tag{1.38}$$

where we have taken E_r from Eq. (1.34) The charge within the Gaussian sphere is given (using Eq. (1.36) by

$$Q = \int_{r \le R} \rho d\tau = q - 4\pi \int_0^R \frac{\mu^2 q e^{-\mu r} r^2 dr}{4\pi r} = q e^{-\mu R}\left(1 + \mu R\right), \tag{1.39}$$

which agrees with Eq. (1.38), and Gauss's law is satisfied. (The last integral above was done using integration by parts.)

Note the importance of the proper treatment of the delta function at the origin.

Problem 1.19:

We want to show that $\mathbf{L}\times\mathbf{L}\phi = i\mathbf{L}\phi$, where $\mathbf{L} = -i\mathbf{r}\times\boldsymbol{\nabla}$. This corresponds to

$$-(\mathbf{r}\times\boldsymbol{\nabla})\times(\mathbf{r}\times\boldsymbol{\nabla}\phi) = \mathbf{r}\times\boldsymbol{\nabla}\phi. \tag{1.40}$$

We introduce $\mathbf{E} = -\boldsymbol{\nabla}\phi$, and then will show that

$$(\mathbf{r}\times\boldsymbol{\nabla})\times(\mathbf{r}\times\mathbf{E}) = -\mathbf{r}\times\mathbf{E}. \tag{1.41}$$

We use

$$\mathbf{a}\times(\mathbf{b}\times\mathbf{c}) = \mathbf{b}(\mathbf{a}\cdot\mathbf{c}) - \mathbf{c}(\mathbf{a}\cdot\mathbf{b}), \tag{1.42}$$

with

$$\mathbf{r}\times\boldsymbol{\nabla} = \mathbf{a}, \quad \mathbf{r} = \mathbf{b}, \quad \mathbf{E} = \mathbf{c}. \tag{1.43}$$

We take the vector derivative twice, first holding $\mathbf{E}$ constant and then holding $\mathbf{r}$ constant. For $\mathbf{E}$ constant,

$$\begin{aligned}(\mathbf{r}\times\boldsymbol{\nabla})\times(\mathbf{r}\times\mathbf{E}) &= [\mathbf{E}\cdot(\mathbf{r}\times\boldsymbol{\nabla})]\mathbf{r} - \mathbf{E}[(\mathbf{r}\times\boldsymbol{\nabla})\cdot\mathbf{r}] \\ &= [(\mathbf{E}\times\mathbf{r})\cdot\boldsymbol{\nabla})]\mathbf{r} - \mathbf{E}[\mathbf{r}\cdot(\boldsymbol{\nabla}\times\mathbf{r})] \\ &= \mathbf{E}\times\mathbf{r}.\end{aligned} \tag{1.44}$$

For $\mathbf{r}$ constant,

$$\begin{aligned}(\mathbf{r}\times\boldsymbol{\nabla})\times(\mathbf{r}\times\mathbf{E}) &= \mathbf{r}[(\mathbf{r}\times\boldsymbol{\nabla})\cdot\mathbf{E}] - [\mathbf{r}\cdot(\mathbf{r}\times\boldsymbol{\nabla})]\mathbf{E} \\ &= \mathbf{r}[\mathbf{r}\cdot(\boldsymbol{\nabla}\times\mathbf{E})] - [(\mathbf{r}\times\mathbf{r})\cdot\boldsymbol{\nabla}]\mathbf{E} \\ &= 0.\end{aligned} \tag{1.45}$$

Adding the results of Eqs. (43) and (44) gives

$$(\mathbf{r}\times\boldsymbol{\nabla})\times(\mathbf{r}\times\mathbf{E}) = -\mathbf{r}\times\mathbf{E}, \tag{1.46}$$

which confirms that $\mathbf{L}\times\mathbf{L}\phi = i\mathbf{L}\phi$.

Chapter 2

Problem 2.1

The electric field just outside the conducting shell equals E_{out}, while the field just inside is 0. Consider the field just outside the small circular section of the shell that will be drilled out as being of two parts: A part E_1 coming from the small circular section, and a part E_2 coming from the rest of the shell. Just outside of the shell, the total field is given by $E_{\text{out}} = E_1 + E_2$. Moving from just outside the shell to a point just inside the shell, E_2 remains about the same, but E_1 will change to the opposite direction because this position is on the other side of the small circular portion. Then the field just inside inside the small circular portion of the shell is given by $E_{\text{in}} = E_2 - E_1 = 0$, which requires $E_1 = E_2 = \frac{1}{2}E_{\text{out}}$. The field in the hole when the circular portion is removed is due only to E_2, which is half of the original field.

Problem 2.3:

(a) The potential energy of a uniformly charged sphere of charge Q and radius R is given by (using the answer to problem 1.13)

$$
\begin{aligned}
U &= \frac{1}{2} \int \rho \phi d\tau = \frac{1}{2} \int_{r \leq R} \left(\frac{3Q}{4\pi R^3} \right) \left[\frac{Q(3R^2 - r^2)}{2R^3} \right] d\tau \\
&= \frac{3Q^2}{16\pi R^6} \int_0^R (3R^2 - r^2) 4\pi r^2 dr \\
&= \frac{3Q^2}{4R^6} \left(\frac{3R^5}{3} - \frac{R^5}{5} \right) = \frac{3Q^2}{5R}.
\end{aligned}
\tag{2.1}
$$

(b) The potential energy is also given by integrating E^2:

$$
\begin{aligned}
U &= \frac{1}{8\pi} \int E^2 d\tau \\
&= \frac{1}{8\pi} \int_{r \leq R} \left(\frac{Qr}{R^3} \right)^2 d\tau + \frac{1}{8\pi} \int_{r \geq R} \left(\frac{Q}{r^2} \right)^2 d\tau \\
&= \frac{Q^2}{2} \left[\int_0^R \frac{r^4 dr}{R^6} + \int_R^\infty \frac{dr}{r^2} \right] = \frac{Q^2}{2R} \left[\frac{1}{5} + 1 \right] \\
&= \frac{3Q^2}{5R},
\end{aligned}
\tag{2.2}
$$

the same as in part (a).

Problem 2.5:

(a) Since the combined system has no charge, the potential on the outer sphere is zero. The potential on the inner sphere is given by

$$\phi(A) = \int_A^B \frac{q\,dr}{r^2} = \frac{q}{A} - \frac{q}{B}. \tag{2.3}$$

Thus the potential energy is

$$U = q\phi(A) = q^2 \left[\frac{1}{A} - \frac{1}{B} \right]. \tag{2.4}$$

(b) The only electric field is between the two spherical shells, where $E = q/r^2$. Then

$$U = \frac{1}{4\pi} \int d\tau\, E^2 = \int_A^B r^2 dr \frac{q^2}{r^4} = q^2 \left[\frac{1}{A} - \frac{1}{B} \right]. \tag{2.5}$$

Problem 2.7:

(a) The electric field a distance $\mathbf{r}$ from a dipole $\mathbf{p_1}$ is

$$\mathbf{E_{p_1}} = \frac{3(\mathbf{p_1}\cdot\hat{\mathbf{r}})\hat{\mathbf{r}} - \mathbf{p_1}}{r^3}. \tag{2.6}$$

The spin torque on a dipole $\mathbf{p_2}$ at the point $\mathbf{r}$ is given by

$$
\begin{aligned}
\boldsymbol{\tau_2} &= \mathbf{p_2}\times\mathbf{E_{p_1}} = \frac{\mathbf{p_2}\times[3(\mathbf{p_1}\cdot\hat{\mathbf{r}})\hat{\mathbf{r}} - \mathbf{p_1}]}{r^3} \\
&= \frac{[3(\mathbf{p_1}\cdot\hat{\mathbf{r}})(\mathbf{p_2}\times\hat{\mathbf{r}}) - \mathbf{p_2}\times\mathbf{p_1}]}{r^3}.
\end{aligned} \tag{2.7}
$$

(b) For the spin torque on the dipole $\mathbf{p_1}$ due to $\mathbf{p_2}$, we interchange $\mathbf{p_1}$ and $\mathbf{p_2}$, and take $\mathbf{r} \to -\mathbf{r}$ to get

$$\boldsymbol{\tau_1} = \frac{[3(\mathbf{p_2}\cdot\hat{\mathbf{r}})(\mathbf{p_1}\times\hat{\mathbf{r}}) - \mathbf{p_1}\times\mathbf{p_2}]}{r^3}. \tag{2.8}$$

(c) The sum of the two spin torques is

$$\boldsymbol{\tau}_{\text{spin}} = \boldsymbol{\tau_1} + \boldsymbol{\tau_2} = \frac{3(\mathbf{p_2}\cdot\hat{\mathbf{r}})(\mathbf{p_1}\times\hat{\mathbf{r}}) + 3(\mathbf{p_1}\cdot\hat{\mathbf{r}})(\mathbf{p_2}\times\hat{\mathbf{r}})}{r^3}. \tag{2.9}$$

The force on dipole $\mathbf{p_2}$ due to dipole $\mathbf{p_1}$ is

$$\mathbf{F_{21}} = \frac{3(\mathbf{p_1}\cdot\hat{\mathbf{r}})\mathbf{p_2} + 3(\mathbf{p_2}\cdot\hat{\mathbf{r}})\mathbf{p_1} + 3(\mathbf{p_1}\cdot\mathbf{p_2})\hat{\mathbf{r}} - 15(\mathbf{p_1}\cdot\hat{\mathbf{r}})(\mathbf{p_2}\cdot\hat{\mathbf{r}})\hat{\mathbf{r}}}{r^4}. \tag{2.10}$$

Since the force $\mathbf{F}_{12}$ on dipole $\mathbf{p}_1$ due to dipole $\mathbf{p}_2$ is equal and opposite to $\mathbf{F}_{21}$, the two forces form a force couple, and we can take moments of the two forces about any fixed point. For simplicity, we take moments of the forces about the position of the dipole $\mathbf{p}_1$, so the torque of $\mathbf{F}_{12}$ vanishes, and the orbital torque on the two dipoles is given by

$$\boldsymbol{\tau}_{\text{orbital}} = \mathbf{r} \times \mathbf{F}_{21} = \frac{3(\mathbf{p}_1 \cdot \hat{\mathbf{r}})(\hat{\mathbf{r}} \times \mathbf{p}_2) + 3(\mathbf{p}_2 \cdot \hat{\mathbf{r}})(\hat{\mathbf{r}} \times \mathbf{p}_1)}{r^3}. \tag{2.11}$$

The sum of the spin torque and the orbital torque is

$$\boldsymbol{\tau}_{\text{net}} = \boldsymbol{\tau}_{\text{spin}} + \boldsymbol{\tau}_{\text{orbital}} = 0. \tag{2.12}$$

Problem 2.9:

(a) The quadrupole moment of a uniformly charged needle of length L and charge q is

$$Q_0 = \frac{1}{2} \int (2z^2 - x^2 - y^2) dq = \frac{1}{2} \frac{q}{L} \int_{-L/2}^{+L/2} 2z^2 dz = \frac{1}{12} q L^2, \tag{2.13}$$

where we have used $dq = (q/L)dz$.

(b) The quadrupole moment of a uniformly charged disk of radius R and charge q is

$$Q_0 = \frac{1}{2} \int (3\cos^2\theta - 1) r^2 dq = -\frac{q}{R^2} \int_0^R r^3 dr = -\frac{1}{4} q R^2, \tag{2.14}$$

where we have used $dq = (q/\pi R^2) 2\pi r dr = (2q/R^2) r dr$.

(c) The quadrupole moment of a spherical shell of radius R with a surface charge distribution $\sigma = (q/4\pi R^2) \cos^2\theta$ is

$$\begin{aligned}
Q_0 &= \frac{1}{2} \int (3\cos^2\theta - 1) r^2 dq \\
&= \frac{qR^4}{8\pi R^2} \oint (3\cos^2\theta - 1) \cos^2\theta d\Omega \\
&= \frac{qR^2}{4} \int_0^\pi (3\cos^4\theta - \cos^2\theta) \sin\theta d\theta \\
&= \frac{2}{15} q R^2, \tag{2.15}
\end{aligned}$$

where we have used $dq = \sigma R^2 d\Omega = \sigma R^2 \sin\theta d\theta d\phi$.

Problem 2.11:

The quadrupole force on a nucleus with quadrupole moment Q_0 a distance $\mathbf{r}$ from an electron of charge $-e$ is given by

$$
\begin{aligned}
\mathbf{F_Q} &= \frac{1}{2}Q_0(\hat{\mathbf{k}}\cdot\nabla)(\hat{\mathbf{k}}\cdot\nabla)\mathbf{E} \\
&= \frac{1}{2}Q_0(\hat{\mathbf{k}}\cdot\nabla)(\hat{\mathbf{k}}\cdot\nabla)\left(\frac{-e\mathbf{r}}{r^3}\right) \\
&= -\frac{1}{2}eQ_0(\hat{\mathbf{k}}\cdot\nabla)\left[\hat{\mathbf{k}}\cdot\left(\frac{\hat{\mathbf{n}}}{r^3}-\frac{3\mathbf{r}\mathbf{r}}{r^5}\right)\right] \\
&= -\frac{1}{2}eQ_0(\hat{\mathbf{k}}\cdot\nabla)\left[\frac{\hat{\mathbf{k}}}{r^3}-\frac{3(\hat{\mathbf{k}}\cdot\mathbf{r})\mathbf{r}}{r^5}\right] \\
&= -\frac{1}{2}eQ_0\hat{\mathbf{k}}\cdot\left[-\frac{3\hat{\mathbf{k}}\mathbf{r}}{r^5}-\frac{3\mathbf{r}\hat{\mathbf{k}}}{r^5}-\frac{3(\hat{\mathbf{k}}\cdot\mathbf{r})\hat{\mathbf{n}}}{r^5}+\frac{15(\hat{\mathbf{k}}\cdot\mathbf{r})\mathbf{r}\mathbf{r}}{r^7}\right] \\
&= \frac{1}{2}eQ_0\left[\frac{3\hat{\mathbf{r}}}{r^4}+\frac{3(\hat{\mathbf{k}}\cdot\hat{\mathbf{r}})\hat{\mathbf{k}}}{r^4}+\frac{3(\hat{\mathbf{k}}\cdot\hat{\mathbf{r}})\hat{\mathbf{k}}}{r^4}-\frac{15(\hat{\mathbf{k}}\cdot\mathbf{r})^2\hat{\mathbf{r}}}{r^4}\right] \\
&= -\frac{3eQ_0}{2r^4}[5(\hat{\mathbf{k}}\cdot\hat{\mathbf{r}})^2\hat{\mathbf{r}}-2(\hat{\mathbf{k}}\cdot\hat{\mathbf{r}})\hat{\mathbf{k}}-\hat{\mathbf{r}}]. \tag{2.16}
\end{aligned}
$$

The quadrupole force on the electron is given by

$$
\mathbf{F}_{el} = -e\mathbf{E_Q} = \frac{3eQ_0}{2r^4}[5(\hat{\mathbf{k}}\cdot\mathbf{r})^2\hat{\mathbf{r}}-2(\hat{\mathbf{k}}\cdot\hat{\mathbf{r}})\hat{\mathbf{k}}-\hat{\mathbf{r}}], \tag{2.17}
$$

showing that it is equal and opposite to the quadrupole force on the nucleus.

Problem 2.13:

(a) The energy of two symmetric quadrupoles, each of magnitude Q_0 with symmetry axes $\hat{\mathbf{k}}$ and $\hat{\mathbf{q}}$ respectively, a distance $\mathbf{r}$ apart, is given by

$$
\begin{aligned}
U &= -\frac{1}{2}Q_0(\hat{\mathbf{k}}\cdot\nabla)(\hat{\mathbf{k}}\cdot\mathbf{E}) = -\frac{3}{4}Q_0^2(\hat{\mathbf{k}}\cdot\nabla)\hat{\mathbf{k}}\cdot\left[\frac{5(\hat{\mathbf{q}}\cdot\mathbf{r})^2\mathbf{r}}{r^7}-\frac{2(\hat{\mathbf{q}}\cdot\mathbf{r})\hat{\mathbf{q}}}{r^5}-\frac{\mathbf{r}}{r^5}\right] \\
&= -\frac{3}{4}Q_0^2(\hat{\mathbf{k}}\cdot\nabla)\left[\frac{5(\hat{\mathbf{q}}\cdot\mathbf{r})^2(\hat{\mathbf{k}}\cdot\mathbf{r})}{r^7}-\frac{2(\hat{\mathbf{q}}\cdot\mathbf{r})(\hat{\mathbf{k}}\cdot\hat{\mathbf{q}})}{r^5}-\frac{(\hat{\mathbf{k}}\cdot\mathbf{r})}{r^5}\right] \\
&= -\frac{3}{4}Q_0^2\hat{\mathbf{k}}\cdot\left[\frac{10\hat{\mathbf{q}}(\hat{\mathbf{q}}\cdot\mathbf{r})(\hat{\mathbf{k}}\cdot\mathbf{r})}{r^7}+\frac{5(\hat{\mathbf{q}}\cdot\mathbf{r})^2\hat{\mathbf{k}}}{r^7}-\frac{35(\hat{\mathbf{q}}\cdot\mathbf{r})^2(\hat{\mathbf{k}}\cdot\mathbf{r})\mathbf{r}}{r^9}\right. \\
&\qquad\qquad \left.-\frac{2\hat{\mathbf{q}}(\hat{\mathbf{k}}\cdot\hat{\mathbf{q}})}{r^5}+\frac{10(\hat{\mathbf{q}}\cdot\mathbf{r})(\hat{\mathbf{k}}\cdot\hat{\mathbf{q}})\mathbf{r}}{r^7}-\frac{\hat{\mathbf{k}}}{r^5}+\frac{5(\hat{\mathbf{k}}\cdot\mathbf{r})\mathbf{r}}{r^7}\right] \\
&= \frac{3Q_0^2}{4r^5}[35(\hat{\mathbf{k}}\cdot\hat{\mathbf{r}})^2(\hat{\mathbf{q}}\cdot\hat{\mathbf{r}})^2-20(\hat{\mathbf{k}}\cdot\hat{\mathbf{r}})(\hat{\mathbf{q}}\cdot\hat{\mathbf{r}})(\hat{\mathbf{k}}\cdot\hat{\mathbf{q}}) \\
&\qquad\qquad -5(\hat{\mathbf{k}}\cdot\hat{\mathbf{r}})^2-5(\hat{\mathbf{q}}\cdot\hat{\mathbf{r}})^2+2(\hat{\mathbf{k}}\cdot\hat{\mathbf{q}})^2+1] \tag{2.18}
\end{aligned}
$$

(b) When $\hat{\mathbf{k}}$ and $\hat{\mathbf{q}}$ are parallel, and each perpendicular to $\mathbf{r}$,

$$U = \frac{9Q_0^2}{4r^5}. \tag{2.19}$$

(c) When $\hat{\mathbf{k}}$ and $\hat{\mathbf{q}}$ are parallel, and each parallel to $\mathbf{r}$,

$$U = \frac{3Q_0^2}{4r^5}[35 - 20 - 5 - 5 + 2 + 1] = \frac{6Q_0^2}{r^5}. \tag{2.20}$$

Chapter 3

Problem 3.1:

For a point charge q a distance d from a grounded plane , the electric field is that of the original point charge and an image charge $-q$ at a distance d directly behind the plane. We take the grounded plane as the x-y plane with the point charge q on the positive z-axis.

(a) The surface charge density on the plane is given by [Eq. (3.21)]:

$$\sigma(\rho) = \frac{1}{4\pi}E_\perp = \frac{-qd}{2\pi(\rho^2 + d^2)^{3/2}}, \tag{3.1}$$

where ρ is the distance in the plane from the z-axis. The total charge induced on the plane is given by

$$Q = \int \sigma dA = \int_0^\infty 2\pi \rho d\rho \left[\frac{-qd}{2\pi(\rho^2 + d^2)^{3/2}} \right] = -q. \tag{3.2}$$

(b) The force on the original charge q is given by $\mathbf{F} = q\mathbf{E}$, where $\mathbf{E}$ is found by integrating Coulomb's law over the surface charge distribution. By symmetry, the force will be in the z direction with

$$\begin{aligned} F_z &= \int_0^\infty 2\pi \rho d\rho \left[\frac{-qd}{2\pi(\rho^2 + d^2)^{3/2}} \right] \left[\frac{d}{(\rho^2 + d^2)^{3/2}} \right] \\ &= -qd^2 \int_0^\infty \frac{\rho d\rho}{(\rho^2 + d^2)^3} = -\frac{q}{4d^2}, \end{aligned} \tag{3.3}$$

the same as the force between the original point charge and the image charge.

Problem 3.3:

A dipole $\mathbf{p}$ is a distance $\mathbf{d}$ from a grounded plane, and makes an angle of 60° with the vector $\mathbf{d}$. The appropriate image dipole to keep the plane at zero potential can be deduced by thinking of the dipole as $(+)$ and $(-)$ point charges as in Figure 2.2. Then, the image dipole $\mathbf{p}'$ will equal $\mathbf{p}$ in magnitude and make an angle of 60° in the opposite direction from the original dipole's angle. The angle between the two dipoles will thus be 120°.

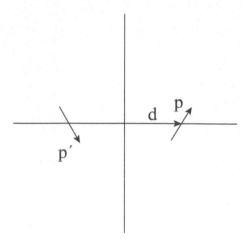

(a) The energy of the system of the original dipole **p** and the grounded plane is given in terms of the image dipole **p'** by

$$U = -\frac{1}{2}\mathbf{p}\cdot\mathbf{E_{p'}} = \frac{1}{2(2d)^3}[\mathbf{p}\cdot\mathbf{p'} - 3(\mathbf{p}\cdot\hat{\mathbf{d}})(\mathbf{p'}\cdot\hat{\mathbf{d}}].\qquad(3.4)$$

Note that this is a factor $\frac{1}{2}$ of the value for U given in Eq. (2.63) for the energy of two real dipoles. The reason for this is the same as discussed in Section 3.2.1 for the energy of a point charge outside a grounded plane. The actual charge distribution on the grounded plane is at zero potential and does not contribute in the integral $U = \frac{1}{2}\int \phi dq$. Thus the energy of the dipole-grounded plane system is

$$U = \frac{p^2}{16d^3}[\cos(120°) - 3\cos^2(60°)] = -\frac{5p^2}{64d^2}.\qquad(3.5)$$

(b) The force on the dipole **p** is given by [Eq. (2.64)]:

$$
\begin{aligned}
\mathbf{F} &= \frac{3(\mathbf{p}\cdot\hat{\mathbf{d}})\mathbf{p'} + 3(\mathbf{p'}\cdot\hat{\mathbf{d}})\mathbf{p} + 3(\mathbf{p}\cdot\mathbf{p'})\hat{\mathbf{d}} - 15(\mathbf{p}\cdot\hat{\mathbf{d}})(\mathbf{p'}\cdot\hat{\mathbf{d}})\hat{\mathbf{d}}}{(2d)^4} \\[4pt]
&= \frac{3}{16d^4}[p\cos(60°)\mathbf{p'} + p'\cos(60°)\mathbf{p} + pp'\cos(120°)\hat{\mathbf{d}} - 5pp'\cos^2(60°)\hat{\mathbf{d}} \\[4pt]
&= \frac{3p^2\hat{\mathbf{d}}}{16d^4}[2\cos^2(60°) + \cos(120°) - 5\cos^2(60°)] \\[4pt]
&= -\frac{15p^2\hat{\mathbf{d}}}{64d^4},\qquad\qquad\qquad\qquad\qquad\qquad\qquad\qquad(3.6)
\end{aligned}
$$

where we have used the fact that the magnitudes p and p' are equal and that
$\mathbf{p} + \mathbf{p'} = 2p\cos(60°)\hat{\mathbf{d}}$.

(c) The spin torque on the dipole is

$$\boldsymbol{\tau} = \mathbf{p} \times \mathbf{E}_{\mathbf{p}'} = \mathbf{p} \times \left[\frac{3(\mathbf{p}' \cdot \hat{\mathbf{d}})\hat{\mathbf{d}} - \mathbf{p}'}{(2d)^3} \right] = \frac{[3(\mathbf{p}' \cdot \hat{\mathbf{d}})(\mathbf{p} \times \hat{\mathbf{d}}) - \mathbf{p} \times \mathbf{p}'}{8d^3}]. \quad (3.7)$$

For the case when $\mathbf{p}$ makes an angle of $60°$ with $\hat{\mathbf{d}}$, the torque on $\mathbf{p}$ is

$$\boldsymbol{\tau} = \frac{p^2 \hat{\mathbf{n}}}{8d^3} [3 \cos(60°) \sin(60°) - \sin(120°)] = \frac{\sqrt{3}p^2 \hat{\mathbf{n}}}{32d^3}, \quad (3.8)$$

where $\hat{\mathbf{n}}$ is a unit vector in the direction of $\mathbf{p} \times \mathbf{d}$. This torque tends to rotate the dipole toward the $\hat{\mathbf{d}}$ direction.

(d) The dipole will accelerate in a straight line toward the grounded plane. The dipole's angle with the $\hat{\mathbf{d}}$ direction will oscillate with increasing frequency and amplitude as the magnitude of d decreases.

Problem 3.5:

(a) The charge induced on the surface of a grounded sphere of radius a by a point charge q at a distance $\mathbf{d}$ $(d > a)$ from the center of the sphere is given by $4\pi\sigma = E_\perp$, where $E_\perp$ is the normal component of the field at the surface of the sphere due to the original charge and an image charge $q' = -aq/d$ at a distance $\mathbf{d}' = (a^2/d)\hat{\mathbf{d}}$ from the center of the sphere. Thus

$$\sigma = \frac{1}{4\pi} \hat{\mathbf{r}} \cdot \left[\frac{q(a\hat{\mathbf{r}} - \mathbf{d})}{|a\hat{\mathbf{r}} - \mathbf{d}|^3} + \frac{(q')(a\hat{\mathbf{r}} - \mathbf{d}')}{|a\hat{\mathbf{r}} - \mathbf{d}'|^3} \right]. \quad (3.9)$$

This is a suitable form in vector notation for the surface charge density, but we can also write it in algebraic form as

$$
\begin{aligned}
\sigma &= \frac{1}{4\pi} \left[\frac{q(a - d\cos\theta)}{(a^2 + d^2 - 2ad\cos\theta)^{3/2}} - \frac{(qa/d)[a - (a^2/d)\cos\theta]}{[a^2 + a^4/d^2 - (2a^3/d)\cos\theta]^{3/2}} \right] \\
&= \frac{1}{4\pi} \left[\frac{q(a - d\cos\theta)}{(a^2 + d^2 - 2ad\cos\theta)^{3/2}} - \frac{(qd/a)(d - a\cos\theta)}{(d^2 + a^2 - 2ad\cos\theta)^{3/2}} \right] \\
&= \frac{q}{4\pi a} \left[\frac{a^2 - d^2}{(a^2 + d^2 - 2ad\cos\theta)^{3/2}} \right], \quad (3.10)
\end{aligned}
$$

where θ is the angle between $\hat{\mathbf{r}}$ and $\hat{\mathbf{d}}$.

(b) The total charge induced on the sphere is given by the integral

$$
\begin{aligned}
Q &= a^2 \oint \sigma d\Omega = 2\pi a^2 \int_0^\pi \sigma \sin\theta d\theta \\
&= \frac{qa(a^2 - d^2)}{2} \int_0^\pi \frac{\sin\theta d\theta}{(a^2 + d^2 - 2ad\cos\theta)^{3/2}} \\
&= \frac{-qa(a^2 - d^2)}{2d} \left[(a^2 + d^2 - 2ad\cos\theta)^{-1/2} \right]_0^\pi \\
&= \frac{q(a^2 - d^2)}{2d} \left[\frac{1}{|d - a|} - \frac{1}{d + a} \right] = -\frac{qa}{d} \quad \text{for } d > a. \quad (3.11)
\end{aligned}
$$

The integral over the actual surface charge thus equals the image charge used to reproduce the boundary condition at the spherical surface, as is required by Gauss's law.

Note: For a charge q within a hollow grounded sphere, with $d < a$, the absolute value $|d - a|$ would be $(a - d)$, and the total surface charge would equal $-q$, the negative of the charge inside, corresponding to zero electric field outside the grounded sphere.

(c) The energy of the charge q and the grounded sphere is given by

$$U = \frac{1}{2}q\phi_{q'} = \frac{1}{2}\frac{qq'}{(d - d')} = \frac{-q^2 a/d}{2(d - a^2/d)} = \frac{-q^2 a}{2(d^2 - a^2)}. \tag{3.12}$$

For $q = -e$, $d=2$ Å, and $a=1$ Å, the energy is

$$U = \frac{-(4.8 \times 10^{-10})(1) \times 300}{(2 \times 10^{-8})(2^2 - 1^2)} = -2.4 \text{ eV}. \tag{3.13}$$

Problem 3.7:

The Fourier cosine expansion for a function defined on the interval $(0, L)$ is

$$f(x) = \sum_n a_n \cos(n\pi x/L). \tag{3.14}$$

The coefficients a_n can be found using the orthogonality relation for cosine functions defined in the interval $(0, L)$:

$$\int_0^L \cos(n\pi x/L)\cos(n'\pi x/L)dx = \frac{L}{2}\delta_{nn'}, \quad n, n' > 0. \tag{3.15}$$

We multiply each side of Eq. (3.14) by $\cos(n'\pi x/L)$, and integrate with respect to x from 0 to L to get

$$\int_0^L f(x)\cos(n'\pi x/L)\,dx = \int_0^L \sum_n a_n \cos(n\pi x/L)\cos(n'\pi x/L)dx$$

$$= \sum_n \frac{L}{2}a_n\delta_{nn'} = \frac{L}{2}a'_n, \quad n' > 0. \tag{3.16}$$

In the above, we interchanged the sum and the integral, integrating before summing. The coefficents for the Fourier cosine series are given by

$$a_n = \frac{2}{L}\int_0^L f(x)\cos(n\pi x/L)dx, \quad n > 0. \tag{3.17}$$

The coefficient a_0 is a special case, given by

$$a_0 = \frac{1}{L}\int_0^L f(x)dx = \langle f \rangle. \tag{3.18}$$

That is, a_0 is the average value of $f(x)$ in the interval $(0, L)$.

Problem 3.9:

The exponential Fourier expansion for a function defined on the interval $(-L, L)$ is

$$f(x) = \sum_{-\infty}^{+\infty} c_n e^{i\pi n x/L}. \tag{3.19}$$

The coefficients c_n can be found by using the complex orthogonality of the exponential functions in the interval $(-L, L)$:

$$\int_{-L}^{L} e^{-i\pi n' x/L} e^{i\pi n x/L} dx = 2L\delta_{nn'}. \tag{3.20}$$

Multiplying each side of Eq. (3.19) by $e^{-i\pi n' x/L}$, and integrating with respect to x from
$-L$ to L, we get

$$\int_{-L}^{L} e^{-i\pi n' x/L} f(x) dx = \int_{-L}^{L} \sum_n c_n e^{-i\pi n' x/L} e^{i\pi n x/L} dx$$
$$= 2L \sum_n c_n \delta_{nn'} = 2L c_{n'}. \tag{3.21}$$

Thus the coefficents for the exponential Fourier series are given by

$$c_n = \frac{1}{2L} \int_{-L}^{L} e^{i\pi n x/L} f(x) dx. \tag{3.22}$$

Note that the integral for the coefficients of the exponential series is divided by the length of the interval and not half the length as for trigonometric series. Also, the function f(x) is integrated with the complex conjugate of the corresponding exponential in the Fourier sum.

We can write the exponential series as

$$f(x) = c_0 + \sum_{n>0}\{c_n e^{i\pi n x/L} + c_{-n} e^{-i\pi n x/L}\}$$
$$= c_0 + \sum_{n>0}\{c_n[\cos(\pi n x/L) + i\sin(2\pi n x/L)] + c_{-n}[\cos(2\pi n x/L) - i\sin(2\pi n x/L)\}$$
$$= c_0 + \sum_{n>0}\{(c_n + c_{-n})[\cos(2\pi n x/L) + i(c_n - c_{-n})\sin(2\pi n x/L)\}. \tag{3.23}$$

Comparing this with the trigonemetric Fourier series in problem 3.7, we see that the coefficients are related by

$$a_0 = c_0, \text{ and for } n > 0 : a_n = c_n + c_{-n}, \quad b_n = i(c_n - c_{-n}). \tag{3.24}$$

Problem 3.11:

(a) The potential inside a cubical cavity of size $L \times L \times L$ with five sides grounded is given by

$$\phi(x, y, z) = \sum_{mn} \alpha_{mn} \sin(m\pi x/L) \sin(n\pi y/L) \sinh(\gamma_{mn} z), \qquad (3.25)$$

with γ_{mn} given by $\gamma_{mn} = (\pi/L)\sqrt{m^2 + n^2}$. For the top face kept at a constant potential $V(x, y) = V_0$, the coefficients α_{mn} are given by

$$
\begin{aligned}
\alpha_{mn} &= \frac{4V_0}{L^2 \sinh(\gamma_{mn}L)} \int_0^L \int_0^L \sin(m\pi x/L) \sin(n\pi y/L) dx dy \\
&= \frac{16V_0}{\pi^2 mn \sinh(\gamma_{mn}L)}, \quad m, n \text{ odd} \qquad (3.26) \\
&= 0, \quad m \text{ or } n \text{ even.} \qquad (3.27)
\end{aligned}
$$

Thus the potential inside the cavity is given by

$$\phi(x, y, z) = \frac{16V_0}{\pi^2} \sum_{mn \text{ odd}} \frac{\sin(m\pi x/L) \sin(n\pi y/L) \sinh(\gamma_{mn} z)}{mn \sinh(\gamma_{mn}L)}. \qquad (3.28)$$

(b) The potential at the center of the cavity is

$$
\begin{aligned}
\phi(L/2, L/2, L/2) &= \frac{16V_0}{\pi^2} \sum_{mn \text{ odd}} \frac{\sin(m\pi/2) \sin(n\pi/2) \sinh(\gamma_{mn}L/2)}{mn \sinh(\gamma_{mn}L)} \\
&= \frac{8V_0}{\pi^2} \sum_{mn \text{ odd}} \frac{(-1)^{(m-n)/2}}{mn \cosh(\gamma_{mn}L/2)}, \qquad (3.29)
\end{aligned}
$$

where we have used the hyperbolic identity

$$\sinh(\gamma L) = 2 \sinh(\gamma L/2) \cosh(\gamma L/2). \qquad (3.30)$$

(c) Putting in numbers for

| (m,n)= | (1,1) | [(1,3)+(3,1)] | (3,3) | [(1,5)+(5,1)] |

$$
\begin{aligned}
\phi &= \frac{8V_0}{\pi^2} \left[\frac{1}{\cosh(\sqrt{2}\pi/2)} - \frac{2}{3\cosh(\sqrt{10}\pi/2)} + \frac{1}{9\cosh(\sqrt{18}\pi/2)} + \frac{2}{5\cosh(\sqrt{26}\pi/2)} \right] \\
&= (0.1738 - 0.0075 + 0.0003 + 0.0003)V_0 = 0.167V_0. \qquad (3.3)
\end{aligned}
$$

This equals $V_0/6$ (the average potential of the six faces) to three significant figures. Higher terms do not contribute to three significant figures.

Problem 3.13:

(a) **Note:** This derivation is the same as that in [Sec. 3.3.3] of the text, with the only change being that cosine is used instead of sine.
The Fourier cosine expansion for a function defined on the interval $(0, L)$ is

$$f(x) = \sum_n a_n \cos(n\pi x/L), \tag{3.32}$$

with the coefficients given by

$$a_n = \frac{2}{L} \int_0^L f(x) \cos(n\pi x/L)dx, \quad n > 0. \tag{3.33}$$

To take the limit as $L \to \infty$, we introduce a discrete variable k_n related to n by

$$k_n = \frac{n\pi}{L}. \tag{3.34}$$

Then the Fourier sum can be written as

$$f(x) = \sum_n a_n \cos(k_n x)\Delta n \tag{3.35}$$

where we have introduced the quantity Δn, which just equals one. From Eq. (3.34), we see that

$$\Delta n = \frac{L}{\pi}\Delta k_n, \tag{3.36}$$

so Eq. (3.35) can be written

$$f(x) = \sum_n \frac{L}{\pi}a_n \cos(k_n x)\Delta k_n. \tag{3.37}$$

A new expansion coefficient

$$F(k_n) = \frac{L}{\pi}a_n \tag{3.38}$$

can be introduced, and then

$$f(x) = \sum_n F(k_n) \cos(k_n x)\Delta k_n. \tag{3.39}$$

Now, we can take the limit as $L \to \infty$, $\Delta k_n \to 0$ in Eq. (3.39). Then the sum in Eq. (3.39) approaches the usual definition of an integral, and we can write

$$f(x) = \int_0^\infty F(k) \cos(kx)dk, \tag{3.40}$$

with k now being a continuous variable. In this limit, equation (3.33) for the expansion coefficient becomes

$$F(k) = \frac{2}{\pi} \int_0^\infty \cos(kx)f(x)dx. \tag{3.41}$$

(b) To extend the exponential Fourier sum for a function defined on the interval (-L,L) to an integral over the infinite interval $(-\infty, \infty)$, we use the same steps as in the part (a), using $e^{2\pi inx/L} \rightarrow e^{ikx}$ instead of $\cos(n\pi x/L) \rightarrow \cos(kx)$. Then the summation coefficients c_n become the Fourier transform function $F(k)$ given by

$$F(k) = \frac{1}{2\pi} \int_{-\infty}^{\infty} f(x)e^{-ikx}dx. \tag{3.42}$$

Problem 3.15:

For this problem, the boundary condition on the left end of the open channel described in problem 3.14 is $\phi(0, y) = V_0 y/H$. for which the expansion coefficients for the potential are given by

$$\alpha_n = \frac{2V_0}{H^2} \int_0^H y\sin(n\pi y/H)dy = \frac{-2V_0}{n\pi}\cos(n\pi) = (-1)^{(n-1)}\frac{2V_0}{n\pi}. \tag{3.43}$$

Then, the potential inside the open channel is

$$\phi(x, y) = \frac{2V_0}{\pi}\sum_n \frac{(-1)^{(n-1)}e^{-n\pi x/H}}{n}\sin(n\pi y/H). \tag{3.44}$$

For $x = H/2$, $y = H/2$, the potential is

$$\phi(H/2, H/2) = \frac{2V_0}{\pi}\sum_n \frac{(-1)^{(n-1)}e^{-n\pi/2}}{n}\sin(n\pi/2)$$

$$= \frac{2V_0}{\pi}\sum_{n \text{ odd}} \frac{(-1)^{(n-1)/2}e^{-n\pi/2}}{n}. \tag{3.45}$$

This is one half the potential $\phi(H/2, H/2)$ found in problem 3.14. This is because the constant potential of V_0 on the left face in problem 3.14 can be written as

$$V_0 = V_0 y/H + V_0(1 - y/H), \tag{3.46}$$

and each of these two boundary conditions leads to the same potential at any point with $y = H/2$. Thus, the potential $\phi(H/2, H/2)$ for this problem is one half the potential $\phi(H/2, H/2)$ for problem 3.14.

Problem 3.17:

The surface Green's function for the $x = 0$ face (the left end surface) of the open channel in Fig. 3.5 is found by the following steps:

1. The potential is written in terms of the surface Green's function g(x,y;0,y')

$$\phi(x, y) = \int_0^H g(x, y; 0, y')\phi(0, y')dy'. \tag{3.47}$$

2. We introduce the delta function for the set of discrete eigenfunctions $\sin(n\pi y/L)$:

$$\delta(y - y') = (2/H) \sum \sin(n\pi y/H) \sin(n\pi y'/H). \qquad (3.48)$$

3. We introduce the appropriate function of x into the sum to make it a solution of Laplace's equation:

$$g(x, y; 0, y') = (2/H) \sum \sin(n\pi y/H) \sin(n\pi y'/H) e^{-n\pi x/H}, \qquad (3.49)$$

which is the surface Green's function.

Chapter 4

Problem 4.1:

(a) In spherical coordinates, the potential of a dipole becomes

$$\psi = \frac{\mathbf{p}\cdot\hat{\mathbf{r}}}{r^2} = \frac{p\cos\theta}{r^2}, \tag{4.1}$$

where we have taken the z-axis in the direction of the dipole.

(b) The spherical coordinates of the electric field are given by

$$E_r = = -\partial_r(p\cos\theta/r^2) = 2p\cos\theta/r^3 \tag{4.2}$$

$$E_\theta = -(1/r)\partial_\theta\psi = -(1/r)\partial_\theta(p\cos\theta/r^2) = p\sin\theta/r^3. \tag{4.3}$$

The vector expression for $\mathbf{E}$ is

$$\mathbf{E} = \frac{3(\mathbf{p}\cdot\hat{\mathbf{r}})\hat{\mathbf{r}} - \mathbf{p}}{r^3}. \tag{4.4}$$

The spherical components of this are

$$E_r = \hat{\mathbf{r}}\cdot\mathbf{E} = \frac{\hat{\mathbf{r}}\cdot[3(\mathbf{p}\cdot\hat{\mathbf{r}})\hat{\mathbf{r}} - \mathbf{p}]}{r^3} = 2p\cos\theta/r^3 \tag{4.5}$$

$$E_\theta = \hat{\boldsymbol{\theta}}\cdot\mathbf{E} = \frac{\hat{\boldsymbol{\theta}}\cdot[3(\mathbf{p}\cdot\hat{\mathbf{r}})\hat{\mathbf{r}} - \mathbf{p}]}{r^3} = \frac{-\hat{\boldsymbol{\theta}}\cdot\mathbf{p}}{r^3} = p\sin\theta/r^3. \tag{4.6}$$

(c) The spherical coordinates of the curl of $\mathbf{E}$ are

$$(\boldsymbol{\nabla}\times\mathbf{E})_r = \frac{\partial_\theta(r\sin\theta E_\phi) - \partial_\phi(rE_\theta)}{r^2\sin\theta} = 0 - \frac{\partial_\phi(2p\cos\theta/r^2)}{r^2\sin\theta} = 0 \tag{4.7}$$

$$(\boldsymbol{\nabla}\times\mathbf{E})_\theta = \frac{\partial_\phi E_r - \partial_r(r\sin\theta E_\phi)}{r\sin\theta} = \frac{\partial_\phi(2p\cos\theta/r^2) - 0}{r\sin\theta} = 0 \tag{4.8}$$

$$(\boldsymbol{\nabla}\times\mathbf{E})_\phi = \frac{\partial_r(rE_\theta) - \partial_\theta E_r}{r} = \frac{\partial_r(p\sin\theta/r^2) - \partial_\theta(2p\cos\theta/r^3}{r}$$

$$= \frac{-2p\sin\theta/r^3 + 2p\sin\theta/r^3}{r} = 0. \tag{4.9}$$

The divergence of $\mathbf{E}$ in spherical coordinates is

$$\boldsymbol{\nabla}\cdot\mathbf{E} = \frac{1}{r^2\sin\theta}\left[\partial_r(r^2\sin\theta E_r) + \partial_\theta(r\sin\theta E_\theta) + \partial_\phi(rE_\phi)\right]$$

$$= \frac{1}{r^2}\partial_r\left(\frac{2p\cos\theta}{r}\right) + \frac{1}{r\sin\theta}\partial_\theta\left(\frac{p\sin^2\theta}{r^3}\right) + 0$$

$$= \frac{-2p\cos\theta}{r^4} + \frac{2p\cos\theta}{r^4} = 0, \quad r \neq 0. \tag{4.10}$$

Problem 4.3:

We start with [Eq. (4.73)]:

$$(1 - 2tx + t^2) \sum_n nP_n(x)t^{n-1} = (x - t) \sum_m P_m(x)t^m, \qquad (4.11)$$

which follows from the generating function. This can be rewritten as

$$\sum_n P_n(x)(nt^{n-1} - 2xnt^n + nt^{n+1}) = \sum_m P_m(x)(xt^m - t^{m+1}). \qquad (4.12)$$

Next, we bring all the terms to the left hand side:

$$\sum_n P_n(x)[nt^{n-1} - (2nx + x)t^n + (n + 1)t^{n+1}] = 0. \qquad (4.13)$$

Now we change the summation index in each term to make each term go like t^n. In the first term, we make the change $n \to n + 1$, and, in the third term we make the change $n \to n - 1$. This leaves

$$\sum_n [(n + 1)P_{n+1} - (2n + 1)xP_n + nP_{n-1}]t^n = 0. \qquad (4.14)$$

Since coefficients of a power series are unique, the combination in square brackets must vanish, and we have

$$P_{l+1}(x) = \frac{(2l + 1)xP_l(x) - lP_{l-1}(x)}{l + 1} \qquad (4.15)$$

for any value of $l > 0$. This is the recursion relation in [Eq. (4.71)] of the text.

Problem 4.5:

(a) For the boundary condition $\psi(R, \theta) = \psi_0 \cos^2 \theta$ on a sphere of radius R, the potential outside the sphere is given by [Eq. (4.87)]:

$$\psi(r, \theta) = \sum_l \frac{b_l}{r^{l+1}} P_l(\cos \theta), \quad r \geq R, \qquad (4.16)$$

with the coefficients b_l given by

$$\begin{aligned} b_l &= \frac{(2l + 1)R^{l+1}}{2} \int_{-1}^{+1} V(\theta)P_l(\cos \theta)d(\cos \theta) \\ &= \frac{(2l + 1)R^{l+1}\psi_0}{2} \int_{-1}^{+1} x^2 P_l(x)d(x), \quad \text{with } x = \cos \theta. \end{aligned} \qquad (4.17)$$

To do this integral, we recognize that

$$x^2 = \frac{3x^2 - 1}{3} + \frac{1}{3} = \frac{2}{3}P_2(x) + \frac{1}{3}P_0(x). \qquad (4.18)$$

Then we can use the orthogonality of the Legendre polynomials to get

$$b_0 = \frac{RV_0}{2}\int_{-1}^{+1}\frac{1}{3}[P_0(x)]^2 dx = \frac{R}{3}V_0 \tag{4.19}$$

$$b_2 = \frac{5R^3 V_0}{2}\int_{-1}^{+1}\frac{2}{3}[P_2(x)]^2 dx = \frac{2R^3}{3}V_0 \tag{4.20}$$

$$b_l = 0, \quad l \neq 0 \text{ or } 2. \tag{4.21}$$

Then, $$\psi(r,\theta) = \frac{V_0}{3}\left[\left(\frac{R}{r}\right) + 2\left(\frac{R}{r}\right)^3 P_2(\cos\theta)\right], \quad r \geq R. \tag{4.22}$$

(b) The surface charge distribution on the sphere is given by [Eq. (4.88)]:

$$\sigma(\theta) = \frac{1}{4\pi}\sum_l \frac{(2l+1)b_l}{R^{l+2}}P_l(\cos\theta) = \frac{V_0}{4\pi R}\left[\frac{1}{3}P_0(\cos\theta) + \frac{10}{3}P_2(\cos\theta)\right]$$

$$= \frac{V_0}{12\pi R}[15\cos^2\theta - 4]. \tag{4.23}$$

(c) The quadrupole moment of the charged sphere is given by:

$$p_2 = \int r^2 P_2(\cos\theta)dq = 2\pi R^4 \int_{-1}^{1} P_2(\cos\theta)\,\sigma(\theta)\,d(\cos\theta)$$

$$= \frac{V_0 R^3}{2}\int_{-1}^{1} P_2(x)\left[\frac{1}{3}P_0(x) + \frac{10}{3}P_2(x)\right] dx = \frac{2}{3}V_0 R^3. \tag{4.24}$$

This is the same as the coefficient of $P_2(\cos\theta)/r^3$ in Eq. (4.22) for the potential.

Problem 4.7:

(a) The potential on the axis (chosen as the z-axis) of a uniformly charged ring of charge Q and radius R is

$$\psi(z) = \frac{Q}{\sqrt{z^2 + R^2}} \tag{4.25}$$

because every point on the ring is the same distance $\sqrt{z^2 + R^2}$ from the point z.

(b) We can write the potential in terms of the variables r and θ as

$$\psi(r,0°) = \frac{Q}{\sqrt{r^2 + R^2}}. \tag{4.26}$$

This can be expanded for $r > R$ as

$$\psi(r,0°) = Q[r^2+R^2]^{-1/2} = \frac{Q}{r}\left[1 + \frac{R^2}{r^2}\right]^{-1/2} = Q\sum_{n=0}^{\infty}\binom{-1/2}{n}\frac{R^{2n}}{r^{2n+1}}, \quad r > R. \tag{4.27}$$

We introduce the index $l = 2n$. Then l must be even and $n = l/2$. The expansion for $\psi(r, 0°)$ can be written as

$$\psi(r, 0°) = Q \sum_{\text{even } l} \binom{-1/2}{l/2} \frac{R^l}{r^{l+1}}, \quad r > R. \tag{4.28}$$

This can now be extended to all angles by multiplying each term in the sum by $P_l(\cos\theta)$:

$$\psi(r, \theta) = Q \sum_{\text{even } l} \binom{-1/2}{l/2} \frac{R^l P_l(\cos\theta)}{r^{l+1}}, \quad r > R. \tag{4.29}$$

For $r < R$ the potential still satifies Laplace's equation because there is no charge inside the ring. We can expand $\psi(r, 0°)$ in powers of r^2/R^2, and following steps similar to those above leads to

$$\psi(r, \theta) = Q \sum_{\text{even } l} \binom{-1/2}{l/2} \frac{r^l P_l(\cos\theta)}{R^{l+1}}, \quad r < R. \tag{4.30}$$

(c) We can identify each multipole moment as the coeffcent of the term $P_l(\cos\theta)/r^{l+1}$ in the sum in Eq. (4.29). This gives

$$p_l = Q\binom{-1/2}{l/2} R^l, \quad l \text{ even.} \tag{4.31}$$

$$= 0 \quad l \text{ odd.} \tag{4.32}$$

For example, the quadrupole moment is

$$p_2 = Q\binom{-1/2}{1} R^2 = -\frac{1}{2}QR^2. \tag{4.33}$$

(d) The multipole moments can also be found by integrating over the charge distribution of the ring:

$$p_l = \int r^l P_l(\cos\theta) dq = QR^l P_l(0), \tag{4.34}$$

The $P_l(0)$ enters because all the charge on the ring is at an angle of 90° from the z-axis. We can evaluate $P_l(0)$ by using the generating function:

$$\frac{1}{\sqrt{1 - 2xt + t^2}} = \sum_l t^l P_l(x), \quad t < 1. \tag{4.35}$$

Setting $x = 0$ results in

$$\frac{1}{\sqrt{1 + t^2}} = [1 + t^2]^{-1/2} = \sum_n \binom{-1/2}{n} t^{2n}$$

$$= \sum_{\text{even } l} \binom{-1/2}{l/2} t^l = \sum_l t^l P_l(0), \quad t < 1, \tag{4.36}$$

where we changed the index n to $l/2$. Equating the coefficients of the same power of t in the last two sums gives

$$P_l(0) = \begin{pmatrix} -1/2 \\ l/2 \end{pmatrix}, \quad l \text{ even.} \tag{4.37}$$

Then Eq. (4.34) gives the same result for the multipole moments as in part (c).

(e) A uniformly charged disk can be considered to be composed of uniformly charged rings. The potential of rings can be integrated to find the potential of the disc. Each ring has a radius x and a charge $dq = [Q/(\pi R^2)]2\pi x dx$, so the potential on the axis of the disk is

$$\phi_{\text{disk}}(z) = \frac{2Q}{R^2} \int_0^R \frac{x dx}{\sqrt{z^2 + x^2}} = \frac{2Q}{R^2} \left[\sqrt{z^2 + R^2} - z \right]. \tag{4.38}$$

To find the potential off the axis of the disk, we first expand the potential on the axis in a binomial expansion in powers of R/z.

$$\begin{aligned}
\phi_{\text{disk}}(z) &= \frac{2Qz}{R^2} \left[\sqrt{1 + R^2/z^2} - 1 \right] \\
&= \frac{2Q}{R} \sum_1^\infty \begin{pmatrix} 1/2 \\ n \end{pmatrix} \left(\frac{R}{z} \right)^{2n-1} \\
&= \frac{2Q}{R} \sum_{\text{even } l}^\infty \begin{pmatrix} 1/2 \\ \frac{l}{2} + 1 \end{pmatrix} \left(\frac{R}{z} \right)^{l+1}.
\end{aligned} \tag{4.39}$$

We have made substitution $l = 2n - 2$ to facilitate the expansion off the axis of the disk.

The potential of the disk satisfies Laplace's equation for the region $r > R$. That means each term in the series that goes like $\frac{1}{z^{l+1}}$ can be extended off the axis by simply multiplying that term by the Legendre polynomial $P_l(\cos\theta)$. This gives

$$\phi_{\text{disk}}(r > R, \theta) = 2Q \sum_{\text{even } l}^\infty \begin{pmatrix} 1/2 \\ \frac{l}{2} + 1 \end{pmatrix} \left[\frac{R^l P_l(\cos\theta)}{r^{l+1}} \right] \tag{4.40}$$

for the potential off the axis of the disk, but only in the region $r > R$.

In order find the potential off the axis of a disk for the region $r < R$, we first find the off-axis potential for a uniformly charged ring for each region, $r > R$ and $r < R$.

$$\phi_{\text{ring}}(z) = \frac{Q}{\sqrt{z^2 + R^2}} = (Q/z)[1 + R^2/z^2]^{-\frac{1}{2}}$$

$$= \frac{Q}{R} \sum_0^\infty \binom{-1/2}{n} \left(\frac{R}{z}\right)^{2n+1}$$

$$= \frac{Q}{R} \sum_{\text{even } l}^\infty \binom{-1/2}{\frac{l}{2}} \left(\frac{R}{z}\right)^{l+1}. \qquad (4.41)$$

$$\text{Then,} \quad \phi_{\text{ring}}(r > R, \theta) = Q \sum_{\text{even } l}^\infty \binom{-1/2}{\frac{l}{2}} \left(\frac{R^l P_l(\cos\theta)}{r^{l+1}}\right) \qquad (4.42)$$

is the off-axis potential for a uniformly charged ring for $r > R$.

There is no charge for $r < R$ in the interior of the ring, so the potential satisfies Laplace's equation there, and will have a Legendre polynomial multipole expansion. The off-axis potential for $r < R$ just interchanges r and R, giving

$$\phi_{\text{ring}}(r < R, \theta) = Q \sum_{\text{even } l}^\infty \binom{-1/2}{\frac{l}{2}} \left(\frac{r^l P_l(\cos\theta)}{R^{l+1}}\right). \qquad (4.43)$$

To get the off-axis potential for a uniformly charged disk, we follow the procedure we used in integrating over rings to get the on-axis potential for a disk, applied now for the off-axis potential.

$$\begin{aligned}
\phi_{\text{disk}}(r < R, \theta) &= \frac{2}{R^2} \int_0^r \phi_{\text{ring}}(r > x, \theta) x \, dx + \frac{2}{R^2} \int_r^R \phi_{\text{ring}}(r < x, \theta) x \, dx \\
&= \frac{2Q}{R^2} \sum_{\text{even } l}^\infty \binom{-1/2}{\frac{l}{2}} P_l(\cos\theta) \left[\int_0^r \frac{x^{l+1} dx}{r^{l+1}} + \int_r^R \frac{r^l dx}{x^l}\right] \\
&= \frac{2Q}{R^2} \sum_{\text{even } l}^\infty \binom{-1/2}{\frac{l}{2}} P_l(\cos\theta) \left[\frac{r}{l+2} + \frac{r}{l-1} - \frac{r^l}{(l-1)R^{(l-1)}}\right] \\
&= \frac{2Qr}{R^2} \sum_{\text{even } l}^\infty \binom{-1/2}{\frac{l}{2}} \frac{(2l+1)P_l(\cos\theta)}{(l+2)(l-1)} \\
&\quad - \frac{2Q}{R} \sum_{\text{even } l}^\infty \binom{-1/2}{\frac{l}{2}} \frac{r^l P_l(\cos\theta)}{(l-1)R^l}. \qquad (4.44)
\end{aligned}$$

Problem 4.9:

(a) If we substitute [Eq. (4.85)] for the coefficents b_l into [Eq. (4.83)] for the Legendre polynomial expansion of the potential $\psi(R, \theta)$, we get the expression

$$\begin{aligned}
\psi(R, \theta) &= \sum_l \frac{b_l}{R^{l+1}} P_l(\cos\theta) \\
&= \sum_l P_l(\cos\theta) \frac{(2l+1)}{2} \int_{-1}^{+1} \psi(R, \theta') P_l(\cos\theta') d(\cos\theta'). \quad (4.45)
\end{aligned}$$

Interchanging the order of summation and integration gives

$$\psi(R,\theta) = \int_{-1}^{+1} \psi(R,\theta')d(\cos\theta') \sum_l \frac{(2l+1)}{2} P_l(\cos\theta)P_l(\cos\theta'). \quad (4.46)$$

Inspection of the integral over $\cos\theta'$ shows that the sum is equal to the delta function in the interval $(-1,+1)$:

$$\delta(x-x') = \sum_l \frac{(2l+1)}{2} P_l(x)P_l(x'), \quad -1 < x < +1, \quad (4.47)$$

where we have introduced the variables $x = \cos\theta$ and $x' = \cos\theta'$.

(b) The potential inside a sphere of radius R with the boundary condition that $\psi(R,\theta)$ is specified is given in terms of a surface Green's function by

$$\psi(r,\theta) = \int_{-1}^{+1} g(r,\theta;R,\theta')\psi(R,\theta')d(\cos\theta'). \quad (4.48)$$

We form the delta function in $\cos\theta$ using the $P_l(\cos\theta)$, as in Eq. (4.47):

$$\delta(\cos\theta - \cos\theta') = \sum_l \frac{(2l+1)}{2} P_l(\cos\theta)P_l(\cos\theta') \quad (4.49)$$

To form the surface Green's function, we introduce a function $f(r,R)$ that makes g a solution of Laplace's equation, and satisfies $f(R,R) = 1$. The appropriate function (for $r < R$) is

$$f(r,R) = (r/R)^l. \quad (4.50)$$

Then the surface Green's function is given by

$$g(r,R;\theta,\theta') = \sum_l \frac{(2l+1)r^l}{2R^l} P_l(\cos\theta)P_l(\cos\theta'). \quad (4.51)$$

Problem 4.11:

(a) For the boundary condition

$$\begin{align} \phi(R,\theta) &= +V_0, \quad 0° < \theta < 180°, \quad (4.52)\\ \phi(R,\theta) &= -V_0, \quad 180° < \theta < 360°, \quad (4.53) \end{align}$$

on the surface of a long hollow cylinder, the potential inside the cylinder is

$$\phi(r,\theta) = \sum_n a_n (r/R)^n \sin(n\theta), \quad r < R, \quad (4.54)$$

with the coefficients given by

$$
\begin{aligned}
a_n &= \frac{1}{\pi}\int_0^{2\pi}\sin(n\theta)\phi(R,\theta)d\theta \\
&= \frac{V_0}{\pi}\left[\int_0^{\pi}\sin(n\theta)d\theta - \int_{\pi}^{2\pi}\sin(n\theta)d\theta\right] \\
&= \frac{V_0}{n\pi}\{[1-\cos(n\pi)] + [\cos(2n\pi)-\cos(n\pi)]\} \\
&= \frac{4V_0}{n\pi}, \quad n \text{ odd}, \\
&= 0, \quad\ n \text{ even}.
\end{aligned}
$$

$$(4.55)$$
$$(4.56)$$

(b) The potential outside the cylinder is

$$
\phi(r,\theta) = \sum_n a_n (R/r)^n \sin(n\theta), \quad r > R, \tag{4.57}
$$

with the same coefficients a_n. The surface charge on the cylinder is

$$
\begin{aligned}
\sigma(\theta) &= -\frac{1}{4\pi}\partial_r(\psi_{\text{out}} - \psi_{\text{in}}) = \sum_{\text{odd }n}\left(\frac{2n}{4\pi R}\right)\left(\frac{4V_0}{n\pi}\right)\sin(n\theta) \\
&= \frac{2V_0}{\pi^2 R}\sum_{\text{odd }n}\sin(n\theta).
\end{aligned}
\tag{4.58}
$$

This sum does not look convergent, but it can be summed by the following method:

$$
\begin{aligned}
\sum_{\text{odd }n}\sin(n\theta) &= \sum_{m=0}^{\infty}\sin[(2m+1)\theta] = \operatorname{Im}\sum_{m=0}^{\infty}e^{i[(2m+1)\theta]} \\
&= \operatorname{Im}\left[e^{i\theta}\sum_{m=0}^{\infty}e^{i2m\theta}\right] = \operatorname{Im}\left[\frac{e^{i\theta}}{1-e^{i2\theta}}\right] \\
&= \operatorname{Im}\left[\frac{1}{e^{-i\theta}-e^{i\theta}}\right] = \operatorname{Im}\left[\frac{i}{2\sin\theta}\right] \\
&= \frac{1}{2\sin\theta}.
\end{aligned}
\tag{4.59}
$$

Then the surface charge density can be written:

$$
\sigma = \frac{V_0}{\pi^2 R\sin\theta}. \tag{4.60}
$$

Problem 4.13:

(a) The potential inside a hollow cylinder of radius R and length $L = 2R$, with both ends grounded and the curved surface kept at a potential V is given by [Eq. 4.264] with no θ dependence:

$$
\psi(r,\theta,z) = \sum_n a_n I_0(n\pi r/L)\sin(n\pi z/L), \tag{4.61}
$$

with the coefficients α_n given by

$$
\begin{aligned}
\alpha_n &= \frac{2V}{LI_0(n\pi R/L)} \int_0^L dz\, \sin(n\pi z/L) \\
&= \frac{2V[1 - \cos(n\pi)]}{n\pi I_0(n\pi/2)} \\
&= \frac{4V}{n\pi I_0(n\pi/2)}, \quad n \text{ odd}, \tag{4.62} \\
&= 0, \quad n \text{ even}. \tag{4.63}
\end{aligned}
$$

The potential at the center of the cylinder is given by

$$
\begin{aligned}
\psi(0, \theta, L/2) &= \sum_{\text{odd } n} \alpha_n I_0(0) \sin(n\pi/2) \\
&= \frac{4V}{\pi} \sum_{\text{odd } n} \frac{(-1)^{(n-1)/2}}{n I_0(n\pi/2)}. \tag{4.64}
\end{aligned}
$$

(b) We use the numerical values:

$$
I_0(\pi/2) = 1.719, \qquad I_0(3\pi/2) = 21.11 \tag{4.65}
$$

for the terms in the summation:

$$
\begin{array}{ccc}
(n) = & (1) & (3)
\end{array}
$$

$$
\psi(0, \theta, L/2) = \quad (0.7408 - 0.0201)V = 0.721\,V. \tag{4.66}
$$

Higher terms do not contribute to three significant figures.

(c) Adding twice the value of the central potential from problem 4.12 to the value from this problem gives $0.279V_0 + 0.721V_0 = V_0$. This is because keeping every surface of the closed cylinder at a constant potential V_0 would make the potential constant inside the cylinder.

Chapter 5

Problem 5.1:

We use Green's reciprocity theorem to find the net charge induced on each of the grounded planes.

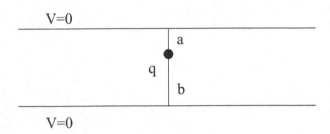

For situation 1, we have the charge q at the position shown, and the two planes at zero potential. For situation 2, we choose to keep the lower plane at potential $\phi_2(0) = 0$, and give the upper plane a constant potential $\phi_2(a+b) = \phi_0$, with no charge between the planes. In case 2, the potential at the point shown in the figure (a distance b above the bottom plane) will be $\phi_2(b) = [b/(a+b)]\phi_0$. Green's reciprocity theorem states

$$\int [\rho_1 \phi_2 - \rho_2 \phi_1] d\tau = \sum_n [Q_2 V_1 - Q_1 V_2]_n. \tag{5.1}$$

For the conditions 1 and 2 of this problem, the theorem reduces to

$$q\phi_2(b) - 0 = 0 - Q_{\text{top}}\phi_0, \tag{5.2}$$

where Q_{top} is the charge induced on the top plane. Solving for Q_{top} gives

$$Q_{\text{top}} = \frac{-q\phi_2(b)}{\phi_0} = -q\left(\frac{b}{a+b}\right). \tag{5.3}$$

A similar derivation gives $Q_{\text{bottom}} = -q[a/(a+b)]$.

Problem 5.3:

(a) For a horizontal infinite plane with the boundary condition

$$V(r, \theta, 0) = V_0, \quad r < a$$
$$V(r, \theta, 0) = 0, \quad r > a, \tag{5.4}$$

the surface Green's function solution for the potential is

$$\psi(\mathbf{r}) = \int g(\mathbf{r}, \mathbf{r}')\psi(\mathbf{r}')dA'$$

$$= \int \frac{z\psi(x', y', 0)dA'}{2\pi[(x' - x)^2 + (y' - y)^2 + z^2]^{\frac{3}{2}}}, \tag{5.5}$$

where the surface Green's function, derived by using an image charge behind the plane, is taken from [Eq. (5.25)]. We use cylindrical coordinates for the boundary condition (using ρ instead of r for the distance from the z-axis), and then the integral becomes

$$\psi(\rho, \theta, z) = \frac{V_0 z}{2\pi} \int_0^a \int_0^{2\pi} \frac{\rho' d\rho' d\theta'}{[(\rho'\cos\theta' - \rho\cos\theta)^2 + (\rho'\sin\theta' - \rho\sin\theta)^2 + z^2]^{\frac{3}{2}}}, \tag{5.6}$$

(b) For $\rho = 0$, the integral for $\psi(0, \theta, z)$ simplifies:

$$\psi(0, \theta, z) = \frac{V_0 z}{2\pi} \int_0^a \int_0^{2\pi} \frac{\rho' d\rho' d\theta'}{[\rho'^2 \cos^2\theta' + \rho'^2 \sin^2\theta' + z^2]^{\frac{3}{2}}}$$

$$= V_0 z \int_0^a \frac{\rho' d\rho'}{[\rho'^2 + z^2]^{\frac{3}{2}}}$$

$$= V_0 - \frac{V_0 z}{\sqrt{z^2 + a^2}}. \tag{5.7}$$

Note: This result can be compared to the potential on the axis of a uniformly charged disk in problem 1.9(b). That potential is $\psi(z) = (V_c/R)[\sqrt{z^2 + a^2} - z]$, where V_c is the potential at the center of the disk. The ratio of the two potentials (with $V_c = V_0$) is

$$\frac{\psi(\text{constant potential})}{\psi(\text{uniform charge})} = \frac{a}{\sqrt{z^2 + a^2}}, \tag{5.8}$$

which is always less than 1.

Problem 5.5:

The triple sum Green's function is given by

$$G(x, y, z; x', y', z') = \frac{32}{\pi L} \sum_{lmn} \left[\frac{\sin\left(\frac{l\pi x}{L}\right) \sin\left(\frac{m\pi y}{L}\right) \sin\left(\frac{n\pi z}{L}\right) \sin\left(\frac{l\pi x'}{L}\right) \sin\left(\frac{m\pi y'}{L}\right) \sin\left(\frac{n\pi z'}{L}\right)}{(l^2 + m^2 + n^2)} \right] \tag{5.9}$$

(a): The triple sum surface Green's function for the top surface is

$$g(x, y, z; x', y', L) = -\frac{1}{4\pi}\partial_z' G(x, y, z; x', y', z')|_{z'=L}$$

$$= -\frac{8}{\pi L^2} \sum_{lmn} \left[\frac{n \sin\left(\frac{l\pi x}{L}\right) \sin\left(\frac{m\pi y}{L}\right) \sin\left(\frac{n\pi z}{L}\right) \sin\left(\frac{l\pi x'}{L}\right) \sin\left(\frac{m\pi y'}{L}\right) \cos(n\pi)}{(l^2 + m^2 + n^2)} \right]. \tag{5.10}$$

(b): The potential inside the cavity is

$$\phi(x,y,z) = V_0 \int_0^L dx' \int_0^L dy' g(x.y.z; x', y', L))$$

$$= -\frac{8V_0}{\pi L^2} \int_0^L dx' \int_0^L dy' \sum_{lmn} \left[\frac{(-1)^n n \sin\left(\frac{l\pi x}{L}\right) \sin\left(\frac{m\pi y}{L}\right) \sin\left(\frac{n\pi z}{L}\right) \sin\left(\frac{l\pi x'}{L}\right) \sin\left(\frac{m\pi y'}{L}\right)}{(l^2 + m^2 + n^2)} \right]$$

$$= -\frac{8V_0}{\pi^3} \sum_{lmn} \left\{ \frac{(-1)^n n [1 - \cos(l\pi)] 1 - \cos(m\pi)] \sin\left(\frac{l\pi x}{L}\right) \sin\left(\frac{m\pi y}{L}\right) \sin\left(\frac{n\pi z}{L}\right)}{lm(l^2 + m^2 + n^2)} \right\}$$

$$= -\frac{32V_0}{\pi^3} \sum_{lmn} \left\{ \frac{(-1)^n n \sin\left(\frac{n\pi z}{L}\right)}{lm(l^2 + m^2 + n^2)} \right\} \quad l \text{ and } m \text{ odd.} \tag{5.11}$$

(c): The potential at the center of the cavity is

$$\phi(x,y,z) = -\frac{32V_0}{\pi^3} \sum_{lmn} \left[\frac{(-1)^n \sin\left(\frac{l\pi x}{L}\right) \sin\left(\frac{m\pi y}{L}\right) \sin\left(\frac{n\pi z}{L}\right)}{lm(l^2 + m^2 + n^2)} \right] \quad l \text{ and } m \text{ odd}$$

$$= \frac{32V_0}{\pi^3} \sum_{lmn} \left[\frac{(-1)^{\frac{l+m+n-3}{2}} n}{lm(l^2 + m^2 + n^2)} \right] \quad l, m, \text{ and } n \text{ odd}$$

$$= 0.167 V_0 \tag{5.12}$$

if the sum is evaluated on a computer. The sum is slowly convergent, but this can be improved by combining adjacent pairs of terms in the sum on n.

Chapter 6

Problem 6.1:

(a) The bound surface charge density in a dielectric lamina perpendicular to an electric field $\mathbf{E_0}$ is given by:

$$\sigma_b = \hat{\mathbf{n}} \cdot \mathbf{P} = \frac{1}{4\pi}(D - E) = \frac{1}{4\pi}(E_0 - E_0/\epsilon) = \frac{(\epsilon - 1)E_0}{4\pi\epsilon}, \tag{6.1}$$

where we have used the fact that $\hat{\mathbf{n}} \cdot \mathbf{D}$ is continuous at the face of the dielectric so $D = E_0$.

(b) In a parallel plate capacitor with charge Q and plate area A, D in the dielectric is given by $D = 4\pi\sigma = 4\pi Q/A$, so the bound surface charge density on the dielectric is

$$\sigma_b = \frac{1}{4\pi}(D - D/\epsilon) = \frac{(\epsilon - 1)}{\epsilon}\frac{Q}{A}. \tag{6.2}$$

Problem 6.3:

(a) In a coaxial cylindrical capacitor with inner radius A and outer radius B filled with a dielectric of permittivity ϵ, the electric field is radial (by symmetry), and Gauss's law shows that $E \sim 1/r$. Then E can be given as

$$E = \frac{AE_A}{r}, \tag{6.3}$$

where E_A is the value of E at the inner surface. The potential difference between the two cylinders is

$$V = \int_A^B \mathbf{E} \cdot d\mathbf{r} = AE_A \int_A^B \frac{dr}{r} = AE_A \ln(B/A). \tag{6.4}$$

The surface charge on the inner cylinder is then given by

$$\sigma = \epsilon E_A/4\pi, \tag{6.5}$$

and the charge per unit length on the cylinder is

$$Q/L = 2\pi A\sigma = \epsilon AE_A/2. \tag{6.6}$$

The capacitance per unit length (C/L) is given by

$$C/L = \frac{Q}{LV} = = \frac{\epsilon AE_A/2}{AE_A \ln(B/A)} = \frac{\epsilon}{2\ln(B/A)}. \tag{6.7}$$

(b) If the dielectric fills only the lower half of the capacitor, we can assume that the electric field will still be radial. This is because, with the dielectric filling just half of the cavity, a radial electric fill will not induce any bound charge on the surface of the dielectric Maxwell's equation $\nabla \times \mathbf{E} = 0$ means that $\mathbf{E}$ cannot change perpendicular to its direction, so the magnitude of E will be independent of angle. Thus, the electric field between the cylinders will still be give by Eq. (6.3), and the potential difference between the two spheres will still be given by Eq. (6.4). The only difference will be that Eq. (6.6) will change to

$$Q/L = (\epsilon + 1)AE_A/4, \tag{6.8}$$

because only half the volume between the cylinders has the dielectric. This gives the capacitance per unit length as

$$C/L = \frac{(1 + \epsilon)}{4 \ln(B/A)}. \tag{6.9}$$

Problem 6.5:

(a) The bound surface charge density on a dielectric sphere placed in a uniform field $\mathbf{E_0}$ is given by

$$\sigma_b = \frac{1}{4\pi}(\epsilon - 1)\hat{\mathbf{r}} \cdot \mathbf{E_I}, \tag{6.10}$$

where $\mathbf{E_I}$ is the electric field just inside the surface of the sphere. This field is given by [Eq. (6.68)]:

$$\mathbf{E_I}(\mathbf{r}) = \frac{3\mathbf{E_0}}{\epsilon + 2}. \tag{6.11}$$

Using this field, the surface charge density is

$$
\begin{aligned}
\sigma_b &= \frac{3}{4\pi}\left(\frac{\epsilon - 1}{\epsilon + 2}\right)\hat{\mathbf{r}} \cdot \mathbf{E_0} \\
&= \frac{3}{4\pi}\left(\frac{\epsilon - 1}{\epsilon + 2}\right)E_0 \cos\theta,
\end{aligned} \tag{6.12}
$$

where θ is the angle from the direction of $\mathbf{E_0}$.

(b) The induced dipole moment on the dielectric sphere (of radius R) is given by the integral

$$
\begin{aligned}
\mathbf{p} &= \int \mathbf{r}dq = \int \mathbf{r}\sigma dA = R^3 \int \hat{\mathbf{r}}\sigma d\Omega \\
&= \frac{3R^3}{4\pi}\left(\frac{\epsilon - 1}{\epsilon + 2}\right)\int \hat{\mathbf{r}}(\hat{\mathbf{r}} \cdot \mathbf{E_0})d\Omega.
\end{aligned} \tag{6.13}
$$

The angular integral is

$$\int \hat{\mathbf{r}}(\hat{\mathbf{r}}\cdot\mathbf{E_0})d\Omega = \mathbf{E_0}\int(\hat{\mathbf{E}}_0\cdot\hat{\mathbf{r}})(\hat{\mathbf{r}}\cdot\hat{\mathbf{E}}_0)d\Omega = \frac{4\pi}{3}\mathbf{E_0}, \tag{6.14}$$

and then

$$\mathbf{p} = \left(\frac{\epsilon-1}{\epsilon+1}\right)\mathbf{E_0}R^3, \tag{6.15}$$

in agreement with [Eq. (6.69)].

Problem 6.7:

(a) The dipole moment induced in a dielectric sphere of radius R by a dipole $\mathbf{p_0}$, a distance $\mathbf{d}$ away, is given by [Eq (6.83)]:

$$\mathbf{p} = \left(\frac{\epsilon-1}{\epsilon+2}\right)\frac{R^3}{d^3}[3(\mathbf{p_0}\cdot\hat{\mathbf{d}})\hat{\mathbf{d}} - \mathbf{p_0}]. \tag{6.16}$$

The force between the two dipoles is given by [Eq. (2.64)]:

$$\begin{aligned}\mathbf{F}_{\mathbf{pp_0}} &= \frac{3(\mathbf{p}\cdot\hat{\mathbf{d}})\mathbf{p_0} + 3(\mathbf{p_0}\cdot\hat{\mathbf{d}})\mathbf{p} + 3(\mathbf{p}\cdot\mathbf{p_0})\hat{\mathbf{d}} - 15(\mathbf{p}\cdot\hat{\mathbf{d}})(\mathbf{p_0}\cdot\hat{\mathbf{d}})\hat{\mathbf{d}}}{d^4} \\ &= \frac{3R^3}{d^7}\left(\frac{\epsilon-1}{\epsilon+2}\right)[f_1 + f_2 + f_3 + f_4], \end{aligned} \tag{6.17}$$

where the f_i are four terms we will evaluate in turn.

$$\begin{aligned}f_1 &= \{[3(\mathbf{p_0}\cdot\hat{\mathbf{d}})\hat{\mathbf{d}} - \mathbf{p_0}]\cdot\hat{\mathbf{d}}\}\mathbf{p_0} = 2(\mathbf{p_0}\cdot\hat{\mathbf{d}})\mathbf{p_0} && (6.18)\\ f_2 &= (\mathbf{p_0}\cdot\hat{\mathbf{d}})[3(\mathbf{p_0}\cdot\hat{\mathbf{d}})\hat{\mathbf{d}} - \mathbf{p_0}] = 3(\mathbf{p_0}\cdot\hat{\mathbf{d}})^2\hat{\mathbf{d}} - (\mathbf{p_0}\cdot\hat{\mathbf{d}})\mathbf{p_0} && (6.19)\\ f_3 &= \{[3(\mathbf{p_0}\cdot\hat{\mathbf{d}})\hat{\mathbf{d}} - \mathbf{p_0}]\cdot\mathbf{p_0})\}\hat{\mathbf{d}} = 3(\mathbf{p_0}\cdot\hat{\mathbf{d}})^2\hat{\mathbf{d}} - p_0^2\hat{\mathbf{d}} && (6.20)\\ f_4 &= -5\{[(3\mathbf{p_0}\cdot\hat{\mathbf{d}})\hat{\mathbf{d}} - \mathbf{p_0}]\cdot\hat{\mathbf{d}}\}(\mathbf{p_0}\cdot\hat{\mathbf{d}})\hat{\mathbf{d}} = -10(\mathbf{p_0}\cdot\hat{\mathbf{d}})^2\hat{\mathbf{d}}. && (6.21)\end{aligned}$$

Adding these terms up gives

$$\mathbf{F} = \frac{-3R^3}{d^7}\left(\frac{\epsilon-1}{\epsilon+2}\right)[4(\mathbf{p_0}\cdot\hat{\mathbf{d}})^2\hat{\mathbf{d}} + p_0^2\hat{\mathbf{d}} - (\mathbf{p_0}\cdot\hat{\mathbf{d}})\mathbf{p_0}], \tag{6.22}$$

which agrees with [Eq. (6.84)].

(b) Averaging over the polarization of $\mathbf{p_0}$ gives

$$\langle(\mathbf{p_0}\cdot\hat{\mathbf{d}})^2\rangle = \frac{1}{4\pi}\int(\mathbf{p_0}\cdot\hat{\mathbf{d}})^2 d\Omega_{\mathbf{p_0}} = \frac{1}{3}p_0^2 \tag{6.23}$$

$$\langle(\mathbf{p_0}\cdot\hat{\mathbf{d}})\mathbf{p_0}\rangle = \langle(\mathbf{p_0}\cdot\hat{\mathbf{d}})(\mathbf{p_0}\cdot\hat{\mathbf{d}})\rangle\hat{\mathbf{d}} = \frac{1}{3}p_0^2\hat{\mathbf{d}}. \tag{6.24}$$

Then, the force between unpolarized dipoles is

$$\mathbf{F} = -6p_0^2\left(\frac{\epsilon-1}{\epsilon+2}\right)\frac{R^3}{d^7}\hat{\mathbf{d}}, \tag{6.25}$$

in agreement with Eq. (6.85).

Problem 6.9:

The average dipole moment of a Boltzmann distribution of dipoles in an electric field is given by [Eq. (6.102)]:

$$\langle \mathbf{p} \rangle = N \oint \mathbf{p_0} \exp\left[\frac{\mathbf{p_0 \cdot E}}{kT}\right] d\Omega_{\mathbf{p_0}} \approx N \oint \mathbf{p_0} \left[1 + \frac{\mathbf{p_0 \cdot E}}{kT}\right] d\Omega_{\mathbf{p_0}}. \qquad (6.26)$$

We have expanded the exponential to first order for energies small compared to kT, The first term averages to zero, leaving

$$\langle \mathbf{p} \rangle = \frac{N}{kT} \oint \mathbf{p_0}(\mathbf{p_0 \cdot E}) d\Omega_{\mathbf{p_0}} = \frac{NE}{kT} \oint (\hat{\mathbf{E}} \cdot \mathbf{p_0})(\mathbf{p_0} \cdot \hat{\mathbf{E}}) d\Omega_{\mathbf{p_0}} = \frac{4\pi N p_0^2}{3kT} \mathbf{E}. \qquad (6.27)$$

To this order in the expansion, the normalization constant is determined by

$$1 = N \oint d\Omega_{\mathbf{p_0}} = 4\pi N, \qquad (6.28)$$

so the average dipole moment is

$$\langle \mathbf{p} \rangle = \frac{p_0^2}{3kT} \mathbf{E}, \quad \text{as given in Eq. (6.103).} \qquad (6.29)$$

Problem 6.11:

(a) The electric field in a dielectric sphere in an electric field $\mathbf{E_0}$ is given by Eq. (6.68):

$$\mathbf{E_I(r)} = \frac{3}{\epsilon + 2} \mathbf{E_0}. \qquad (6.30)$$

To get the field inside a spherical cavity in a dielectric that has an electric field $\mathbf{E_M}$, we just have to let $\epsilon \to 1/\epsilon$. Then

$$\mathbf{E_I(r)} = \frac{3\epsilon}{1 + 2\epsilon} \mathbf{E_M} \qquad (6.31)$$

is the field inside the cavity.

(b) The microscopic electric field in a dielectric that has a macroscopic field $\mathbf{E_M}$ is given by Eq. (6.89):

$$\mathbf{E_\mu(r)} = \mathbf{E_0} + \frac{4\pi}{3} \mathbf{P} = \mathbf{E_M} + \frac{1}{3}(\epsilon - 1)\mathbf{E_M} = \frac{\epsilon + 2}{3} \mathbf{E_M}. \qquad (6.32)$$

The difference of the two fields is

$$E_\mu - E_I = \frac{(\epsilon + 2)(2\epsilon + 1) - 9\epsilon}{3(2\epsilon + 1)} = \frac{2\epsilon^2 - 4\epsilon + 2}{3(2\epsilon + 1)} = \frac{2(\epsilon - 1)^2}{3(2\epsilon + 1)}, \qquad (6.33)$$

which is positive for all ϵ. The field in a cavity is weaker because the electric field in the medium gets weaker near the cavity.

Problem 6.13:

(a) For a spherical electrode of radius a in an extended medium, the electric field in the medium is $\mathbf{E} = aV\hat{\mathbf{r}}/r^2$. This produces a current density in the medium $\mathbf{j} = \sigma\mathbf{E} = \sigma aV\hat{\mathbf{r}}/r^2$, and the current leaving the electrode is given by $I = \oint \mathbf{j}\cdot d\mathbf{A} = 4\pi\sigma aV$. The resistance of the electrode is $R = V/I = 1/4\pi\sigma a$. The surface charge density on the electrode is

$$\sigma_{\text{charge}} = \frac{1}{4\pi}\hat{\mathbf{r}}\cdot\mathbf{D} = \frac{\epsilon}{4\pi}\hat{\mathbf{r}}\cdot\mathbf{E} = \frac{\epsilon aV}{4\pi a^2}. \tag{6.34}$$

Consequently, the charge on the electrode is

$$Q = 4\pi a^2 \sigma_{\text{charge}} = \epsilon aV. \tag{6.35}$$

The capacitance is $C = Q/V = \epsilon a$.
The product $RC = \epsilon/(4\pi\sigma) = \tau$, the relaxation time.

(b) For a spherical capacitor with inner radius A and outer radius B filled with a dielectric of permittivity ϵ, the capacitance found in problem 6.2(a) was

$$C = \frac{Q}{V} = \epsilon\left[\frac{1}{A} - \frac{1}{B}\right]^{-1} = \frac{\epsilon AB}{(B-A)}. \tag{6.36}$$

For a conductivity σ between the electrodes, the current density is

$$\mathbf{j} = \sigma\mathbf{E} = \frac{\sigma Q}{\epsilon r^2}, \tag{6.37}$$

and the current leaving the inner electrode will be

$$I = \oint \mathbf{j}\cdot d\mathbf{A} = \frac{4\pi A^2\sigma Q}{\epsilon A^2} = \frac{4\pi\sigma}{\epsilon}\left[\frac{V\epsilon AB}{(B-A)}\right] = \frac{4\pi\sigma VAB}{(B-A)}. \tag{6.38}$$

Then, the resistance is

$$R = \frac{V}{I} = \frac{(B-A)}{4\pi\sigma AB}. \tag{6.39}$$

The product $RC = \epsilon/(4\pi\sigma) = \tau$.

(c) When the dielectric fills only the lower half of the capacitor in part (b), the capacitance found in problem 6.2(b) was

$$C = \frac{Q}{V} = \frac{(1+\epsilon)}{2}\left[\frac{1}{A} - \frac{1}{B}\right]^{-1} = \frac{(1+\epsilon)AB}{2(B-A)}. \tag{6.40}$$

The current leaving the inner electrode is given by

$$I = \oint \mathbf{j}\cdot d\mathbf{A} = 2\pi A^2\sigma E_A, \tag{6.41}$$

where E_A is the electric field at the surface of the inner electrode. The effective area for this integral is only $2\pi A^2$ because there is no current in the upper half of the capacitor. The field E_A is related to V by

$$V = \int_A^B \frac{A^2 E_A dr}{r^2} = A^2 E_A \left[\frac{1}{A} - \frac{1}{B}\right] = \frac{A E_A (B - A)}{B}. \tag{6.42}$$

Then the effective resistance is given by

$$R = \frac{V}{I} = \frac{(B - A)}{2\pi\sigma AB}, \tag{6.43}$$

and, the product RC is

$$RC = \frac{(\epsilon + 1)}{4\pi\sigma}. \tag{6.44}$$

Note: The product RC does not equal $\epsilon/(4\pi\sigma)$ here because ϵ and σ are not constant throughout the capacitor.

Chapter 7

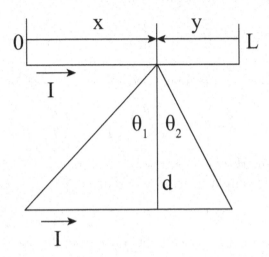

(a) Two parallel straight wires, each of length L and carrying current I, are a distance d apart as shown in the figure. The magnetic field at the top wire due to the bottom wire is given by

$$\mathbf{B} = \frac{\mathbf{I} \times \hat{\mathbf{d}}}{cd}[\sin\theta_2 - \sin\theta_1]. \tag{7.1}$$

The function $\sin\theta_1$ can be written in terms of the distance x from the left end of the top wire as

$$\sin\theta_1 = \frac{-x}{\sqrt{x^2 + d^2}}, \tag{7.2}$$

where the minus sign follows from the sign convention in the text that an angle to the left of vertical in the figure is negative. The function $\sin\theta_2$ can be written in terms of the distance y from the right end of the top wire as

$$\sin\theta_2 = \frac{y}{\sqrt{y^2 + d^2}}, \tag{7.3}$$

The magnitude of force on the upper wire is given by integrating over the length of the wire as follows:

$$\mathbf{F} = \frac{I}{c}\int \mathbf{dl} \times \mathbf{B} = \frac{-I^2\hat{\mathbf{d}}}{c^2 d}\left[\int_0^L \frac{y\,dy}{\sqrt{y^2 + d^2}} + \int_0^L \frac{x\,dx}{\sqrt{x^2 + d^2}}\right]. \tag{7.4}$$

525

We have integrated the first term over the variable y and the second term over the variable x. The two integrals are the same, so the result for **F** is

$$\mathbf{F} = \frac{-2I^2\hat{\mathbf{d}}}{c^2d}\left[\int_0^L \frac{y\,dy}{\sqrt{y^2+d^2}}\right] = \frac{-2I^2\hat{\mathbf{d}}}{c^2d}\left[\sqrt{L^2+d^2}-d\right]. \qquad (7.5)$$

(b) The difference between this force and the result for long wire is

$$\Delta F = \frac{2I^2}{c^2d}\left[\sqrt{L^2+d^2}-d\right] - \frac{2I^2L}{c^2d} = \frac{2I^2L}{c^2d}\left[\sqrt{1+d^2/L^2}-d/L-1\right]. \qquad (7.6)$$

For $d \ll L$, this can be expanded to lowest order in d/L:

$$\Delta F \approx \left(\frac{2I^2L}{c^2d}\right)\left(\frac{d}{L}\right) \approx \left(\frac{d}{L}\right)F_\infty. \qquad (7.7)$$

This means the the force on the wires of length L will be 1% lower than the long wire formula when $d/L = 0.01$.

Problem 7.3:

(a) The magnitude of the magnetic field a perpendicular distance r from the midpoint of a straight wire segment of length $2a$ carrying a current I is given by

$$B = \frac{2aI/c}{r\sqrt{a^2+r^2}}. \qquad (7.8)$$

For a configuration of four such wires forming a square loop, the distance r from the center of each wire to to a point a distance d above the plane of the loop, and on its axis, is $r = \sqrt{d^2+a^2}$. This means that the magnitude of B at the height d will be

$$B = \frac{2aI/c}{\sqrt{d^2+a^2}\sqrt{d^2+2a^2}} \qquad (7.9)$$

for each wire. The components of **B** perpendicular to the axis cancel, and the sum of the four components along the axis is given by $4B\cos\alpha$ where α is the angle that **B** makes with the loop's axis. Thus the magnetic field due to the square current loop is

$$B_\square = 4B\cos\alpha = \frac{4Ba}{\sqrt{d^2+a^2}} = \frac{8a^2I/c}{(d^2+a^2)\sqrt{d^2+2a^2}}. \qquad (7.10)$$

(b) The magnetic field of a circular loop of radius a, carrying a current I is given by

$$B_\bigcirc = \frac{2\pi a^2 I/c}{(d^2+a^2)^{\frac{3}{2}}}. \qquad (7.11)$$

(c) At the center of each loop ($d = 0$), the fields are

$$B_\square(0) = 4\sqrt{2}\frac{I}{ca} = 5.66\frac{I}{ca}, \qquad B_\bigcirc(0) = \frac{2\pi I}{ca} = 6.28\frac{I}{ca}, \qquad (7.12)$$

so $B_\bigcirc(0 > B_\square(0)$. This is because the circular loop is inscribed within the square which makes its wires closer to the center of the loop than for the square loop.

(ii) For large $d \gg a$, we can expand each form for B to get

$$B_\square(d \gg a) = 8\frac{a^2 I}{cd^3}, \qquad B_\bigcirc(d \gg a) = \frac{2\pi a^2 I}{cd^3} = 6.28\frac{Ia^2}{cd^3}, \qquad (7.13)$$

so $B_\bigcirc(d \gg a) < B_\square(d \gg a)$. This is because the large d limit of the magnetic field is determined by the dipole moment of each loop. The dipole moment is proportional to the area of the loop, which is smaller for the inscribed circle.

Problem 7.5:

(a) The magnetic field required to bend electrons with kinetic energy 200 eV into a circular path with radius 5 cm is given by

$$B = \frac{mvc}{eR} = \frac{\sqrt{(mv^2)(mc^2)}}{eR} = \frac{\sqrt{2mc^2 T}}{eR}$$

$$= \frac{\sqrt{(2)(511 \times 10^3)(200)}}{5} = 2,900 \text{ gauss.} \qquad (7.14)$$

Note: We took the c inside the square root to make $mc^2 = 511$ keV, we used $mv = \sqrt{2mT}$ where T is the kinetic energy, and we did not use the magnitude e of the electron charge because the energies were given in electron volt units.

(b) The velocity of 200 eV electrons is given by solving $T = mv^2/2$ for v:

$$v = \sqrt{\frac{2T}{m}} = c\sqrt{\frac{2T}{mc^2}} = c\sqrt{\frac{2 \times 200}{511 \times 10^3}} = 0.028\,c = 8.4 \times 10^8 \text{ cm/sec.} \qquad (7.15)$$

The angular velocity of the 200 eV electrons is given by

$$\omega = \frac{v}{R} = \frac{8.4 \times 10^8}{5} = 1.7 \times 10^8 \text{ radians/sec.} \qquad (7.16)$$

Note: The energies were in eV, and all other units were in cgs.

Problem 7.7:

(a) The force on an electron (with charge $q = -e$) in an electric field $\mathbf{E} = -E\hat{\mathbf{j}}$ and a magnetic field $\mathbf{B} = -B\hat{\mathbf{k}}$ is given by the Lorentz force

$$\mathbf{F} = q\mathbf{E} + \frac{q}{c}\mathbf{v}\times\mathbf{B} = eE\hat{\mathbf{j}} + \frac{eB}{c}(\mathbf{v}\times\hat{\mathbf{k}}). \tag{7.17}$$

The equations of motion for the electron are

$$\dot{v}_x = F_x/m = (e/mc)Bv_y \tag{7.18}$$
$$\dot{v}_y = F_y/m = (e/m)E - (e/mc)Bv_x. \tag{7.19}$$

Taking the time derivative of $\dot{v}_y$ leads to

$$\ddot{v}_y = -(eB/mc)^2 v_y \tag{7.20}$$

which has the solution [for $v_y(0) = 0$, since the electron starts at rest]

$$v_y = A\sin(\omega t), \quad \text{with } \omega = \frac{eB}{mc}. \tag{7.21}$$

Then, v_x is given by solving Eq. (7.19) for v_x

$$v_x = \frac{cE}{B} - \frac{mc}{eB}\dot{v}_y = \frac{cE}{B} - A\cos(\omega t). \tag{7.22}$$

Since $v_x(0) = 0$, the constant $A = cE/B$. Now x and y can be found by integration:

$$x = \frac{cE}{\omega B}[\omega t - \sin(\omega t)], \tag{7.23}$$

$$y = \frac{cE}{\omega B}[1 - \cos(\omega t)]. \tag{7.24}$$

(b) Letting $R = (cE/\omega B)$, we can solve for $R\sin(\omega t)$ and $R\cos(\omega t)$:

$$R\sin(\omega t) = \omega R t - x, \tag{7.25}$$
$$R\cos(\omega t) = R - y. \tag{7.26}$$

Squaring and adding these equations, we get

$$(x - \omega R t)^2 + (y - R)^2 = R^2. \tag{7.27}$$

(c) The first three cycles of the motion are plotted below.

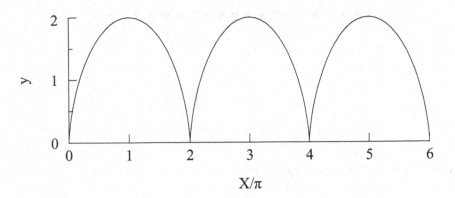

Problem 7.9

(a) For a circular current loop, the magnetic field on the z-axis is given by [Eq. (7.24)]:

$$B_z = \frac{2\pi I a^2}{c(z^2 + a^2)^{\frac{3}{2}}}.$$ (7.28)

Assuming that the integral along a semicircle of infinite radius vanishes, the Ampere's law integral is given by

$$\int_{-\infty}^{\infty} B_z dz = \frac{2\pi I a^2}{c} \int_{-\infty}^{\infty} \frac{dz}{(z^2 + a^2)^{\frac{3}{2}}} = \frac{4\pi I}{c},$$ (7.29)

in accord with Ampere's law.

(b) For a square current loop, the magnetic field on the z-axis is given by [Eq. (7.10)] in problem 7.3:

$$B_z = \frac{8R^2 I / c}{(z^2 + R^2)\sqrt{z^2 + 2R^2}}.$$ (7.30)

The Ampere's law integral is given by

$$\int_{-\infty}^{\infty} B_z dz = \frac{8R^2 I}{c} \int_{-\infty}^{\infty} \frac{dz}{(z^2 + R^2)\sqrt{z^2 + 2R^2}}.$$ (7.31)

This integral can be simplified by the substitution $z = \sqrt{2}R \tan\theta$, yielding

$$\int_{-\infty}^{\infty} B_z dz = \frac{8I}{c} \int_{-\pi/2}^{\pi/2} \frac{\cos\theta \, d\theta}{(\sin^2\theta + 1)} = \frac{8I}{c} \int_{-1}^{1} \frac{dx}{(x^2 + 1)} = \frac{4\pi I}{c},$$ (7.32)

in accord with Ampere's law.

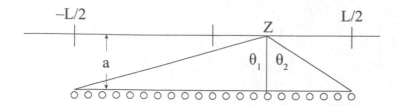

(c) For a solenoid, the magnetic field on the z-axis is given by

$$B = \frac{2\pi n I}{c}[\sin\theta_2 - \sin\theta_1], \tag{7.33}$$

where the angles θ_1 and θ_2 are defined in the figure. The angles are positive if the point z is to the right of vertical at the point of measurement (z) of B, and negative if to the left. (In the figure, θ_1 is negative and θ_2 is positive.)

If the point $z = 0$ is taken at the middle of the solenoid, then z is related to each angle by

$$\sin\theta_1 = \frac{-(z+L/2)}{\sqrt{(z+L/2)^2 + a^2}}, \quad \sin\theta_2 = \frac{L/2-z}{\sqrt{(z-L/2)^2 + a^2}}. \tag{7.34}$$

and the Ampere's law integral is given by

$$\int_{-\infty}^{\infty} B_z dz = \frac{2\pi n I}{c} \int_{-\infty}^{\infty} \left[\frac{L/2-z}{\sqrt{(z-L/2)^2 + a^2}} + \frac{z+L/2}{\sqrt{(z+L/2)^2 + a^2}} \right]. \tag{7.35}$$

The two separate integrals diverge, but their sum is convergent, so we integrate from $-R$ to $+R$, and then take the limit as $R \to \infty$:

$$\int_{-\infty}^{\infty} B_z dz = \frac{2\pi n I}{c} \lim_{R\to\infty} \left[-\sqrt{(z-L/2)^2 + a^2} + \sqrt{(z+L/2)^2 + a^2} \right]_{-R}^{R} \tag{7.36}$$

$$= \frac{4\pi n I}{c} \lim_{R\to\infty} \left[\sqrt{(R+L/2)^2 + a^2} - \sqrt{(R-L/2)^2 + a^2} \right]. \tag{7.37}$$

We expand the square roots for large R, to get

$$\int_{-\infty}^{\infty} B_z dz = \frac{4\pi n I}{c} \lim_{R\to\infty} [R\sqrt{1 + L/R + (a^2 + L^2/4)/R^2}$$

$$- R\sqrt{1 - L/R + (a^2 + L^2/4)/R^2}]$$

$$= \frac{4\pi n I}{c} \lim_{R\to\infty} [R + L/2 - (R - L/2)]$$

$$= \frac{4\pi n L I}{c} = \frac{4\pi N I}{c}, \tag{7.38}$$

in accordance with Ampere's law.

Problem 7.11

We can consider a long straight wire of radius R carrying a current $\mathbf{I}$ with a circular portion of radius a cut out at a distance d from the center, as shown in the figure, to be two separate currents: one uniform current, $\mathbf{I_1}$, in the original wire, in the direction of $\mathbf{I}$, and centered at the origin; and a second uniform current, $\mathbf{I_2}$, in the cut-out portion, in the opposite direction and centered at the point d.

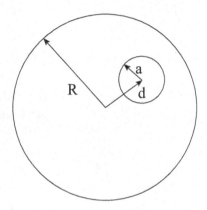

The magnetic field for $r < R$ due to current $\mathbf{I_1}$ is given by

$$\mathbf{B_1} = \left(\frac{2\pi r^2}{c}\right)\frac{\mathbf{j} \times \mathbf{r}}{r^2} = \left(\frac{2\pi}{c}\right)\mathbf{j} \times \mathbf{r}, \tag{7.39}$$

where $\pi r^2 \mathbf{j}$ is the current within radius r. The current density $\mathbf{j}$ is given by

$$\mathbf{j} = \frac{\mathbf{I}}{\pi(R^2 - a^2)}. \tag{7.40}$$

Similarly, the magnetic field due to $\mathbf{I_2}$ inside the cut-out area is

$$\mathbf{B_2} = \left(\frac{-2\pi}{c}\right)\mathbf{j} \times (\mathbf{r} - \mathbf{d}). \tag{7.41}$$

We take the current densities for I_1 and I_2 equal in magnitude, so that there will be no net current in the cut-out portion. Adding the two fields gives

$$\mathbf{B} = \mathbf{B_1} + \mathbf{B_2} = \left(\frac{2\pi}{c}\right)\mathbf{j} \times \mathbf{d} = \frac{2\mathbf{I} \times \mathbf{d}}{c(R^2 - a^2)}. \tag{7.42}$$

This is a constant field in the cut-out area.

Problem 7.13:

(a) The magnetic scalar potential at a height z on the axis of of a square current loop of dimension $2a \times 2a$ carrying a current I is given by

$$\phi_m = \frac{I}{c}\Omega. \tag{7.43}$$

The solid angle subtended by the square current loop is given by the integral

$$\Omega = \int_{-a}^{a}\int_{-a}^{a} \frac{zdxdy}{(x^2 + y^2 + z^2)^{3/2}}. \tag{7.44}$$

We make the substitution $y = \sqrt{x^2 + z^2}\tan\theta$, and integrate over θ from $-\theta_a$ to $+\theta_a$, where the angle θ_a is shown in the figure:

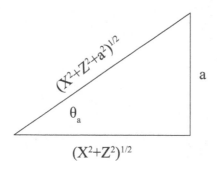

We make the substitution $y = \sqrt{x^2 + z^2}\tan\theta$

$$\Omega = z\int_{-a}^{a} dx \int_{-\theta_a}^{\theta_a} \frac{\sqrt{x^2 + z^2}\sec^2\theta d\theta}{(x^2 + z^2)^{3/2}\sec^3\theta} = z\int_{-a}^{a} \frac{dx}{(x^2 + z^2)}\int_{-\theta_a}^{\theta_a}\cos\theta d\theta$$

$$= 2z\int_{-a}^{a} \frac{\sin\theta_a dx}{(x^2 + z^2)} = 2z\int_{-a}^{a} \frac{adx}{(x^2 + z^2)\sqrt{x^2 + z^2 + a^2}}, \tag{7.45}$$

where we have used $\sin\theta_a = a/\sqrt{x^2 + z^2 + a^2}$, as seen in the figure for θ_a. We do this integral with the substitution $x = \sqrt{z^2 + a^2}\tan\theta$, and integrate over θ from $-\theta_z$ to $+\theta_z$, where $\sin\theta_z = a/\sqrt{z^2 + 2a^2}$. (See the figure for θ_a with $x \to a$.) This gives

$$\Omega = 2az\int_{-\theta_z}^{\theta_z} \frac{\sqrt{z^2 + a^2}\sec^2\theta d\theta}{[(z^2 + a^2)\tan^2\theta + z^2]\sqrt{z^2 + a^2}\sec\theta}$$

$$= 2az\int_{-\theta_z}^{\theta_z} \frac{\cos\theta d\theta}{[(z^2 + a^2)\sin^2\theta + z^2\cos^2\theta]}$$

$$= 2az\int_{-\theta_z}^{\theta_z} \frac{\cos\theta d\theta}{a^2\sin^2\theta + z^2}. \tag{7.46}$$

We do this integral by the substitution $u = (a/z)\sin\theta$, and integrate from $-u_z$ to $+u_z$ with $u_z = (a/z)\sin\theta_z = a^2/(z\sqrt{z^2 + 2a^2})$. Then

$$\Omega = 2az \int_{-u_z}^{u_z} \frac{du}{az(u^2 + 1)} = 4\tan^{-1} u_z$$

$$= 4\tan^{-1}\left[\frac{a^2}{z\sqrt{z^2 + 2a^2}}\right]. \tag{7.47}$$

Finally,

$$\phi_m = \frac{I}{c}\Omega = \frac{4I}{c}\tan^{-1}\left[\frac{a^2}{z\sqrt{z^2 + 2a^2}}\right]. \tag{7.48}$$

Note: The magnetic scalar potential can also be found by starting with B_z as calculated for the square current loop in problem 7.3:

$$B_z = \frac{8a^2 I/c}{(z^2 + a^2)\sqrt{z^2 + 2a^2}}. \tag{7.49}$$

The magnetic scalar potential on the z-axis is then given by

$$\phi_m(z) = \int_z^\infty B_z dz = \int_z^\infty \frac{8a^2 I}{c(z'^2 + a^2)\sqrt{z'^2 + 2a^2}}. \tag{7.50}$$

This integral can be done by making the substitution $z' = \sqrt{2}a\tan\theta$, which leads to

$$\phi_m(z) = \frac{8I}{c}\int_{\theta_z}^{\pi/2} \frac{\cos\theta d\theta}{(\sin^2\theta + 1)}, \tag{7.51}$$

where the lower limit satisfies $\tan\theta_z = (z/\sqrt{2}a)$. Now making the substitution
$x = \sin\theta$, we get

$$\phi_m(z) = \frac{8I}{c}\int_{x_z}^1 \frac{dx}{(x^2 + 1)}, \tag{7.52}$$

where $x_z = z/\sqrt{z^2 + 2a^2}$. This integration gives

$$\phi_m(z) = \frac{8I}{c}[\arctan(1) - \arctan(x_z)] = \frac{8I}{c}\left[\frac{\pi}{4} - \arctan\left(\frac{z}{\sqrt{z^2 + 2a^2}}\right)\right]$$

$$= \frac{4I}{c}\left[\frac{\pi}{2} - 2\arctan\left(\frac{z}{\sqrt{z^2 + 2a^2}}\right)\right]. \tag{7.53}$$

The two answers for ϕ_m appear different, but they can be shown to be equal using some more trigonometry. We write

$$\phi_m(z) = \frac{4I}{c}\left[\frac{\pi}{2} - 2\alpha\right], \quad \text{with } \alpha = \arctan\left(\frac{z}{\sqrt{z^2 + 2a^2}}\right). \tag{7.54}$$

Then

$$\tan\left[c\phi_m(z)/4I\right] = \cot(2\alpha) = \frac{\cos^2\alpha - \sin^2\alpha}{2\sin^2\alpha\cos^2\alpha}$$

$$= \frac{z^2 + 2a^2 - z^2}{2z\sqrt{z^2 + 2a^2}} = \frac{a^2}{z\sqrt{z^2 + 2a^2}}, \qquad (7.55)$$

which shows that the two formulas for ϕ_m are equivalent.

(b) The magnetic field on the axis of the square current loop is given by

$$B_z = -\partial_z\phi_m = -\partial_z\left[\frac{4I}{c}\tan^{-1}\left(\frac{a^2}{z\sqrt{z^2+2a^2}}\right)\right]$$

$$= -\frac{4I}{c}\left[\frac{z^2(z^2+2a^2)}{z^2(z^2+2a^2)+a^4}\right]\left[\frac{-a^2}{z^2\sqrt{z^2+2a^2}} - \frac{a^2}{(z^2+2a^2)^{3/2}}\right]$$

$$= \frac{4Ia^2}{c}\left[\frac{z^2(z^2+2a^2)}{(z^2+a^2)^2}\right]\left[\frac{z^2+2a^2+z^2}{z^2(z^2+2a^2)^{3/2}}\right]$$

$$= \frac{8a^2I/c}{(z^2+a^2)\sqrt{z^2+2a^2}}. \qquad (7.56)$$

I agree that the magnetic scalar potential is not the easiest way to find the magnetic field in this case.

Problem 7.15:

(a) A circular current loop of radius R (in the x-y plane) carrying a current I has a magnetic moment given by

$$\boldsymbol{\mu} = \pi R^2 I \hat{\mathbf{k}}/c. \qquad (7.57)$$

The magnetic field due to this dipole moment is

$$\mathbf{B}\boldsymbol{\mu} = \frac{[3(\boldsymbol{\mu}\cdot\hat{\mathbf{r}})\hat{\mathbf{r}} - \boldsymbol{\mu}]}{r^3} \qquad (7.58)$$

The force on a similar current loop a distance $L = 4R$ above the first loop depends on the cross product $\mathbf{I}\times\mathbf{B}$. The horizontal components of the force, coming from the vertical component of $\mathbf{B}_{\boldsymbol{\mu}}$, will cancel. Thus, the force on the upper loop will depend only on the horizontal component of $\mathbf{B}_{\boldsymbol{\mu}}$, which produces a vertical force. The horizontal component of $\mathbf{B}_{\boldsymbol{\mu}}$ at the location of the top loop is

$$B_\rho = \frac{3\mu z\rho}{[\rho^2 + z^2]^{5/2}}, \qquad (7.59)$$

Here, we use ρ for the radial coordinate in cylindrical coordinates. In cylindrical coordinates, the current is given by $I\hat{\boldsymbol{\theta}}$, and the force on the top loop is

$$\mathbf{F} = \frac{1}{c}\int_0^{2\pi} I\hat{\boldsymbol{\theta}}\times\mathbf{B}_{\boldsymbol{\mu}}Rd\theta = -2\pi RIB_\rho\hat{\mathbf{k}}/c = \frac{-6\pi RIz\rho\boldsymbol{\mu}}{c[\rho^2+z^2]^{5/2}}. \quad (7.60)$$

putting in the values $\rho = R$, $z = L$, we get for the force

$$\mathbf{F} = \frac{-6\pi LR^2 I\mu\hat{\mathbf{k}}}{c[R^2+L^2]^{5/2}} = \frac{-6\pi^2 LR^4 I^2\hat{\mathbf{k}}}{c^2[R^2+L^2]^{5/2}}. \quad (7.61)$$

The force is attractive for the two currents in the same direction. For $L = 4R$, the force on the upper loop is

$$\mathbf{F} = -\frac{24\pi^2 I^2\hat{\mathbf{k}}}{17^{5/2}c^2} = -0.20\frac{I^2\hat{\mathbf{k}}}{c^2}. \quad (7.62)$$

This is an approximation to the force because we approximated the lower loop by a magnetic dipole. The full calculation would require an expansion of the magnetic field in Legendre polynomials.

(b) A simpler approximation is to replace both loops by magnetic dipoles. Then the force on the upper loop is given by the same formula as the electric dipole-dipole force given by [Eq. (2.64)]:

$$\begin{aligned}
\mathbf{F}_{\mu'\mu} &= \frac{3(\boldsymbol{\mu}\cdot\hat{\mathbf{k}})\boldsymbol{\mu}' + 3(\boldsymbol{\mu}'\cdot\hat{\mathbf{k}})\boldsymbol{\mu} + 3(\boldsymbol{\mu}\cdot\boldsymbol{\mu}')\hat{\mathbf{k}} - 15(\boldsymbol{\mu}\cdot\hat{\mathbf{k}})(\boldsymbol{\mu}'\cdot\hat{\mathbf{k}})\hat{\mathbf{k}}}{L^4} \\
&= \frac{[3\mu^2+3\mu^2+3\mu^2-15\mu^2]\hat{\mathbf{k}}}{L^4} = \frac{-6\mu^2\hat{\mathbf{k}}}{L^4} \\
&= \frac{-6\pi^2 R^4 I^2\hat{\mathbf{k}}}{c^2 L^4} = \frac{-6\pi^2 I^2\hat{\mathbf{k}}}{256c^2} = -0.23\frac{I^2\hat{\mathbf{k}}}{c^2}. \quad (7.63)
\end{aligned}$$

Problem 7.17:

(a) The magnetic moment of a sphere with a uniform surface charge density σ, rotating with angular velocity $\boldsymbol{\omega}$ is given by

$$\begin{aligned}
\boldsymbol{\mu} &= \frac{1}{2c}\int(\mathbf{r}\times\sigma\mathbf{v})dS = \frac{\sigma}{2c}\int\mathbf{r}\times(\boldsymbol{\omega}\times\mathbf{v})dS = \frac{\sigma}{2c}\int[\boldsymbol{\omega}r^2 - (\boldsymbol{\omega}\cdot\mathbf{r})\mathbf{r}]dS \\
&= \frac{\sigma R^4}{2c}\int[\boldsymbol{\omega}-(\boldsymbol{\omega}\cdot\hat{\mathbf{r}})\hat{\mathbf{r}}]d\Omega = \frac{4\pi\sigma R^4\boldsymbol{\omega}}{3c} = \frac{QR^2\boldsymbol{\omega}}{3c}, \quad (7.64)
\end{aligned}$$

where we have used $\sigma = Q/4\pi R^2$.

(b) The spin angular momentum of the rotating sphere is given by

$$\mathbf{L} = \int (\mathbf{r} \times \rho_m \mathbf{v}) d\tau = \rho_m \int \mathbf{r} \times (\boldsymbol{\omega} \times \mathbf{v}) d\tau = \rho_m \int [\boldsymbol{\omega} r^2 - (\boldsymbol{\omega} \cdot \mathbf{r}) \mathbf{r}] d\tau$$

$$= \rho_m \int [\boldsymbol{\omega} - (\boldsymbol{\omega} \cdot \hat{\mathbf{r}}) \hat{\mathbf{r}}] r^2 d\tau = \frac{8\pi \rho_m \boldsymbol{\omega}}{3} \int_0^R r^4 dr = \frac{8\pi \rho_m R^5 \boldsymbol{\omega}}{15}$$

$$= \frac{2MR^2\boldsymbol{\omega}}{5}, \tag{7.65}$$

where we have used $\rho_m = 4\pi R^3/3$. The gyromagnetic ratio of the rotating sphere is

$$G = \frac{\mu}{L} = \frac{2QR^2\omega/3c}{2MR^2\omega/5} = \frac{5}{6} \frac{Q}{Mc}. \tag{7.66}$$

Its g factor is

$$g = \frac{G}{Q/2Mc} = \frac{5}{3}. \tag{7.67}$$

Problem 7.19: The magnitude of the Fermi-Breit interaction energy between an electron and a proton is given by

$$U_{\text{FB}}(\mathbf{r}) = (8\pi/3)|\mu_e\mu_p|\delta(\mathbf{r}), \tag{7.68}$$

where $\mu_e = \mu_{\text{B}}$, $\mu_p = 2.79\mu_{\text{N}}$, and $\mathbf{r}$ is the distance from the proton to the electron. For a probability distribution $P(r) = e^{-2r/a}/\pi a^3$, the F-B energy is given by

$$U = \int P(r) U_{\text{FB}} d\tau = (8\pi/3)|\mu_e\mu_p|P(0) = \frac{8\mu_e\mu_p}{3a^3}. \tag{7.69}$$

Because of the delta function interaction, the integral depended only on $P(0)$. Putting in numbers (taking μ_{B} and μ_N from Problem 7.20.):

$$U = \frac{8|\mu_e\mu_p|}{3a^3} = \frac{8(5.79 \times 10^{-9})(2.79 \times 3.15 \times 10^{-12})}{3 \times (.53 \times 10^{-8})^3}$$

$$= 911 \times 10^3 \frac{(\text{eV/gauss})^2}{\text{cm}^3}. \tag{7.70}$$

To simplify the units, we write them as

$$1\frac{(\text{eV/gauss})^2}{\text{cm}^3} == \left(\frac{e}{\text{gauss} - \text{cm}^2}\right)\left(\frac{\text{Volt}}{\text{gauss} - \text{cm}}\right) \text{eV}, \tag{7.71}$$

and use the conversions

$$1 \text{ gauss} - \text{cm} = 1 \text{ statvolt} = 300 \text{ Volts}, \tag{7.72}$$

$$e = 4.8 \text{ statcoulomb} = 4.8 \times 10^{-8} \text{ gauss} - \text{cm}^2. \tag{7.73}$$

to get

$$\frac{(\text{eV/gauss})^2}{\text{cm}^3} = \left(\frac{4.8 \times 10^{-8}}{300}\right) \text{eV} = 1.6 \times 10^{-10} \text{ eV}. \tag{7.74}$$

Then $\qquad U = (911 \times 10^3)(1.6 \times 10^{-10}) = 1.46 \times 10^{-4} \text{ eV}. \tag{7.75}$

Chapter 8

Problem 8.1:

For a disk of permeabilty μ ($>> 1$) with a radius a and thickness t, whose axis makes an angle θ with a uniform magnetic field $\mathbf{B_0}$, the induced magnetic field $\mathbf{B}$ in the disk will be mainly in the plane of the disk. Its magnitude will be determined by the boundary condition that $\mathbf{H_T}$ is continuous at the face of the disk. This means that $\mathbf{H_T}$ in the disk will be

$$\mathbf{H_T} = B_0 \sin \theta \hat{\mathbf{t}}, \tag{8.1}$$

where $\hat{\mathbf{t}}$ is a unit vector in the direction of the projection of $\mathbf{B_0}$ onto the disk. The magnetic field in the disk will be $\mathbf{B} = \mu \mathbf{H_T} = \mu B_0 \sin \theta \hat{\mathbf{t}}$. If $\mu >> 1$, the $\mathbf{H}$ field will be negligible compare to $\mathbf{B}$ in the disk. Then the magnetization in the disk will be

$$\mathbf{M} = (\mathbf{B} - \mathbf{H})/4\pi \approx \mathbf{B}/4\pi = \mu B_0 \sin \theta \hat{\mathbf{t}}/4\pi. \tag{8.2}$$

(a) The magnetic moment of the disk is given be $\mathbf{M}$ times the volume of the disk:

$$\mathbf{m} = \pi a^2 t \mathbf{M} = \frac{1}{4} a^2 t \mu B_0 \sin \theta \hat{\mathbf{t}}. \tag{8.3}$$

We have used $\mathbf{m}$ for the magnetic moment to avoid confusion with the permeability μ.

(b) The torque on the disk is given by

$$\boldsymbol{\tau} = \mathbf{m} \times \mathbf{B_0} = \frac{1}{4} a^2 t \mu B_0 \sin \theta \hat{\mathbf{t}} \times \mathbf{B_0} = \frac{1}{8} \mu a^2 t B_0^2 \sin(2\theta) \hat{\boldsymbol{\tau}}, \tag{8.4}$$

where $\hat{\boldsymbol{\tau}}$ is a unit vector in the direction of $\hat{\mathbf{t}} \times \mathbf{B_0}$. This torque will tend to rotate the disk so that the plane of the disk will be parallel to $\mathbf{B_0}$.

Problem 8.3:

(a) The bound magnetic charge density is

$$
\begin{aligned}
\rho_m &= -\boldsymbol{\nabla}\cdot\mathbf{M} = -\boldsymbol{\nabla}\cdot[a\hat{\mathbf{k}}(\hat{\mathbf{k}}\cdot\mathbf{r})^2] = -a\hat{\mathbf{k}}\cdot\boldsymbol{\nabla}[(\hat{\mathbf{k}}\cdot\mathbf{r})^2] \\
&= -2a(\hat{\mathbf{k}}\cdot\mathbf{r})\hat{\mathbf{k}}\cdot\boldsymbol{\nabla}(\hat{\mathbf{k}}\cdot\mathbf{r}) = -2a(\hat{\mathbf{k}}\cdot\mathbf{r}).
\end{aligned} \tag{8.5}
$$

(b) the bound magnetic surface charge density is

$$\sigma_m = \hat{\mathbf{r}}\cdot\mathbf{M} = a(\hat{\mathbf{r}}\cdot\hat{\mathbf{k}})(\hat{\mathbf{k}}\cdot\mathbf{r})^2 = ar^2(\hat{\mathbf{k}}\cdot\hat{\mathbf{r}})^3. \tag{8.6}$$

(c) The magnetic moment of the sphere will come from integrals over the volume and the surface charge densities:

$$\boldsymbol{\mu} = \boldsymbol{\mu}_{\text{volume}} + \boldsymbol{\mu}_{\text{surface}}. \tag{8.7}$$

$$\begin{aligned}
\boldsymbol{\mu}_{\text{volume}} &= \int \mathbf{r}\rho_m d^3r = -2a\int \mathbf{r}(\hat{\mathbf{k}}\cdot\mathbf{r})d^3r \\
&= -2a\hat{\mathbf{k}}\int(\hat{\mathbf{k}}\cdot\mathbf{r})^2 d^3r = -\frac{8\pi a}{3}\int_0^R r^4 dr \\
&= -\frac{8\pi aR^5}{15}.
\end{aligned} \tag{8.8}$$

$$\begin{aligned}
\boldsymbol{\mu}_{\text{surface}} &= a\oint dS\sigma_{\text{surface}} \\
&= a\oint dSrr^2(\hat{\mathbf{k}}\cdot\hat{\mathbf{r}})^3 = aR^5\hat{\mathbf{k}}\int dS(\hat{\mathbf{k}}\cdot\hat{\mathbf{r}})^4 \\
&= aR^5\hat{\mathbf{k}}\oint d\Omega(\hat{\mathbf{k}}\cdot\hat{\mathbf{r}})^4 = aR^5\hat{\mathbf{k}}\oint d\Omega\cos^4\theta \\
&= \frac{4\pi aR^5}{5}.
\end{aligned} \tag{8.9}$$

Then

$$\begin{aligned}
\boldsymbol{\mu} &= \boldsymbol{\mu}_{\text{volume}} + \boldsymbol{\mu}_{\text{surface}} \\
&= \frac{12\pi aR^5}{15} - \frac{8\pi aR^5}{15} = \frac{4\pi aR^5}{15}.
\end{aligned} \tag{8.10}$$

Problem 8.5:

(a) The bound current density is

$$\begin{aligned}
\mathbf{j}_b &= c\nabla\times\mathbf{M} = c\nabla\times[a\hat{\mathbf{k}}(\hat{\mathbf{k}}\cdot\mathbf{r})^2] = -ac\hat{\mathbf{k}}\times\nabla[(\hat{\mathbf{k}}\cdot\mathbf{r})^2] \\
&= -2ac\hat{\mathbf{k}}\times\hat{\mathbf{k}}(\hat{\mathbf{k}}\cdot\mathbf{r}) = 0.
\end{aligned} \tag{8.11}$$

(b) The bound surface current density is

$$\mathbf{K}_b = c\mathbf{M}\times\hat{\mathbf{r}} = ca(\hat{\mathbf{k}}\times\hat{\mathbf{r}})(\hat{\mathbf{k}}\cdot\mathbf{r})^2. \tag{8.12}$$

(c) The magnetic moment is

$$
\begin{aligned}
\boldsymbol{\mu} &= \frac{1}{2c} \oint dS\mathbf{r} \times \mathbf{K} = \frac{1}{2c} \oint dS\mathbf{r} \times \left[ca(\hat{\mathbf{k}} \times \hat{\mathbf{r}})(\hat{\mathbf{k}} \cdot \mathbf{r})^2 \right] \\
&= \frac{a}{2} \oint dS R^3 \left[\hat{\mathbf{k}} - (\hat{\mathbf{k}} \cdot \hat{\mathbf{r}})\hat{\mathbf{r}} \right] (\hat{\mathbf{k}} \cdot \hat{\mathbf{r}})^2 \\
&= \frac{aR^5}{2} \oint d\Omega \hat{\mathbf{k}} \left[(\hat{\mathbf{k}} \cdot \hat{\mathbf{r}})^2 - (\hat{\mathbf{k}} \cdot \hat{\mathbf{r}})^4 \right] \\
&= \pi a R^5 \hat{\mathbf{k}} \int_0^\pi \sin\theta d\theta \left[\cos^2\theta - \cos^4\theta \right] \\
&= \frac{4\pi}{15} a R^5 \hat{\mathbf{k}}
\end{aligned}
\tag{8.13}
$$

Problem 8.7:

(a) The $\mathbf{H}$ field on the axis of a bar magnet can be found by treating the magnet as having a sheet of magnetic surface charge located at each end. This corresponds to the electrostatic problem of uniformly charged disks located at $z = +L/2$ with $\sigma_m = M$, and at $z = -L/2$ with $\sigma_m = -M$, for a magnet of length L and polarization $\mathbf{M}$.

The electric field on the axis of a uniformly charged disk was found for positive z in Problem 1.6(b) to be

$$
E_z = 2\pi\sigma - \frac{2\pi\sigma z}{\sqrt{z^2 + a^2}}, \quad z > 0.
\tag{8.14}
$$

For negative z, this becomes

$$
E_z = -2\pi\sigma - \frac{2\pi\sigma z}{\sqrt{z^2 + a^2}}, \quad z < 0.
\tag{8.15}
$$

The $\mathbf{H}$ field in the magnet is given by uniformly charged disks at $z = +L/2$ ($\sigma_m = M$) and $z = -L/2$ ($\sigma_m = -M$). Outside the magnet, H_z on the axis is given by

$$
H_z(\text{outside}) = 2\pi M \left[\frac{L/2 - z}{\sqrt{(z - L/2)^2 + a^2}} + \frac{z + L/2}{\sqrt{(z + L/2)^2 + a^2}} \right], \quad |z| > L/2.
\tag{8.16}
$$

The terms $2\pi\sigma$ cancelled here because of their opposite charges. Inside the magnet, the $2\pi\sigma$ terms add (both are negative), and H_z is given by

$$
H_z(\text{inside}) = H_z(\text{outside}) - 4\pi M, \quad |z| < L/2.
\tag{8.17}
$$

(b) The $\mathbf{B}$ field of a bar magnet with uniform magnetization $\mathbf{M}$ is equivalent to the $\mathbf{B}$ field of a solenoid, with the relation $nI = K = Mc$ between

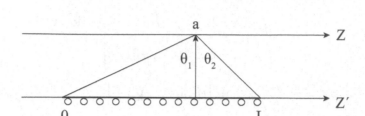

the current in the solenoid and the magnetization of the magnet. The magnetic field on the axis of a solenoid is given by

$$B_z = \frac{2\pi n I}{c}[\sin \theta_2 - \sin \theta_1], \tag{8.18}$$

where the angles θ_1 and θ_2 are defined in the figure.

The angles are positive if the point z is to the right of vertical at the point of measurement of B, and negative if to the left. (In the figure, θ_1 is negative and θ_2 is positive.) To compare this field to the H field, we relate the angles θ_1 and θ_2 to the variable z, taking the point $z = 0$ at the midpoint of the solenoid. From the figure, we see that

$$\sin \theta_1 = \frac{-(z + L/2)}{\sqrt{(z + L/2)^2 + a^2}}; \qquad \sin \theta_2 = \frac{L/2 - z}{\sqrt{(L/2 - z)^2 + a^2}}. \tag{8.19}$$

Then, B for the bar magnet, considered as a solenoid is given by

$$B_z = 2\pi M \left[\frac{L/2 - z}{\sqrt{(L/2 - z)^2 + a^2}} + \frac{L/2 + z}{\sqrt{(L/2 - z)^2 + a^2}} \right]. \tag{8.20}$$

This is the same result as in part (a) for H_z outside the magnet, but this formula for B_z also holds inside the magnet

(c) Comparing our results for B_z and H_z shows that $H_z = B_z - 4\pi M$ inside the magnet.

Problem 8.9:

(a) The force required to separate two identical bar magnets is the same as the force between two sheets of charge each having a surface charge density of $\sigma_m = M$:

$$F = 2\pi M^2 A. \tag{8.21}$$

For one iron magnet to lift another, this force must be greater than the weight of one magnet. This means that

$$F = 2\pi M^2 A \geq \rho g L A, \tag{8.22}$$

where $\rho = 7.9$ gram/cm^3 is the density of iron. Solving for M, we get (for $L=10$ cm)

$$M = \sqrt{\rho g L/(2\pi)} = \sqrt{(7.9)(980)(10)/(2\pi)} = 110 \, \text{gauss}. \qquad (8.23)$$

(b) If the other iron bar were not a permanent magnet, but had permeability $\mu = 4$, then there would be an image magnet (just as with a dielectric) with magnetization $M' = (\mu - 1)/(\mu + 1)M$, and the force required to separate them would be

$$F = 2\pi M^2 A \left(\frac{\mu - 1}{\mu + 1}\right). \qquad (8.24)$$

Then the minimum magnetization required for the permanent magnet to pick up the iron bar is

$$M = 110\sqrt{\frac{\mu + 1}{\mu - 1}} = 140 \, \text{gauss}. \qquad (8.25)$$

(c) Place the second bar at the middle of the first bar, and perpendicular to it. If the second bar is not a magnet, there will be no force on it.

Chapter 9

Problem 9.1:

(a) A copper ring of radius a with a current $I = I_0 t/\tau$ has a magnetic dipole moment of

$$\boldsymbol{\mu} = \frac{I}{c} \times \textbf{area} = \frac{\pi a^2 I_0 t \hat{\textbf{d}}}{c\tau}, \tag{9.1}$$

where $\hat{\textbf{d}}$ is a unit vector along the axis of the ring.

(b) The magnetic flux through a similar ring a distance d above the first ring can be approximated by the dipole magnetic field on the axis of the first ring times the area of the upper ring:

$$\Phi = \pi a^2 B = \frac{2\mu \pi a^2}{d^3} = \frac{2\pi^2 a^4 I_0 t}{c\tau d^3}. \tag{9.2}$$

(c) The induced EMF in the upper ring is

$$\mathcal{E} = \frac{1}{c}\frac{d\Phi}{dt} = \frac{2\pi^2 a^4 I_0}{c^2 \tau d^3}. \tag{9.3}$$

The current in the ring is

$$I' = \frac{\mathcal{E}}{R} = \frac{2\pi^2 a^4 I_0}{Rc^2 \tau d^3}, \tag{9.4}$$

in a direction opposite to the current in the lower ring (by Lenz's law).

(d) The induced dipole moment of the upper ring is

$$\boldsymbol{\mu}' = -\frac{\pi a^2 I' \hat{d}}{c} = -\frac{2\pi^3 a^6 I_0 \hat{d}}{Rc^3 \tau d^3}. \tag{9.5}$$

(e) The force on the upper ring can be approximated by the dipole-dipole force given by (with the two dipoles aligned with $\hat{\textbf{d}}$)

$$\textbf{F} = \frac{-6\boldsymbol{\mu} \cdot \boldsymbol{\mu}' \hat{d}}{d^4} = \frac{12\pi^4 a^8 I_0^2 t \hat{d}}{Rc^4 \tau^2 d^7} \tag{9.6}$$

Problem 9.3:
   To find the electric field, the length of the wire is needed.
It can be deduced using $L = \pi a^2 R / \rho(\text{copper})$.

(a) The electric field at the surface of the wire is

$$\mathbf{E} = \frac{IR}{L} = \frac{IR\rho}{\pi a^2 R} = \frac{I\rho}{\pi a^2}. \tag{9.7}$$

The magnetic field at the surface is

$$\mathbf{B} = \frac{2\mathbf{I}\times\hat{\mathbf{r}}}{(ca)}. \tag{9.8}$$

(b) The Poynting vector is given by

$$\mathbf{S} = \frac{c}{4\pi}\mathbf{E}\times\mathbf{B} = \frac{c}{4\pi}\left(\frac{IR}{L}\right)\times\left(\frac{2\mathbf{I}\times\hat{\mathbf{r}}}{ca}\right) = \frac{-I^2 R\hat{\mathbf{r}}}{2\pi La} \tag{9.9}$$

The magnitude of the Poynting vector is constant on the surface, so the power flux out of the wire is given by

$$P = \int \hat{\mathbf{r}}\cdot\mathbf{S}dA = -2\pi aLS = -I^2 R. \tag{9.10}$$

The minus sign means that the power is entering the wire.

(c) E is constant in the wire, and $B = 2Ir/(ca^2)$ (using Ampere's law). The electromagnetic energy in the wire is

$$
\begin{aligned}
U &= \frac{1}{8\pi}\int (E^2 + B^2)d\tau \\
&= \frac{1}{8\pi}\left[\left(\frac{IR}{L}\right)^2(\pi a^2 L) + 2\pi L\int_0^a\left(\frac{2Ir}{ca^2}\right)^2 rdr\right] \\
U &= \frac{1}{8\pi}\left[\frac{\pi a^2 I^2 R^2}{L} + \frac{2\pi LI^2}{c^2}\right] \tag{9.11}
\end{aligned}
$$

$$\text{or} \quad U = \frac{I^2 R}{8\pi}\left[\rho + \frac{2\pi^2 a^2}{c^2\rho}\right]. \tag{9.12}$$

We have written the energy in the wire in terms of L, and then in terms of ρ. Since ρ has the unit of second in the Gaussian system, we see that the U has the proper unit of power×time.

The electromagnetic momentum in the wire is given by

$$
\begin{aligned}
\mathbf{P_{EB}} &= \frac{1}{4\pi c}\int \mathbf{E}\times\mathbf{B}d\tau \\
&= \frac{1}{4\pi c}\int\left(\frac{IR}{L}\right)\times\left(\frac{2\mathbf{I}\times\mathbf{r}}{ca^2}\right)d\tau \\
&= \frac{-2I^2 R}{4\pi c^2 a^2 L}\int \mathbf{r}d\tau = 0 \tag{9.13}
\end{aligned}
$$

since the vector $\mathbf{r}$ averages to zero in the angular integral.

Problem 9.5:

(a) The electric part of the Maxwell stress tensor $[\mathbf{T_{DE}}]$ is given by

$$[\mathbf{T_{DE}}] = \frac{1}{4\pi}\left[\mathbf{DE} - \frac{1}{2}\hat{\mathbf{n}}(\mathbf{D}\cdot\mathbf{E})\right] \qquad (9.14)$$

We have used $[\mathbf{T_{DE}}]$ here, because it is the appropriate form to find the force on the charges in part (d). That is, integrating the form $[\mathbf{T_{DE}}]$ will find the force on free charge only, and not on the dielectric. For two point charges, each of charge q, one located at $\mathbf{d}$, and the other at $-\mathbf{d}$, the electric field on the plane surface midway between the charges is

$$\mathbf{E} = \frac{2q\rho}{\epsilon(\rho^2 + d^2)^{3/2}}, \qquad (9.15)$$

where $\boldsymbol{\rho}$ is the radius vector in the plane. We have included the case where the charges are immersed in a simple dielectric of infinite extent [as in part (d)]. The Maxwell stress tensor on the plane surface is

$$[\mathbf{T_{DE}}] = \frac{q^2}{\pi\epsilon}\left[\frac{\boldsymbol{\rho\rho} - \frac{1}{2}\hat{\mathbf{n}}\rho^2}{(\rho^2 + d^2)^3}\right]. \qquad (9.16)$$

(b) & (d) To find the force on the charge at $-\mathbf{d}$, we integrate $[\mathbf{T}]\cdot d\mathbf{S}$ over a surface composed of the plane between the charges, and a hemisphere of radius $R \to \infty$ closing the surface around the charge at $-\mathbf{d}$. Since $TdS \sim (\rho^2/\rho^4)$ for large ρ the integral on the hemisphere vanishes in the limit $R \to \infty$. The integral over the plane surface is

$$\int [\mathbf{T_{DE}}]\cdot d\mathbf{S} = 2\pi\int\left\{[\mathbf{T_{DE}}]\cdot\hat{\mathbf{d}}\right\}\rho d\rho = \frac{2q^2}{\epsilon}\int_0^\infty\left\{\frac{[\boldsymbol{\rho\rho} - \frac{1}{2}\hat{\mathbf{n}}\rho^2]\cdot\hat{\mathbf{d}}}{(\rho^2 + d^2)^3}\right\}\rho d\rho$$

$$= \frac{-q^2\hat{\mathbf{d}}}{\epsilon}\int_0^\infty\frac{\rho^3 d\rho}{(\rho^2 + d^2)^3} = \frac{-q^2\hat{\mathbf{d}}}{4\epsilon d^2}. \qquad (9.17)$$

This is the same repulsive force given by Coulomb's law. For the force on the charge at $\mathbf{d}$, the vector surface would be $d\mathbf{S} = -\hat{\mathbf{d}}dS$, and the force would come out in the $+\hat{\mathbf{d}}$ direction. We have included the permittivity in the calculation, so the force above is the answer to part (d). The answer to part (b) just has $\epsilon=1$.

(c) If the charge at position $\mathbf{d}$ has charge $-q$, then the electric field on the plane surface midway between the charges is

$$\mathbf{E} = \frac{2q\mathbf{d}}{\epsilon(\rho^2 + d^2)^{3/2}}. \qquad (9.18)$$

This makes the integral over the plane surface

$$\int [\mathbf{T_{DE}}] \cdot d\mathbf{S} \; = \; 2\pi \int \left\{ [\mathbf{T_{DE}}] \cdot \hat{\mathbf{d}} \right\} \rho d\rho = \frac{2q^2}{\epsilon} \int_0^\infty \left\{ \frac{[\mathbf{dd} - \frac{1}{2}\hat{n}d^2] \cdot \hat{\mathbf{d}}}{(\rho^2 + d^2)^3} \right\} \rho d\rho$$

$$= \; \frac{+q^2 d^2 \hat{\mathbf{d}}}{\epsilon} \int_0^\infty \frac{\rho d\rho}{(\rho^2 + d^2)^3} = \frac{+q^2 \hat{\mathbf{d}}}{4\epsilon d^2}. \tag{9.19}$$

This is an attractive force on the charge at $-\mathbf{d}$.

Problem 9.7:

(a) For a magnetic dipole $\boldsymbol{\mu}$ located at the center of a uniform electric charge distribution of radius R, charge e, and mass m, the electromagnetic angular momentum is

$$\mathbf{L_{em}} \; = \; \frac{1}{4\pi c} \int \mathbf{r} \times (\mathbf{E} \times \mathbf{B}) d\tau = \int \mathbf{r} \times \left\{ [\hat{\mathbf{r}} E(r)] \times \left[\frac{3(\boldsymbol{\mu} \cdot \hat{\mathbf{r}})\hat{\mathbf{r}} - \boldsymbol{\mu}}{4\pi c r^3} \right] \right\} d\tau$$

$$= \; -\int \frac{[\hat{\mathbf{r}} \times (\hat{\mathbf{r}} \times \boldsymbol{\mu})] E(r) d\tau}{4\pi c r^2} = -\int \frac{[\hat{\mathbf{r}}(\hat{\mathbf{r}} \cdot \boldsymbol{\mu}) - \boldsymbol{\mu}] E(r) d\tau}{4\pi c r^2}$$

$$= \; \frac{8\pi \boldsymbol{\mu}}{3} \int_0^\infty \frac{E(r) r^2 dr}{4\pi c r^2} = \frac{2e\boldsymbol{\mu}}{3c} \left[\int_0^R \frac{r dr}{R^3} + \int_R^\infty \frac{dr}{r^2} \right]$$

$$= \; \frac{2e\boldsymbol{\mu}}{3c} \left[\frac{1}{2R} + \frac{1}{R} \right] = \frac{e\boldsymbol{\mu}}{cR}. \tag{9.20}$$

(b) The gyromagnetic ratio for this distribution is

$$G = \frac{\mu}{L} = \frac{cR}{e}. \tag{9.21}$$

The g factor is given by

$$g = \frac{G}{e/(2mc)} = \left(\frac{2mc}{e} \right) \left(\frac{cR}{e} \right) = \frac{2mc^2 R}{e^2}. \tag{9.22}$$

The g factor will equal 2 if

$$R = \frac{e^2}{mc^2} = \frac{\alpha \hbar c}{mc^2} = \frac{197\,\mathrm{MeV} - \mathrm{fm}}{(137)(.511)\,\mathrm{MeV}} = 2.8 \;\mathrm{fm}. \tag{9.23}$$

Note we have used the combination $\hbar c$ to simplify the numerical calculation.

Chapter 10

Problem 10.1:

(a) The intensity of light is given by

$$\overline{\mathbf{S}} = \frac{c\hat{\mathbf{k}}}{8\pi}\sqrt{\frac{\epsilon}{\mu}}|\mathbf{E_0}|^2. \tag{10.1}$$

Solving for E_0 (with $\epsilon = \mu = 1$), we get (for the intensity of sunlight at the Earth's surface of $12\times10^5 \mathrm{ergs/cm^2}$-sec.)

$$
\begin{aligned}
E_0 &= \sqrt{\frac{8\pi\overline{S}}{c}} = \sqrt{\frac{(8\pi)(12\times10^5)}{3\times10^{10}}} = 3.17\times10^{-2}\mathrm{gauss\ (stavolt/cm)} \\
&= 3.17\times10^{-2}\times300(V/sV)\times100(\mathrm{cm/m}) = 950\,\mathrm{V/m}. \tag{10.2}
\end{aligned}
$$

(b) The radiation pressure from light reflected at normal incidence, is given by

$$p_{\mathrm{rad}} = \frac{2\overline{S}\sqrt{\epsilon\mu}}{c} = \frac{2\times12\times10^5}{3\times10^{10}} = 8\times10^{-5}\,\mathrm{dynes/cm^2} \tag{10.3}$$

Atmospheric pressure is $p_{\mathrm{atm}} = 10^6$ dynes/cm^2, so $p_{\mathrm{rad}} \sim 10^{-10}p_{\mathrm{atm}}$.

(c) **(i)** If the sunlight is absorbed, the radiation pressure will be cut in half: $p_{\mathrm{rad}} = 4\times10^{-5}$ dynes/cm^2.

(ii) If the sunlight is diffusely reflected, the normal component of the reflected momentum will vary like $\hat{\mathbf{k}}\cdot\hat{\mathbf{n}}$. This will lead to a factor $\int_0^{\pi/2}\cos\theta\sin\theta d\theta = 1/2$ in the reflection contribution to p_{rad}, and make $p_{\mathrm{rad}} \to (3/4)p_{\mathrm{rad}}$:
$$p_{\mathrm{rad}} = 6\times10^{-5}\ \mathrm{dynes/cm^2}$$

(d) The intensity of solar radiation varies like $1/r^2$. The ratio of the Earth's orbital radius to the Sun's radius is $R = (150)/(.7) = 2,140$. Then the intensity of the solar radiation at the surface of the sun is

$$
\begin{aligned}
\overline{S}_{\mathrm{sun}} &= 12\times10^5\times2140^2 = 5.5\times10^{12}\ \mathrm{ergs/cm^2-sec.} \\
&= 5.5\times10^5\,\mathrm{kW/cm^2}. \tag{10.4}
\end{aligned}
$$

The electric field at the Sun is

$$E_{\text{sun}} = E_{\text{earth}} \times 2140 = 0.032 \times 2140 = 70 \text{ gauss}. \quad (10.5)$$

The radiation pressure (for 100% reflection) at the sun would be

$$p_{\text{sun}} = p_{\text{earth}} \times 2140^2 = 8 \times 10^{-5} \times 2140^2 = 370 \text{ dynes/cm}^2. \quad (10.6)$$

Problem 10.3:

In elliptically polarized light, the planar components of the electric field are given by

$$E'_x = (E_+/\sqrt{2})(1+r)\cos(\omega t - \alpha/2) \quad (10.7)$$
$$E'_y = (E_+/\sqrt{2})(1-r)\sin(\omega t - \alpha/2). \quad (10.8)$$

In this problem, the maximum beam intensity ($9I_0$) is along the x'-axis at 30° above the x-axis, and the minimum intensity (I_0) is along the y'-axis, 30° above the y-axis.

(a) The angle $\alpha/2$ is the angle above the original x-axis for maximum intensity. If maximum intensity occurs at 30°, then $\alpha = 60° = \pi/3$. The ratio $(1+r)/(1-r)$ is the ratio of maximum to minimum electric field strength as a polarizer is rotated in the beam. The intensity of an EM wave with electric field amplitude E_0 is $I = (c/8\pi)|E_0|^2$. The intensity $\sim E^2$, so

$$\left(\frac{1+r}{1-r}\right)^2 = 9, \quad \text{and } r = \frac{1}{2}. \quad (10.9)$$

From Eq. (10.8) above, we see that the minimum intensity is given by

$$I_{\text{min}} = I_0 = \frac{c}{16\pi}[E_+(1-r)]^2 = \frac{c|E_+|^2}{64\pi}. \quad (10.10)$$

Then, $E_+ = 8\sqrt{\pi I_0/c},$ \quad (10.11)

and $E_- = re^{i\alpha}E_+ = \frac{1}{2}e^{i\pi/3}E_+ = 4e^{i\pi/3}\sqrt{\pi I_0/c}.$ \quad (10.12)

(b) The fields E_x and E_y in the planar basis are given by [Eq. (10.50)] (with $\alpha = \pi/3$ and $r = 1/2$):

$$\begin{pmatrix} E_x \\ E_y \end{pmatrix} = \frac{E_+}{\sqrt{2}}\begin{bmatrix} \sqrt{3}/2 - 1/2 \\ 1/2 + \sqrt{3}/2 \end{bmatrix}\begin{pmatrix} (3/2)\cos(\omega t - \pi/6) \\ (1/2)\sin(\omega t - \pi/6) \end{pmatrix}. \quad (10.13)$$

After matrix multiplication:

$$E_x = \frac{E_+}{4\sqrt{2}}[3\sqrt{3}\cos(\omega t - 30°) - \sin(\omega t - 30°)]$$

$$= \frac{E_+}{4\sqrt{2}}[5\cos(\omega t) + \sqrt{3}\sin(\omega t)] \qquad (10.14)$$

$$E_y = \frac{E_+}{4\sqrt{2}}[3\cos(\omega t - 30°) + \sqrt{3}\sin(\omega t - 30°)]$$

$$= \frac{E_+}{4\sqrt{2}}[\sqrt{3}\cos(\omega t) + 3\sin(\omega t)]. \qquad (10.15)$$

The transmitted intensity if a polarizer is placed along the x-axis will be

$$I_x = \frac{c|E_x|^2}{8\pi} = \frac{c|E_+|^2}{8\pi}\left(\frac{27+1}{32}\right) = 7\frac{|E_+|^2}{64\pi} = 7I_0. \qquad (10.16)$$

Similarly, $$I_y = \left(\frac{3+9}{4}\right)I_0 = 3I. \qquad (10.17)$$

Problem 10.5:

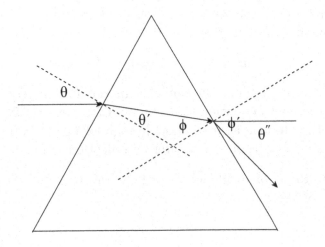

The angles are a bit tricky in this problem. We have to proceed in steps, angle by angle.

1. The incident angle is $\theta = 30°$, and (by Snell's law) the refracted angle of the ray in the prism is given by

$$\sin\theta' = (1/n)\sin\theta = (1/1.5)(1/2) = 1/3, \rightarrow \theta' = 19.5°. \qquad (10.18)$$

2. The angle θ' is the angle between the direction of the ray in the prism and the normal to the first face of the prism. It is related to the angle of incidence ϕ for the ray leaving the prism by $\theta' + \phi = 60°$. This is because the angles θ'

and ϕ are in the same triangle, whose third angle is 120°. (120° is the angle between perpendiculars to two sides of an equilateral triangle.) Thus

$$\phi = 60° - \theta' = 40.5°. \tag{10.19}$$

3. Applying Snell's law again gives

$$\sin \phi' = n \sin \phi = 1.5 \sin 40.5° = 0.974, \rightarrow \phi' = 77°. \tag{10.20}$$

4. The angle ϕ' is the angle between the emerging ray and the normal to the second face of the prism. It is related to the angle θ'', the angle below horizontal of the emerging ray by $\theta'' = \phi' - 30°$. This is because the angle between the normal to the second face of the equilateral triangle and the horizontal is 30°. Thus

$$\theta'' = \phi' - 30° = 47°. \tag{10.21}$$

This is the angle below horizontal of the emerging beam.

Problem 10.7:

(a) For a rectangular fish tank, the angle of refraction θ' as a ray enters the water (from the back) is equal to the angle of incidence ϕ as the ray leaves the water because the back face is parallel to the front face. Applying Snell's law at each face gives, for the final emerging ray ϕ',

$$\sin \phi' = n \sin \phi = n \sin \theta' = \sin \theta, \tag{10.22}$$

so that the entering angle (θ) and the emerging angle (ϕ') are equal. We see from the above equation that any ray entering the water at the back of the tank will have a small enough angle of incidence leave the water at the front end without total internal reflection.

(b) The largest angle of incidence for a light ray to enter the water at the right face of the fish tank is $\theta = 90°$.

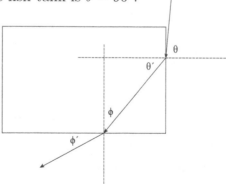

This means that the largest angle of refraction entering the water is given by

$$\sin \theta' = (n_{\text{air}}/n_{\text{water}}) \sin 90° = 3/4 \rightarrow \theta' = 48.6°. \tag{10.23}$$

This is the largest angle of refraction for any ray entering the water. (It equals the critical angle.) The angle θ' is related to the angle of incidence ϕ as the ray leaves the front of the fish tank (so you can see it) by $\phi = 90° - \theta' = 41.4°$. This is because the front face and the right face of the fish tank are perpendicular. The angle at which this ray would leave the fish tank's front face is given by Snell's law:

$$\sin \phi' = n \sin \phi = 1.33 \sin 41.4° = 0.882, \rightarrow \phi' = 62°. \tag{10.24}$$

This is the largest angle from the normal to the front face from which you could see through the right face. For any angle closer to the normal direction, the right face would look like a mirror.

(c) There would always be total internal reflection from the right face if n were large enough the the maximum angle in part (b) were 90°. This means that both the angle ϕ at which the light is incident on the front face (on the way to your eye) and the angle θ' with which the light enters the water from the right face are the critical angle θ_c. But the two angles add up to 90°. This means that the critical angle must equal 45°. Then

$$\sin \theta_c = \frac{1}{n}, \quad \text{so } n = \frac{1}{\sin \theta_c} = \frac{1}{\sin 45°} = \sqrt{2} = 1.4. \tag{10.25}$$

(d) The glass does not effect this problem because the angle at which light leaves the glass equals the angle at which light enters the glass. This was shown in part (a).

Problem 10.9:

(a) The boundary conditions for a coated surface (with $\mu = 1$) are given by

$$E_1 + E_1' = E_2 + E_2' e^{ik_2 d} \tag{10.26}$$
$$n_1(E_1 - E_1') = n_2(E_2 - E_2' e^{ik_2 d}) \tag{10.27}$$
$$E_2 e^{ik_2 d} + E_2' = E_3 \tag{10.28}$$
$$n_2(E_2 e^{ik_2 d} - E_2') = n_3 E_3. \tag{10.29}$$

To solve these equations for the transmitted field E_3 and the reflected field E_1' in terms of the incident field E_1, we first solve the last two equations for E_2 and E_2' in terms of E_3. We can do this by multiplying Eq. (10.28) by n_2, and then adding and subtracting the last two equations. This gives

$$E_2 = \frac{(n_2 + n_3)e^{-ik_2 d}}{2n_2} E_3 \tag{10.30}$$

$$E_2' = \frac{(n_2 - n_3)}{2n_2} E_3. \tag{10.31}$$

Substituting these two equations into the first two equations gives

$$E_1 + E_1' = \left[\frac{(n_2 + n_3)e^{-ik_2d} + (n_2 - n_3)e^{ik_2d}}{2n_2}\right] E_3 \quad (10.32)$$

$$n_1(E_1 - E_1') = n_2\left[\frac{(n_2 + n_3)e^{-ik_2d} - (n_2 - n_3)e^{ik_2d}}{2n_2}\right] E_3. (10.33)$$

We rewrite these two equations as

$$n_1 n_2 (E_1 + E_1') = [n_1 n_2 \cos(k_2 d) - i n_1 n_3 \sin(k_2 d)] E_3 \quad (10.34)$$
$$n_1 n_2 (E_1 - E_1') = [n_2 n_3 \cos(k_2 d) - i n_2^2 \sin(k_2 d)] E_3. \quad (10.35)$$

Adding these two equations gives

$$2 n_1 n_2 E_1 = [n_2(n_1 + n_3)\cos(k_2 d) - i(n_1 n_3 + n_2^2)\sin(k_2 d)] E_3, \quad (10.36)$$

which can be solved for E_3:

$$E_3 = \left[\frac{2n_1 n_2}{n_2(n_1 + n_3)\cos(k_2 d) - i(n_1 n_3 + n_2^2)\sin(k_2 d)}\right] E_1. \quad (10.37)$$

Adding Eqs. (10.34) and (10.35) gives

$$2 n_1 n_2 E_1' = [n_2(n_1 - n_3)\cos(k_2 d) - i(n_1 n_3 - n_2^2)\sin(k_2 d)] E_3. \quad (10.38)$$

We use this equation and Eq. (10.37 to give E_1' in terms of E_1:

$$E_1' = \left[\frac{n_2(n_1 - n_3)\cos(k_2 d) - i(n_1 n_3 - n_2^2)\sin(k_2 d)}{n_2(n_1 + n_3)\cos(k_2 d) - i(n_1 n_3 + n_2^2)\sin(k_2 d)}\right] E_1, \quad (10.39)$$

which is [Eq. (10.122)] in the text.

(b) The transmission coefficent is given by

$$T = \frac{n_3|E_3|^2}{n_1|E_1|^2} = \frac{4n_1 n_2^2 n_3}{n_2^2(n_1 + n_3)^2 \cos^2(k_2 d) + (n_1 n_3 + n_2^2)^2 \sin^2(k_2 d)}. \quad (10.40)$$

The reflection coefficient is given by

$$R = \frac{|E_1'|^2}{|E_1|^2} = \frac{n_2^2(n_1 - n_3)^2 \cos^2(k_2 d) + (n_1 n_3 - n_2^2)^2 \sin^2(k_2 d)}{n_2^2(n_1 + n_3)^2 \cos^2(k_2 d) + (n_1 n_3 + n_2^2)^2 \sin^2(k_2 d)}. \quad (10.41)$$

These are the same as [Eqs. (10.123) and (10.124)] in the text.

Note: In addition to the conditions $n_2^2 = n_1 n_3$ and $k_2 d = \pi/2$ which make $R = 0$ (as pointed out in the text), there is another possibility: $n_1 = n_3$ and $k_2 d = \pi$. This might be achieved in optical systems by a thin air space between two glass plates, but would require an extremely small separation d.

(c) The index of refraction for the interface is

$$n_2 = \sqrt{n_1 n_3} = \sqrt{1.5} = 1.22. \tag{10.42}$$

The thickness of the interface for 5,000Å incident light is

$$d = \frac{\lambda_1}{4} \sqrt{\frac{n_1}{n_3}} = \frac{5000}{4} \sqrt{\frac{1}{1.5}} = 1,020 \text{Å}. \tag{10.43}$$

Chapter 11

Problem 11.1:

(a)

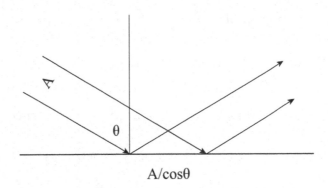

A/cosθ

The radiation pressure deduced for a plane wave using conservation of momentum is given for normal incidence by [Eq. (10.33)]:

$$p_{\text{rad}} = \frac{\epsilon}{4\pi} E_0^2, \quad \text{normal incidence.} \tag{11.1}$$

For a plane wave incident at an angle θ onto the flat surface of a perfect conductor, two factors of $\cos\theta$ enter. One factor of $\cos\theta$ is because only the normal component of the wave produces the force. The other factor of $\cos\theta$ enters because a portion of the beam of area A will reflect off an area $(A/\cos\theta)$ of the surface. Thus the radiation pressure is given by

$$p_{\text{rad}} = \frac{\epsilon}{4\pi} E_0^2 \cos^2\theta. \tag{11.2}$$

(b) For a wave that is polarized perpendicular the plane of incidence, the electric field is parallel to the surface and does not produce a force. The magnetic component exerts a force on the induced current. The radiation pressure due to this is derived in [Sec. 11.2.2], resulting, for normal incidence, in [Eq. (11.34)]:

$$p_{\text{rad}} = \frac{1}{4\pi\mu} |B_0|^2, \quad \text{normal incidence.} \tag{11.3}$$

For incidence at an angle θ, the only change in the derivation is in [Eq. (11.32)] relating the total **B** field to the incident field $(\mathbf{B_0})$ and the reflected field. This equation now becomes

$$\mathbf{B} = \mathbf{B_0} + \mathbf{B}(\text{reflected}) = 2\cos\theta\, B_0 \hat{\mathbf{t}}, \tag{11.4}$$

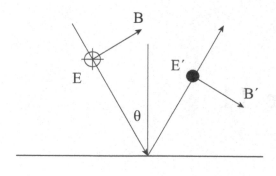

where $\hat{\mathbf{t}}$ is a unit vector perpendicular to the scattering plane (parallel to the surface). This makes the final result for the radiation pressure

$$p_{\text{rad}} = \frac{1}{4\pi\mu}B_0^2\cos^2\theta = \frac{\epsilon}{4\pi}E_0^2\cos^2\theta. \tag{11.5}$$

(c) For a wave that is polarized parallel to the plane of incidence, The magnetic

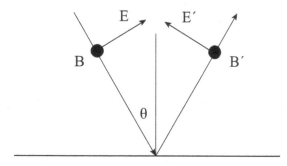

field will be parallel to the surface, and will have the same contribution to the radiation pressure as for normal incidence. The incident and reflected electric fields will each make an angle θ with the surface, and each will have a component normal to the surface given by $E_\perp = E\sin\theta$. Since the reflected and incident fields are equal in magnitude, the total $E_\perp$ at the surface will be $E_\perp = \mathbf{E}\cdot\hat{\mathbf{n}} = 2E_{\text{inc}}\sin\theta$, where $\hat{\mathbf{n}}$ is a normal unit vector out of the surface. This will induce a surface charge of $\sigma = -2E_{\text{inc}}\sin\theta/4\pi$. (The minus sign is because $\hat{\mathbf{n}}$ is an outward normal from the surface.) There will be a force per unit area (pressure) on this surface charge given by $p = \frac{1}{2}\sigma E_\perp$, the factor of $\frac{1}{2}$ arising because one-half of $E_\perp$ is caused by the surface charge itself. (This is discussed more fully, for the $\mathbf{H}$ field, in [Sec. 11.2.2].) Thus there is a force on the induced surface charge of

$$\begin{aligned} p_{\sigma E} &= \frac{1}{2}\sigma E_\perp = \frac{1}{2}(-2E_{\text{inc}}\sin\theta/4\pi)(2E_{\text{inc}}\sin\theta) \\ &= -\frac{\epsilon}{2\pi}E_0^2\sin^2\theta\cos^2(\omega t). \end{aligned} \tag{11.6}$$

Here we have included the oscillating time dependence of $\mathbf{E}$. One more factor of $\frac{1}{2}$ enters due in the time average of E^2, because $\overline{\cos^2(\omega t)} = \frac{1}{2}$. Finally, the electric field pressure is

$$\overline{p}_{\sigma E} = -\frac{\epsilon}{4\pi} E_{\text{inc}}^2 \sin^2 \theta. \tag{11.7}$$

The negative sign signifies that this electric force is actually a pull on the surface charge. Adding the forces due to **B** and **E** gives the total radiation pressure

$$p_{\text{rad}} = \frac{\epsilon}{4\pi} E_0^2 - \frac{\epsilon}{4\pi} E_0^2 \sin^2 \theta = \frac{\epsilon}{4\pi} E_0^2 \cos^2 \theta. \tag{11.8}$$

We see that all three calculations give the same p_{rad}.

Problem 11.3:

(a) The permittivity is given by [Eq. (11.52)]:

$$\epsilon = 1 + \frac{4\pi N \sum_i z_i \eta_i}{1 - \frac{4\pi}{3} N \sum_i z_i \eta_i}, \qquad \eta_i = \frac{e^2}{m[\omega_i^2 - \omega^2 - i\omega\gamma_i]}. \tag{11.9}$$

The imaginary part of ϵ will only be appreciable when ω is near a resonance, so we break the sum into one nearby resonance and a remaining sum over the distant resonances. Then

$$\begin{aligned}
\epsilon &= 1 + \frac{3A + 3B/(\omega_i^2 - \omega^2 - i\omega\gamma_i)}{1 - A - B/(\omega_i^2 - \omega^2 - i\omega\gamma_i)} \\
&= 1 + \frac{3A(\omega_i^2 - \omega^2 - i\omega\gamma_i) + 3B}{(1 - A)(\omega_i^2 - \omega^2 - i\omega\gamma_i) - B},
\end{aligned} \tag{11.10}$$

where

$$A = \frac{4\pi N}{3} \sum_{j \neq i} z_j \eta_j, \qquad B = \frac{4\pi}{3} N z_i e^2 / m. \tag{11.11}$$

We can rewrite ϵ as

$$\begin{aligned}
\epsilon &= 1 + \frac{3A[(\omega_i^2 - \omega^2 - i\omega\gamma_i) + B/A - B/(1 - A) + B/(1 - A)]}{(1 - A)[(\omega_i^2 - \omega^2 - i\omega\gamma_i) - B/(1 - A)]} \\
&= 1 + \frac{3A}{1 - A} + \frac{3B + 3AB/(1 - A)}{(1 - A)(\omega_i^2 - \omega^2 - i\omega\gamma_i) - B} \\
&= \frac{1 + 2A}{1 - A} + \frac{3B/(1 - A)^2}{\omega_i^2 - \omega^2 - i\omega\gamma_i - B/(1 - A)}.
\end{aligned} \tag{11.12}$$

Letting $C = B/(1 - A)$, we can write

$$\begin{aligned}
\epsilon &= \frac{1 + 2A}{1 - A} + \frac{3C/(1 - A)}{\omega_i^2 - \omega^2 - C - i\omega\gamma_i} \tag{11.13} \\
&= \frac{1 + 2A}{1 - A} + \frac{3C(\omega_i^2 - \omega^2 - C + i\omega\gamma_i)}{(1 - A)[(\omega_i^2 - \omega^2 - C)^2 + (\omega\gamma_i)^2]} \tag{11.14}
\end{aligned}$$

This form displays the real and imaginary parts of ϵ.

Note: This problem is easier if A is assumed to be negligible from the start. Then $A = 0$ and $C = B$. This approximation is used in [Eq. (11.109)].

(b) To locate the singularities of ϵ, we find the zeros of the denominator in [Eq. (11.13)]:

$$\omega_\pm = -i(\gamma/2) \pm \sqrt{-(\gamma/2)^2 + \omega_i^2 - C}. \tag{11.15}$$

Both zeros are in the lower half plane. It is unlikely that the argument of the square root could be negative enough to change this. That would require $C > \omega_i$, which would make $\epsilon < 1$ at low frequencies. (See Eq. (11.10).

Problem 11.5: **(a)** Plot of $|f(x)|^2$ for $f(x) = e^{im\pi x/L}$ $0 < x < L$:

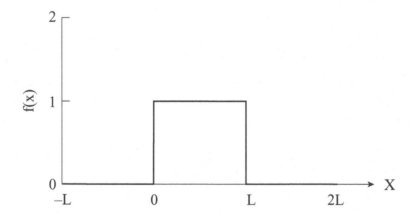

(b) The Fourier transform of $f(x)$ is

$$
\begin{aligned}
A(k) &= \frac{1}{2\pi} \int_0^L e^{im\pi x/L} e^{-ikx} dx = \frac{1}{2\pi} \int_0^L e^{-ix(k - m\pi/L)} dx \\
&= \frac{1}{2\pi} \left[\frac{e^{-ix(k - m\pi/L)}}{-i(k - m\pi/L)} \right]_0^L = \frac{1 - e^{-i(kL - m\pi)}}{2\pi i(k - m\pi/L)} = \frac{[1 - (-1)^m e^{-ikL}]}{2\pi i(k - m\pi/L)} \\
&= \frac{e^{-ikL/2} \sin(kL/2)}{\pi(k - m\pi/L)}, \quad m \text{ even}, \tag{11.16} \\
&= \frac{e^{-ikL/2} \cos(kL/2)}{i\pi(k - m\pi/L)}, \quad m \text{ odd} \tag{11.17}
\end{aligned}
$$

The absolute value of $A(k)$ is

$$|A(k)|^2 = \frac{\sin^2(kL/2)}{[\pi(k - m\pi/L)]^2}, \quad m \text{ even} \tag{11.18}$$

$$|A(k)|^2 = \frac{\cos^2(kL/2)}{[\pi(k - m\pi/L)]^2}, \quad m \text{ odd}. \tag{11.19}$$

For m either even or odd, $|A|^2$ has a maximum at $k = m\pi/L$. Double application of L'Hospital's rule gives the maximum value of $|A|^2$ as $A_0^2 = L^2/4\pi^2$. The first zeros of $|A|^2$ occur at $k = m\pi/L \pm 2\pi/L$, with recurring zeros and smaller maxima as k increases or decreases. The plot of $|A|^2$ is:

Fourier Transforms

(c) The range in x of $f(x)$ is from 0 to L, so a reasonable Δx is $\Delta x = L/2$. A reasonable Δk for $|A(k)|^2$ is when $|A|^2$ is half its maximum. From the plot of $|A|^2$, this is about $\Delta k = \pi/L$. Then $\Delta x \Delta k = \pi/2$ for this case.

Problem 11.7:
(a) We assume that the given red (4,500 Å) and blue (6,500 Å) wavelengths are for vacuum, and that the formula $n = 1.350 - 100\text{Å}/\lambda$ holds for the wavelength in the glass. We first solve for λ in the glass in terms of λ_0, the wave length in vacuum (or air):

$$\lambda = \frac{c}{\nu n} = \frac{\lambda_0}{n} = \frac{\lambda_0}{1.35 - 100/\lambda} = \frac{\lambda_0 + 100}{1.35}. \tag{11.20}$$

The wavelengths in the glass are

$$\lambda_{\text{red}} = \frac{4500 + 100}{1.35} = 3410\text{Å}; \qquad \lambda_{\text{blue}} = \frac{6500 + 100}{1.35} = 4890\text{Å}. \tag{11.21}$$

The angle in the glass is determined by Snell's law:

$$\sin\theta = \frac{\sin 30°}{n} = \frac{1}{2(1.35 - 100/\lambda)}, \tag{11.22}$$

$$\sin\theta_{\text{red}} = \frac{1}{2(1.35 - 100/3410)} = 0.379 \to \theta_{\text{red}} = 22.25° \tag{11.23}$$

$$\sin\theta_{\text{blue}} = \frac{1}{2(1.35 - 100/4890)} = 0.0.376 \to \theta_{\text{blue}} = 22.09°. \tag{11.24}$$

The angular difference is $\Delta\theta = 22.25 - 22.09 = 0.16°$.

(b) The phase velocities in the glass are

$$v_p = \frac{c}{n} = \frac{c}{1.35 - 100/\lambda}, \tag{11.25}$$

$$v_p(\text{red}) = \frac{c}{1.35 - 100/3410} = 0.757\,c, \tag{11.26}$$

$$v_p(\text{blue}) = \frac{c}{1.35 - 100/4890} = 0.752\,c. \tag{11.27}$$

The group velocities are given by [Eq. (11.104)]:

$$v_g = \frac{c}{n}\left(1 + \frac{\lambda}{n}\frac{dn}{d\lambda}\right). \tag{11.28}$$

The derivative is

$$\frac{dn}{d\lambda} = \frac{d}{d\lambda}(1.35 - 100/\lambda) = 100/\lambda^2, \tag{11.29}$$

$$\text{so}\quad v_g = \frac{c}{n}\left(1 + \frac{100}{n\lambda}\right) = \frac{c}{n^2}(1.35 - 100/\lambda + 100/\lambda) = \frac{1.35c}{n^2} = \frac{1.35v_p}{n}, \tag{11.30}$$

$$v_g(\text{red}) = \frac{1.35c}{(1.35 - 100/3410)^2} = 0.774\,c, \tag{11.31}$$

$$v_g(\text{blue}) = \frac{1.35c}{(1.35 - 100/4890)^2} = 0.764\,c. \tag{11.32}$$

Problem 11.9:
Setting the numerator of [Eq. (11.110)] to zero gives the quadratic equation

$$\omega^2 + i\omega\gamma - \omega_i^2 - 2b/3 = 0, \tag{11.33}$$

with the solution
$$\omega = -i\gamma/2 \pm \sqrt{\gamma^2/4 + \omega_i^2 + 2b/3}. \tag{11.34}$$

Setting the denominator of [Eq. (11.110)] to zero gives the quadratic equation

$$\omega^2 + i\omega\gamma - \omega_i^2 + b/3 = 0, \tag{11.35}$$

with the solution
$$\omega = -i\gamma/2 \pm \sqrt{\gamma^2/4 + \omega_i^2 - b/3}. \tag{11.36}$$

The argument of the square root is positive if ϵ is positive at low frequencies. **Note:** The zeros and poles of $\epsilon(\omega)$ are in the lower half plane, so we see that the square root singularities of $n(\omega)$ in [Eq. (11.112)] are all in the lower half plane.

Chapter 12

Problem 12.1:

For a coaxial wave guide with copper cylinders of inner radius $A = 0.10$ cm and outer radius $B = 0.40$ cm with the outer cylinder grounded and the inner cylinder at a potential $V=120$ Volts (V=120/300=0.40 statvolts) oscillating at a frequency $\nu=12$ MHz,

(a) the electric field between the cylinders is given by (omitting time dependence for now)

$$\mathbf{E(r)} = \frac{AE_A}{r}\hat{\mathbf{r}}, \tag{12.1}$$

where the constant E_A is the mgnitude of the electric field at the surface of the inner cylinder. E_A is related to the potential of the inner cylinder by

$$V = \int_A^B \mathbf{E \cdot dr} = \int_A^B \frac{AE_A dr}{r} = AE_A \ln\left(\frac{B}{A}\right), \quad \text{so} \quad \mathbf{E}$$

$$= \frac{V}{r\ln(B/A)}\hat{\mathbf{r}}. \tag{12.2}$$

The electric field at the inner cylinder is

$$\mathbf{E} = \frac{V\hat{\mathbf{r}}e^{-i\omega t}}{A\ln(B/A)} = \frac{(.4)\hat{\mathbf{r}}\exp[-(2\pi i)(12 \times 10^6 t)]}{(.1)(\ln 4)} = 2.9\hat{\mathbf{r}}\exp[-24\pi i \times 10^6 t] \text{ gauss.} \tag{12.3}$$

The magnetic field between the cylinders is given by [Eq. (12.34) (with $\epsilon=2.0$ and $\mu=0$):

$$\mathbf{H} = \sqrt{\frac{\epsilon}{\mu}}\hat{\mathbf{k}}\times\mathbf{E} = \frac{\sqrt{2}V\hat{\theta}e^{-i\omega t}}{r\ln(B/A)}, \tag{12.4}$$

and the magnetic field at the inner cylinder is

$$\mathbf{H} = 2.9\sqrt{2}\hat{\theta}\exp[-24\pi i \times 10^6 t] \text{ gauss.} \tag{12.5}$$

(b) The transmitted power is given by [Eq. (12.70)]:

$$P = \frac{c}{8\pi}\int(\mathbf{E^*}\times\mathbf{H})\cdot d\mathbf{A} = \frac{c}{8\pi}\sqrt{\frac{\epsilon}{\mu}}\int|\mathbf{E}|^2 dA$$

561

$$= \frac{c}{4}\sqrt{\frac{\epsilon}{\mu}}\left(\frac{V}{\ln(B/A)}\right)^2 \int_A^B \frac{r\,dr}{r^2}$$

$$= \frac{cV^2\sqrt{\epsilon/\mu}}{4\ln(B/A)} = \frac{(3\times 10^{10})(.4^2)(\sqrt{2})}{(4)\ln(4)}$$

$$= 0.12\times 10^{10}\text{ ergs/sec} = 120\text{ Watts.} \tag{12.6}$$

(c) The rate of power loss is given by [Eq. (12.82)]:

$$\frac{dP}{dz} = -\frac{c}{8\pi}\sqrt{\frac{\omega\mu_c}{8\pi\sigma}}\oint |\mathbf{H}_{\text{surface}}|^2 dl = -\frac{c}{8\pi}\sqrt{\frac{\omega\mu_c}{8\pi\sigma}}\left[\frac{(\epsilon/\mu)V^2}{\ln^2(B/A)}\right] 2\pi\left(\frac{1}{A}+\frac{1}{B}\right)$$

$$= -\sqrt{\frac{\omega\mu_c}{8\pi\sigma}}\left[\frac{c(\epsilon/\mu)V^2}{4\ln^2(B/A)}\right]\left(\frac{A+B}{AB}\right)$$

$$= -\sqrt{\frac{\omega\mu_c}{8\pi\sigma}}\left[\frac{\sqrt{\epsilon/\mu}}{4\ln(B/A)}\right]\left(\frac{A+B}{AB}\right)\times P. \tag{12.7}$$

The attenuation length for a coaxial wave guide is

$$L_{\text{atten}} = \frac{P}{-dP/dz} = \sqrt{\frac{8\pi\sigma}{\omega\mu_c}}\left[\frac{4\ln(B/A)}{\sqrt{\epsilon/\mu}}\right]\left(\frac{AB}{A+B}\right). \tag{12.8}$$

Putting in numbers (using $\mu_c = 1$ and $\sigma_c = 54\times 10^{16}$ sec^{-1}:

$$L_{\text{atten}} = \sqrt{\frac{(4)(54\times 10^{16})}{12\times 10^6}}\left[\frac{4\ln(4)}{\sqrt{2}}\right]\left(\frac{4}{5}\right) = 1.3\times 10^6\text{ cm} = 13\text{ km.} \tag{12.9}$$

Problem 12.3:

(a) For a rectangular wave guide of dimensions 0.40×0.60 cm, the TM cutoff frequencies are given by [Eq. (12.57)]. The TM cutoff frequencies in Hz are

$$\nu_{lm} = \frac{c}{2\sqrt{\epsilon\mu}}\left(\frac{l^2}{A^2}+\frac{m^2}{B^2}\right)^{\frac{1}{2}} = \frac{c}{2}\left(\frac{l^2}{.16}+\frac{m^2}{.36}\right)^{\frac{1}{2}} = \frac{c}{2.4}\sqrt{9l^2+4m^2}. \quad (12.10)$$

The first five TM cutoff frequencies are

$$
\begin{array}{llll}
(1) & \nu_{11} & = \frac{3\times10^{10}}{2.4}\sqrt{9+4} & = 45\,\text{GHz} & (12.11)\\
(2) & \nu_{12} & = 12.5\times10^9\sqrt{9+16} & = 63\,\text{GHz} & (12.12)\\
(3) & \nu_{21} & = 12.5\times10^9\sqrt{36+4} & = 79\,\text{GHz} & (12.13)\\
(4) & \nu_{13} & = 12.5\times10^9\sqrt{9+36} & = 84\,\text{GHz} & (12.14)\\
(5) & \nu_{22} & = 12.5\times10^9\sqrt{36+16} & = 90\,\text{GHz} & (12.15)
\end{array}
$$

(c) The first five TE cutoff frequencies are given by the same formula, but either l or m can equal zero. The first five TE frequencies are

$$
\begin{array}{llll}
(1) & \nu_{01} & = 12.5\times10^9\sqrt{0+4} & = 25\,\text{GHz} & (12.16)\\
(2) & \nu_{10} & = 12.5\times10^9\sqrt{9+0} & = 37\,\text{GHz} & (12.17)\\
(3) & \nu_{11} & = 12.5\times10^9\sqrt{9+4} & = 45\,\text{GIIz} & (12.18)\\
(4) & \nu_{02} & = 12.5\times10^9\sqrt{0+16} & = 50\,\text{GHz} & (12.19)\\
(5) & \nu_{12} & = 12.5\times10^9\sqrt{9+16} & = 63\,\text{GHz} & (12.20)
\end{array}
$$

(**b,d**) Nodal planes (shown as heavy lines) for 4 mm×6 mm rectangular wave guide:

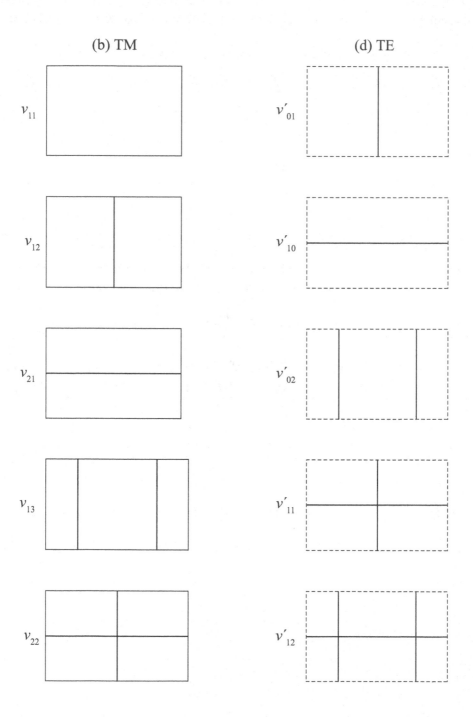

Problem 12.5:

(a) The first two TE cutoff frequencies of the .4cm×.6cm waveguide of Prob. 12.3 are 25 GHz ($=\nu_c$) and 37 GHz, so the frequency midway between them is $\nu = (25 + 37)/2 = 31$ GHz$=1.24\nu_c$, The power for the lowest TE mode in a rectangular wave guide is given by

$$P = 2 \times \frac{\mu \omega k H_0^2 AB}{32\pi\gamma^2} = \sqrt{\frac{\mu}{\epsilon}} \frac{cH_0^2 AB}{32\pi}(\nu/\nu_c)\sqrt{(\nu/\nu_c)^2 - 1}. \qquad (12.21)$$

Putting in numbers:

$$\begin{aligned} P &= \frac{(1)(3 \times 10^{10})}{32\pi}(.2)^2(.4)(.6)(1.24)\sqrt{1.24^2 - 1} \\ &= 0.23 \times 10^7 \text{ ergs/sec} = 0.23 \text{ Watts.} \end{aligned} \qquad (12.22)$$

(b) The geometric factor for the lowest TE mode in a rectangular (.4 cm×.6 cm) wave guide is given by [Eq. (12.88)]:

$$G_{01} = \frac{A^3 B}{A^2 B/2 + (\nu_c/\nu)^2 A^3} = \frac{AB}{B/2 + (\nu_c/\nu)^2 A} = \frac{(.4)(.6)}{.6/2 + (1/1.24)^2 \times .4} = 0.348 \text{ cm.}$$

$$(12.23)$$

The attenuation length is given by [Eq. (12.86)]:

$$\begin{aligned} L_{\text{atten}} &= \left[\frac{\mu\sigma_c}{4\epsilon\mu_c}\right]^{\frac{1}{2}} \left[\frac{\sqrt{1 - \nu_c^2/\nu^2}}{\sqrt{\nu}}\right] G_{lm}, \\ &= \left[\frac{(1)(.5 \times 10^{18})}{(4)(1)(1)}\right]^{\frac{1}{2}} \left[\frac{\sqrt{1 - 1/1.24^2}}{\sqrt{2.5 \times 10^{10})}}\right] \times .446 \\ &= 0.059 \times 10^4 \text{ cm} = 5.9 \text{ meters.} \end{aligned} \qquad (12.24)$$

Problem 12.7

(a) For a copper circular wave guide of radius $R=0.40$ cm filled with air, the cutoff frequencies are given by

$$
\begin{aligned}
\nu_{ml} &= \frac{\omega_{ml}}{2\pi} = \frac{c\gamma_{ml}}{2\pi\sqrt{\epsilon\mu}} = \frac{cj_{ml}}{2\pi R} = \frac{3\times 10^{10}}{2\pi \times .4}\, j_{ml} \\
&= 11.94\times 10^{9}\, j_{ml} = 11.94\, j_{ml} \text{ GHz},
\end{aligned}
\tag{12.25}
$$

where j_{ml} is the lth zero of the Bessel function J_m, defined by $J_m(j_{ml}) = 0$.

Using Table [12.2] in the text for the j_{ml}, the first five TM cutoff frequencies are

$$
\begin{aligned}
\nu_{01} &= 11.94\times 2.4048 &=& \quad 29 \text{ GHz} & (12.26)\\
\nu_{11} &= 11.94\times 3.8317 &=& \quad 46 \text{ GHz} & (12.27)\\
\nu_{21} &= 11.94\times 5.1356 &=& \quad 61 \text{ GHz} & (12.28)\\
\nu_{02} &= 11.94\times 5.5200 &=& \quad 66 \text{ GHz} & (12.29)\\
\nu_{12} &= 11.94\times 7.0156 &=& \quad 84 \text{ GHz}. & (12.30)
\end{aligned}
$$

(c) The TE cutoff frequencies for this wave guide are given by

$$
\nu'_{ml} = \frac{cj'_{ml}}{2\pi R} = 11.94\times 10^{9}\, j'_{ml},
\tag{12.31}
$$

where j'_{ml} is the lth zero of the derivative of the Bessel function J_m, defined by $J'_m(j'_{ml}) = 0$.
Using Table [12.2] for the j'_{ml}, the first five TE cutoff frequencies are

$$
\begin{aligned}
\nu'_{11} &= 11.94\times 1.8412 &=& \quad 22 \text{ GHz} & (12.32)\\
\nu'_{21} &= 11.94\times 3.0542 &=& \quad 36 \text{ GHz} & (12.33)\\
\nu'_{02} &= 11.94\times 3.8317 &=& \quad 46 \text{ GHz} & (12.34)\\
\nu'_{12} &= 11.94\times 5.3314 &=& \quad 64 \text{ GHz} & (12.35)\\
\nu'_{22} &= 11.94\times 6.7061 &=& \quad 80 \text{ GHz}. & (12.36)
\end{aligned}
$$

Nodal surfaces (shown as heavy lines) for circular wave guide:

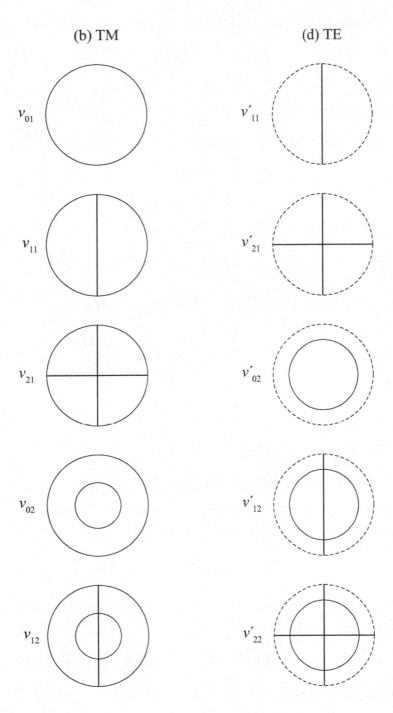

(b) TM

v_{01}

v_{11}

v_{21}

v_{02}

v_{12}

(d) TE

v'_{11}

v'_{21}

v'_{02}

v'_{12}

v'_{22}

Problem 12.9:

For a TM mode in a cavity of length L the field E_z satifies the end conditions $\partial_z E_z = 0$, so

$$E_z(\mathbf{r_T}, z) = \cos(n\pi z/L)\psi(\mathbf{r_T}), \qquad (12.37)$$

where $\psi(\mathbf{r_T})$ is a solution of the transverse eigenvalue equation. Using [Eq. (12.91)] (with $H_z{=}0$), we can write

$$\boldsymbol{\nabla}_{\mathbf{T}}{\times}(\boldsymbol{\nabla}_{\mathbf{T}}{\times}\mathbf{E_T}) = 0 = \boldsymbol{\nabla}_{\mathbf{T}}(\boldsymbol{\nabla}_{\mathbf{T}}{\cdot}\mathbf{E_T}) - \nabla_T^2\mathbf{E_T} = \boldsymbol{\nabla}_{\mathbf{T}}(\boldsymbol{\nabla}_{\mathbf{T}}{\cdot}\mathbf{E_T}) + \gamma^2\mathbf{E_T}.$$
$$(12.38)$$

Then, using [Eq. (12.89)]:

$$\begin{aligned}\mathbf{E_T} &= -\frac{1}{\gamma^2}\boldsymbol{\nabla}_{\mathbf{T}}(\boldsymbol{\nabla}_{\mathbf{T}}{\cdot}\mathbf{E_T}) = \frac{1}{\gamma^2}\boldsymbol{\nabla}(\partial_z E_z) \\ &= -\frac{n\pi}{L\gamma^2}\sin(n\pi z/L)\boldsymbol{\nabla}_{\mathbf{T}}\psi \quad [\text{Eq. (12.102)}].\end{aligned} \qquad (12.39)$$

From [Eq. (12.92) and (12.90)] (with $H_z = 0$), we can write

$$\begin{aligned}\boldsymbol{\nabla}_{\mathbf{T}}{\times}(\boldsymbol{\nabla}_{\mathbf{T}}{\times}\mathbf{H_T}) &= \boldsymbol{\nabla}_{\mathbf{T}}(\boldsymbol{\nabla}_{\mathbf{T}}{\cdot}\mathbf{H_T}) - \nabla^2\mathbf{H_T} \\ &= 0 + \gamma^2\mathbf{H_T} = \frac{-i\epsilon\omega}{c}\boldsymbol{\nabla}_{\mathbf{T}}{\times}(\hat{\mathbf{k}}E_z),\end{aligned} \qquad (12.40)$$

and then

$$\begin{aligned}\mathbf{H_T} &= \frac{i\epsilon\omega}{c\gamma^2}\hat{\mathbf{k}}{\times}\boldsymbol{\nabla}_{\mathbf{T}}(E_z) \\ &= \frac{i\epsilon\omega}{c\gamma^2}\cos(n\pi z/L)\hat{\mathbf{k}}{\times}\boldsymbol{\nabla}_{\mathbf{T}}\psi \quad [\text{Eq. (12.103)}].\end{aligned} \qquad (12.41)$$

For a TE mode, H_z satifies the end conditions $H_z{=}0$, so

$$H_z(\mathbf{r_T}, z) = \sin(n\pi z/L)\psi(\mathbf{r_T}), \qquad (12.42)$$

where $\psi(\mathbf{r_T})$ is a solution of the transverse eigenvalue equation. Using [Eq. (12.92)] (with $E_z{=}0$), we can write

$$\boldsymbol{\nabla}_{\mathbf{T}}{\times}(\boldsymbol{\nabla}_{\mathbf{T}}{\times}\mathbf{H_T}) = 0 = \boldsymbol{\nabla}_{\mathbf{T}}(\boldsymbol{\nabla}_{\mathbf{T}}{\cdot}\mathbf{H_T}) - \nabla_T^2\mathbf{H_T} = \boldsymbol{\nabla}_{\mathbf{T}}(\boldsymbol{\nabla}_{\mathbf{T}}{\cdot}\mathbf{H_T}) + \gamma^2\mathbf{H_T}.$$
$$(12.43)$$

Then, using [Eq. (12.90)]:

$$\begin{aligned}\mathbf{H_T} &= -\frac{1}{\gamma^2}\boldsymbol{\nabla}_{\mathbf{T}}(\boldsymbol{\nabla}_{\mathbf{T}}{\cdot}\mathbf{H_T}) = \frac{1}{\gamma^2}\boldsymbol{\nabla}(\partial_z H_z) \\ &= \frac{n\pi}{L\gamma^2}\cos(n\pi z/L)\boldsymbol{\nabla}_{\mathbf{T}}\psi \quad [\text{Eq. (12.105)}].\end{aligned} \qquad (12.44)$$

From [Eq. (12.91) and (12.89)] (with $E_z = 0$), we can write

$$\boldsymbol{\nabla_T} \times (\boldsymbol{\nabla_T} \times \mathbf{E_T}) \;=\; \boldsymbol{\nabla_T}(\boldsymbol{\nabla_T} \cdot \mathbf{E_T}) - \nabla^2 \mathbf{E_T}$$

$$= \; 0 + \gamma^2 \mathbf{E_T} = \frac{i\mu\omega}{c} \boldsymbol{\nabla_T} \times (\hat{\mathbf{k}} E_z), \qquad (12.45)$$

and then

$$\mathbf{E_T} \;=\; \frac{-i\mu\omega}{c\gamma^2} \hat{\mathbf{k}} \times \boldsymbol{\nabla_T}(E_z)$$

$$= \; \frac{-i\mu\omega}{c\gamma^2} \sin(n\pi z/L)\hat{\mathbf{k}} \times \boldsymbol{\nabla_T}\psi \quad \text{[Eq. (12.106)]}. \qquad (12.46)$$

Problem 12.11: For a cubical cavity of dimensions $2 \times 2 \times 2$ cm, the resonant frequencies are given by

$$\nu_{lmn} = \frac{c}{2}\left[\left(\frac{l}{2}\right)^2 + \left(\frac{m}{2}\right)^2 + \left(\frac{n}{2}\right)^2\right]^{\frac{1}{2}}. \qquad (12.47)$$

There is a threefold degeneracy for the fundamental frequency:

$$\nu_{110} = \nu_{101} = \nu_{011} = \frac{3 \times 10^{10}\sqrt{2}}{4} = 10.6 \text{ GHz.} \qquad (12.48)$$

To find the Q factor, we need to know the energy in the cavity U and the rate of energy loss dU/dt. For the lowest mode, with $n = 0$, [Eq. (12.119)] for the energy in the cavity simplifies to

$$U_{110} = \frac{\epsilon E_0^2 ABC}{32\pi}. \qquad (12.49)$$

This equation holds for the energy of the lowest mode for any rectangular cavity.

The fields in the cavity can be found as in the preceding problem (12.10) for a TM mode with $n = 0$ (with ϵ and $\mu = 1$). $\mathbf{E_T}$ and H_z are zero, and the other components are given by Eqs. (**??**) and (**??**):

$$E_z \;=\; E_0 \sin(\pi x/2) \sin(\pi y/2) \qquad (12.50)$$

$$H_x \;=\; \frac{-iAE_0 \sin(\pi x/A) \sin(\pi y/B)}{\sqrt{A^2 + B^2}} \qquad (12.51)$$

$$H_y \;=\; \frac{iBE_0 \cos(\pi x/A) \sin(\pi y/B)}{\sqrt{A^2 + B^2}} \qquad (12.52)$$

The power loss is given by [Eq. (12.124)] (for dimensions $A \times B \times C$):

$$\frac{dU}{dt} = -\frac{c}{8\pi}\sqrt{\frac{\omega\mu_c}{8\pi\sigma_c}} \int |\mathbf{H}_{\text{surface}}|^2 dA$$

$$= -\frac{c}{16\pi}\sqrt{\frac{\nu}{\sigma_c}}\left\{ 2\int_0^A dx \int_0^B dy \left[|H_x(x,y,0)|^2 + |H_y(x,y,0)|^2\right]\right.$$

$$\left. +2\int_0^A dx \int_0^C dz |H_x(x,0,z)|^2 + 2\int_0^B dy \int_0^C dz |H_y(x,y,0)|^2\right\}$$

$$= -\frac{cE_0^2}{8\pi(A^2+B^2)}\sqrt{\frac{\nu}{\sigma_c}}$$

$$\left\{ \int_0^A dx \int_0^B dy \left[A^2\sin^2(\pi x/A)\cos^2(\pi y/B) + B^2\cos^2(\pi x/A)\sin^2(\pi y/B)\right]\right.$$

$$\left. +\int_0^A dx \int_0^C dz\, A^2\sin^2(\pi x/A) + \int_0^B dy \int_0^C dz\, B^2\sin^2(\pi y/B)\right\}$$

$$= -\frac{cE_0^2}{8\pi(A^2+B^2)}\sqrt{\frac{\nu}{\sigma_c}}\left[\frac{1}{4}A^3B + \frac{1}{4}AB^3 + \frac{1}{2}A^3C + \frac{1}{2}B^3C\right]$$

$$= -\frac{cE_0^2}{32\pi}\sqrt{\frac{\nu}{\sigma_c}}\left[AB + 2C\left(\frac{A^3+B^3}{A^2+B^2}\right)\right]. \qquad (12.53)$$

Then, the quality factor is

$$Q = \frac{2\pi\nu U}{-\frac{dU}{dt}} = \frac{2\pi\sqrt{\sigma_c\nu}}{c}\left[\frac{ABC(A^2+B^2)}{AB(A^2+B^2) + 2C(A^3+B^3)}\right] \qquad (12.54)$$

This equation holds for the lowest mode of any rectangular cavity of dimensions $A \times B \times C$ (with C the shortest dimension and with ϵ and $\mu = 0$). For the cubical $2 \times 2 \times 2$ cavity of this problem, we get

$$Q = \frac{2\pi\sqrt{(5\times10^{17})(10.6\times10^9)}}{3\times10^{10}}\left\lceil\frac{64}{96}\right\rceil = 10,000. \qquad (12.55)$$

Problem 12.13:

(a) The resonant frequencies of a circular cavity are given by [Eq. (12.112)]:

$$\text{TM}: \quad \nu_{lmn} = \frac{c}{2\sqrt{\epsilon\mu}}[(j_{ml}/\pi R)^2 + (n/L)^2]^{\frac{1}{2}} = \frac{c}{2\pi R\sqrt{\epsilon\mu}}[(j_{ml}^2 + (n\pi R/L)^2]^{\frac{1}{2}}. \quad (12.56)$$

$$\text{TE}: \quad \nu'_{lmn} = \frac{c}{2\sqrt{\epsilon\mu}}[(j'_{ml}/\pi R)^2 + (n/L)^2]^{\frac{1}{2}} = \frac{c}{2\pi R\sqrt{\epsilon\mu}}[(j'^2_{ml} + (n\pi R/L)^2]^{\frac{1}{2}}. \quad (12.57)$$

For $R = 2.0$ cm and $L = 3.0$ cm, and $\epsilon = \mu = 1$,

$$\text{TM}: \quad \nu_{lmn} = 2.39\sqrt{j_{ml}^2 + 4.39n^2} \quad \text{GHz}. \qquad (12.58)$$

$$\text{TE}: \quad \nu'_{lmn} = 2.39\sqrt{j'^2_{ml} + 4.39n^2} \quad \text{GHz}. \qquad (12.59)$$

Using Table [12.3] in the text, the first six resonant frequencies are

$$
\begin{aligned}
&(1) \quad \nu_{100} && = 2.39 \times 2.40 && = 5.7\,\text{GHz} && (12.60)\\
&(2) \quad \nu'_{111} && = 2.39\sqrt{1.84^2 + 4.39} && = 6.7\,\text{GHz} && (12.61)\\
&(3) \quad \nu_{101} && = 2.39\sqrt{2.40^2 + 4.39} && = 7.6\,\text{GHz} && (12.62)\\
&(4) \quad \nu'_{121} && = 2.39\sqrt{3.05^2 + 4.39} && = 8.8\,\text{GHz} && (12.63)\\
&(5) \quad \nu_{110} && = 2.39 \times 3.83 && = 9.2\,\text{GHz} && (12.64)\\
&(6) \quad \nu_{111} && = 2.39\sqrt{3.83^2 + 4.39} && = 10.4\,\text{GHz} && (12.65)\\
&(6') \quad \nu'_{201} && = 2.39\sqrt{3.83^2 + 4.39} && = 10.4\,\text{GHz}. && (12.66)
\end{aligned}
$$

The TE modes are distinguished by the notation ν'. The sixth mode is degenerate.

(b) The nodal surfaces for each mode are shown (as heavy lines) below. The r and θ nodes are shown on the circle and the z nodes are shown on the rectangle.

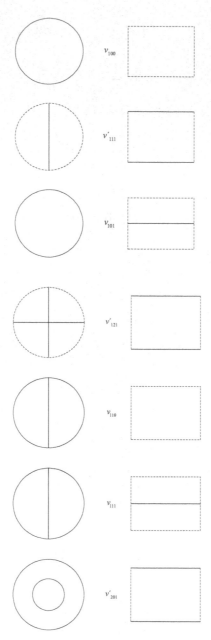

Chapter 13

Problem 13.1:

(a) The radiation current for a center fed antenna is given by [Eq. (13.53)]:

$$\boldsymbol{J} = I_0 k \hat{\mathbf{z}} \int_{-L/2}^{+L/2} \sin(kL/2 - k|z|) e^{-ikz\cos\theta} dz, \tag{13.1}$$

The sine function here is an even function of z because it depends on $|z|$. Consequently, we can replace $e^{-ikz\cos\theta}$ by $\cos(kz\cos\theta)$ in the integral from $-L/2$ to $+L/2$. Then, the integral becomes

$$
\begin{aligned}
\boldsymbol{J} &= I_0 k \hat{\mathbf{z}} \int_{-L/2}^{+L/2} \sin(kL/2 - k|z|) \cos(kz\cos\theta) dz \\
&= 2 I_0 k \hat{\mathbf{z}} \int_0^{+L/2} \sin(kL/2 - kz) \cos(kz\cos\theta) dz \\
&= I_0 k \hat{\mathbf{z}} \int_0^{+L/2} [\sin(kL/2 - kz + kz\cos\theta) + \sin(kL/2 - kz - kz\cos\theta) dz] \\
&= I_0 k \hat{\mathbf{z}} \left[\frac{\cos(kL/2 - kz + kz\cos\theta)}{k(1 - \cos\theta)} + \frac{\cos(kL/2 - kz - kz\cos\theta)}{k(1 + \cos\theta)} \right]_0^{+L/2} \\
&= I_0 k \hat{\mathbf{z}} \left[\frac{\cos\left(\frac{kL}{2}\cos\theta\right) - \cos\left(\frac{kL}{2}\right)}{k(1 - \cos\theta)} + \frac{\cos\left(\frac{-kL}{2}\cos\theta\right) - \cos\left(\frac{kL}{2}\right)}{k(1 + \cos\theta)} \right] \\
&= 2 I_0 \hat{\mathbf{z}} \left[\frac{\cos\left(\frac{kL}{2}\cos\theta\right) - \cos\left(\frac{kL}{2}\right)}{\sin^2\theta} \right], \tag{13.2}
\end{aligned}
$$

which is the same as [Eq. (13.54)].

(b) The radiation pattern for a center fed antenna is given by [Eq. (13.56)]:

$$\frac{dP}{d\Omega} = \frac{I_0^2}{2\pi c} \left[\frac{\cos\left(\frac{kL}{2}\cos\theta\right) - \cos\left(\frac{kL}{2}\right)}{\sin\theta} \right]^2. \tag{13.3}$$

For a half wave antenna with $kL = \pi$, this reduces to

$$\frac{dP}{d\Omega} = \frac{I_0^2}{2\pi c} \left[\frac{\cos\left(\frac{\pi}{2}\cos\theta\right) - \cos\left(\frac{\pi}{2}\right)}{\sin\theta} \right]^2$$

$$= \frac{I_0^2}{2\pi c} \left[\frac{\cos\left(\frac{\pi}{2}\cos\theta\right)}{\sin\theta} \right]^2, \tag{13.4}$$

which is the same as [Eq. (13.57)].

(c) For a full wave antenna with $kL = 2\pi$, the radiation pattern reduces to

$$\frac{dP}{d\Omega} = \frac{I_0^2}{2\pi c} \left[\frac{\cos\left(\pi\cos\theta\right) - \cos\pi}{\sin\theta} \right]^2$$

$$= \frac{I_0^2}{2\pi c} \left[\frac{1 - \cos\left(\pi\cos\theta\right)}{\sin\theta} \right]^2$$

$$= \frac{2I_0^2}{\pi c} \left[\frac{\cos^2\left(\frac{\pi}{2}\cos\theta\right)}{\sin\theta} \right]^2, \tag{13.5}$$

which is the same as [Eq. (13.58)].

(d) For $kL \ll 1$, we expand the cosines and the radiation pattern of Eq. (13.1) reduces to

$$\frac{dP}{d\Omega} = \frac{I_0^2}{2\pi c} \left[\frac{1 - \frac{1}{2}\left(\frac{kL}{2}\cos\theta\right)^2 - 1 + \frac{1}{2}\left(\frac{kL}{2}\right)^2}{\sin\theta} \right]^2$$

$$= \frac{I_0^2(kL)^4}{128\pi c} \left[\frac{1 - \cos^2\theta}{\sin\theta} \right]^2$$

$$= \frac{I_0^2(kL)^4 \sin^2\theta}{128\pi c}, \tag{13.6}$$

which is the same as [Eq. (13.59)].

Problem 13.3:

(a) The current distribution

$$I(z,t) = I_0 \sin(2\pi|z|/L)e^{-i\omega t}, \quad |z| < L/2, \qquad (13.7)$$

is plotted below (for $t = 0$):

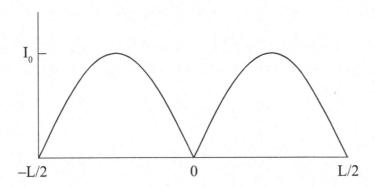

(b) The radiation current for this current distribution is given by the integral in [Eq. (13.39)]:

$$
\begin{aligned}
\boldsymbol{\mathcal{J}}(k\hat{\mathbf{r}}) &= k\int d^3r' e^{-ik\mathbf{r}'\cdot\hat{\mathbf{r}}}\mathbf{j}(\mathbf{r}') \\
&= kI_0\hat{\mathbf{z}}\int_{-L/2}^{+L/2} \sin(2\pi|z|/L)e^{-ikz\hat{\mathbf{z}}\cdot\hat{\mathbf{r}}}dz \\
&= kI_0\hat{\mathbf{z}}\int_{-L/2}^{+L/2} \sin(2\pi|z|/L)\cos(kz\cos\theta)dz \\
&= 2kI_0\hat{\mathbf{z}}\int_{0}^{+L/2} \sin(2\pi z/L)\cos(kz\cos\theta)dz \\
&= kI_0\hat{\mathbf{z}}\int_{0}^{+L/2} [\sin(2\pi z/L + kz\cos\theta) + \sin(2\pi z/L - kz\cos\theta)]dz \\
&= -kI_0\hat{\mathbf{z}}\left[\frac{\cos(2\pi z/L + kz\cos\theta)}{k\cos\theta + 2\pi/L} - \frac{\cos(2\pi z/L - kz\cos\theta)}{k\cos\theta - 2\pi/L}\right]_{0}^{L/2} \\
&= -kI_0\hat{\mathbf{z}}\left\{\frac{\cos[\pi + (kL/2)\cos\theta] - 1}{k\cos\theta + 2\pi/L} - \frac{\cos[\pi - (kL/2)\cos\theta] - 1}{k\cos\theta - 2\pi/L}\right\} \\
&= I_0\hat{\mathbf{z}}\left\{\frac{\cos[(kL/2)\cos\theta] + 1}{\cos\theta + 2\pi/kL} - \frac{\cos[(kL/2)\cos\theta] + 1}{\cos\theta - 2\pi/kL}\right\} \\
&= \frac{8\pi I_0\hat{\mathbf{z}}}{kL}\left\{\frac{\cos^2[(kL/4)\cos\theta]}{\cos^2\theta - (2\pi/kL)^2}\right\}. \qquad (13.8)
\end{aligned}
$$

The radiation pattern is given by:

$$\frac{dP}{d\Omega} = \frac{|\boldsymbol{\mathcal{J}}\sin\theta|^2}{8\pi c} = \frac{8\pi I_0^2}{c(kL)^2}\left\{\frac{\sin\theta\cos^2[(kL/4)\cos\theta]}{\cos^2\theta - (2\pi/kL)^2}\right\}^2. \qquad (13.9)$$

The radiation pattern for $kL = 2\pi$, using Eq. (13.9) above, is given by

$$\frac{dP}{d\Omega} = \frac{2I_0^2}{\pi c} \left\{ \frac{\cos^2[(\pi/2)\cos\theta]}{\sin\theta} \right\}^2, \tag{13.10}$$

which agrees with [Eq. (13.58)] for the full wave case in the text.

The choice $kL = \pi$ has the radiation pattern:

$$\frac{dP}{d\Omega} = \frac{8I_0^2}{\pi c} \left\{ \frac{\sin\theta \cos^2[(\pi/4)\cos\theta]}{4 - \cos^2\theta} \right\}^2. \tag{13.11}$$

The radiation patterns of Eqs. (13.10) and (13.11) are shown below: The curves have been normalized to have the same maximum intensity.

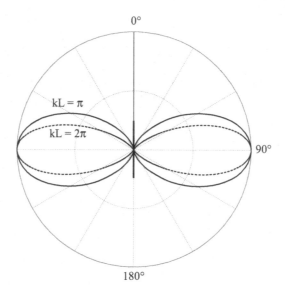

The radiation pattern for $kL = 2\pi$ (dashed curve) is identical to that for $kL = 2\pi$ in [Fig. 13.4]. The pattern for $kL = \pi$ (solid curve) is almost the same as that for $kL = \pi$ in [Fig 13.4], even though the current distributions and the algebraic forms for $\frac{dP}{d\Omega}$ are quite different.

Problem 13.5:

(a) The Coulomb force on a charge e a distance $\mathbf{r}$ from the center of a uniform spherical charge distribution of radius R is $\mathbf{F} = -e^2\mathbf{r}/R^3$ for $r \leq R$. That is the force on the heavy proton in this model. By Newton's third law, the force on the electron cloud will be equal and opposite to this force. Because the proton is much heavier than the electron, it is the electron cloud that actually oscillates. The force on the electron cloud is a linear restoring force with spring constant $k = e^2/R^3$. Thus the angular frequlency of small oscillations will be

$$\omega = \sqrt{\frac{k}{m}} = \sqrt{\frac{e^2}{mR^3}} = \frac{c}{R}\sqrt{\frac{e^2/R}{mc^2}}. \tag{13.12}$$

(b) The ratio e^2/R is the magnitude of the electron's potential energy in hydrogen (twice the binding energy), given by $e^2/R = (4.8 \times 10^{-10} \times 300/(.53 \times 10^{-8}) = 27$ eV. Then,

$$\omega = \frac{(3 \times 10^{10})}{(.53 \times 10^{-8})} \sqrt{\frac{27}{.51 \times 10^6}} = 41 \times 10^{15} \text{ sec}^{-1}. \qquad (13.13)$$

The period of oscillation is $T = 2\pi/\omega = 0.153 \times 10^{-15}$ sec.

(c) The electric dipole moment is $\mathbf{p} = -eA_0\hat{\mathbf{r}}e^{-i\omega t}$.
(d) The power radiated (for amplitude A) is given by [Eq. (13.72)]:

$$P = \frac{\omega^4}{3c^3}|\mathbf{p}|^2 = \left(\frac{e^4}{3m^2R^6c^3}\right)(e^2A^2) = \frac{e^6A^2}{3m^2R^6c^3}. \qquad (13.14)$$

The initial power, with $A = R$, is

$$
\begin{aligned}
P_0 &= \frac{e^6}{3R^4m^2c^3} = \left(\frac{c}{3R}\right)\frac{(e^2/R)^3}{(mc^2)^2} \\
&= \left[\frac{(3 \times 10^{10} \text{ cm/sec})}{(3)(.53 \times 10^{-8} \text{ cm})}\right]\left[\frac{(27 \text{ eV})^3}{(.51 \times 10^6 \text{ eV})^2}\right] = 1.4 \times 10^{11} \text{ eV/sec}. \quad (13.15)
\end{aligned}
$$

(e) The potential energy of a point charge $-e$ in a uniformly charged sphere of radius R and charge $+e$ is found from the answer to Prob. 1.13(b):

$$U = \frac{e^2r^2}{2R^3} - \frac{3e^2}{2R^3} = \frac{e^2A^2\cos^2(\omega t)}{2R^3} - \frac{3e^2}{2R^3}, \qquad (13.16)$$

where we have put in the oscillating time dependence of r. We can neglect the motion and kinetic energy of the heavy proton, and the kinetic energy of the oscillating electron is

$$T = \frac{1}{2}mv^2 = \frac{mA^2\omega^2\sin^2(\omega t)}{2} = \frac{e^2A^2\sin^2(\omega t)}{2R^3}, \qquad (13.17)$$

where we have substituted for ω from Eq. (13.12) above. The total energy of this system is

$$W = U + T = \frac{e^2A^2}{2R^3} - \frac{3e^2}{2R^3}. \qquad (13.18)$$

This depends on time only through the time dependence of the amplitude A, due to the radiation. We can equate $\frac{dW}{dt}$ to the power lost due to radiation given by eq. (13.14):

$$\frac{dW}{dt} = \left(\frac{e^2A}{R^3}\right)\frac{dA}{dt} = -\frac{e^6A^2}{3R^6m^2c^3}. \qquad (13.19)$$

This is a differential equation for $A(t)$, with the solution

$$A = A_0 \exp\left[\frac{-e^4}{3R^3m^2c^3}\right] = A_0 e^{-t/\tau}, \tag{13.20}$$

where the lifetime τ is

$$
\begin{aligned}
\tau &= \frac{3R^3m^2c^3}{e^4} = \frac{(e^2/R)}{P_0} \\
&= \frac{(27\,\text{eV})}{(1.4 \times 10^{11}\,\text{eV/sec})} = 2 \times 10^{-10}\,\text{sec}.
\end{aligned} \tag{13.21}
$$

We wrote τ in terms of the initial power using Eq. (13.15).

Problem 13.7:

(a) The Larmor formula for radiation by an accelerating electron is

$$P = \frac{2e^2|\mathbf{a}|^2}{3c^3}. \tag{13.22}$$

[For the purposes of this problem, we assume that the deceleration is constant, but this assumption is not a good one. It is just made to give a simpler problem with a constant radiated power. The actual deceleration increases in magnitude as the electron slows down. This can be deduced from [Eq. (15.59)] in the text. This problem could be repeated with the actual deceleration, but that calculation is quite complicated, and the final numerical answer would not be much different.]

A constant acceleration is related to the stopping distance by $a = v_0^2/2L$, where v_0 is the initial velocity. Then the total energy radiated (for 2 keV electrons stopping in 0.01 cm) is

$$
\begin{aligned}
W_{\text{rad}} &= Pt = \left[\frac{2e^2|\mathbf{a}|^2}{3c^3}\right]\sqrt{\frac{2L}{a}} = \left[\frac{e^2v_0^4}{6L^2c^3}\right]\sqrt{\frac{4L^2}{v_0^2}} \\
&= \frac{c^2v_0^3}{3Lc^3} = \left(\frac{e^2}{3L}\right)\left(\frac{mv_0^2}{mc^2}\right)^{\frac{3}{2}} \\
&= \left[\frac{(4.8 \times 10^{-10})(300)}{(3)(.01)}\right]\left[\frac{(2 \times 2\,\text{keV})}{511\,\text{keV}}\right]^{\frac{3}{2}} \\
&= 0.33 \times 10^{-8}\,\text{eV} \\
&= \left(\frac{.33 \times 10^{-8}}{2 \times 10^3}\right)W_0 = 17 \times 10^{-13}W_0.
\end{aligned} \tag{13.23}
$$

For an initial beam power of 300 Watts, the radiated power is

$$P_{\text{rad}} = 17 \times 10^{-13} \times 300 = 5.1 \times 10^{-10}\,\text{Watts}. \tag{13.24}$$

(c) The maximum intensity of the radiation 20 cm away is given, using [Eq. (13.93)], by

$$\frac{dP}{dA} = \frac{1}{r^2}\frac{dP}{d\Omega} = \frac{e^2|\mathbf{a}|^2}{4\pi c^3 r^2} = \frac{e^2 v_0^4}{16\pi c^3 L^2 r^2} = \left[\frac{e^2 c}{16\pi L^2 r^2}\right]\left[\frac{mv_0^2}{mc^2}\right]^2$$

$$= \left[\frac{(4.8\times10^{-10})^2(3\times10^{10})}{(16)(\pi)(.01)^2(20)^2}\right]\left[\frac{(4)}{(511)}\right]^2$$

$$= 2.1\times10^{-15}\,\text{ergs/sec}-\text{cm}^2 = 2.1\times10^{-18}\,\text{Watts/meter.}\quad(13.25)$$

Problem 13.9:

(a) The power radiated by a magnetic moment $\boldsymbol{\mu}$, rotating with angular velocity $\boldsymbol{\omega}$ about an axis perpendicular to $\boldsymbol{\mu}$ is given by the same equation as that for a rotating electric dipole by [Eq. (13.80)]:

$$P = \frac{2\mu^2\omega^4}{3c^3}.\quad(13.26)$$

(b) The energy radiated in one rotation is

$$W_{\text{rad}} = \frac{2\pi}{\omega}P = \frac{4\pi\mu^2\omega^3}{3c^3}.\quad(13.27)$$

The rotational energy of the star is

$$W_{\text{rot}} = \frac{1}{2}I\omega^2 = \frac{1}{5}MR^2\omega^2,$$

$$\text{so}\quad \frac{W_{\text{rad}}}{W_{\text{rot}}} = \frac{20\pi\mu^2\omega}{3MR^2}\quad(13.28)$$

(c) The radiative power loss equals the rate of rotational energy loss:

$$\frac{dW_{\text{rot}}}{dt} = \frac{2}{5}MR^2\omega\frac{d\omega}{dt}$$

$$= P = \frac{2\mu^2\omega^4}{3c^3}.\quad(13.29)$$

$$\text{and then}\quad \frac{d\omega}{\omega} = \frac{5\mu^2\omega^2}{3MR^2c^3}dt.\quad(13.30)$$

The change in ω in one year will be small, so the result above would give the fractional change in ω for $dt = $ one year.

Problem 13.11:

(a) If a conducting sphere of radius R is placed in a oscillating, but uniform, magnetic field $\mathbf{B_0}$, we can use the magnetic scalar potential to find the field outside the conducting sphere. This works because there is no free current outside the sphere. For $r > R$, the magnetic scalar potential is

$$\phi_m = \sum_l \frac{a_l P_l(\cos\theta)}{r^2} - B_0\,r\cos\theta, \quad r \geq R. \tag{13.31}$$

The normal component of **B** must vanish at the conductor's surface as discussed in [Sec. 11.2.1]. This means that

$$\partial_r\phi_m|_{r=R} = 0 = \sum_l \frac{-2a_l P_l(\cos\theta)}{R^3} - B_0\cos\theta. \tag{13.32}$$

All the a_l must vanish except for $a_1 = -B_0 R^3/2$, and then

$$\phi_m = \frac{-B_0 R^3\cos\theta}{2r^2} - B_0\,r\cos\theta. \tag{13.33}$$

Note: At this point we already can identify the magnetic moment of the sphere as $\boldsymbol{\mu} = -\mathbf{B_0}R^3/2$, but the purpose of the problem is to find $\boldsymbol{\mu}$ from the induced current. In any event, we can write ϕ_m in vector form as

$$\phi_m = \frac{R^3\mathbf{B_0}\cdot\hat{\mathbf{r}}}{2r^2} - \mathbf{B_0}\cdot\mathbf{r}, \tag{13.34}$$

and then the magnetic field outside the sphere is

$$\mathbf{B} = -\boldsymbol{\nabla}\phi_m = -\left(\frac{R^3}{2}\right)\left[\frac{3(\mathbf{B_0}\cdot\hat{\mathbf{r}})\hat{\mathbf{r}} - \mathbf{B_0}}{r^3}\right] + \mathbf{B_0} \tag{13.35}$$

The surface current induced on the sphere follows from the boundary condition [in Eq. (11.29)] on $\hat{\mathbf{r}}\times\mathbf{B}$ at the surface of the sphere:

$$\hat{\mathbf{r}}\times(\mathbf{B}_{\text{out}} - \mathbf{B}_{\text{in}}) = \frac{4\pi}{c}\mathbf{K}. \tag{13.36}$$

Since $\mathbf{B}_{\text{in}} = 0$ for the conducting sphere, the surface current is given by

$$\mathbf{K} = \frac{c}{4\pi}\hat{\mathbf{r}}\times\mathbf{B}_{\text{out}}(R) = \frac{3c}{8\pi}\hat{\mathbf{r}}\times\mathbf{B_0}. \tag{13.37}$$

(b) The magnetic moment is given by adapting [Eq. (7.125)] to a surface current:

$$\begin{aligned}
\boldsymbol{\mu} &= \frac{1}{2c}\int\mathbf{r}\times\mathbf{K}\,dS = \frac{3}{16\pi}\int\mathbf{r}\times(\hat{\mathbf{r}}\times\mathbf{B_0}]R^2\,d\Omega \\
&= \frac{3}{16\pi}\int[\mathbf{r}(\hat{\mathbf{r}}\cdot\mathbf{B_0}) - \mathbf{B_0}]R^2\,d\Omega = \frac{3}{16\pi}\left(\frac{-8\pi}{3}\right)R^3\mathbf{B_0} \\
&= -\frac{1}{2}R^3\mathbf{B_0}.
\end{aligned} \tag{13.38}$$

Chapter 14

Problem 14.1:

(a) The relativistic addition of velocities for velocity **c** at an angle θ to a velocity **v**, is

$$u_\parallel = \frac{c\cos\theta + v}{1 + v\cos\theta/c} \tag{14.1}$$

$$u_\perp = \frac{c\sin\theta\sqrt{1 - v^2/c^2}}{(1 + v\cos\theta/c)}. \tag{14.2}$$

The squared magnitude of the resultant velocity is

$$u_\parallel^2 + u_\perp^2 = \frac{\left[(c\cos\theta + v)^2 + c^2\sin^2\theta - v^2\sin^2\theta\right]}{(1 + v\cos\theta/c)^2}$$

$$= \frac{\left[c^2 + 2vc\cos\theta + v^2\cos^2\theta\right]}{(1 + v\cos\theta/c)^2} = c^2. \tag{14.3}$$

(b) The relativistic addition of any two velocities **u'** and **v** is given by

$$u_\parallel = \frac{u'\cos\theta + v}{1 + u'v\cos\theta/c^2} \tag{14.4}$$

$$u_\perp = \frac{u'\sin\theta\sqrt{1 - v^2/c^2}}{(1 + u'v\cos\theta/c^2)}. \tag{14.5}$$

The squared magnitude of the resultant velocity is

$$u_\parallel^2 + u_\perp^2 = \frac{\left[(u'\cos\theta + v)^2 + u'^2\sin^2\theta - u'^2v^2\sin^2\theta/c^2\right]}{(1 + u'v\cos\theta/c^2)^2}$$

$$= \frac{\left[u'^2 + 2u'v\cos\theta + v^2 - u'^2v^2\sin^2\theta/c^2\right]}{(1 + u'v\cos\theta/c^2)^2} = \frac{N}{D}. \tag{14.6}$$

We want to show that this result is less than c^2, that is that the numerator N and denominator D are related by $c^2 D - N > 0$.

$$c^2 D - N = c^2 + 2u'v\cos\theta + u'^2v^2\cos^2\theta/c^2 - u'^2 - 2u'v\cos\theta - v^2 + u'^2v^2\sin^2\theta/c^2$$

$$= c^2 + u'^2v^2/c^2 - u'^2 - v^2 = \frac{(c^2 - v^2)(c^2 - u'^2)}{c^2}, \tag{14.7}$$

which is greater than zero for all u' and v less than c.

Problem 14.3:

To find the relativistic transformation of acceleration, we use [Eqs. (14.24) and (14.25)] for **u** and calculate $d\mathbf{u}/dt$. For $\mathbf{a}_{\parallel}$, we use

$$
\begin{aligned}
\mathbf{a}_{\parallel} &= \frac{d\mathbf{u}_{\parallel}}{dt} = \frac{d\left[\frac{\mathbf{u}'_{\parallel}+\mathbf{v}}{1+\mathbf{u}'\cdot\mathbf{v}/c^2}\right]}{\gamma_v(dt' + d\mathbf{r}'\cdot\mathbf{v}/c^2)} \\
&= \frac{\left[(1+\mathbf{u}'\cdot\mathbf{v}/c^2)d\mathbf{u}'_{\parallel} - (\mathbf{u}'_{\parallel}+\mathbf{v})d\mathbf{u}'\cdot\mathbf{v}/c^2\right]}{\gamma_v(dt' + d\mathbf{r}'\cdot\mathbf{v}/c^2)(1+\mathbf{u}'\cdot\mathbf{v}/c^2)^2} \\
&= \frac{\left[(1+\mathbf{u}'\cdot\mathbf{v}/c^2)\mathbf{a}'_{\parallel} - (\mathbf{u}'_{\parallel}+\mathbf{v})\mathbf{a}'\cdot\mathbf{v}/c^2\right]}{\gamma_v(1+\mathbf{u}'\cdot\mathbf{v}/c^2)(1+\mathbf{u}'\cdot\mathbf{v}/c^2)^2} \\
&= \frac{\mathbf{a}'_{\parallel}}{\gamma_v^3(1+\mathbf{u}'\cdot\mathbf{v}/c^2)^3},
\end{aligned} \tag{14.8}
$$

which agrees with [Eq. (14.24)]. For $\mathbf{a}_{\perp}$, we use

$$
\begin{aligned}
\mathbf{a}_{\perp} &= \frac{d\mathbf{u}_{\perp}}{dt} = \frac{d\left[\frac{\mathbf{u}'_{\perp}}{\gamma_v(1+\mathbf{u}'\cdot\mathbf{v}/c^2)}\right]}{\gamma_v(dt' + d\mathbf{r}'\cdot\mathbf{v}/c^2)} \\
&= \frac{\gamma_v(1+\mathbf{u}'\cdot\mathbf{v}/c^2)d\mathbf{u}'_{\perp} - \gamma_v\mathbf{u}'_{\perp}d\mathbf{u}'\cdot\mathbf{v}/c^2}{\gamma_v(dt' + d\mathbf{r}'\cdot\mathbf{v}/c^2)\gamma_v^2(1+\mathbf{u}'\cdot\mathbf{v}/c^2)^2} \\
&= \frac{\mathbf{a}'_{\perp} + (\mathbf{u}'\cdot\mathbf{v})\mathbf{a}'_{\perp}/c^2 - \mathbf{u}'_{\perp}(\mathbf{a}'\cdot\mathbf{v})/c^2}{\gamma_v^2(1+\mathbf{u}'\cdot\mathbf{v}/c^2)^3} \\
&= \frac{\mathbf{a}'_{\perp} + (\mathbf{u}'\times\mathbf{a}')\times\mathbf{v}/c^2}{\gamma_v^2(1+\mathbf{u}'\cdot\mathbf{v}/c^2)^3},
\end{aligned} \tag{14.9}
$$

which agrees with [Eq. (14.25)].

Problem 14.5:

An object with an acceleration g in its rest system has an acceleration given by [Eq. (14.26)] with $\mathbf{u}' = \mathbf{0}$. The object's acceleration in a system in which the object has a velocity v will be

$$
a = g/\gamma_v^3 = g(1-v^2)^{\frac{3}{2}} = \frac{dv}{dt} \tag{14.10}
$$

[**Note:** We will use natural units with c absent, or you could consider that the velocity is in units of LY/year, in which case $c = 1$.]

We can solve the differential equation for the velocity after falling from rest for a time t

$$
gt = \int_0^v \frac{dv}{(1-v^2)^{\frac{3}{2}}} = \frac{v}{\sqrt{1-v^2}} = v/\gamma_v, \tag{14.11}
$$

where we integrated using the substitution $v = \sin\theta$. Solving for v, we get

$$v = \frac{gt}{\sqrt{1 + g^2 t^2}} = gt/\gamma_e. \qquad (14.12)$$

(b) To find the velocity after one year, it is convenient to use the distance unit LY (light year) and time unit Y (year). In these units, c obviously equals one, and

$$g = \frac{980\left(\frac{cm}{sec^2}\right) \times (3.16 \times 10^7)^2 \left(\frac{sec}{year}\right)^2}{(3.16 \times 10^7)(3 \times 10^{10})\left(\frac{cm}{LY}\right)} = 1.03 \text{ LY}/\text{Y}^2. \qquad (14.13)$$

We will use the approximate value $g=1$ LY/Y^2. Then the speed of the object after one year is $v = 1 \times 1/\sqrt{1+1} = 1/\sqrt{2}$ (or $c/\sqrt{2}$ if you're not ready for natural units).

(c) The momentum of an object of mass m is

$$p = mv\gamma_e = mgt. \qquad (14.14)$$

We plot v and p/m below:

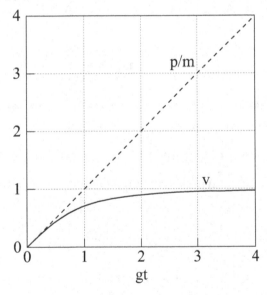

Problem 14.7:

The velocity of the space ship was calculated in Prob. 14.5, and we will use that result here: $v = gt/\sqrt{1 + g^2 t^2}$. We will also use the units LY for distance and Y (years) for time. In these units $c = 1$ by definition, and $g = 1$ by chance (or intelligient design?), as calculated in Prob. 14.4.

(a) The distance the space ship travels from rest in a time $t = 5$ years is

$$x = \int_0^t v\,dt = \int_0^t \frac{gt\,dt}{\sqrt{1 + g^2 t^2}}$$

$$= \frac{\sqrt{1+g^2t^2}-1}{g} = \frac{\sqrt{1+5^2}-1}{1} = 4.1LY. \tag{14.15}$$

The distance traveled for 5 years with a deceleration g will be the same as for the acceleration, so the distance traveled in the 10 year trip is 8.2 LY.

(b) The astronauts will age at a rate $dt' = dt/\gamma$, where dt is the time differential at rest on Earth. Thus, in a time $T{=}5$ Earth years, the astronauts will age a time T', given by

$$T' = \int_0^T \frac{dt}{\gamma} = \int_0^T \frac{dt}{\sqrt{1+g^2t^2}} = \frac{1}{g}\ln\left(gT+\sqrt{1+g^2T2}\right) \tag{14.16}$$

$$= \ln(5+\sqrt{26}) = 2.3 \text{ years.} \tag{14.17}$$

After traveling 20 Earth years, the astronauts will be $4{\times}2.3{=}9.2$ years older.

Problem 14.9:

The relativistic transformation of velocity to a Lorentz system moving with velocity $\mathbf{v}$ is given by [Eqs. (14.24) and (14.25)] (with the primes interchanged and $\mathbf{v} \to -\mathbf{v}$):

$$\mathbf{u}'_{\parallel} = \frac{\mathbf{u}_{\parallel}-\mathbf{v}}{1-\mathbf{u}{\cdot}\mathbf{v}/c^2}, \qquad \mathbf{u}'_{\perp} = \frac{\mathbf{u}_{\perp}}{\gamma_v(1-\mathbf{u}{\cdot}\mathbf{v}/c^2)}. \tag{14.18}$$

We consider two space ships with two parallel, but unequal, velocities $\mathbf{u}(1)$ and $\mathbf{u}(2)$, and take the cross product $\mathbf{u}'(1){\times}\mathbf{u}'(2)$:

$$\begin{aligned} \mathbf{u}'(1){\times}\mathbf{u}'(2) &= \mathbf{u}'_{\parallel}(1){\times}\mathbf{u}'_{\perp}(2) + \mathbf{u}'_{\perp}(1){\times}\mathbf{u}'_{\parallel}(2) \\ &= \frac{\left[\mathbf{u}_{\parallel}(1)-\mathbf{v}\right]{\times}\mathbf{u}_{\perp}(2) + \mathbf{u}_{\perp}(1){\times}\left[\mathbf{u}_{\parallel}(2)-\mathbf{v}\right]}{\gamma_v\left[1-\mathbf{u}(1){\cdot}\mathbf{v}/c^2\right]\left[1-\mathbf{u}(2){\cdot}\mathbf{v}/c^2\right]} \\ &= \frac{\mathbf{u}(1){\times}\mathbf{u}(2) + \mathbf{v}{\times}\left[\mathbf{u}_{\perp}(1)-\mathbf{u}_{\perp}(2)\right]}{\gamma_v\left[1-\mathbf{u}(1){\cdot}\mathbf{v}/c^2\right]\left[1-\mathbf{u}(2){\cdot}\mathbf{v}/c^2\right]}. \end{aligned} \tag{14.19}$$

The first term in the numerator $\mathbf{u}(1){\times}\mathbf{u}(2) = 0$, because $\mathbf{u}(1)$ and $\mathbf{u}(2)$ are parallel. But the second term will not vanish if $\mathbf{u}_{\perp}(1)$ and $\mathbf{u}_{\perp}(2)$ have different magnitudes. Thus, the velocities $\mathbf{u}'(1)$ and $\mathbf{u}'(2)$ are generally not parallel. **(b)** Since the two spaceships cannot collide in the original system, they cannot collide in any Lorentz system. This is because the outcome of any experiment must be the same in all Lorentz systems.

Problem 14.11:

A star moving with velocity $\mathbf{v}$ at an angle θ, as in [Fig. (14.6)], has a transverse velocity of $v\sin\theta$. In a pre-relativistic model, if a telescope were moving with this transverse velocity and were pointed directly at the star, the light from the star would pass straight through the telescope. A telescope at rest would have a relative velocity of $-v \sin \theta$ with respect to that (hypothetical) moving telescope. Consequently, the light would hit the back of the telescope a distance $\delta = vt \sin \theta$ away from where it would have hit if the star had no transverse velocity. The time t for the light to pass through the telescope is given by $t = L/c$, so the distance $\delta = (vL/c) \sin \theta$. The small angle at which the telescope should be deflected away from directly toward the star is given by

$$\phi \approx \tan \phi = \frac{\delta}{L} = \frac{v}{c} \sin \theta. \tag{14.20}$$

The relativistic result for stellar aberration is given by [Eq. (14.109)]:

$$\tan \theta' = \frac{k'_\perp}{k'_\parallel} = \frac{\sin \theta}{\gamma(\cos \theta + v/c)}, \tag{14.21}$$

The angle ϕ, which is is the angle by which the telescope is to be deflected from pointing directly at the star, is given by $\phi = \theta - \theta'$. Then

$$\tan \phi = \tan(\theta - \theta') = \frac{\tan \theta - \tan \theta'}{1 + \tan \theta \tan \theta'} = \frac{\gamma(\cos \theta + v/c) \tan \theta - \sin \theta}{\gamma(\cos \theta + v/c) + \tan \theta \sin \theta}. \tag{14.22}$$

We expand to lowest order in v/c for $v \ll c$ by setting $\gamma = 1$, and neglecting the v/c in the denominator. This gives

$$\phi \approx \tan \phi = \frac{(v/c) \tan \theta}{\cos \theta + \tan \theta \sin \theta} = \frac{v}{c} \sin \theta, \tag{14.23}$$

which is the same as the non-relativistic calculation.

Problem 14.13:

(a) A proton of mass m=938 MeV and kinetic energy T=469 MeV has a total energy of $E = T + m$=469+938=1,407 MeV.

(b) Its momentum is $p = \sqrt{E^2 - m^2} = \sqrt{1407^2 - 938^2}$=1,049 MeV.

(c) Its velocity is $v = p/E$=1049/1407=0.745.

(d) Its Lorentz γ is given by $\gamma = E/m$=1407/938=1.50.

Problem 14.15:

(a) In the process $\pi^+(140) \to \mu^+(106) + \nu(0)$, the π^+ is at rest in the center of mass system. Conservation of energy can be written as

$$E_\nu = p_\nu = p_\mu = m_\pi - E_\mu \tag{14.24}$$

where, because the neutrino is massless here, we used $E_\nu = p_\nu$, and then $p_\nu = p_\mu$, because the two momenta have equal magnitude in the barycentric system. Squaring this equation, we get

$$p_\mu^2 = m_\pi^2 + m_\mu^2 + p_\mu^2 - 2m_\pi E_\mu, \tag{14.25}$$

and solving for E_μ gives

$$E_\mu = \frac{m_\pi^2 + m_\mu^2}{2m_\pi} = \frac{(140^2) + (106)^2}{(2)(140)} = 110 \text{ MeV}. \tag{14.26}$$

The muon momentum is

$$\begin{aligned} p_\mu &= \sqrt{(E_\mu^2 - m_\mu^2)^2} = \frac{\sqrt{(m_\pi^2 + m_\mu^2)^2 - 4m_\pi^2 m_\mu^2}}{2m_\pi} \\ &= \frac{m_\pi^2 - m_\mu^2}{2m_\pi} = \frac{(140^2) - (106)^2}{(2)(140)} = 30 \text{ MeV}. \end{aligned} \tag{14.27}$$

The muon's velocity is

$$v_\mu = \frac{p_\mu}{E_\mu} = \frac{30}{110} = 0.27. \tag{14.28}$$

(b) If there were no neutrino emitted in the beta decay process, $n(939.6) \to p(938.3) + e(0.51) + \nu(0)$, the energy conservation equation in the neutron rest system (in which $E_n = m_n$ and $\mathbf{p_p} = -\mathbf{p_e}$) would be

$$\sqrt{p^2 + m_p^2} = m_n - E_e = m_n - \sqrt{p^2 + m_e^2}. \tag{14.29}$$

Squaring both sides, we get

$$p^2 + m_p^2 = m_n^2 + p^2 + m_e^2 - 2m_n E_e, \tag{14.30}$$

and then

$$\begin{aligned} E_e &= \frac{m_n^2 + m_e^2 - m_p^2}{2m_n} \\ &= \frac{(939.6)^2 + (.51)^2 - (938.3)^2}{(2)(939.6)} = 1.30 \text{ MeV}. \end{aligned} \tag{14.31} \tag{14.32}$$

With a neutrino being emitted, this is the upper limit on the electron energy.

Problem 14.17:

The distance a muon travels in one lifetime is

$$x = vt = \frac{p}{E}\gamma\tau = \frac{p}{E}\frac{E}{m}\tau = \frac{p}{m}\tau. \tag{14.33}$$

For a muon to travel 80 km in one lifetime of $\tau = 2.2 \times 10^{-6}$ sec, its energy must be

$$E \approx p = \frac{mx}{\tau} = \frac{(206)(8 \times 10^6)}{(2.2 \times 10^{-6})}$$

$$= 7500 \times 10^{12} \left(\frac{\text{Mev} - \text{cm}}{\text{sec}}\right) \frac{1}{3 \times 10^{10}} \left(\frac{\text{sec}}{\text{cm}}\right) = 250 \, \text{GeV} \quad (14.34)$$

Note: The original answer was not in the units we wanted, so we divided by the appropriate conversion factor.

Problem 14.19:

The electromagnetic field tensor transforms as

$$[F'] = [L][F][\tilde{L}]$$

$$= \begin{pmatrix} \gamma & -\beta\gamma & 0 & 0 \\ -\beta\gamma & \gamma & 0 & 0 \\ 0 & 0 & 1 & 0 \\ 0 & 0 & 0 & 1 \end{pmatrix} \begin{pmatrix} 0 & -E_x & -E_y & -E_z \\ E_x & 0 & -B_z & B_y \\ E_y & B_z & 0 & -B_x \\ E_z & -B_y & B_x & 0 \end{pmatrix} \begin{pmatrix} \gamma & -\beta\gamma & 0 & 0 \\ -\beta\gamma & \gamma & 0 & 0 \\ 0 & 0 & 1 & 0 \\ 0 & 0 & 0 & 1 \end{pmatrix}$$

$$= \begin{pmatrix} \gamma & -\beta\gamma & 0 & 0 \\ -\beta\gamma & \gamma & 0 & 0 \\ 0 & 0 & 1 & 0 \\ 0 & 0 & 0 & 1 \end{pmatrix} \begin{pmatrix} \beta\gamma E_x & -\gamma E_x & -E_y & -E_z \\ \gamma E_x & -\beta\gamma E_x & -B_z & B_y \\ \gamma E_y - \beta\gamma B_y & -\beta\gamma E_y + \gamma B_z & 0 & -B_x \\ \gamma E_z + \beta\gamma B_y & -\beta\gamma E_z - \gamma B_y & B_x & 0 \end{pmatrix}$$

$$= \begin{pmatrix} \beta\gamma^2 E_x - \beta\gamma^2 E_x & -\gamma^2 E_x + \beta^2\gamma^2 E_x & -\gamma E_y + \beta\gamma B_z & -\gamma E_z - \beta\gamma B_y \\ -\beta^2\gamma^2 E_x + \gamma^2 E_x & \beta\gamma^2 E_x - \beta\gamma^2 E_x & \beta\gamma E_y - \gamma B_z & \beta\gamma E_z + \gamma B_y \\ \gamma E_y - \beta\gamma B_z & -\beta\gamma E_y + \gamma B_z & 0 & -B_x \\ \gamma E_z + \beta\gamma B_y & -\beta\gamma E_z - \gamma B_y & B_x & 0 \end{pmatrix}$$

$$= \begin{pmatrix} 0 & -E_x & -\gamma E_y + \beta\gamma B_z & -\gamma E_z - \beta\gamma B_y \\ E_x & 0 & \beta\gamma E_y - \gamma B_z & \beta\gamma E_z + \gamma B_y \\ \gamma E_y - \beta\gamma B_z & -\beta\gamma E_y + \gamma B_z & 0 & -B_x \\ \gamma E_z + \beta\gamma B_y & -\beta\gamma E_z - \gamma B_y & B_x & 0 \end{pmatrix} \quad (14.35)$$

Comparison of this final matrix with the matrix [F] (in the first step above) shows agreement with [Eq. (14.140] for the transformed **E** and **B** fields.

Chapter 15

Problem 15.1:

The covariant equation of motion for a charged particle is given by

$$\frac{dp^\mu}{d\tau} = qF^{\mu\nu}U_\nu. \tag{15.1}$$

In vector form, this becomes

$$\frac{\gamma dp^0}{dt} = qF^{0\nu}U_\nu \rightarrow \frac{\gamma dW}{dt} = q\gamma\mathbf{v}\cdot\mathbf{E} \rightarrow \frac{dW}{dt} = q\mathbf{v}\cdot\mathbf{E}, \tag{15.2}$$

$$\frac{\gamma dp^1}{dt} = qF^{10}U_0 + qF^{1j}U_j \rightarrow \frac{\gamma dp_x}{dt} = q\gamma[E_x + B_z v_y - B_y v_z]$$

$$\frac{\gamma dp^2}{dt} = qF^{20}U_0 + qF^{2j}U_j \rightarrow \frac{\gamma dp_y}{dt} = q\gamma[E_y + B_x v_z - B_z v_x]$$

$$\frac{\gamma dp^3}{dt} = qF^{30}U_0 + qF^{3j}U_j \rightarrow \frac{\gamma dp_z}{dt} = q\gamma[E_z + B_y v_x - B_x v_y].$$

These last three equations can be written in vector form as

$$\frac{d\mathbf{p}}{dt} = q\mathbf{E} + q\mathbf{v}\times\mathbf{B}. \tag{15.3}$$

(b) To write $d\mathbf{p}/dt$ in terms of the acceleration, we write $\mathbf{p}$ in terms of the velocity:

$$\frac{d\mathbf{p}}{dt} = \frac{d(m\mathbf{v}\gamma)}{dt} = m\left[\gamma\frac{d\mathbf{v}}{dt} + \mathbf{v}\frac{d\gamma}{dt}\right]. \tag{15.4}$$

$$\frac{d\gamma}{dt} = \frac{d}{dt}\left[\frac{1}{\sqrt{1-\mathbf{v}\cdot\mathbf{v}}}\right] = (1-v^2)^{-\frac{3}{2}}\mathbf{v}\cdot\frac{d\mathbf{v}}{dt} = \gamma^3\mathbf{v}\cdot\mathbf{a}. \tag{15.5}$$

Then

$$\frac{d\mathbf{p}}{dt} = m\left[\gamma\mathbf{a} + \gamma^3\mathbf{v}(\mathbf{v}\cdot\mathbf{a})\right], \tag{15.6}$$

Problem 15.3:

The acceleration is related to $\frac{d\mathbf{p}}{dt}$ by Eq. (15.6):

$$\frac{d\mathbf{p}}{dt} = m\left[\gamma\mathbf{a} + \gamma^3\mathbf{v}(\mathbf{v}\cdot\mathbf{a})\right], \tag{15.7}$$

If $d\mathbf{p}/dt = -mg\hat{\mathbf{j}}$, and $\mathbf{v} = v_x\hat{\mathbf{i}} + v_y\hat{\mathbf{j}}$, then

$$
\begin{aligned}
-mg\hat{\mathbf{j}} &= m\left[\gamma\mathbf{a} + \gamma^3(v_x\hat{\mathbf{i}} + v_y\hat{\mathbf{j}})(v_x a_x + v_y a_y)\right] \\
-g &= \left[\gamma a_y + \gamma^3 v_y(v_x a_x + v_y a_y)\right] & (15.8) \\
0 &= \left[\gamma a_x + \gamma^3 v_x(v_x a_x + v_y a_y)\right]. & (15.9)
\end{aligned}
$$

We solve Eq. (15.9) for v_y:

$$
a_y = \frac{-a_x(\gamma + \gamma^3 v_x^2)}{\gamma^3 v_x v_y} = \frac{-a_x(1 - v_y^2)}{v_x v_y}, \tag{15.10}
$$

and substitute into Eq. (15.8):

$$
-g = a_x\gamma^3 v_x v_y - a_x(\gamma + \gamma^3 v_x^2)\frac{(1 - v_y^2)}{v_x v_y} = \frac{\gamma^3 a_x}{v_x v_y}[v_x^2 v_y^2 - (1 - v_x^2)(1 - v_y^2)] = \frac{-\gamma a_x}{v_x v_y}. \tag{15.11}
$$

Then

$$
a_x = \frac{v_x v_y g}{\gamma}, \tag{15.12}
$$

and

$$
a_y = -\left(\frac{1 - v_y^2}{v_x v_y}\right)\left(\frac{v_x v_y g}{\gamma}\right) = \frac{-(1 - v_y^2)g}{\gamma}. \tag{15.13}
$$

As a check, this has the right non-relativistic limit.

Problem 15.5:

(a) For the nonrelativistic Lagrangian

$$
\begin{aligned}
\boldsymbol{\nabla}L &= \boldsymbol{\nabla}\left[\frac{1}{2}mv^2 - q\phi + q\mathbf{v}\cdot\mathbf{A}\right] = q[-\boldsymbol{\nabla}\phi + \mathbf{v}\times(\boldsymbol{\nabla}\times\mathbf{A}) + (\mathbf{v}\cdot\boldsymbol{\nabla})\mathbf{A}] \\
&= q[-\boldsymbol{\nabla}\phi + \mathbf{v}\times\mathbf{B} + (\mathbf{v}\cdot\boldsymbol{\nabla})\mathbf{A}], & (15.14) \\
\boldsymbol{\nabla}_{\mathbf{v}}L &= \boldsymbol{\nabla}_{\mathbf{v}}\left[\frac{1}{2}mv^2 - q\phi + q\mathbf{v}\cdot\mathbf{A}\right] = m\mathbf{v} + q\mathbf{A} & (15.15)
\end{aligned}
$$

Then Lagrange's equation is

$$
\begin{aligned}
\boldsymbol{\nabla}L &= \frac{d}{dt}\boldsymbol{\nabla}_{\mathbf{v}}L & (15.16) \\
q[\mathbf{v}\times\mathbf{B} - \boldsymbol{\nabla}\phi + (\mathbf{v}\cdot\boldsymbol{\nabla})\mathbf{A}] &= \frac{d}{dt}[m\mathbf{v} + q\mathbf{A}] = m\dot{\mathbf{v}} + q\frac{d\mathbf{A}}{dt} \\
m\dot{\mathbf{v}} &= q\left[\mathbf{v}\times\mathbf{B} - \boldsymbol{\nabla}\phi - \partial_t\mathbf{A}\right] \\
m\dot{\mathbf{v}} &= q\mathbf{E} + q\mathbf{v}\times\mathbf{B}, & (15.17)
\end{aligned}
$$

the nonrelativistic equation of motion.

(b) The relativistic Lagrangian is given by [Eq. (15.28)]:

$$L = \frac{-m}{\gamma} - q\phi + q\mathbf{v}\cdot\mathbf{A}. \tag{15.18}$$

This only differs from the nonrelativistic Lagrangian by the replacement $\frac{1}{2}mv^2 \to mv/\gamma$. Thus, $\boldsymbol{\nabla}L$ is unchanged, but the term $m\mathbf{v}$ in $\boldsymbol{\nabla}_{\mathbf{v}}L$ will change to

$$\boldsymbol{\nabla}_{\mathbf{v}}\left(\frac{-m}{\gamma}\right) = -m\boldsymbol{\nabla}_{\mathbf{v}}\sqrt{1-v^2} = \frac{m\mathbf{v}}{\sqrt{1-v^2}} = m\mathbf{v}\gamma = \mathbf{p}. \tag{15.19}$$

Then Lagrange's equation leads to the relativistic equation of motion

$$\frac{d}{dt}(m\mathbf{v}\gamma) = \frac{d\mathbf{p}}{dt} = q\mathbf{E} + q\mathbf{v}\times\mathbf{B}. \tag{15.20}$$

(c) The relativistic Lagrangian in tensor notation is given by [Eq. (15.28)]:

$$L = \frac{-m}{\gamma} - \left(\frac{q}{\gamma}\right)U_\mu A^\mu. \tag{15.21}$$

The gauge transformation $A^\mu \to A'^\mu = A^\mu + \partial^\mu\Lambda$ changes the Lagrangian by

$$L \to L' = L - (q/\gamma)U_\mu\partial^\mu\Lambda = L - q\partial_t\Lambda - q\mathbf{v}\cdot\boldsymbol{\nabla}\Lambda = L - q\frac{d\Lambda}{dt}. \tag{15.22}$$

To see if this change in L changes the equation of motion, we put the additional term into Lagrange's equation:

$$\frac{d}{dt}\boldsymbol{\nabla}_{\mathbf{v}}[-q\partial_t\Lambda - q\mathbf{v}\cdot\boldsymbol{\nabla}\Lambda] - \boldsymbol{\nabla}[-q\partial_t\Lambda - q\mathbf{v}\cdot\boldsymbol{\nabla}\Lambda] = -q\frac{d(\boldsymbol{\nabla}\Lambda)}{dt} + q\boldsymbol{\nabla}\left[\frac{d\Lambda}{dt}\right] = 0, \tag{15.23}$$

so the additional term has no effect on the equation of motion.

(d) The Hamiltonian for a charged particle is given by

$$\sqrt{(\boldsymbol{\mathcal{P}} - q\mathbf{A})^2 + m^2} + q\phi. \tag{15.24}$$

Hamilton's canonical equations are:

$$\boldsymbol{\nabla}_{\boldsymbol{\mathcal{P}}}H = \mathbf{v}, \qquad \boldsymbol{\nabla}H = -\frac{d\boldsymbol{\mathcal{P}}}{dt}. \tag{15.25}$$

From the first one, we get

$$\mathbf{v} = \boldsymbol{\nabla}_{\boldsymbol{\mathcal{P}}}\left[\sqrt{(\boldsymbol{\mathcal{P}} - q\mathbf{A})^2 + m^2} + q\phi\right] = \frac{(\boldsymbol{\mathcal{P}} - q\mathbf{A})}{\sqrt{(\boldsymbol{\mathcal{P}} - q\mathbf{A})^2 + m^2}}. \tag{15.26}$$

Solving this for $(\mathcal{P} - q\mathbf{A})$, we get

$$\mathbf{v}^2 \left[(\mathcal{P} - q\mathbf{A})^2 + m^2 \right] = (\mathcal{P} - q\mathbf{A})^2$$
$$(\mathcal{P} - q\mathbf{A}) = m\mathbf{v}\gamma. \tag{15.27}$$

The other Hamilton equation gives

$$\frac{d\mathcal{P}}{dt} = -\boldsymbol{\nabla} \left[\sqrt{(\mathcal{P} - q\mathbf{A})^2 + m^2} + q\phi \right] = \frac{-\boldsymbol{\nabla}[(\mathcal{P} - q\mathbf{A}){\cdot}(\mathcal{P} - q\mathbf{A})}{2\sqrt{(\mathcal{P} - q\mathbf{A})^2 + m^2}} - q\boldsymbol{\nabla}$$

$$= \frac{\{q(\mathcal{P} - q\mathbf{A})\times(\boldsymbol{\nabla}\times\mathbf{A}) + q[(\mathcal{P} - q\mathbf{A}){\cdot}\boldsymbol{\nabla}]\mathbf{A}\}}{\sqrt{(\mathcal{P} - q\mathbf{A})^2 + m^2}} - \boldsymbol{\nabla}\phi$$

$$\frac{d}{dt}(m\mathbf{v}\gamma + q\mathbf{A}) = \frac{\{qm\gamma\mathbf{v}\times\mathbf{B} + qm\gamma(\mathbf{v}{\cdot}\boldsymbol{\nabla})\mathbf{A}\}}{\sqrt{m^2v^2\gamma^2 + m^2}} - \boldsymbol{\nabla}\phi$$

$$\frac{d}{dt}(m\mathbf{v}\gamma) = q\mathbf{v}\times\mathbf{B} + q(\mathbf{v}{\cdot}\boldsymbol{\nabla})\mathbf{A} - q\frac{d\mathbf{A}}{dt} - \boldsymbol{\nabla}\phi = q\mathbf{v}\times\mathbf{B} + q\mathbf{E}, \tag{15.28}$$

the same as from the relativistic Lagrangian.

Problem 15.7:

(a) The electric field at a distance $\mathbf{r} = -vt\hat{\mathbf{i}} + y\hat{\mathbf{j}}$ from a particle of charge q moving with velocity $\mathbf{v} = v\hat{\mathbf{i}}$ is given by Eqs. (15.53) and (15.54) in the text. The force on a particle of charge q' due to this electric field is given by $\mathbf{F} = q\mathbf{E}$:

$$F'_x = \frac{-qq'\gamma vt}{(y^2 + \gamma^2 v^2 t^2)^{\frac{3}{2}}} \tag{15.29}$$

$$F'_y = \frac{qq'\gamma y}{(y^2 + \gamma^2 v^2 t^2)^{\frac{3}{2}}}. \tag{15.30}$$

The force on particle q due to the electric field from the article q' at rest is

$$F_x = \frac{qq'vt}{(y^2 + v^2 t^2)^{\frac{3}{2}}} \tag{15.31}$$

$$F_y = \frac{-qq'}{(y^2 + v^2 t^2)^{\frac{3}{2}}}. \tag{15.32}$$

(b) The x component of the impulse on either particle will vanish because the $\int_{-\infty}^{\infty} E_x dt = 0$ since $E_x(-t) = -E_x(t)$. The y impulse on particle q' is

$$I'_y = \int_{-\infty}^{\infty} F'_y dt = qq'\gamma y \int_{-\infty}^{\infty} \frac{dt}{(y^2 + \gamma^2 v^2 t^2)^{\frac{3}{2}}}$$

$$= qq'\gamma y \int_{-\pi/2}^{\pi/2} \frac{(y/\gamma v)\sec^2\theta d\theta}{y^3 \sec^3\theta} = \frac{2qq'}{vy}. \tag{15.33}$$

The time integral of F_y will be the same (except for a minus sign) as that for F'_y because the factor γ cancelled in the time integral of F'_y, and that factor was the only difference in the two integrals. Thus

$$I_y = -I'_y = \frac{-2qq'}{vy}, \tag{15.34}$$

and the total impulse on both charges $\mathbf{I} + \mathbf{I}' = \mathbf{0}$.

Problem 15.9:

(a) For this problem, the acceleration term for the electric field is negligible compared to the velocity field until the distance is so great that both terms are negligible for finding the electron's energy. For two electrons, a distance x apart, the magnitude of the force on either electron is given Eq. (15.54)] in the text:

$$\frac{dp}{dt} = qE = \frac{e^2}{\gamma^2 x^2}. \tag{15.35}$$

We can rewrite this equation as

$$\frac{dp}{dt} = \frac{dp}{dx}\frac{dx}{dt} = v\frac{dp}{dx} = \frac{p}{E}\frac{dp}{dx} = \frac{m^2 e^2}{E^2 x^2}. \tag{15.36}$$

The last equality in Eq. (15.36) can be written as,

$$\frac{m^2 e^2 dx}{x^2} = Epdp = E^2 dE. \tag{15.37}$$

and we integrate this equality for electrons starting from rest ($E = m$) a distance d apart:

$$\int_d^\infty \frac{m^2 e^2 dx}{x^2} = \int_m^E E'^2 dE'$$
$$\frac{m^2 e^2}{d} = \frac{1}{3}(E^3 - m^3). \tag{15.38}$$

Solving for E, we get

$$E = \left[\frac{3m^2 e^2}{d} + m^3\right]^{\frac{1}{3}} = m\left[1 + \frac{3e^2}{md}\right]^{\frac{1}{3}}. \tag{15.39}$$

(b) To evaluate the kinetic energy for each electron for $d = 1$ Å we expand the cube root to lowest order in e^2/md, getting

$$T = E - m \approx \frac{e^2}{d} = \frac{(4.8 \times 10^{-10})(300)}{(1 \times 10^{-8})} = 14 \text{ eV}. \tag{15.40}$$

This is the same as the nonrelativistic result. For $d=1$ fm, we cannot expand and the electron's kinetic energy is

$$
\begin{aligned}
E &= m\left[1+\frac{3e^2}{md}\right]^{\frac{1}{3}} \\
&= (.51\text{ MeV})\left[1+\frac{(3)(4.8\times10^{-10})(300)}{(.51\times10^6)(1\times10^{-13})}\right]^{\frac{1}{3}} = 1.1\text{ MeV}. \quad (15.41)
\end{aligned}
$$

For this case, the nonrelativistic calculation would give

$$
T_{\text{NR}} = \frac{e^2}{d} = \frac{(4.8\times10^{-10})(300)}{(1\times10^{-13})} = 1.4\text{ MeV}. \quad (15.42)
$$

Problem 15.11:

(a) Putting numbers into Eq. (15.125) in the text for the energy loss in one circular orbit, we get

$$
\begin{aligned}
\Delta W &= \frac{4\pi}{3}\frac{q^2}{R}\left(\frac{W}{m}\right)^4 = \frac{4\pi}{3}\left[\frac{(4.8\times10^{-10})(300)}{100\text{ cm/meter}}\right]\left[\frac{W}{.511\times10^{-3}}\right]^4 \\
&= 88.5\text{ keV}\left\{\frac{[W(\text{GeV})]^4}{R(\text{meters})}\right\} \quad (15.43)
\end{aligned}
$$

(b) To keep 50 GeV electrons in a storage ring of radius $R=200$ meters, the energy input per revolution must be

$$
\begin{aligned}
\Delta W &= 88.5\text{ keV}\left\{\frac{[W(\text{GeV})]^4}{R(\text{meters})}\right\} \\
&= 88.5\text{ keV}\left[\frac{(50)^4}{(200)}\right] = 2.8\text{ GeV}. \quad (15.44)
\end{aligned}
$$

(c) To keep 1 TeV protons in a storage ring of radius $R=40$ meters requires an energy input per revolution of

$$
\begin{aligned}
\Delta W &= 88.5\text{ keV}\left\{\frac{[W(\text{GeV})]^4}{R(\text{meters})}\right\}\left[\frac{m_e}{m_p}\right]^4 \\
&= 88.5\text{ keV}\left[\frac{(1\times10^3)^4}{(40)}\right]\left[\frac{.511}{938}\right]^4 = 0.2\text{ keV}. \quad (15.45)
\end{aligned}
$$

Index